AA004431

International Workshop on Flavour Changing and Conserving Processes 2015 (FCCP 2015)

EPJ Web of Conferences Volume 118 (2016)

Anacapri, Italy
10 - 12 September 2015

Editors:

Giancarlo D'Ambrosio
Massimo Passera
Paolo Massarotti
Michele Iacovacci
Graziano Venanzoni
Stefano Mastroianni

ISBN: 978-1-5108-2331-0

Printed from e-media with permission by:

Curran Associates, Inc.
57 Morehouse Lane
Red Hook, NY 12571

Some format issues inherent in the e-media version may also appear in this print version.

This work is licensed under a Creative Commons Attribution license:
http://creativecommons.org/licenses/by/2.0/
You are free to:
Share – copy and redistribute the material in any medium or format.
Adapt – remix, transform, and build upon the material for any purpose, even commercial.
The licensor cannot revoke these freedoms as long as you follow the license terms.
Under the following terms:
You must give appropriate credit, provide a link to the license, and indicate if changes were made. You may do so in any reasonable manner, but not in any way that suggests the licensor endorses you or your use. The copyright is retained by the corresponding authors.

Printed by Curran Associates, Inc. (2016)

For additional information, please contact EDP Sciences – Web of Conferences at the address below.

EDP Sciences – Web of Conferences
17, Avenue du Hoggar
Parc d'Activité de Courtabœuf
BP 112
F-91944 Les Ulis Cedex A
France

Phone: +33 (0) 1 69 18 75 75
Fax: +33 (0) 1 69 28 84 91

contact-edps@webofconferences.org

Additional copies of this publication are available from:

Curran Associates, Inc.
57 Morehouse Lane
Red Hook, NY 12571 USA
Phone: 845-758-0400
Fax: 845-758-2633
Email: curran@proceedings.com
Web: www.proceedings.com

TABLE OF CONTENTS

Workshop on "Flavour changing and conserving processes" 2015 (FCCP2015)

10-12 September 2015 *Villa Orlandi, Anacapri, Capri Island, Italy*
Europe/Rome timezone

Web page: http://fccp2015.na.infn.it
E-mail: loc_fccp2015(at)na.infn.it
Tel. +39 081 18891222/223
Fax +39 081 18891224

FCCP2015

"International Workshop on Flavour changing and conserving processes"

Villa Orlandi, Anacapri – Capri Island, Italy

10 - 12 September 2015

BULLETIN

1. Place and Date.

The "International Workshop on Flavour changing and conserving processes" (FCCP2015), is organized by the Physics Department of the University of Naples "Federico II" and by INFN (Istituto Nazionale di Fisica Nucleare).
It will take place in Anacapri, Capri Island, from Thursday, September 10 to Saturday, September 12, 2015.
This is the first workshop of a series which aims to discuss the status and prospects of the physics of "Flavour changing and conserving processes" and addresses the scientific community involved in experiments like g-2, mu2e, meg, meee, COMET, EDM and more generally to the physics of K and B mesons. It wants to be a "thinking effort" on the outlook, both theoretical and experimental, as well as a picture of the "state of the art".

2. International Advisory Committee

F. Bedeschi (INFN, Pisa- Italy)
P. Campana (INFN, Frascati- Italy)
A. Czarnecki (University of Alberta- Canada)
E. De Rafael (CPT, Marseille- France)
G. La Rana (University of Naples - Italy)
W. J. Marciano (Brookhaven- USA)
A. Masiero (University of Padova- Italy)
J. Miller (University of Boston -USA)
M. Nio (RIKEN Nishina center - Japan)
B. L. Roberts (University of Boston - USA)

3. Local Organizing Committee

G. D'Ambrosio (INFN, Naples)
G. Carlino (INFN, Naples)
M. Iacovacci (University of Naples)
P. Massarotti (University of Naples)
S. Mastroianni (INFN, Naples)
M. Passera (INFN, Padova)
G. Venanzoni (INFN, Frascati)

4. Secretariat

Studiocongress
Via dei Mille 16
80121 Napoli
http://www.studiocongress.it
Tel.+39 081 18891224
Fax +39 081 18891224

5. Preliminary schedule of the workshop
(http://people.na.infn.it/~mastroia/FCCP2015_programme.pdf)

Thursday 10th		
08:00-09:00	Registration	
Morning session		
09:00 -09:20	*Welcome*	
09:20 -09:30	*Introduction*	
09:30 -09:55	*Theory review on the muon g-2*	Kirill Melnikov (Th g-2 Rev)
09:55 -10:20	*Status of muon g-2 experiment at FNAL*	David Herzog (g-2 Experiment)
10:20 - 10:45	*Status of g-2 and edm experiment at JPARC*	speaker TBA (g-2 Experiment)
10:45 – 11:10	*Leading-order hadronic contribution to the electron and muon g-2*	Fred Jegerlehner (Th g-2 (e+mu) Hadronic LO)
11:10-11:30	**Coffee break**	
11:30 -11:55	*Perspectives on the hadronic contribution to the muon g-2*	Marc Knecht (Th g-2 Hadronic contribution)
11:55 -12:20	*Review of recent calculations of the hadronic vacuum polarization contribution to a_mu*	Zhiqing Zhang (Th g-2 Hadronic LO)
12:20 -12:45	*Leading-order hadronic contributions to the lepton anomalous magnetic moments from the lattice*	Marcus Petschlies (Th g-2 Hadronic LO Lattice)
12:45 -13:10	*VMD/HLS approach to the muon (g-2): a solution to the tau--e+e- puzzle*	Maurice Benayoun (Th g-2 tau puzzle)
13:10 -15:00	**Lunch time**	
Afternoon session		

15:00 -15:25	*A new approach to evaluate the leading hadronic corrections to the muon g-2*	Luca Trentadue (Th g-2 Hadronic LO - new approach)
15:25 -15:50	*Recent developments in the calculation of the hadronic light-by-light contribution to the muon g-2*	Massimiliano Procura (Th g-2 Hadonic LBL Recent develop)
15:50 -16:15	*Recent developments in the calculation of the hadronic light-by-light contribution to the muon g-2*	Marc Vanderhaeghen (Th g-2 Hadonic LBL Recent develop)
16:15 -16:40	*What does holographic QCD predicts for the light-by-light scattering contribution to (g-2)_mu*	Luigi Cappiello (Th g-2 Hadronic LBL Holograph)
16:40-17:00	**Coffee break**	
17:00 -17:25	*Hadronic light-by-light contribution to a_mu: extended Nambu-Jona-Lasinio and chiral quarks models*	Johan Bijnens (Th g-2 Hadronic LBL)
17:25 -17:50	*Hadronic contributions to the muon anomaly in the Constituent Chiral Quark Model*	David Greynat (Th Hadronic g-2 muon)
17:50 -18:15	*Review on gamma-gamma physics at KLOE*	Dario Moricciani (Exp gamma-gamma KLOE Rev)
18:15 -18:40	*On the precision of a data-driven estimate of the pseudoscalar-pole contribution to hadronic light-by-light scattering in the muon g-2*	Andreas Nyffeler (Th g-2 Hadronic LBL Recent develop)
18:40 -19:05	*Lattice calculation of the hadronic contribution to the muon g-2*	Christoph Lehner (Th g-2 Hadronic LBL Lattice)
20:00	**Welcome party**	
	Friday 11th	
Morning session		
09:00 -09:25	*Higher-order hadronic and heavy-lepton contributions to the muon g-2*	Matthias Steinhauser (Th g-2 Hadronic HO HeLep)
09:25 -09:50	*Recent developments on the MSSM contribution to the muon g-2*	Dominik Stockinger (Th g-2 Supersymm contrib)
09:50 -10:15	*The muon g-2 in two-Higgs-doublet Models*	Eung Jin Chun (Th g-2 Two Higgs doublets)
10:15 – 10:40	*The muon g-2 and degenerate supersymmetry*	Ketan Patel
11:00 -15:30	***Boat excursion around Capri Island***	
Afternoon session		
15:30 -15:55	*Review on measurement of the R ratio at Novosibirsk and ISR, and radiative corrections*	Simon Eidelman (Exp g-2 R Rev)
15:55 -16:20	*Radiative corrections and Monte Carlo generators for low energies*	Guido Montagna (MC g-2 Rad Corr (low energies)
16:20 -16:45	*Review on transition form factors*	Andrzej Kupsc (Exp Trans Form Fact Rev)

16:45 -17:10	*Pseudoscalar transition form factors: rare decays, g-2, and mixing angles*	Pere Masjuan
17:10 -17:35	*Review on Dark Photon*	Francesca Curciarello
17:35-17:55	**Coffee break**	
17:55 -18:20	*On the interrelationship among leptonic g−2, EDMs and lepton flavor violation*	Paride Paradisi (Th Flavor EDM)
18:20 -18:45	*The EDM in the g-2 experiment*	Rebecca Chislett (g-2 Experiment EDM)
18:45 -19:10	*Radiative mu and tau leptonic decays and the possibility to determine the tau dipole moments*	Matteo Fael (Th Flavor – EDM)
19:10-19:35	*Review on the fine-structure constant and the electron g-2*	Saida Guellati
21:00	**Social dinner**	
	Saturday 12th	
Morning session		
09:00 -09:25	*Experimental review on Lepton Universality and Lepton Flavour Violation tests at the B-factories*	Alberto Lusiani (Exp Flavour – Rev)
09:25 -09:50	*Charged lepton flavor violation – recent developments*	Francesca Borzumati (Th Flavor Rev)
09:50 -10:15	*Status of the Mu2e experiment*	Stefano Miscetti (Exp Flavour)
10:15 - 10:40	*Status of the Mu3e experiment at PSI*	Ann-Kathrin Perrevoort (Exp Flavour)
10:40 – 11:05	*Status of the MEG experiment at PSI*	Fabrizio Cei (Exp Flavour)
11:05-11:20	**Coffee break**	
11:20-11:45	*The SHIP Facility at CERN*	Giovanni De Lellis (Exp Flavour)
11:45 -12:10	*Status of the COMET experiment*	Peter Dornan (Exp Flavour)
12:10-12:35	*Lepton-flavour-violating decays in theories with Dim-6 operators*	Marco Pruna (Th Flavour)
12:35 -13:00	*Two-Higgs-doublet model facing experimental hints*	Julian Heeck (Th Flavour)
13:00 -13:25	*New Physics effects in flavour physics*	Andreas Crivellin (Th Flavour)
13:25 -14:50	**Lunch time**	
Afternoon session		
14:50 -15:15	*Status of the NA62 Experiment*	TBA (Exp Flavour)
15:15 -15:40	*Search for edm at storage ring*	Yannis Semertzidis (EDM Experiment Review)
15:40 -16:30	*Outlook*	Arkady Vainshtein

6. Social Events

- Thursday, September 10: Welcome cocktail at 20:00;
- Friday, September 11: Boat tour of the island (starting at 11:00)
- Friday, September 11: Social Dinner (21:00 h) at Restaurant "Da Gelsomina" Via Migliara, 72 – Anacapri (http://www.dagelsomina.com)

7. Workshop Site

The Workshop will be held in Villa Orlandi, Anacapri, Capri Island.
All information can be accessed online through the web page:

http://fccp2015.na.infn.it

A valid passport is required for entry in Italy from outside of the european Schengen area. Participants from some countries will also need a visa. Please see http://vistoperitalia.esteri.it/home/en for details.

8. Registration and Accommodation

People willing to participate should carefully fill in the Registration Form on the web page of the workshop (https://agenda.infn.it/confRegistrationFormDisplay.py?confId=9430) not later than July 31, 2015.

9. Accomodation and Fee

All participants are kindly requested to reserve their accommodation on their own.
Special deals have been with the following hotels.
Please fill this form for Hotel reservation and payment fee (docx, pdf).

For any questions on Hotel reservation contact:

Studiocongress
Nicoletta de Pertis
Via dei Mille 16
80121 Napoli
tel 081 18891222 / 223 fax 081 18891224
http://www.studiocongress.it
postmaster@studiocongress.it

The registration fee is 250 euros for participant and 80 euros for accompanying person.

The registration fee includes proceedings, lunches, coffee breaks and social events.
Accompanying people will have to pay 80 euros for social program, namely boat tour on Friday 11.th and social dinner.

Please fill the form you find in the Accommodation and fee page (https://agenda.infn.it/internalPage.py?pageId=4&confId=9430).

10. Travel and information

You can get a fast boat to Capri from Naples' main port *Molo Beverello.* There are also some slower boats (car ferries) leaving from *Porta di Massa*, which is 500m from Molo Beverello (picture). Information about timetables can be found at the following link http://www.capri.net/salsa/lang/en/page/transport.html Please note that the timetables usually change end of March.

Arriving by plane: From Napoli Capodichino-Airport it takes 45 min. to reach the Port of Naples. The company *Alibus* runs a bus that leaves in front of terminal 1 every 20 minutes (approximately at .00, .20, .40). The bus makes two stops, one at the central train station, the second at Piazza Municipio. The port (Molo Beverello) is at the bottom of the square. The ride costs 3 euro and the ticket can be bought on the bus. Taxis are also available (ride to Molo Beverello about 25 euro, verify price BEFORE boarding the car!)

Arriving by train: Stazione Centrale (15 minutes from the Molo Beverello) is connected with the port by bus and tram.

Anacapri is one of the two main villages in the island of Capri (the other one is called "Capri", like the whole island). The hotels where we are staying are located in Anacapri. To reach them, you can choose one of these ways:

1. From Marina Grande (the port where the boat leaves you) take the direct bus to Anacapri (runs on average every hour; cost=1,80 Euro; it takes 15 min.; buy a ticket at the ticket-desk). Tell the driver the name of your hotel.
2. At Marina Grande (the port where the boat leaves you) buy at the ticket-desk a ticket for the "Funicolare"=cable-car to Capri village (runs every 15 min.; it takes 4 min.); then, from Capri take the bus to Anacapri (runs every 10 minutes; it takes 15 min.). You need a second ticket for this ride. Tell the driver the name of your hotel.
3. From Marina Grande (the port where the boat leaves you) take a taxi (it costs about 25 Euros; it takes 15 min.; verify the price before boarding the taxi).

How to reach Villa Orlandi: from the Piazza Vittoria bus stop in Anacapri, follow the suggested route to Villa Orlandi.

Useful links:

www.capritourism.com
www.capri.net

11. Proceedings

The Proceedings of the Workshop will be published on-line by the Open Access Journal "EPJ Web of Conferences" (http://www.epj-conferences.org).
All contributing authors will be asked to prepare a contribution to be included in the proceedings and submit it within the deadline of October 31, 2015. More information can be found on the web page of the Workshop.
(https://agenda.infn.it/internalPage.py?pageId=7&confId=9430)

12. Internet Connection

A wifi connection will be available for participants.

13. Contact

If you have any questions or comments, please send an e-mail message
to the following address: loc_fccp2015(at)na.infn.it

14. Important dates

Deadline of the registration:	July 31, 2015
Arrival:	September 9, 2015
Beginning of the meeting:	September 10, 2015
End of the meeting:	September 12, 2015
Deadline for the proceedings:	October 31, 2015

PREFACE

The International Workshop on "Flavour changing and conserving processes" 2015 (FCCP2015) took place on September 10-12, 2015 in Anacapri, Capri Island, Italy. It was organized by the Physics Department of the University of Naples Federico II and by INFN (Istituto Nazionale di Fisica Nucleare). This was the first workshop of a series which aims to discuss the status and prospects of the physics of Flavour changing and conserving processes, and addresses the scientific community involved in experiments like $g-2$, Mu2e, MEG, Mu3e, COMET, EDM and more generally to the physics of K and B mesons. Forty talks were presented at the Workhop, covering the theoretical status as well as the latest results of experiments in various Laboratories of the world.

We wish to thank everybody who contributed to the success of the Workshop. The advice of the International Advisory Committee was very useful in defining the scientific program. The session chairs greatly contributed to the smooth operation of the meeting. The speakers did an excellent job in presenting the latest results.

We look forward to the next edition in 2017 when new exciting results on the physics of Flavour changing and conserving processes are expected!

Giancarlo D'Ambrosio
Michele Iacovacci
Massimo Passera
Graziano Venanzoni
Paolo Massarotti
Stefano Mastroianni

Editors
FCCP2015

© The Authors, published by EDP Sciences. This is an open access article distributed under the terms of the Creative Commons Attribution License 4.0 (http://creativecommons.org/licenses/by/4.0/).

edp sciences EPJ Web of Conferences

Statement of Peer review

In submitting conference proceedings to EPJ Web of Conferences, I certify to the Publisher that I adhere to the **Policy on Publishing Integrity** of the journal in order to safeguard good scientific practice in publishing.

1. All articles have been subjected to peer review administered by the proceedings editors.
2. Reviews have been conducted by expert referees, who have been requested to provide unbiased and constructive comments aimed, whenever possible, at improving the work.
3. Proceedings editors have taken all reasonable steps to ensure the quality of the materials they publish and their decision to accept or reject a paper for publication has been based only on the merits of the work and the relevance to the journal.

Title, date and place of the conference

International Workshop on "Flavour changing and conserving processes" 2015 (FCCP2015), Villa Orlandi, Anacapri – Capri Island, Italy, 10 - 12 September 2015

Proceedings editor(s):

Giancarlo D'Ambrosio; Michele Iacovacci;
Massimo Passera; Graziano Venanzoni;
Paolo Massarotti; Stefano Mastroianni .

Date and editor's signature

19 - MARCH - 2016

17, avenue du Hoggar - PA de Courtabœuf – BP 112 - 91944 Les Ulis Cedex A (France)
Tél. : 33 (0)1 69 18 75 75 - Fax : 33(0)1 69 07 45 17 – www.edpsciences.org

The HLS approach to $(g-2)_\mu$: A Solution to the "τ versus e^+e^-" Puzzle

Maurice Benayoun[1,a,b]

[1] *LPNHE des Universités Paris VI et Paris VII, IN2P3/CNRS, 4 Place Jussieu, 75252 Paris, France*

Abstract. The Hidden Local Symmetry (HLS) Model provides a framework able to encompass several physical processes and gives a unfied description of these in an energy range extending up to the ϕ mass. Supplied with appropriate symmetry breaking schemes, the HLS Model gives a broken Effective Lagrangian (BHLS). The BHLS Lagrangian gives rise to a fit procedure in which a simultaneous description of the e^+e^- annihilations to $\pi^+\pi^-$, $\pi^0\gamma$, $\eta\gamma$, $\pi^+\pi^-\pi^0$, K^+K^-, K_LK_S and of the dipion spectrum in the decay $\tau^\pm \to \pi^\pm\pi^0\nu$ can be performed. Supplemented with a few pieces of information on the $\rho^0-\omega-\phi$ system, the τ dipion spectrum is shown to predict accurately the pion form factor in e^+e^- annihilations. Physics results derived from global fits involving or excluding the τ dipion spectra are found consistent with each others. Therefore, no obvious mismatch between the τ and e^+e^- physics properties arises and the $\tau-e^+e^-$ puzzle vanishes within the broken HLS Model.

1 Introduction

The pion form factor in the $e^+e^- \to \pi^+\pi^-$ annihilation ($F_\pi^{ee}(s)$) and in the the $\tau^\pm \to \pi^\pm\pi^0\nu$ decay ($F_\pi^\tau(s)$) are expected to differ only by isospin symmetry breaking (IB) terms. Understanding the relationship between $F_\pi^{ee}(s)$ and $F_\pi^\tau(s)$ is important as it can allow for 2 different evaluations of the dipion contribution to $a_\mu(\pi\pi)$, the muon Hadronic Vacuum Polarization (HVP) which could be merged together if consistent with each other. However, this relationship supposes a good understanding of isospin symmetry breaking and an appropriate modelling.

For a long time [1, 2], the comparison between $|F_\pi^{ee}(s)|^2$ and $|F_\pi^\tau(s)|^2$ was not satisfactory and the mismatch [3] was severe enough that one started to speak of a "e^+e^- vs τ" puzzle. This has continued up to very recently [4]. However, some works [5, 6] indicated that this puzzle could well be a modelling issue of the isospin symmetry breaking phenomenon. On the other hand, it was also shown [7] that, numerically, the $e^+e^- - \tau$ discrepancy sensitively depends on the $e^+e^- \to \pi^+\pi^-$ sample considered. Therefore the so–called $e^+e^- - \tau$ puzzle may carry several components.

The Hidden Local Symmetry (HLS) Model provides a framework and a procedure able to address this puzzle in the various aspects just sketched. The HLS model encompasses several physical processes and gives a unfied description of these in an energy range extending up to the ϕ mass. However, in order to account precisely for experimental data, it should be supplied with several symmetry breaking schemes. Among these, an energy dependent mixing mechanism of the neutral vector meson system ($\rho^0-\omega-\phi$) is generated via loop effects and allows to define an effective broken HLS (BHLS) model. Within this framework, the e^+e^- annihilations to $\pi^+\pi^-$, $\pi^0\gamma$, $\eta\gamma$, $\pi^+\pi^-\pi^0$, K^+K^-, K_LK_S and the dipion spectrum in the decay $\tau^\pm \to \pi^\pm\pi^0\nu$ are *simultaneously* accounted for with the same set of parameters derived from global fits in procedures involving all the existing data samples covering the channels listed above. One can also define a variant – named τ + PDG – where the $e^+e^- \to \pi^+\pi^-$ data are replaced by tabulated ρ^0, ω and ϕ particle properties[18]; with such a tool, one can compare the τ predictions for the $e^+e^- \to \pi^+\pi^-$ annihilation and the data. If the $e^+e^- - \tau$ puzzle is a relevant concept, this comparison should enhance the issue.

After a brief review of the BHLS Model and its breaking in Sections 2, 3 and 4, the various aspects of the global fit method are outlined in Section 5. Section 6 studies the τ +PDG predictions and their comparison with the existing $e^+e^- \to \pi^+\pi^-$ data samples. In Section 7, one reports on global fits mixing $e^+e^- \to \pi^+\pi^-$ and τ dipion data. In Section 8, the effects of τ data on the muon HVP are discussed. Section 9 is devoted to the conclusions.

2 Basics of the Hidden Local Symmetry Model

The Hidden Local Symmetry Model (HLS) Model is a framework which encompasses simultaneously several different physics processes covered by a large number of already available data samples. A comprehensive review of the HLS Model is given in [8] and a brief account can be found in [9]; however, in order to really deal with experimental data at their present level of accuracy, breaking procedures need to be implemented. As these are tightly connected with the HLS Model structure, it is worth giving a brief outline of its main features.

[a] e-mail: benayoun@in2p3.fr
[b] e-mail: maurice.benayoun@gmail.com

© The Authors, published by EDP Sciences. This is an open access article distributed under the terms of the Creative Commons Attribution License 4.0 (http://creativecommons.org/licenses/by/4.0/).

Beside its non–anomalous sector, which allows to address some e^+e^- annihilation channels and some τ decays, the HLS Model also contains an anomalous (FKTUY) sector [8, 10] which provides couplings of the form VVP, $VPPP$, γPPP, $VP\gamma$ or $P\gamma\gamma$ among the light flavor mesons[1]. Intrisically, the HLS validity range does not extend much beyond the ϕ mass.

The $e^+e^- \to \pi^+\pi^-/K\overline{K}$ annihilations or the $\tau^\pm \to \pi^\pm\pi^0\nu$ decay clearly proceed from only the non–anomalous sector of the HLS model. However, the radiative meson decays ($VP\gamma$ or $P\gamma\gamma$ couplings) as well as the annihilation channels like $e^+e^- \to \pi^0\gamma$, $e^+e^- \to \eta\gamma$ or $e^+e^- \to \pi^0\pi^+\pi^-$ involve couplings only available from the anomalous (FKTUY) sector of the HLS model.

The construction of the HLS Lagrangian starts by defining the (right and left) ξ fields :

$$\xi_{R,L} = \exp\left[\pm iP/f_\pi\right] \tag{1}$$

where f_π is the pion decay constant and $P = P_8 + P_0$ is the U(3) matrix of the pseudoscalar fields which includes the octet and singlet field components [9].

The HLS non–anomalous Lagrangian is defined by[2] :

$$\left\{\begin{array}{lll} \mathcal{L}_{HLS} = & \mathcal{L}_A + a\mathcal{L}_V & \\ \mathcal{L}_A = & -\frac{f_\pi^2}{4}\mathrm{Tr}[(D_\mu\xi_L\xi_L^\dagger - D_\mu\xi_R\xi_R^\dagger)^2] & \equiv -\frac{f_\pi^2}{4}\mathrm{Tr}[L-R]^2 \\ \mathcal{L}_V = & -\frac{f_\pi^2}{4}\mathrm{Tr}[(D_\mu\xi_L\xi_L^\dagger + D_\mu\xi_R\xi_R^\dagger)^2] & \equiv -\frac{f_\pi^2}{4}\mathrm{Tr}[L+R]^2 \end{array}\right. \tag{2}$$

These expressions involve the covariant derivatives of the $\xi_{R,L}$ fields :

$$\left\{\begin{array}{l} D_\mu\xi_L = \partial_\mu\xi_L - igV_\mu\xi_L + i\xi_L\mathcal{L}_\mu \\ D_\mu\xi_R = \partial_\mu\xi_R - igV_\mu\xi_R + i\xi_R\mathcal{R}_\mu \end{array}\right. \tag{3}$$

which introduce the usual bare vector field matrix[3] V; the other gauge bosons of the Standard Model (A, $W^\pm$ and Z) are hidden inside $\mathcal{L}_\mu$ and $\mathcal{R}_\mu$; neglecting the influence of the Z boson field absent from the physics we address, these are written as :

$$\left\{\begin{array}{l} \mathcal{L}_\mu = eQA_\mu + \frac{g_2}{\sqrt{2}}(W_\mu^+T_+ + W_\mu^-T_-) \\ \mathcal{R}_\mu = eQA_\mu \end{array}\right. \tag{4}$$

The quark charge matrix Q is standard and the matrix $T_+ = [T_-]^\dagger$ is constructed out of matrix elements of the Cabibbo–Kobayashi–Maskawa matrix [8]. Concerning the physics parameters, the above expressions exhibit the electric charge e, the universal vector coupling g and the weak coupling g_2 (related to the Fermi constant by $g_2 = 2m_W\sqrt{G_F\sqrt{2}}$). Finally, a is a specific HLS parameter not fixed by the model and expected of the order of 2.

[1] In the following, V and P denote generically any of the resp. vector and pseudoscalar light flavor mesons and, also, the corresponding field matrices without ambiguities.

[2] Within the present one page reminder, we do not discuss the anomalous sectors and refer the interested reader to [8] or [9].

[3] Which involves the so–called ideal combinations ρ_I^0, ω_I and ϕ_I for the neutral fields.

3 Usual symmetry-breaking schemes of the HLS Model

The HLS model obviously provides an elegant unified framework which covers an important set of annihilation and decay processes. However, as such, it cannot produce a satisfactory account of the real experimental data falling into its scope. A simple illustration is given by the pion and kaon decay constants found of equal magnitudes within the unbroken HLS Lagrangian.

This clearly indicates that symmetry breaking mechanisms should be supplied. The authors of the HLS Model were aware of this difficulty and soon proposed a simple (BKY) mechanism to break the flavor SU(3) symmetry [11] of the model; this has been later extended to include isospin breaking [12]; for practical purpose, we use the BKY mechanism as reformulated in [13]. This turns out to modify Eqs. (2) by introducing breaking matrices in the following way :

$$\left\{\begin{array}{lll} \mathcal{L}_A & = & -\frac{f_\pi^2}{4}\mathrm{Tr}[(L-R)X_A]^2 \\ \mathcal{L}_V & = & -\frac{f_\pi^2}{4}\mathrm{Tr}[(L+R)X_V]^2 \end{array}\right. \tag{5}$$

where X_A and X_V are real diagonal matrices :

$$\left\{\begin{array}{l} X_A = \mathrm{Diag}(q_A, y_A, z_A) \\ X_V = \mathrm{Diag}(q_V, y_V, z_V) \quad . \end{array}\right. \tag{6}$$

which should be derived from data.

The departures of $q_{A/V}$ and $y_{A/V}$ from 1 measure the isospin symmetry breaking, while $z_{A/V}$ carry the flavor SU(3) symmetry breaking. Together with determinant terms [14] which permit to break nonet symmetry in the pseudoscalar sector, this provides a reliable description of the light meson radiative decays and of the $e^+e^- \to P\gamma$ annihilations.

4 Vector field mixing, a new symmetry breaking mechanism

The coupling of the neutral vector mesons carrying no open strangeness to a pseudoscalar meson pair is given by the following piece of the $\mathcal{L}_V$ HLS Lagrangian[4] [5] :

$$\begin{array}{l} \frac{iag}{2}\rho_I^0\pi^- \overleftrightarrow{\partial} \pi^+ + \frac{iag}{4z_A}(\rho_I^0 + \omega_I - \sqrt{2}z_V\phi_I)K^- \overleftrightarrow{\partial} K^+ \\ + \frac{iag}{4z_A}(\rho_I^0 - \omega_I + \sqrt{2}z_V\phi_I)K^0 \overleftrightarrow{\partial} \overline{K}^0 \end{array} \tag{7}$$

The last two terms obviously give rise to pseudoscalar kaon loops which modify the vector mass matrix by s–dependent terms. Moreover, the kaon loops generate transitions among the ideal ρ_I, ω_I and ϕ_I of the original Lagrangian and, thus, give *non – diagonal* entries inside the vector meson squared mass matrix [5, 9]. Stated otherwise, at one–loop order, the ideal ρ_I, ω_I and ϕ_I fields are no longer mass eigenstates as expected for the physical

[4] The isospin breaking effects generated by the X_A and X_V matrices have been removed for clarity; they can be found in [9].

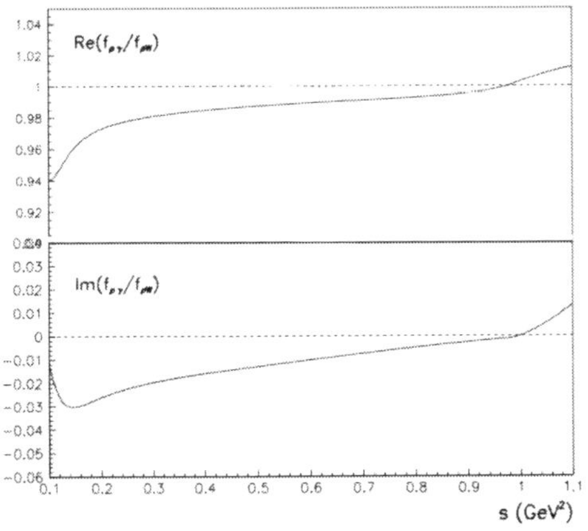

Figure 1. The ratio $f_{\rho^0\gamma}(s)/f_{\rho^\pm W}$ derived from a BHLS global fit over the region from the two-pion threshold to slightly above the ϕ mass.

vector-meson fields. The one-loop order mass squared matrix can be written as :

$$\begin{cases} M^2(s) &= M_0^2(s) + \delta M^2(s) \quad \text{with :} \\ M_0^2(s) &= \text{Diag}(m^2 + \Pi_{\pi\pi}(s),\ m^2,\ z_V m^2). \end{cases} \tag{8}$$

where $\delta M^2(s)$ is a non–diagonal perturbation matrix depending on the kaon loops to the (otherwise) diagonal matrix $M_0^2(s)$. The entries of $M_0^2(s)$ depend on m^2 – the squared ρ (or ω) meson mass, as it occurs in the original HLS Lagrangian – and on $\Pi_{\pi\pi}(s)$, the dipion loop to which only ρ_I couples.

As $\delta M^2(s)$ is small compared to $M_0^2(s)$ within the range of validity of the HLS model (*i.e.* up to the ϕ mass region), the eigenvalue problem Eq. (8) can be solved perturbatively. The relation between the ideal fields (V_I) and the physical fields (V_R) can be written :

$$\begin{pmatrix} \rho_I \\ \omega_I \\ \phi_I \end{pmatrix} = \begin{pmatrix} 1 & -\alpha & \beta \\ \alpha & 1 & \gamma \\ -\beta & -\gamma & 1 \end{pmatrix} \begin{pmatrix} \rho_R^0 \\ \omega_R \\ \phi_R \end{pmatrix} \tag{9}$$

where α, β and γ are functions of s, the energy flowing through the vector meson line. These functions essentially[5] depend on the sum and the difference of the charged and neutral kaon loops and are small compared to 1 [5, 9]. Therefore, the vector field mixing mechanism introduces breaking terms which are s–dependent, s being the running vector meson mass.

This change of fields is what mostly generates an isospin 1 component inside the physical ω and ϕ mesons and, then, their couplings to a pion pair as, using Eq. (9), one gets at first order in breaking parameters[6] :

$$\frac{iag}{2}\rho_I \cdot \pi^- \overleftrightarrow{\partial} \pi^+ \Rightarrow$$
$$\frac{iag(1+\Sigma_V)}{2}\left[\rho_R^0 + [(1-h_V)\Delta_V - \alpha(s)\omega + \beta(s)\phi\right] \cdot \pi^- \overleftrightarrow{\partial} \pi^+ \tag{10}$$

At this order, the ρ_R^0 coupling to a pion pair is unchanged and remains identical to those of the $\rho^\pm$ meson. On the other hand, the change of fields Eq. (9) modifies the Lagrangian coupling of the neutral ρ meson to the photon, while leaving unchanged the charged ρ coupling to the W boson. Therefore, the vector field mixing makes the ratio of these two couplings s–dependent :

$$\frac{f_{\rho^0\gamma}(s)}{f_{\rho^\pm W}} = \left[1 + \frac{h_V\Delta_V}{3} + \frac{\alpha(s)}{3} + \frac{\sqrt{2}z_V}{3}\beta(s)\right] . \tag{11}$$

In order to substantiate this specific breaking of the HLS model, let us quote a result derived from a global fit involving all the channels listed in the Introduction; the ratio $f_{\rho^0\gamma(s)}/f_{\rho^\pm W}$ shown in Figure 1 exhibits significant variations over the HLS energy range of interest. It is the main mechanism which allows to reconcile the τ dipion spectrum and the $e^+e^- \to \pi^+\pi^-$ annihilation cross section. These couplings are supplemented by loop corrections [5, 9] which also play an important role in defining the effective γV mixings[7].

On the other hand, the τ decay process also undergoes several specific breaking effects : The short range [15] and long range [16] corrections are included when fitting the τ dipion spectrum, as for the specific $\pi^\pm\pi^0$ phase space factor. These τ breaking effects are accounted for within fits as usually done; they are clearly independent of the HLS breaking mechanisms and only come supplementing them.

5 The various aspects of the global fit method

Gradually equipped since [5] with the various breaking procedures briefly outlined above, the HLS Model has evolved toward a broken version (BHLS) [9] able now to cope simultaneously with several physics processes, namely the e^+e^- annihilations to $\pi^+\pi^-$, $\pi^0\gamma$, $\eta\gamma$, $\pi^+\pi^-\pi^0$, K^+K^-, K_LK_S, the dipion spectrum in the decay $\tau^\pm \to \pi^\pm\pi^0\nu$ and, additionally, some more radiative decays of light flavor mesons. It involves 25 parameters to be extracted from data and they come intricated simultaneously within the various amplitudes.

Therefore, BHLS is a global model permitting a global account of all the processes just listed. As the *same* model parameters are involved in the *different* physics processes, this implies the existence of *physics* correlations among the various spectra or partial widths entering the BHLS realm.

[5]Actually, $K^*\overline{K}$ and $K^*\overline{K^*}$ loops are provided by resp. the VVP anomalous Lagrangian and by the Yang–Mills term; however, up the ϕ mass region they are real and effectively absorbed inside the subtraction polynomials of the kaon loop combinations.

[6]The Δ_V comes from an additional (minor) breaking process [9]; Δ_V and Σ_V are combinations of the breaking parameters q_V and y_V generated by the BKY mechanism.

[7]See also [6].

One obviously expects herefrom an improvement of the knowledge of the physics model parameters involved in each given channel $\mathcal{I}$, as the experimental data covering the other (different) channels contribute to improve the uncertainties of each of the model parameters and, hence, the statistical quality of the account of $\mathcal{I}$. Of course, the fit quality is expected to reflect that the physics constraints exhibited by the (BHLS) model are well accepted by the data.

The number of independent data samples covering the various quoted channels is of the order of 50; they are listed and discussed in [9]. In this Reference, a simultaneous fit of all available e^+e^- annihilation scan data and of the published τ dipion spectra is performed. A very good fit quality is reached and no noticeable issue is observed. As the model is global, it also represents a new tool to examine precisely several issues [17] :

1. The discrepancy between the dipion spectrum in the τ decay and in the e^+e^- annihilation;
2. The relative compatibility of the various available $e^+e^- \to \pi^+\pi^-$ cross section measurements up to the ϕ mass;
3. The compatibility of the τ and e^+e^- based estimates of the hadronic vacuum polarization (HVP) contribution to the muon $g-2$.

The following sections outline the BHLS analysis of these issues.

6 The BHLS prediction of the pion form factor in e^+e^- annihilations

Since 2002 several measurements of the pion form factor in e^+e^- annihilations have been published. Beside the data samples collected in scan mode by CMD–2 [19] and SND [20], the KLOE Collaboration has produced three spectra collected in the ISR mode under different conditions, namely KLOE08 [21], KLOE10 [22] and recently the KLOE12 data sample [23] – strongly correlated with KLOE08. BaBar has also produced a $\pi^+\pi^-$ spectrum [24] extending up to 1.8 GeV. Finally, very recently, the BES-III Collaboration has published a new spectrum [25] limited to the energy interval 0.6 – 0.9 GeV. Except for the BES-III data sample presented in [27], these data samples have been examined in either of [17] or [26] – specifically for the KLOE12 sample. The present study outlines the treatment of the BES-III sample within the BHLS fitter.

A priori, BHLS can predict the pion form factor in the $e^+e^- \to \pi^+\pi^-$ annihilation relying on the measured τ dipion spectra [28–30], provided it is also fed with the appropriate isospin breaking (IB) information. However, despite the intricacy phenomenon noted above, some of the specific IB effects occuring in the $e^+e^- \to \pi^+\pi^-$ annihilation are marginally constrained by the other annihilation channels included within the BHLS realm. So, specific data directly reflecting these IB effects should be provided to the BHLS fitter. Such pieces of information are obviously related with the $\omega/\phi \to \pi^+\pi^-$ couplings. Reference [17] proposed to use the corresponding tabulated [18] partial widths and the (Orsay) phases between the ω/ϕ amplitudes and the underlying coherent $\pi^+\pi^-$ background; this phase information can be replaced by the relevant tabulated products $\Gamma(V \to e^+e^-) \times \Gamma(V \to \pi^+\pi^-)$. Additionally, as the $\rho^0 \to e^+e^-$ coupling is marginally constrained by the non–$\pi^+\pi^-$ annihilation data, one should include the tabulated $\rho^0 \to e^+e^-$ decay width. This method is named τ +PDG for obvious reasons[8].

For reasons which will become clear soon, it deserves noting that none of the 5 pieces of information supplementing the τ spectra in the τ+PDG approach is influenced by any of the KLOE, BaBar or BES-III $\pi^+\pi^-$ spectra. Actually, they are almost 100% determined by the data collected by the CMD–2 and SND Collaborations.

The red curve in Figure 2 displays the τ +PDG distribution and, superimposed, all the available $e^+e^- \to \pi^+\pi^-$ data samples. When MINUIT has converged, one can compute the χ^2 distance of each of the available $\pi^+\pi^-$ spectra to the τ +PDG best fit function; the average χ^2 per data point is then calculated for each data sample and all are reported inside the plots as comments[9].

Obviously, the left–hand side pannel indicates that, overall, the agreement between the τ +PDG *prediction* and the data is satisfactory, and the inset indicates that this agreement extends to the close spacelike region[10]. Therefore, the accepted (PDG) values for the IB pieces of information listed above allow to recover the gross features of the pion form factor with a noticeable precision; this already indicates that the IB mechanism as plugged within BHLS is appropriate.

The right–hand side pannel, however, indicates that we are faced with a contrasting picture, depending on the data samples examined. As a good tag of the agreement between the BHLS τ +PDG and the (secure) NSK data[11], the relevant subpannel in the right–hand side of Figure 2 displays the average χ^2 distance per data point of the NSK samples; one gets $\overline{\chi^2}_{NSK} = 1.2$, close to the best fit value in a fit where the PDG information is replaced by the CMD–2 and SND $e^+e^- \to \pi^+\pi^-$ data samples [9]. This also gives a hint about the range of acceptable values for the $\overline{\chi^2}$ associated with any given sample.

Therefore, the BHLS fit in the τ+PDG mode provides IB parameter values which allow the underlying (BHLS) IB framework to exhibit a full consistency of all non $\pi^+\pi^-$ data with the $\pi^+\pi^-$ NSK (*i.e.* CMD–2 & SND) data. The

[8] It should nevertheless be kept in mind that the annihilation cross–sections to $\pi^0\gamma$, $\eta\gamma$, $\pi^+\pi^-\pi^0$, K^+K^- and K_LK_S are always fully involved within the global fit. They mainly serve to fit the γV transition amplitudes also involved in the $e^+e^- \to \pi^+\pi^-$ annihilation process.

[9] Quite generally, the CMD–2 and SND data samples are grouped together and denoted NSK; to avoid energy calibration issues around the ω mass, the χ^2 value shown for BaBar is calculated by amputating the spectrum from its part falling between 0.76 and 0.80 GeV.

[10] It is nevertheless premature to include this region inside the BHLS fit [27].

[11] The NSK data are implicitly (or explicitly) considered as a reference, as they can accomodate *separately* almost all the other data samples with, however, various qualities.

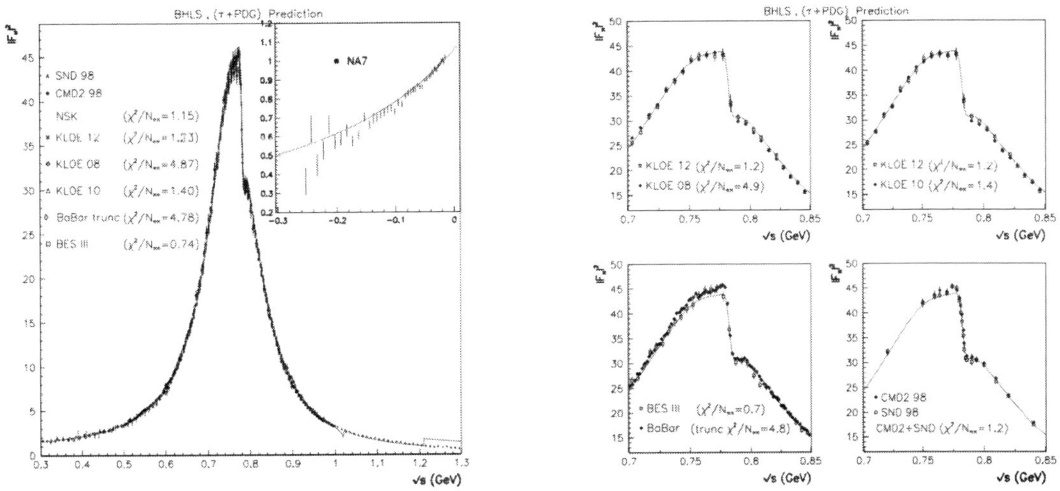

Figure 2. Predictions for the pion form factor in the $e^+e^- \to \pi^+\pi^-$ annihilation derived using the τ + PDG method. None of the $\pi^+\pi^-$ spectra is used in the fitting procedure. The leftmost pannel displays the overall picture with data superimposed, the inset showing the form factor continuation into the spacelike ($s < 0$) region. The rightmost pannel displays, enlarged, the behavior of the various data samples in the $\rho - \omega$ interference region.

picture is clearly alike for the KLOE10 ($\overline{\chi}^2_{KLOE10} = 1.4$), KLOE12 ($\overline{\chi}^2_{KLOE12} = 1.2$) and BES-III ($\overline{\chi}^2_{BESS} = 0.7$) data samples. In contrast, one observes that the KLOE08 ($\overline{\chi}^2_{KLOE08} = 4.9$) or BaBar ($\overline{\chi}^2_{BaBar\ trunc} = 4.8$) samples are farther than could be expected.

The upper left–hand plot in the right pannel of Figure 2 is also quite informative; indeed it shows that the twin samples KLOE08 and KLOE12 [21, 23] carry central values very close to each other and that both follow almost exactly the τ +PDG predicted curve. Nevertheless, KLOE12 exhibits a $\overline{\chi}^2$ value in close agreement with the τ +PDG expectations while KLOE08 does not. This should be due to the estimates and structure of the reported correlated systematic uncertainties, seemingly better understood for the KLOE12 sample.

Finally, there is a clear contradiction between the KLOE and BaBar samples, as already reported by other authors (see for instance [7]) but, also, BaBar does not fit well with the BHLS τ +PDG predictions.

Therefore, as in the comparison between the τ +PDG predictions and the various $\pi^+\pi^-$ data samples, five out of the seven available independent $\pi^+\pi^-$ samples do not exhibit any kind of mismatch, one is obviously tempted to conclude that there is no evidence for a $\tau - e^+e^-$ puzzle. The BHLS approach would rather indicate that the reported puzzle comes from a non–adequate IB modelling.

7 The BHLS global fits

So, the comparison between the τ predictions – also based on commonly accepted PDG information – and the various $\pi^+\pi^-$ data samples indicates various behaviors. More precisely, the τ + PDG method gives a well–founded indication that the CMD–2, SND, KLOE10, KLOE12 and BES-III $e^+e^- \to \pi^+\pi^-$ data samples should be quite consistent with the τ dipion spectra; in contrast, one may expect that KLOE08 and BaBar should exhibit some difficulty to accomodate the τ spectra within the BHLS framework. A step further is to perform global fits within the BHLS framework including the τ spectra and the various $e^+e^- \to \pi^+\pi^-$ data samples, each in isolation or in combinations. The results obtained should allow for more conclusive statements.

The first data line in Table 1 reports some fit information derived using the various $\pi^+\pi^-$ samples in isolation, namely their various $\overline{\chi}^2$ and their global fit probabilities[12]; in these fits, the PDG information previously referred to should be removed. One observes a significant gap between KLOE08 and BaBar, on the one hand and the five other data samples, on the other hand[13].

In the same Table, one also displays the fit results associated with different combinations of the existing data samples. This Table shows that the largest set where each data sample has an average χ^2 per point close to its value in its single mode fit is Combination 1; this combination is our reference for the following. For this combination, the $\overline{\chi}^2$ of the whole set of $\pi^+\pi^-$ data sample is 0.98 with an associated large probability as indicated in the first column.

The pion form factor (FF) in e^+e^- annihilations and in the τ decay derived from fitting with Combination 1 are

[12] As illustrated by Table 3 in [9] and reminded in [17, 27], the global fit probabilities are enhanced towards 1 because several groups of data samples – especially those collected in the $\pi^0\gamma$ and $\eta\gamma$ channels – benefit from very favorable partial χ^2. Under these conditions, small global probabilities indicate suspicious behaviors.

[13] The quantity denoted $\chi^2_{\pi^+\pi^-}$ is, of course, the contribution to the total χ^2 of the $\pi^+\pi^-$ data.

Table 1. Fit mixing the indicated $e^+e^- \to \pi^+\pi^-$ samples *and* the τ spectra in single mode or combined. The values are the ratios $\chi^2/N_{\pi^+\pi^-}$ returned by the fits and the global fit probability are given for each data sample or for the selected sample combinations.

Fit Cond. $\chi^2/N_{\pi^+\pi^-}$	KLOE08	KLOE10	KLOE12	NSK	BES-III	BaBaR (trunc)	BaBaR (full)
Single	1.64	0.96	1.02	0.96	0.56	1.15	1.25
(prob)	(59%)	(97%)	(97%)	(97%)	(99%)	(74%)	(40%)
Comb. 1 (0.98 [99%])	–	1.00	1.05	1.11	0.61	–	–
Comb. 2 (1.06 [97%])	–	1.02	1.05	1.10	–	–	–
Comb. 3 (1.21 [22%])	–	1.01	1.54	1.18	0.56	1.36	–

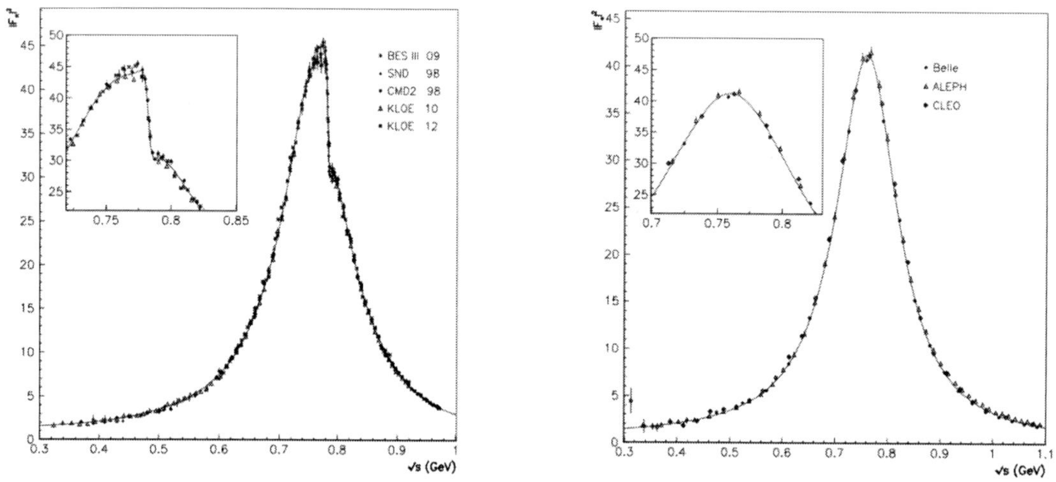

Figure 3. The pion form factor in the $e^+e^- \to \pi^+\pi^-$ annihilation (left–hand pannel) and in the τ decay (right–hand pannel) derived from a BHLS fit involving the CMD–2, SND, KLOE10, KLOE12 and BES-III data samples, on the one hand, and the ALEPH, CLEO and Belle dipion spectra, on the other hand.

displayed in Figure 3; they are clearly satisfactory. The older $\pi^+\pi^-$ data reported in [31] are also included in the fit and, on the whole, one yields $\chi^2_{\pi^+\pi^-}/N_{\pi^+\pi^-} = 361.5/404$ and $\chi^2_{\pi^\pm\pi^0}/N_{\pi^\pm\pi^0} = 86.8/85$. Therefore the picture looks satisfactory and this should be reflected by the residual plots.

Figure 4 displays the e^+e^- pion FF residuals appropriately corrected for the reported scale uncertainty effects as discussed in the Appendix of [17] and in [27]; it looks reasonably flat.

The pion FF in the τ decay is flat for the CLEO and Belle spectra[14]; the ALEPH data sample tends to exhibit a small growth starting at $\simeq$ 850 MeV. However, Figure 3 in [32] indicates that this distribution, thanks to a bug fix, should be scaled down just in this energy region and, then, it behaves like the others.

Therefore, the global fits mixing the τ spectra and the $e^+e^- \to \pi^+\pi^-$ data samples confirm the conclusions reached in the previous Section with the τ + PDG method. Stated otherwise, global fits do not indicate any mismatch between the e^+e^- and τ spectra within BHLS.

8 Including & excluding the τ spectra : Hadronic HVP issues

Let us first examine the contributions to the muon HVP provided by the pion loop in the energy region [0.630, 0.958] GeV; Figure 5 displays our results. The point at top of this Figure displays the τ+PDG prediction for $a_\mu(\pi\pi, [0.63, 0.958])$. The data points in red display the corresponding information directly reconstructed from the samples provided by the indicated experiments; some combinations of these are also shown. So, one observes a quite good correspondance between the "experimental" values and the prediction derived by the τ+PDG method; of course, the experimental values are not influenced at all by BHLS or the τ data.

In the same Figure, one also displays the results derived when merging the $e^+e^- \to \pi^+\pi^-$ data samples and the τ dipion spectra in the minimization procedure. The empty black symbols show the fit results derived by using the iterative method defined in [27]. The points in green

[14] Actually, the 3 residual distributions shown in the lower pannel of Figure 4 look quite similar to the Belle plots displayed in Figure 12 of [30].

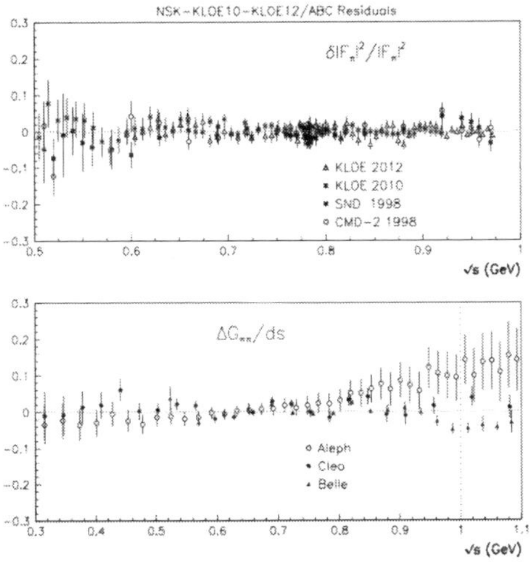

Figure 4. Residuals for the pion form factor in the $e^+e^- \to \pi^+\pi^-$ annihilation (upper pannel) and in the τ decay (lower pannel). The residuals in the upper pannel are corrected for the global scale uncertainty effects (see [17, 27]).

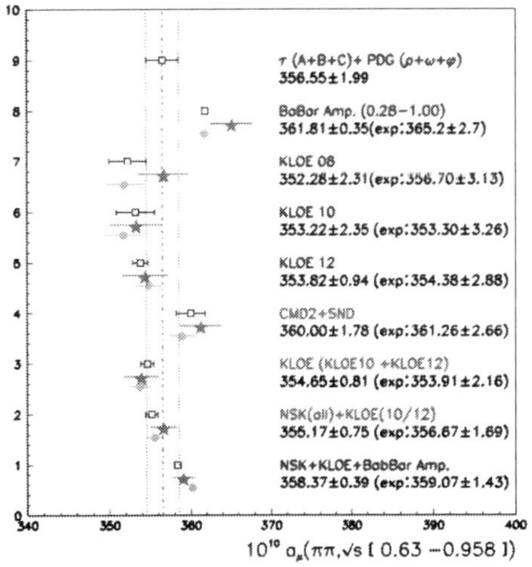

Figure 5. Values for $a_\mu(\pi\pi, [0.63, 0.958])$ in units of 10^{-10} derived from global fits using the indicated $e^+e^- \to \pi^+\pi^-$ data samples or combinations; the τ dipion spectra are always used. The full green circles are the results obtained from the $A = m$ fit (no iteration) and the black empty squares are the results obtained from the $A = M_0$ fit (first iteration) s explained in [27]. The values derived by directly integrating the experimental spectra are indicated by red stars. See Section 8 for comments.

are the corresponding results derived from the same fits but performed without iterating. The motivation for an iterative method are emphasized in [27] and aims at cancelling out possible biases affecting the channels dominated by samples subject to dominant global scale uncertainties

Figure 5 shows that the τ+PDG prediction as well as the fits merging τ and e^+e^- data are consistent with each others and also with the experimental data, except for BaBar which has difficulties to accomodate the BHLS framework as shown in Table 1. Because of the energy boundaries of its spectrum, a BES-III experimental value for $a_\mu(\pi\pi, [0.63, 0.958])$ cannot be produced.

Let us go a step further and examine the contributions to the muon HVP accessible through the BHLS Lagrangian and fitting procedure which is, as already stated, limited upward slightly above the ϕ mass; we chose 1.05 GeV. The results are displayed in Table 2. The numbers have been derived using the iterated fit method already referred to [27].

Let us focus on the $\pi^+\pi^-$ contribution which is actually the main aim of the present study. One thus observes that including the τ spectra shifts the central value by about 1.5×10^{-10} and improves the uncertainty by about 0.3×10^{-10} in both cases. The 1.5×10^{-10} difference between excluding and including the τ spectra looks rather small ($\simeq 1\ \sigma$ or less), similar to those obtained in [6], but much smaller than those in [4]. Therefore, within BHLS, the contribution of the $\pi^+\pi^-$ channel to the HVP does not exhibit any singular behavior : Using or not the τ spectra does not change this picture but improves the results as expected from having a larger statistics (e.g. the $\pi^+\pi^-$ *and* the τ data).

9 Conclusion

The analysis developped above leads to conclude that, actually, one does not observe any mismatch between the e^+e^- and the τ data. To be as precise as possible, Table 1 indicates that the τ spectra collected by ALEPH, CLEO and Belle are in perfect agreement with the CMD–2, SND, KLOE10, KLOE12 and BES-III data samples[15] each taken in isolation or considered together within a sample combination. This leads us to conclude that the so–called $\tau - e^+e^-$ puzzle is only due to the way the implementation of isospin symmetry breaking is performed within some models. In contrast, the BHLS approach and its way to account for IB effects seem to reflect correctly the expected relationship and the expected closeness of the e^+e^- annihilation and τ decay processes. Indeed, BHLS provides successful τ predictions of the e^+e^- pion form factor and a good simultaneous fit of both kinds of data.

However, we are left with a significant tension between the KLOE08 and BaBar (up to 1 GeV) samples on the one hand and the τ spectra on the other hand, as well reflected by the τ+PDG information collected in Figure 2 and by the BHLS global fit results displayed in Table 1. This is indeed an issue but, seemingly, external to the so–called $\tau - e^+e^-$ puzzle which motivates this work.

[15] As, for example, five out of the seven high statistics existing data samples.

Table 2. The various contributions to the muon HVP a_μ in units of 10^{-10} using the BHLS fit excluding the τ data (first data column), including the τ spectra (second data column), compared with the direct integration of the experimental spectra (last data column). The last line displays the total contribution accessible through the BHLS Model. The e^+e^- data samples involved are those from CMD–2, SND, KLOE10, KLOE12 and BES-III.

Channel	Excl. τ	Incl. τ	Direct Estim.
$\pi^+\pi^-$	493.02 ± 1.16	494.59 ± 0.89	492.98 ± 3.38
$\pi^0\gamma$	4.50 ± 0.05	4.54 ± 0.04	3.67 ± 0.11
$\eta\gamma$	0.64 ± 0.01	0.64 ± 0.01	0.56 ± 0.02
$\pi^+\pi^-\pi^0$	40.84 ± 0.62	40.84 ± 0.57	43.54 ± 1.29
$K_L K_S$	11.53 ± 0.09	11.53 ± 0.08	12.21 ± 0.33
K^+K^-	16.88 ± 0.22	16.90 ± 0.20	17.72 ± 0.52
Total (< 1.05 GeV)	567.00 ± 1.63	569.04 ± 1.08	570.68 ± 3.67

References

[1] Davier M, Eidelman S, Hoecker A and Zhang Z 2003 *Eur.Phys.J.* C **27** 497

[2] Davier M, Eidelman S, Hoecker A and Zhang Z 2003 *Eur.Phys.J.* C **31** 503

[3] Davier M 2007 *Nucl. Phys. Proc. Suppl.* **169** 288

[4] Davier M, Hoecker A, Malaescu B and Zhang Z 2011 *Eur.Phys.J.* C **71** 1515 (preprint arxiv:1010.4180)

[5] Benayoun M, David P, DelBuono L, Leitner O and O'Connell H B 2008 *Eur.Phys.J.* C **55** 199-236 (preprint hep-ph/0711.4482)

[6] Jegerlehner F and Szafron 2011 *Eur. Phys. J.* C **71** 1632 (preprint arXiv:1101.2872)

[7] Davier M, Hoecker A, Lopez Castro G, Malaescu B, Mo, X H, Toledo Sanchez G, Wang P,Yuan C Z and Zhang Z 2010 *Eur. Phys. J.* C **66** 127–136 (preprint arxiv:0906.5443)

[8] Harada M and Yamawaki K 2003 *Phys. Rept.* **381** 1-233 (preprint hep-ph/0302103)

[9] Benayoun M, David P, DelBuono, L and Jegerlehner F 2012 *Eur.Phys.J.* C **72** 1848 (preprint arXiv:1106.1315)

[10] Fujiwara T, Kugo T, Terao, H, Uehara S and Yamawaki K 1985 *Prog. Theor. Phys.* **73** 926–941

[11] Bando M, Kugo T and Yamawaki, K 1985 *Nucl. Phys.* B **259** 493–502

[12] Hashimoto M 1996 *Phys. Rev.* D **54** 5611-5619

[13] Benayoun M and O'Connell H B 1998 *Phys. Rev.* D 074006

[14] 't Hooft G 1986 *Phys. Rept.* **142** 357-387

[15] Marciano W J and Sirlin A 1993 *Phys. Rev. Lett.* **71** 3629-3632

[16] Cirigliano V, Ecker G and Neufeld H 2001 *Phys. Lett* B **513** 361-370 (preprint hep-ph/0104267)
Cirigliano V, Ecker G and Neufeld H 2002 *JHEP* **08** 002 (preprint hep-ph/027310)

[17] Benayoun M, David P, DelBuono, L and Jegerlehner F 2013 *Eur.Phys.J.* C **73** 2453 (preprint arXiv:1210.7184)

[18] Beringer J *et al.* 2012 Review of Particle Physics (RPP) *Phys.Rev.* D **86** 010001

[19] Aulchenko V M *et al.* 2002 *Phys. Lett.* B **527** 161-172 (preprint hep-ex/0112031)
Akhmetshin R R *et al.* 2006 *Phys. Lett.* B **648** 28-38 (preprint hep-ex/0610021)
Akhmetshin R R *et al.* 2006 *JETP Lett.* **84** 413-417 (preprint hep-ex/0610016)

[20] Achasov M N *et al.* 2006 *JJ. Exp. Theor. Phys.* **103** 380-384 (preprint hep-ex/0605013)

[21] Venanzoni G *et al.* 2009 *AIP Conf. Proc.* **1182** 665 (preprint arXiv:0906.4331)

[22] Ambrosino F *et al.* 2011 *Phys. Lett.* B **700** 102-110 (preprint arXiv:1006.5313)

[23] Babusci D *et al.* 2013 *Phys. Lett.* B **720** 336-343 (preprint arXiv:1212.4524)

[24] Aubert B *et al.* 2009 *Phys. Rev. Lett.* **103** 231801 (preprint arXiv:0908.3589)
Lees J P *et al.* 2012 *Phys. Rev.* D **86** 032013 (preprint arXiv:1205.2228)

[25] Ablikim M *et al.* 2015 (preprint arXiv:1507.08188)

[26] Benayoun M 2014 *Int.J.Mod.Phys.Conf.Ser.* **35** 1460416; Benayoun M 2013 *PoS* **Photon2013** 048

[27] Benayoun M, David P, DelBuono, L and Jegerlehner F 2015 (preprint arXiv:1507.02943)

[28] Schael S *et al.* 2005 *Phys. Rept.* **421** 191-284

[29] Anderson S *et al.* 2000 *Phys. Rev.* D **61** 112002 (preprint hep-ex/9910046)

[30] Fujikawa M *et al.* 2008 *Phys. Rev.* D **78** 072006 (preprint arXiv:0805.3773)

[31] Barkov L M *et al.* 1985 *Nucl. Phys.* B **256** 365-384

[32] Davier M, Hoecker A, Malaescu B, Yuan C–Z and Zhang Z 2014 *Eur.Phys.J.* C **74** 2803 (preprint arXiv:1312.1501)

EPJ Web of Conferences **118**, 01002 (2016)
FCCP2015
DOI: 10.1051/epjconf/201611801002

Hadronic light-by-light contribution to a_μ: extended Nambu-Jona-Lasinio, chiral quark models and chiral Lagrangians

Johan Bijnens[1,a]

[1] *Department of Astronomy and Theoretical Physics, Lund University, Sölvegatan 14A, SE22362 Lund, Sweden*

Abstract. This talk discusses our old work on the hadronic light-by-light contribution to the muon anomalous magnetic moment and some more recent contributions. I discuss the various contributions starting with pseudo-scalar meson exchange, the quark- and pion-loop, as well as scalar and a_1-exchange. For the π^0-exchange I point out a possible large enhancement when only connected contributions are included. For the quark-loop I include some comments about the more recent estimates of this contribution. The pion-loop is discussed in more detail, in particular I discuss our unpublished work on including effects from a_1 and the polarizability.

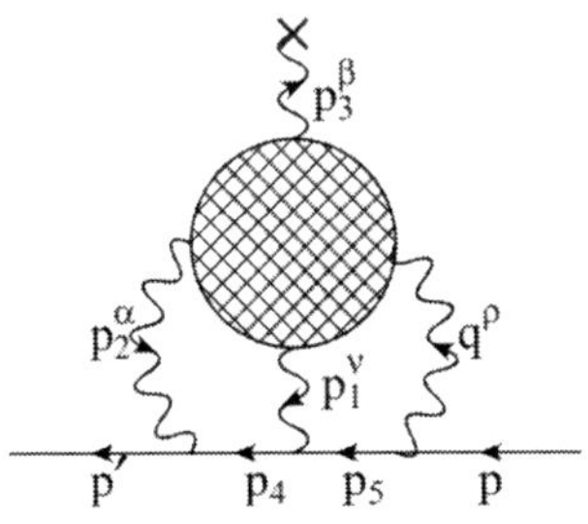

Figure 1. HLbL contribution to the muon $g-2$. The crossed blob indicates the strong interaction part.

1 Introduction

There were many talk on the muon anomalous magnetic moment and hadronic contributions to it at this conference. This manuscript should be read together with those. A general introduction to the theory of the hadronic contributions to the muon anomaly $a_\mu = (g_\mu - 2)/2$ was given in the talks by Melnikov [1] and Knecht [2]. The hadronic light-by-light-contribution (HLbL) shown in Fig. 1 was discussed in the talks by Procura [3], Cappiello [4], Greynat [5], Nyffeler [6], Lehner [7] and Vanderhaeghen [8]. There were also several talks on the underlying form-factors, both theoretically and experimentally.

The main reason is the measurement of the muon anomalous magnetic moment of [9] and the discrepancy with the satandard model prediction. Reviews of the theory can be found in [10–12] but more references and reviews can be found in the remainder of this talk and the talks mentioned above. The main conclusion was that the present best estimate of the HLbL is $(11 \pm 4) \times 10^{-10}$ [10, 12] or $(10.5 \pm 2.6) \times 10^{-10}$ [11]. The main difference is an estimate of the errors which is always somewhat subjective.

In this talk I will concentrate on the work done a long time ago [13–15] as well as some newer work on the pion loop. I will also discuss more recent contributions about the pseudo-scalar exchange and quark-loop. I do not present a new final overall number but will argue that a good estimate for the pion-loop contribution is $-(2.0 \pm 0.5) \times 10^{-10}$.

A short overview of general properties of the underlying four-point functions is Sect. 2 followed by a reminder of the ENJL model used for a large part of [13–15] in Sect. 3. Sect. 4 discusses the numerically largest contribution, pseudo-scalar meson exchange. The contribution with rather large theoretical errors, the quark-loop, is discussed in Sect. 5. Other leading large N_c exchanges are scalar, Sect. 6, and a_1-exchange, Sect. 7. I spent a large amount of space on the π-loop contribution since that is where I have some new results to present. Details are in Sect. 8. I present some conclusions and some possible future directions in the last section.

2 General properties

The problem is that the integration over photon momenta p_1, p_2 in the diagram in Fig. 1 contains both high and low moimenta and mixed cases. Double counting is thus a serious issue when using both quark and hadron contributions. Ref. [16] suggested using chiral p and large N_c counting to distinguish different contributions. This does not fully solve the double counting issue but it is a good start. This suggestion was followed by two groups doing a more or less full evaluation of the HLbL, the one I was involved in [13–15] (BPP) and Kinsohita and collaborators [17–19] (HKS). In fact, these are still the only full calculations that exist.

The underlying object is the four-point function $\Pi^{\rho\nu\alpha\beta}(p_1, p_2, p_3)$ of four electromagnetic vector currents. In fact what we really need is a derivative w.r.t. p_3 at

[a] e-mail: bijnens@thep.lu.se

© The Authors, published by EDP Sciences. This is an open access article distributed under the terms of the Creative Commons Attribution License 4.0 (http://creativecommons.org/licenses/by/4.0/).

$p_3 = 0$,

$$\left.\frac{\delta \Pi^{\rho\nu\alpha\beta}(p_1, p_2, p_3)}{\delta p_{3\lambda}}\right|_{p_3=0} . \tag{1}$$

$\Pi^{\rho\nu\alpha\beta}(p_1, p_2, p_3)$ has in general 138 Lorentz structures which reduces to 43 gauge-invariant structures. Note that in four dimensions there really are 2 less, 136 and 41 [20]. Of the 138 more general structures 28 [21] actually contribute (improving the 32 estimate of [14]). Each of these functions depends on p_1^2, p_2^2, q^2 and before the derivative also on $p_3^2, p_1{\cdot}p_3, p_2{\cdot}p_3$. This should be compared with the lowest order hadronic vacuum polarization where there is one function of one variable. There are two groups using dispersive methods to to establish a link to experiment as close as possible. These were covered in the talks by Procura [3] and Vanderhaeghen [8].

After setting $p_3 \to 0$ the loop integrals over the photon momenta is 8 dimensional. Three of these integrations are trivial and using Gegenbauer polynomial methods two more can be done [12, 21, 22]. So, after having a model or a computation of $\Pi^{\rho\nu\alpha\beta}(p_1, p_2, p_3)$ there is a triple integral over p_1^2, p_2^2, q^2 left. The components and their derivatives then become mulitplied with functions of p_1^2, p_2^2, q^2 examples of which are in [12, 22] and the full results can be found in [21]. In the work I have been involved in we have done the relevant integrations in Euclidean space, i.e. with $P_1^2, P_2^2, Q^2 = -p_1^2, -p_2^2, -q^2$ always positive.

How models actually contribute to the muon anomaly a_μ can be studied by rewriting the integral over $P_1 P_2^2, Q^2$ in the form [10]

$$a_\mu = \int dl_{P_1} dl_{P_2}\, a_\mu^{\mathrm{LL}} = \int dl_{P_1} dl_{P_2} dl_Q\, a_\mu^{\mathrm{LLQ}} \tag{2}$$

with $l_P = (1/2)\ln\left(P^2/\mathrm{GeV}^2\right)$. The reason for choosing the logarithm is that this way it is easiest to see which momentum region contributes. Alternatively one can integrate each momentum up to a cut-off Λ.

One should remember that the different contributions are usually defined within a given model or approach. What is included under the same name can therefore differ and one should be careful when drawing conclusions from comparing calculations.

3 The ENJL model

The main model underlying the work of [13–15] is the extended Nambu-Jona-Lasinio (ENJL) as introduced in [23, 24]. The Lagrangian is given by

$$\begin{aligned}\mathcal{L}_{\mathrm{ENJL}} = &\overline{q}^\alpha \left\{ i\gamma^\mu \left(\partial_\mu - iv_\mu - ia_\mu\gamma_5\right) - (\mathcal{M} + s - ip\gamma_5)\right\} q^\alpha \\ &+ 2\, g_S \left(\overline{q}_R^\alpha q_L^\beta\right)\left(\overline{q}_L^\beta q_R^\alpha\right) \\ &- g_V \left[\left(\overline{q}_L^\alpha \gamma^\mu q_L^\beta\right)\left(\overline{q}_L^\beta \gamma_\mu q_L^\alpha\right) + \left(\overline{q}_R^\alpha \gamma^\mu q_R^\beta\right)\left(\overline{q}_R^\beta \gamma_\mu q_R^\alpha\right)\right]\end{aligned}$$

with $\overline{q} \equiv \left(\overline{u}, \overline{d}, \overline{s}\right)$. v_μ, a_μ, s, p are the usual external vector, axial-vector, scalar and pseudoscalar matrix sources as used in Chiral Perturbation Theory. $\mathcal{M}$ is the quark-mass matrix. This model has no confinement but spontaneous symmetry breaking and has good pion, vector meson and OK axial vector-meson phenomenology. The usual NJL model does not have the g_V term and does not include (axial) vector mesons. The states are dynamically realized via bubble resummation. The simplicity of the model and its reasonably good phenomenology is why we used it as a basis for the calculation.

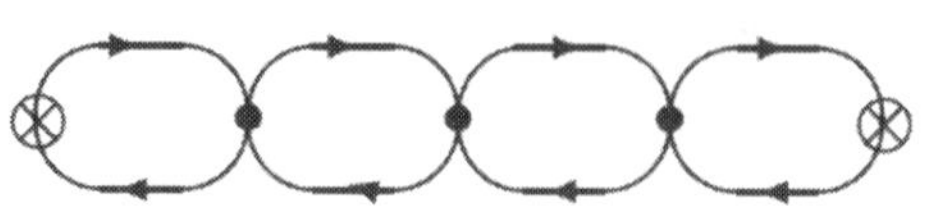

Figure 2. The bubble sum producing meson poles.

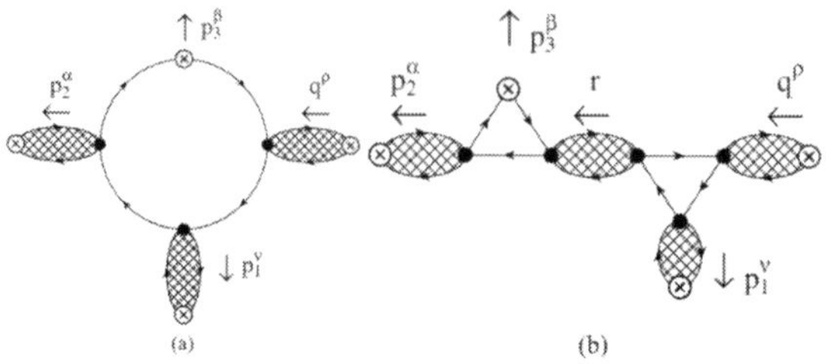

Figure 3. The two calsses of diagrams contributing to HLbL in the ENJL model. The crossed regions indicate the bubble sum "propagators." (a) Quark-loop type (b) Resonance exchange type

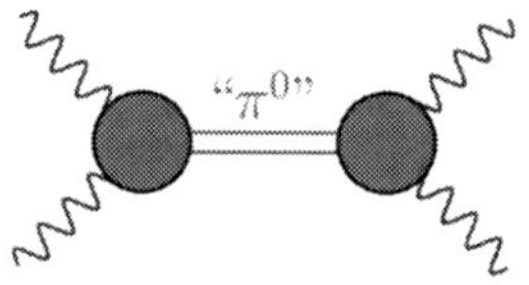

Figure 4. The π^0 exchange contribution. The blobs and the propagator need modeling.

The bubble resumation shown in Fig. 2 produces meson poles. The model with the parameters fitted in [23, 24] has a constituent quark mass of $M_Q = 263$ MeV. It has a number decent matchings to QCD short distance, e.g. for $\Pi_V - \Pi_A$ but fails in others and it always generates a vector meson dominance (VMD) type of behaviour in couplings to external photons. Processes are constructed by one-loop "vertices" coupled together with bubble chain "propagators." The relevant diagrams for HLbL are depicted in Fig. 3. Note that the exhange includes both the so-called pole and off-shell parts as calculated within the model. The vertices have also a nontrivial momentum dependence.

4 π^0-exchange

The single largest numerical contribution is given by "π^0" exchange, depicted in Fig. 4. The blobs need modeling and the propagator in the ENJL model also has corrections to the $1/(p^2 - m_{\pi^0}^2)$. The pointlike vertex has a logarithmic divergence which is uniquely predicted [25, 26]. The VMD form-factor in the $\pi^0\gamma^*\gamma^*$ form-factor, the blobs, were modeled in [14] with a variety of form-factors and as a function of the cut-off Λ (corrected for the overall sign error discovered by [22]). We took the form-factor that was made to fit the then existing data integrated up to

Table 1. The π^0 exchange results of [14].

Λ GeV	$a_\mu \times 10^{10}$ Point-like	ENJL-VMD	Point-VMD	Transv. VMD	CELLO-VMD
0.5	4.92(2)	3.29(2)	3.46(2)	3.60(3)	3.53(2)
0.7	7.68(4)	4.24(4)	4.49(3)	4.73(4)	4.57(4)
1.0	11.15(7)	4.90(5)	5.18(3)	5.61(6)	5.29(5)
2.0	21.3(2)	5.63(8)	5.62(5)	6.39(9)	5.89(8)
4.0	32.7(5)	6.22(17)	5.58(5)	6.59(16)	6.02(10)

2 GeV as our main result with a guesstimate of the error. This result was in quite good agreement with [18] which used the pointlike-VMD approach. This contribution has since been reevaluated many times using different models and approaches. A partial list is:
BPP [14]: $5.9(0.9) \times 10^{-10}$
Nonlocal quark model [27]: 6.27×10^{-10}
DSE (Dyson-Schwinger modeling)[28]: 5.75×10^{-10}
LMD+V [22]: $(5.8 - 6.3) \times 10^{-10}$
Formfactor inspired by AdS/QCD [4, 29]: 6.54×10^{-10}
Chiral Quark Model [30]: 6.8×10^{-10}
Constraint via magnetic susceptibility [31]: 7.2×10^{-10}
$VV'P$ model [32]: 6.66×10^{-10}
All of these are in reasonable agreement, within the errors. Future improvements will come when more experimental results are included as discussed in the talk by Nyffeler [6].

Two more comments are needed. The above numbers are for the π^0. One needs to take into account the η and η' exchange as well. The latter is enhanced due to the charge combinations in the $\eta'\gamma^*\gamma^*$ vertex. In large N_c models like the ENJL model, the pseudoscalar spectrum is not like QCD, one has a π^0, a $\bar\pi$ ($\bar uu + \bar dd$ quark content) and a π_s ($\bar ss$). The $\bar\pi$ has the same mass as the π^0 and due to the quark charges is contributes 25/9 times the π^0 contribution. Lattice QCD calculations with only connected diagrams included will have the $\bar\pi$ contribution as well so there will be an unphysical enhancement compared to the QCD result for the pseudoscalar exchange part. In [14] we used pointlike-VMD to estimate the ratio of π^0, η, η' contributions as 5.58, 1.38, 1.04. Models that include large N_c-breaking effects and fit the mixings to data typically end up with very similar numbers. The total pseudoscalar exchange contribution I thus estimate to be

$$a_\mu^{PS} = (8 - 10) \times 10^{-10} \qquad (3)$$

An example of a specific calculation is the AdS/QCD result of $a_\mu^{PS} = 10.7 \times 10^{-10}$ [33] which also includes excited pseudoscalars.

The other comment is that the short-distance behaviour of the four-point function is known in several limits. In particular when $P_1^2 \approx P_2^2 \gg Q^2$ the four point function is related to the axial-vector-vector-vector three-point function [34]. This three point function has a number of exact properties in QCD and we thus know how it behaves. The above models for π^0-exchange do not exhibit this behaviour. It can be implemented via making one of the blobs in Fig. 4 pointlike [34] and one then obtains 7.7×10^{-10} for the π^0-exchange contribution. Plots how this

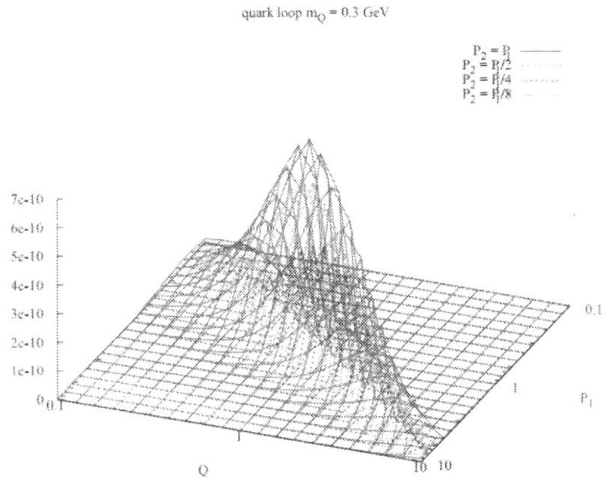

Figure 5. The quantity a_μ^{LLQ} defined in (2) as a function of P_1 and Q for several ratios P_1/P_1. The lines indicate the surface.

affects the contribution of different momentum regions are in [10]. The above behaviour of the four-point function must be obeyed in a full calculation, however whether one implements it via π^0-exchange is a choice. Models incorporating a short-distance quark-loop contribution have the short-distance part of this included [10, 27]. One can see this when comparing quark-loop plus pseudo-scalar exchange of [14] with pseudo-scalar exchange of [34].

5 Quark-loop

The pure quark-loop contribution with a constant mass is known analytically. One of the surprises is that it converges rather slowly. A significant portion is from high momentum regions. With a constituent quark mass of 300 MeV and a cut-off of 1(2) GeV 50(25)% of the full contribution is still missing. A more visual illustration of this is the plot of a_μ^{LLQ} defined in (2) of this contribution. The contribition is plotted in Fig. 5 as a function of P_1 and Q for several ratios of P_2/P_1. The volume under the curve is proportional to a_μ. The contribution peaks for $P_1 \approx P_2 \approx Q$ and around 1 GeV. In [14] we used the ENJL up to a cut-off Λ and added a short-distance quark-loop where we used the quark-mass $M_H = \Lambda$ as a lower cut-off. The estimate used by HKS was a quark-loop damped by VMD factors in the photon legs. The results are given in Tab. 2. Notice especially the stability when we add the ENJL and the short-distance contribution in the region Λ = 1-8 GeV. The conclusion is that the quark-loop is about 2×10^{-10}. In the ENJL model the quark-loop and scalar exchange are needed together to have correct chiral symmetry. The sum of both is very similar to the quark-loop estimate of HKS.

There are a number of estimates of the quark-loop that lead to much larger numbers. These have all in common that there is a momentum region with a fairly small (constituent) quark mass that is not shielded by a VMD-like mechanism. The most prominent example of this is the DSE estimate of [35] $10.7(0.2) \times 10^{-10}$. The present status

Table 2. The quark-loop contribution with VMD damping, the ENJL model and with a heavy quark mass as cut-off. The numbers are $a_\mu \times 10^{10}$.

Cut-off Λ GeV	VMD	ENJL	mass-cut	sum ENJL masscut
0.5	0.48	0.78	2.46	3.2
0.7	0.72	1.14	1.13	2.3
1.0	0.87	1.44	0.59	2.0
2.0	0.98	1.78	0.13	1.9
4.0	0.98	1.98	0.03	2.0
8.0	0.98	2.00	.005	2.0

of this calculation is given in [20]. It not yet a full calculation but includes an estimate of some of the missing parts. This DSE model describes a lot of low-energy phenomenology in a way very similar to the ENJL model. I am quite puzzled by the difference in results.

Similar size numbers are obtained in models with a low constituent quark mass where no VMD-like dynamical effects are included. Examples are the nonlocal chiral quark model [36] with $11.0(0.9) \times 10^{-10}$ and a number of estimates within the chiral quark model $(7.6-8.9) \times 10^{-10}$ [30], $(11.8-14.8) \times 10^{-10}$ [37] and $(7.6-12.5) \times 10^{-10}$[38]. The interpretation varies from an estimate to the full HLbL or just a part that needs to be added to other contributions.

6 Scalar exchange

The estimate of the scalar exchange contribution in the ENJL model is -0.7×10^{-10}. Similar size estimates have been obtained when exchanging a sigma-like particle. It should be pointed out that the scalar in the ENJL model has a phenomenology similar to the sigma but is quite a different underlying object.

A problem here is to distinguish scalar exchange from two-pion or pion-loop contributions. This is one of the areas where the method of [3, 39] will allow progress.

7 a_1-exchange

The exchange of axial vectors in the ENJL model was estimated in [14] to be about 0.6×10^{-10}, but due to the high mass involved, even with a cut-off of 2 GeV only half the contribution was there. The ENJL part also includes some pseudo-scalar meson exchange due to the structure of the calculation.

Axial-vector meson exchange in a more phenomenological way was done using two multiplets in [34] who obtained 2.2×10^{-10}. It was later found that when correct antisymmetrization is included, this becomes smaller by a significant factor and is again in the ballpark of the ENJL result. This was noticed by F. Jegerlehner. He obtains about $(0.76 \pm 0.27) \times 10^{-10}$ for the axial-vector exchange [40, 41]. The evaluation of [42] is also in reasonable agreement with the ENJL estimate.

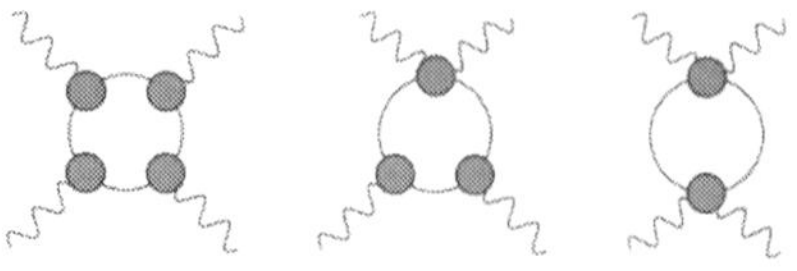

Figure 6. The charged pion loop contribution.

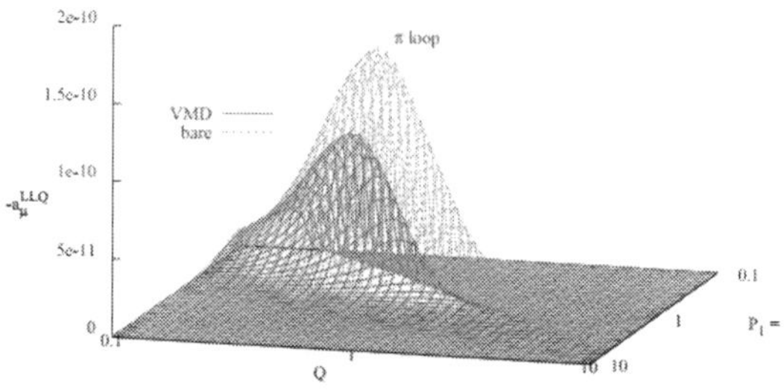

Figure 7. The momentum dependence of the pion loop contribution. Plotted is a_μ^{LLQ} of (2) as a function of $P_1 = P_2$ and Q. Top surface: sQED, bottom surface:full VMD.

8 π-loop

The π-loop contribution to the four-point function is depicted in Fig. 6. The leftmost diagram is the naive one, the other two are required by gauge-invariance. In more general models also a diagram with three photons in one vertex and one with all four in the same vertex might be needed. These have been included in the calculations mentioned below when needed.

The simplest model is a point-like pion or scalar QED (sQED). This gives a contribution of about -4×10^{-10}.

The single photon vertex is in all determinations used as including the pion form-factor. For this one can use either the VMD expression or a more model/experimental inspired version. For the $\pi\pi\gamma^*\gamma^*$ vertex there were originally two main approaches used, full VMD (BPP) and the hidden local symmetry model with vector mesons (HKS). The former is essentially using sQED and putting a VMD-like form-factor in all the photon legs. This was proven to be a consistent procedure in [14]. We obtained there a result of -1.9×10^{-10} using an ENJL inspired pion form-factor. Using a simple VMD typically gives about -1.6×10^{-10}. This version is exactly what is called the model-independent part of the two-pion contribution in [3, 39, 43]. The reason for the lower number compared to the point-like pion loop is obvious in Fig. 7 where we show a_μ^{LLQ} of (2) as a function of $P_1 = P_2$ and Q.

HKS [17, 18] used a different approach. Due to the then existing arguments against full VMD they used the hidden local symmetry model with only vector mesons (HLS) and obtained -0.45×10^{-10}. The difference between this and the previous numbers was the reason for the large error quoted on the pion-loop. This difference was rather puzzling, one reason could be that the HLS model does not have the correct QCD short distance constraint when looking at the two-photon vertex with the same and large virtuality for both photons, the full VMD model has the

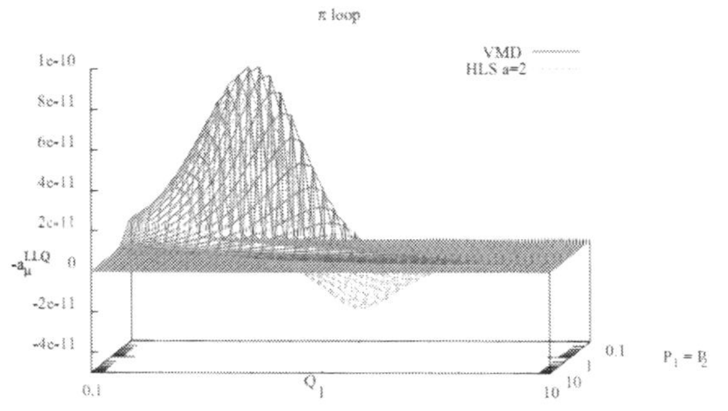

Figure 8. $-a_\mu^{LLQ}$ of (2) as a function of $P_1 = P_2$ and Q. Top surface: full VMD, bottom surface: HLS.

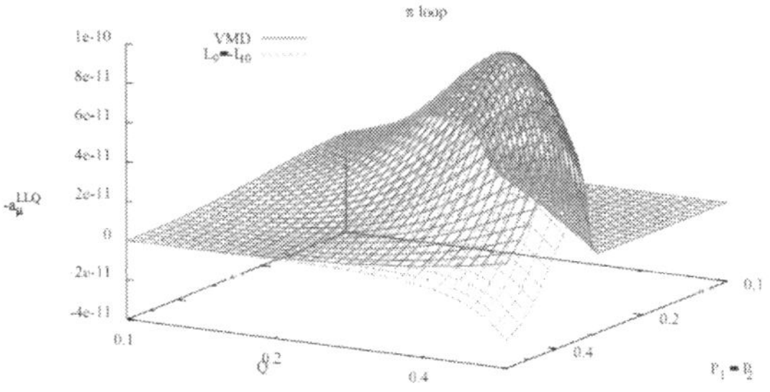

Figure 10. $-a_\mu^{LLQ}$ of (2) as a function of $P_1 = P_2$ and Q. Top surface: full VMD, bottom surface: ChPT with $L_9 = -L_{10}$ so the charge radius is included but no polarizability.

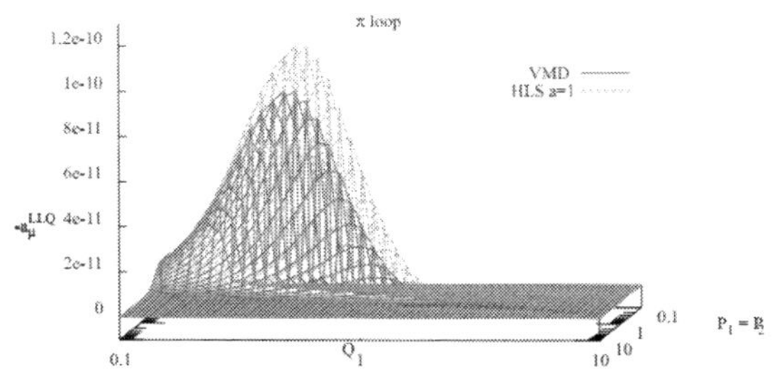

Figure 9. The momentum dependence of the pion loop contribution. $-a_\mu^{LLQ}$ of (2) as a function of $P_1 = P_2$ and Q. Top surface: HLS a=1, bottom surface: full VMD.

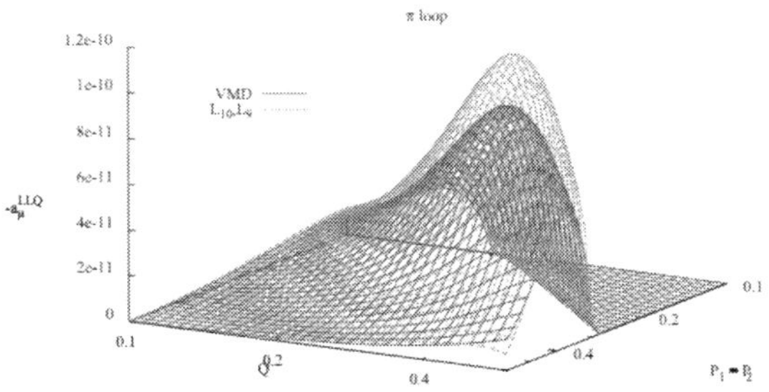

Figure 11. $-a_\mu^{LLQ}$ of (2) as a function of $P_1 = P_2$ and Q. Bottom surface: full VMD, top surface: ChPT with $L_9 \neq -L_{10}$ so the charge radius and the polarizability are included.

correct behaviour. This version of the HLS model also does not give a finite prediction for the π^+-π^0 mass difference. The reason for the large numerical difference is indeed the short distance behaviour. The low momentum behviour is very close but the negative contribution above 1 GeV, clearly visible in Fig. 8, is the main reason for the difference [21, 44]. A comparison as a function of the cut-off can be found in [45]. In fact, using the HLS with an unphysical value of the parameter $a = 1$, which then satisfies the abovementioned short-distance constraint gives very similar numbers as full VMD. This is shown in Fig. 9 From this we conclude that a number in the range $-(1.5$-$1.9)\times 10^{-10}$ is more appropriate with an error of half to 1/3 that.

More recently, it was pointed out that the effect of pion polarizability was neglected in these calculations and a first estimate of this effect given using the Euler-Heisenberg four photon effective vertex produced by pions [46] within Chiral Perturbation Theory. This approximation is only valid below the pion mass. In order to check the size of the pion radius effect and the polarizability we have implemented the low energy part of the four-point function and computed a_μ^{LLQ} for these cases. Partial results are in [44, 45]. Full results will be published in [21]. The effect of the charge radius is shown in Fig. 10 compared to the VMD, notice the different momentum scales compared to the earlier figures. As expected, the charge radius effect is included in the VMD result since the latter gives a good description of the pion form-factor. Including the effect of the polarizability can be done in ChPT by using experimentally determined values for L_9 and L_{10}. The latter can be determined from $\pi^+ \to e\nu\gamma$ or the hadronic vector two-point functions. Both are in good agreement and lead to a prediction of the pion polarizability confirmed by the compass experiment [47]. The effect of including this in ChPT on a_μ^{LLQ} is shown in Fig. 11 [21, 44, 45]. An increase of 10-15% over the VMD estimate can be seen.

ChPT at lowest order or p^4 for a_μ is just the pointlike pion loop or sQED. At NLO pion exchange with pointlike vertices and the pionloop calculated at NLO in ChPT are needed. Both gives divergent contributions to a_μ, so pure ChPT is of little use in predicting a_μ. If we want to see the full effect of the polarizability we need to include a model that can be extended all the way, or at least to a cut-off of about 1 GeV. For the approach of [46] this was done in [48] by including a propagator description of a_1 and choosing it such that the full contribution of the pion-loop to a_μ is finite. They obtained a range of $-(1.1$-$7.1)\times 10^{-10}$ for the pion-loop contribution. I find this range much too broad. One reason is that the range of polarizabilities used in [48] is simply not compatible with ChPT. The pion polarizability is an observable where ChPT should work and indeed the convergence is excellent. The ChPT prediction has also recently been confirmed by experiment. Our work discussed below indicates that $-(2.0\pm0.5)\times 10^{-10}$ is a more appropriate range for the pion-loop contribution.

The work described below will be publsihed in [21]. Preliminary results have been reported at several confer-

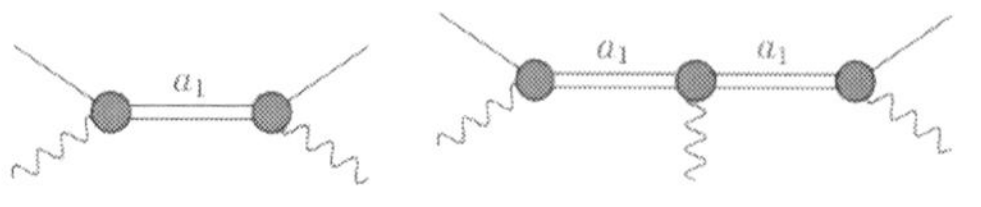

Figure 12. Left: the a_1-exchange that produces the pion polarizability. Right: an example of a diagram that is required by gauge invariance.

ences, see e.g. [49, 50] and will be fully published in [21]. The polarizability comes from $L_9 + L_{10}$ in ChPT. Using [51], we notice that the polarizability is produced by a_1-exchange depicted in Fig. 12. This is depicted pictorially in the left diagram of Fig. 12. However, once such an exchange is there, diagrams like the right one in Fig. 12 lead to effective $\pi\pi\gamma\gamma\gamma$ vertices and are required by electromagnetic gauge invariance. This was done in [48] via the propagator modifications. We deal with them via effective Lagrangians incorporating vector and axial-vector mesons.

If one looks at Fig. 12 one could raise the question "Is including a π-loop but no a_1-loop consistent?" The answer is yes with the following argument. We can fisrt look at a tree level Lagrangian including pions ρ and a_1. We then integrate out the ρ and a_1 and calculate the one-loop pion diagrams wth the resulting Lagrangian. In the diagrams of the original lagrangian this corresponds to only including loops with at least one pion propagator present. Numerical results for cases including full a_1 loops are presented as well below and in [21]. As a technicality, we use anti-symmetric vector fields for the vector and axial-vector mesons. This avoids complications due to π-a_1 mixing. We add vector $V_{\mu\nu}$ and axial-vector $A_{\mu\nu}$ nonet fields. The kinetic terms are given by [51]

$$-\frac{1}{2}\left\langle \nabla^\lambda V_{\lambda\mu}\nabla_\nu V^{\nu\mu} - \frac{M_V^2}{2} V_{\mu\nu}V^{\mu\nu}\right\rangle + V \leftrightarrow A\,. \quad (4)$$

First we add the terms that contribute to the L_i [51]

$$\frac{F_V}{2\sqrt{2}}\left\langle f_{+\mu\nu}V^{\mu\nu}\right\rangle + \frac{iG_V}{\sqrt{2}}\left\langle V^{\mu\nu}u_\mu u_\nu\right\rangle + \frac{F_A}{2\sqrt{2}}\left\langle f_{-\mu\nu}A^{\mu\nu}\right\rangle \quad (5)$$

with $L_9 = \frac{F_V G_V}{2M_V^2}$, $L_{10} = -\frac{F_V^2}{4M_V^2} + \frac{F_A^2}{4M_A^2}$. The Weinberg sum rules imply in the chiral limit $F_V^2 = F_A^2 + F_\pi^2$, $F_V^2 M_V^2 = F_A^2 M_A^2$ and requiring VMD behaviour for the pion formfactor $F_V G_V = F_\pi^2$.

First, look at the model with only π and ρ. The one-loop contributions to $\Pi^{\rho\nu\alpha\beta}$ are not finite. They were also not finite for the HLS model of HKS, but the relevant $\delta\Pi^{\rho\nu\alpha\beta}/\delta p_{3\lambda}$ was. However, in the present model it is only finite for $G_V = F_V/2$ and then the result for a_μ is identical to the HLS model. The same comments as made for the HLS model thus also apply.

Next we do add the a_1 and require $F_A \neq 0$. After a lot of work we find that $\delta\Pi^{\rho\nu\alpha\beta}/\delta p_{3\lambda}|_{p_3=0}$ is finite only for $G_V = F_V = 0$ and $F_A^2 = -2F_\pi^2$ or, if including a full a_1-loop $F_A^2 = -F_\pi^2$. These solutions are clearly unphysical.

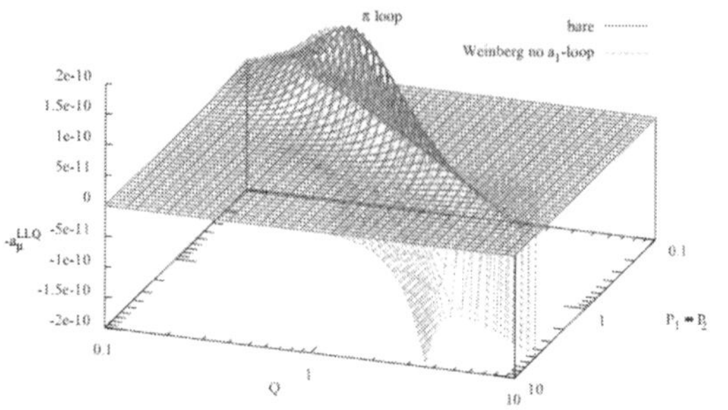

Figure 13. $-a_\mu LLQ$ as defined in (2) as a function of $P_1 = P_2$ and Q with a_1 but no full a_1-loop. Parameters determined by the Weinberg sum rules.

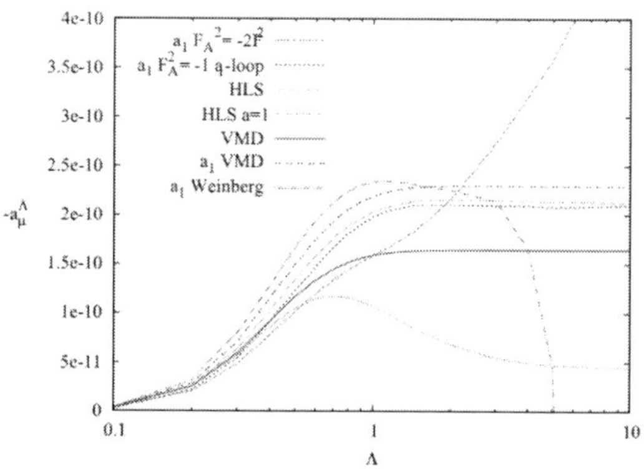

Figure 14. $-a_\mu$ using a variety of models for the pion loop as a functoin of Λ, the cut-off on the photon momenta.

We then add all $\rho a_1 \pi$ vertices given by

$$\begin{aligned}
&\lambda_1\left\langle\left[V^{\mu\nu}, A_{\mu\nu}\right]\chi_-\right\rangle + \lambda_2\left\langle\left[V^{\mu\nu}, A_{\nu\alpha}\right]h_\mu{}^\nu\right\rangle \\
&+ \lambda_3\left\langle i\left[\nabla^\mu V_{\mu\nu}, A_{\nu\alpha}\right]u_\alpha\right\rangle + \lambda_4\left\langle i\left[\nabla_\alpha V_{\mu\nu}, A_{\alpha\nu}\right]u^\mu\right\rangle \\
&+ \lambda_5\left\langle i\left[\nabla^\alpha V_{\mu\nu}, A_{\mu\nu}\right]u_\alpha\right\rangle + \lambda_6\left\langle i\left[V^{\mu\nu}, A_{\mu\nu}\right]f_-{}^\alpha{}_\nu\right\rangle \\
&+ \lambda_7\left\langle iV_{\mu\nu}A^{\mu\rho}A^\nu{}_\rho\right\rangle\,. \qquad (6)
\end{aligned}$$

These are not all independent due to the constraints on $V_{\mu\nu}$ and $A_{\mu\nu}$ [52], there are three relations. After a lot of work [21] we found that no solutions with $\delta\Pi^{\rho\nu\alpha\beta}/\delta p_{3\lambda}|_{p_3=0}$ exists except those already obtained without Λ_i terms. The same conclusions holds if we look at the combination that shows up in the integral over P_1^2, P_2^2, Q^2. We thus find no reasonable model that has a finite prediction for a_μ for the pion-loop including a_1. If we choose the parameters as fixed by the Weinberg sum rules and the VMD behaviour of the pion-form factor we obtain $-a_\mu^{LLQ}$ as shown in Fig. 13. Adding a full a_1-loop changes the plot only marginally. As long as we require the correct polarizability and a VMD-like form-factor behaviour, the plots look quite similar for all cases below 1 GeV. The integrated value up to Λ for a number of cases is hown in Fig. 14. We see that all models end up with a value of $a_\mu = -(2.0 \pm 0.5) \times 10^{-10}$ when integrated up-to a cut-off of order 1-2 GeV. We conclude that that is a resonable estimate for the pion-loop contribution.

9 Conclusions

The present number for the HLbL contribution to the muon anomaly, $a_\mu = (g_\mu - 2)/2$, is (11 ± 4) or $(10.5 \pm 2.6) \times 10^{-10}$ [10–12] depending somewhat on which error estimates and which contributions are taken into account. In this talk I have given an overview of a number of model estimates with the emphasis on my old work [13–15] as well as a number of newer developments. For the latter I have spent quite some time on our reevaluation of the pion loop contribution [21, 44, 45, 49, 50], as well as given a number of arguments why the HLS number of [17, 18] should be considered obsolete. The conclusion is that the pion-loop contributes with $-(2.0 \pm 0.5) \times 10^{-10}$.

One of the remaining problems in the model approach is that the class of models with an "unshielded" quark-loop at relatively low-energies for the photns tend to obtain larger numbers. Whether this is a real phenomenon or not is a question which needs to be settled. My own opion there is that I see no counterpart of it in $\gamma\gamma \to$ hadrons at low to intermediate energies beyond the already included single meson and two-pion exchanges.

For contributions of different mechanisms, progress can be expected both from the dispersive approaches mentioned and experiment restricting the couplings of off-shell or virtual photons to meson that go into the modeling. Alternatively, a full new model calculation that includes phenomenology beyond what the ENJL does, is very desirable. For more deatils about the other approaches and the lattice I refer to the other talks at this conference.

Acknowledgements

I thank the organizers for providing a very nice atmosphere and a good opportunity to discuss many issues regarding HLbL. This work is supported in part by the Swedish Research Council grants 621-2011-5080 and 621-2013-4287.

References

[1] K. Melnikov, *Theory review on the muon g-2*

[2] M. Knecht, *Perspectives on the hadronic contribution to the muon g-2*

[3] M. Procura, *Recent developments in the calculation of the hadronic light-by-light contribution to the muon g-2*

[4] L. Capiello, *What does holographic QCD predicts for the light-by-light scattering contribution to* $(g-2)_\mu$

[5] D. Greynat, *Hadronic contributions to the muon anomaly in the Constituent Chiral Quark Model*

[6] A. Nyffeler, *On the precision of a data-driven estimate of the pseudoscalar-pole contribution to hadronic light-by-light scattering in the muon g-2*

[7] C. Lehner, *Hadronic light-by-light contribution to* $(g-2)_\mu$ *from lattice QCD*

[8] M. Vanderhaeghen (by proxy due to a Lufthansa strike), *Hadronic Light by Light corrections to (g-2)*

[9] G. W. Bennett *et al.* [Muon g-2 Collaboration], Phys. Rev. D **73**, 072003 (2006) [hep-ex/0602035].

[10] J. Bijnens and J. Prades, Mod. Phys. Lett. A **22**, 767 (2007) [hep-ph/0702170].

[11] J. Prades, E. de Rafael and A. Vainshtein, "Hadronic Light-by-Light Scattering Contribution to the Muon Anomalous Magnetic Moment," (Advanced series on directions in high energy physics. 20) [arXiv:0901.0306 [hep-ph]].

[12] F. Jegerlehner and A. Nyffeler, Phys. Rept. **477**, 1 (2009) [arXiv:0902.3360 [hep-ph]].

[13] J. Bijnens, E. Pallante and J. Prades, Phys. Rev. Lett. **75**, 1447 (1995) [Phys. Rev. Lett. **75**, 3781 (1995)] [hep-ph/9505251].

[14] J. Bijnens, E. Pallante and J. Prades, Nucl. Phys. B **474**, 379 (1996) [hep-ph/9511388].

[15] J. Bijnens, E. Pallante and J. Prades, Nucl. Phys. B **626**, 410 (2002) [hep-ph/0112255].

[16] E. de Rafael, Phys. Lett. B **322**, 239 (1994) [hep-ph/9311316].

[17] M. Hayakawa, T. Kinoshita and A. I. Sanda, Phys. Rev. Lett. **75**, 790 (1995) [hep-ph/9503463].

[18] M. Hayakawa, T. Kinoshita and A. I. Sanda, Phys. Rev. D **54**, 3137 (1996) [hep-ph/9601310].

[19] M. Hayakawa and T. Kinoshita, Phys. Rev. D **57**, 465 (1998) [Phys. Rev. D **66**, 019902 (2002)] [hep-ph/9708227].

[20] G. Eichmann, C. S. Fischer, W. Heupel and R. Williams, arXiv:1411.7876 [hep-ph].

[21] J. Bijnens and J. Relefors, work in progress.

[22] M. Knecht and A. Nyffeler, Phys. Rev. D **65**, 073034 (2002) [hep-ph/0111058].

[23] J. Bijnens, C. Bruno and E. de Rafael, Nucl. Phys. B **390**, 501 (1993) [hep-ph/9206236].

[24] J. Bijnens, Phys. Rept. **265**, 369 (1996) [hep-ph/9502335].

[25] M. Knecht, A. Nyffeler, M. Perrottet and E. de Rafael, Phys. Rev. Lett. **88**, 071802 (2002) [hep-ph/0111059].

[26] M. J. Ramsey-Musolf and M. B. Wise, Phys. Rev. Lett. **89**, 041601 (2002) [hep-ph/0201297].

[27] A. E. Dorokhov and W. Broniowski, Phys. Rev. D **78**, 073011 (2008) [arXiv:0805.0760 [hep-ph]].

[28] T. Goecke, C. S. Fischer and R. Williams, Phys. Rev. D **83**, 094006 (2011) [Phys. Rev. D **86**, 099901 (2012)] [arXiv:1012.3886 [hep-ph]].

[29] L. Cappiello, O. Cata and G. D'Ambrosio, Phys. Rev. D **83**, 093006 (2011) [arXiv:1009.1161 [hep-ph]].

[30] D. Greynat and E. de Rafael, JHEP **1207**, 020 (2012) [arXiv:1204.3029 [hep-ph]].

[31] A. Nyffeler, Phys. Rev. D **79**, 073012 (2009) [arXiv:0901.1172 [hep-ph]].

[32] P. Roig, A. Guevara and G. López Castro, Phys. Rev. D **89**, no. 7, 073016 (2014) [arXiv:1401.4099 [hep-ph]].

[33] D. K. Hong and D. Kim, Phys. Lett. B **680**, 480 (2009) [arXiv:0904.4042 [hep-ph]].

[34] K. Melnikov and A. Vainshtein, Phys. Rev. D **70**, 113006 (2004) [hep-ph/0312226].

[35] T. Goecke, C. S. Fischer and R. Williams, Phys. Rev. D **87**, no. 3, 034013 (2013) [arXiv:1210.1759 [hep-ph]].

[36] A. E. Dorokhov, A. E. Radzhabov and A. S. Zhevlakov, Eur. Phys. J. C **75**, no. 9, 417 (2015) [arXiv:1502.04487 [hep-ph]].

[37] R. Boughezal and K. Melnikov, Phys. Lett. B **704**, 193 (2014) [arXiv:1104.4510 [hep-ph]].

[38] P. Masjuan and M. Vanderhaeghen, arXiv:1212.0357 [hep-ph].

[39] G. Colangelo, M. Hoferichter, M. Procura and P. Stoffer, JHEP **1409**, 091 (2014) [arXiv:1402.7081 [hep-ph]].

[40] F. Jegerlehner, talk presented at the workshop on Hadronic contributions to the muon anomalous magnetic moment: strategies for improvements of the accuracy of the theoretical prediction, 1-5 April 2014, Waldthausen Castle near Mainz.

[41] F. Jegerlehner, this conference.

[42] V. Pauk and M. Vanderhaeghen, Eur. Phys. J. C **74**, no. 8, 3008 (2014) [arXiv:1401.0832 [hep-ph]].

[43] G. Colangelo, M. Hoferichter, B. Kubis, M. Procura and P. Stoffer, Phys. Lett. B **738**, 6 (2014) [arXiv:1408.2517 [hep-ph]].

[44] J. Bijnens and M. Z. Abyaneh, EPJ Web Conf. **37**, 01007 (2012) [arXiv:1208.3548 [hep-ph]].

[45] M. Z. Abyaneh, arXiv:1208.2554 [hep-ph].

[46] K. T. Engel, H. H. Patel and M. J. Ramsey-Musolf, Phys. Rev. D **86**, 037502 (2012) [arXiv:1201.0809 [hep-ph]].

[47] C. Adolph *et al.* [COMPASS Collaboration], Phys. Rev. Lett. **114**, 062002 (2015) [arXiv:1405.6377 [hep-ex]].

[48] K. T. Engel and M. J. Ramsey-Musolf, Phys. Lett. B **738**, 123 (2014) [arXiv:1309.2225 [hep-ph]].

[49] K. Kampf *et al.*, arXiv:1308.2575 [hep-ph].

[50] M. Benayoun *et al.*, arXiv:1407.4021 [hep-ph].

[51] G. Ecker, J. Gasser, A. Pich and E. de Rafael, Nucl. Phys. B **321**, 311 (1989).

[52] S. Leupold, private communication.

What does Holographic QCD predicts for anomalous $(g-2)_\mu$?

Luigi Cappiello[1,2,a]

[1] *Dipartimento di Fisica "Ettore Pancini", Universitá di Napoli, Federico II*
[2] *Istituto Nazionale di Fisica Nucleare - Sezione di Napoli*

Abstract. In this talk we discuss how holographic models of QCD have been applied to the study of the light-by-light contribution to the muon anomalous magnetic moment. After a review of the holographic procedure, we discuss the approach we followed in a previous work, in which, using a certain set of holographic model as a "Theory Space", we gave our estimate for the π^0 exchange diagram. Our result depended also on the value of the quark magnetic susceptibility. Thus, in the last part of the talk, we concentrate on the attempts which have been done in the literature to apply holographic methods to the evaluation of this important order parameter of QCD.

1 Introduction

Holographic QCD (HQCD) is a set of *simple* models, aimed at reproducing relevant aspect of Large-N_c QCD in non-perturbative regime. By simple, we mean that they provide explicit analytic expressions for correlation functions of QCD operators, covering the full energy range, with surprisingly good results at lowest energy, where their predictions can be compared with those of Chiral Lagrangians, Vector Meson Dominance, etc.

HQCD models are based on classical Lagrangians, although defined in a (curved) five-dimensional space, contain a small number of parameters, and approximate well the almost conformal behavior of QCD at large momenta. They naturally encode some non-perturbative features of Large-N_c QCD, such as, for instance, the appearance of an infinite number of narrow resonance, saturating amplitude channels.

Many features are, however, model-dependent. Chiral symmetry breaking (ChSB) is differently realized in different HQCD models, each realization having its problematic issues. The mass spectrum of vector mesons do not follow a Regge behavior in some models. Subleading terms in the large momentum expansion of two-point correlation functions, related to QCD condensates, are badly reproduced or completely absent in some models. Moreover, the simplicity of HQCD models can be easily be spoiled, when further aspect of real QCD are tried to be enforced into them.

The simplicity of HQCD makes them useful for the discussion of the Hadronic Ligh-by-Light scattering contribution to the muon anomalous magnetic moment, as was also emphasized by M.Knecht in this workshop [1].

Indeed, it is possible to extract predictions for HLbL scattering using HQCD models, which turn out to be compatible with those obtained using different models and techniques. This strongly suggests that those (partial) aspects of QCD which have been encoded in the holographic models are really (among) the ones relevant for HLbL scattering contributions to $(g-2)_\mu$. Although similar conclusions had been obtained using other models and methods, HQCD offers an additional independent confirmation.

In the following, we briefly review the basics of holographic models in Sect.2, introduce the basic tools for an holographic calculation of the pion-pole contribution in Sect. 3 and illustrate the strategy we used in [2] *(i)* to work with several holographic models as a "Theory Space" and *(ii)* to take into account a new high energy constraint involving the quark magnetic susceptibility χ_0. In Sect.4, we review the various attempts to use or to extend holographic models in order to obtain predictions on χ_0. Conclusions are drawn at the end.

2 Holographic models of QCD

HQCD models have been constructed along the lines of the original conjectured AdS/CFT duality (equivalence) between a four dimensional (conformal) gauge theory at strong coupling and a (classical) five-dimensional field theory in a curved gravitational background with Anti-de-Sitter metric [3–5]. The conjecture can be boldly summarized as follows: for any 4D operator O_Δ of QCD, there exists a corresponding field $\Phi(x,z)$, living in a curved 5D space with (asymptotically) AdS metric, whose value on the 4D boundary of the 5D space, $\Phi(x,0)$, coincide with the classical source $s(x)$ coupled to O_Δ. The generating functional of correlation functions of the 4D theory is equal to the 5D action of $\Phi(x,z)$, evaluated *on-shell*

$$\begin{aligned}\exp(iW[s(x)]) &\equiv \langle\exp\left(i\int d^4x\, s(x)O_\Delta(x)\right)\rangle_{QCD}\\ &= \exp\left(iS_5(\Phi_0(z,x))\right),\end{aligned} \qquad (1)$$

[a] e-mail: luigi.cappiello@na.infn.it

© The Authors, published by EDP Sciences. This is an open access article distributed under the terms of the Creative Commons Attribution License 4.0 (http://creativecommons.org/licenses/by/4.0/).

where $\Phi_0(z, x)$ is the solution of the 5D equation of motion satisfying the boundary condition $\Phi(x, 0) = s(x)$.

The operator-field correspondence can be further specified leading to an holographic dictionary, relating 4D and 5D quantities as exemplified in Tab. 1

Table 1. Example of holographic dictionary. The last three lines contain the operators relevant for HQCD (a is a flavour index).

4D operator	5D dual field
source $s(x)$ coupled to $O_\Delta(x)$	on-shell $\Phi_0(x, z)$
conformal dimension Δ	mass m_Φ
global symmetry	gauge symmetry
conserved current $\bar{q}\gamma_\mu t^a q$	gauge field V^a_M
axial current $\bar{q}\gamma_\mu\gamma_5 t^a q$	gauge field A^a_M
quark bilinear $\bar{q}t^a q$	scalar field X^a

with the following relation between the 5D mass m_Φ of a 5D p-form field and the conformal dimension Δ of the corresponding 4D operator:

$$m^2_\Phi = (\Delta - p)(\Delta + p - 4). \quad (2)$$

One can easily check Eq.(2) for the case of 5D gauge fields, while the tachyonic mass obtained for the 5D scalar field is above the unitarity lower bound valid in an AdS space.

In the case of holographic models of QCD, additional ingredients have to be added in order to encode the effects of confinement and chiral symmetry breaking (ChSB). Different proposals have led to different models. Confinement is realized by effectively compactifying the 5D extradimension. As a consequence of the compactification, vector and axial vector meson resonances appear, corresponding to an infinite set of normalizable 5D states. One HQCD model was proposed by Sakai and Sugimoto [6, 7], who actually derived it from a more sophisticated string theory setting. The SS model is characterized by a curved but not AdS_5 metric background, contains 5D gauge fields invariant under a local chiral symmetry, which reduces to global chiral flavour transformation on the 4D boundary. In the Hard Wall (HW) models of [8, 9] and [10] the extradimension of the AdS_5 space is cut-off at a finite size, while in the Soft-Wall model of [11] the extradimension is infinite, but one still recovers infinite numbers of 4D vector and axial vector states corresponding to 5D normalizable modes in an effective potential generated by an additional 5D dilaton background. Figs. 1 and 2 show cartoons of HW and SW models.

The relevant 5D Lagrangian of an HQCD model contains vector and axial vector gauge fields, V^a_M and A^a_M dual to vector and axial quark currents $\bar{q}\gamma_\mu t^a q$and t $\bar{q}\gamma_\mu\gamma_5 t^a q$ respectively. In some models, in particular in the HW model considered in [8] and [9], one has also the scalar field X^a corresponding to the quark bilinear $\bar{q}t^a q$.

The relevant 5D Lagrangian is the following

$$S_5 = S_{\rm YM} + S_X + S_{\rm CS}\,, \quad (3)$$

where

$$S_{\rm YM} = -{\rm tr}\int d^4x \int_0^{z_0} dz\, e^{-\Phi(z)} \frac{1}{8g_5^2} w(z)\left[\mathcal{F}_L^2 + \mathcal{F}_R^2\right], \quad (4)$$

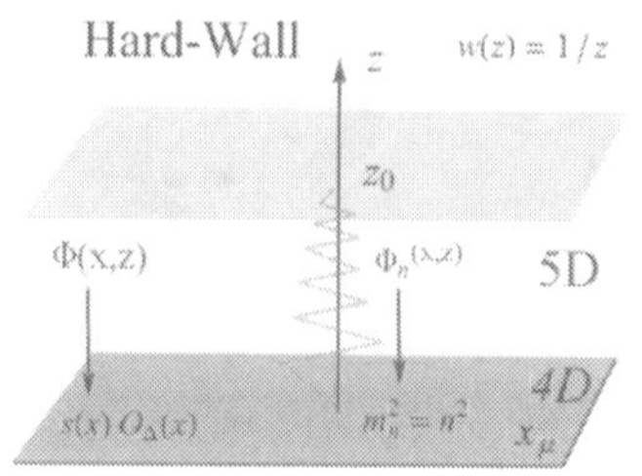

Figure 1. A pictorial representation of a Hard-Wall model. The 5D metric is AdS_5 with warp factor $w(z) = 1/z$. The extradimension z is sharply cut-off at a value z_0. This compactification generates an infinite series of 5D normalizable modes $\Phi_n(x, z)$ corresponding to an infinite tower of 4D resonances with masses $m^2_n \propto n^2$.The mass scale is fixed by the value of the dimensional parameter z_0.

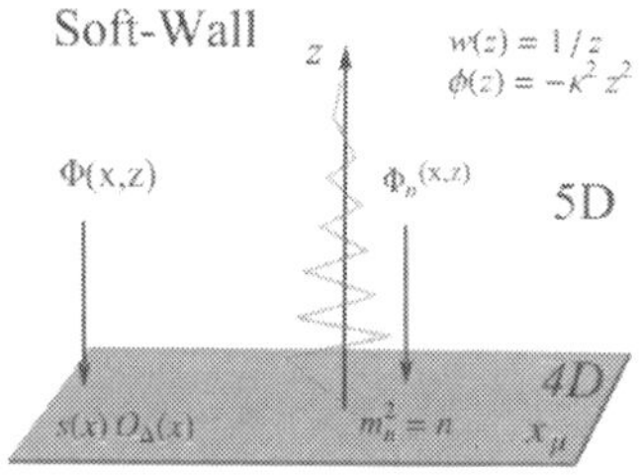

Figure 2. A pictorial representation of the Soft-Wall model.The 5D metric is AdS_5 with warp factor $w(z) = 1/z$ and with unbounded extradimension z. In presence of the dilaton background $\phi(z) = -\kappa^2 z^2$, an infinite series of 5D normalizable modes $\Phi_n(x, z)$ is produced, and hence 4D resonances with masses, $m^2_n \propto n$, following a Regge behavior. The mass scale is fixed by the value of the dimensional parameter κ.

$$S_X = {\rm tr}\int d^4x \int_0^{z_0} dz\, e^{-\Phi(z)} w(z)^3 \left[D^M X D_M X^\dagger + V(X^\dagger X)\right], \quad (5)$$

$$S_{\rm CS} = \frac{N_c}{24\pi^2}\int {\rm tr}\left(\mathcal{A}_L\mathcal{F}_L^2 - \frac{i}{2}\mathcal{A}_L^3\mathcal{F}_L - \frac{1}{10}\mathcal{A}_L^5\right) - (L \to R). \quad (6)$$

In these equations:

- $w(z)$ is the warp factor of the 5D metric $ds^2_5 = w(z)^2\left(dx^2_\mu - dz^2\right)$. For HW and SW models the metric is AdS with $w(z) = 1/z$.
- $\mathcal{F}_{MN} = \partial_M\mathcal{A}_N - \partial_N\mathcal{A}_M - i[\mathcal{A}_M, \mathcal{A}_N]$ and $\mathcal{A}_{L,R} = V \mp A$,
- In HW models the extradimension is IR cut-off at a finite value of z_0, whilst or $z_0 = \infty$ in SW models.
- In SW models, there is also a background dilaton field $\Phi(z) = -\kappa^2 z^2$, absent in the HW models.
- In the HW model of [8] and [9], (from here on dubbed HW1), the 5D scalar field $X(x, z)$, dual to $\bar{q}q$, induces ChSB, by acquiring a non trivial 5D profile $X = v(z)$. Such scalar field, and its action S_X, do not appear in the HW of [10] (which we shall call HW2), where chiral symmetry between vector and axial vector fields is bro-

ken by assigning different boundary conditions for V_μ and A_μ on the IR wall z_0.

- One may notice that the SS model [6] can be reformulated as an HW2 model, with a 5D metric which is not even asymptotically AdS for $z \to 0$.
- Both in the SS and in the HW2 model, the chiral field $U(x)$ realization of Goldstone bosons of the spontaneous ChSB is obtained. Indeed, in these models $U(x)$ is the remnant of non trivial 5D Wilson lines, induced by the ChSB boundary conditions. This makes it possible to derive predictions for the coefficients of the low-energy chiral Lagrangian in a straightforward way.

3 HLbL pion exchange diagram in HQCD

The main motivation of this conference was the forthcoming improvement in the precision of the experimental measure of the muon anomalous magnetic moment, scheduled at Fermilab, as illustrated by D. Herzog's talk [15]. At present, there is a discrepancy between the experimental value [16] and the prediction of the Standard Model. To asses whether this discrepancy is real or not is of paramount importance to understand if new physics is required to fill the gap.

Hadronic light-by-light scattering (HLbL) is the hadronic contribution to the muon anomaly which is less under control from the theoretical side, as illustrated in many reviews on the subject and in the talks by D. Melnikov [17] and M. Knecht [1] at this conference. To-date, the best estimates of the HLbL contributions are given in Ref.[18, 19] as being $(11\pm4)\times10^{-10}$ or in Ref.[20] as being $(10.5\pm)\times10^{-10}$. The two different values and errors show the difficulty to produce a world average from the plethora of different theoretical methods that have been employed in the literature, some of them illustrated in talks at this conference (this one included, of course !).

HQCD models can be used to extract a prediction for the neutral pion exchange contribution to the HLbL diagram of Fig.3. In this diagram the key role is played by the pion form factor $F_{\pi^0\gamma^*\gamma^*}(Q_1^2, Q_2^2)$, with Q_1 and Q_2 the two photon momenta. Using the HQCD action (3) explicit

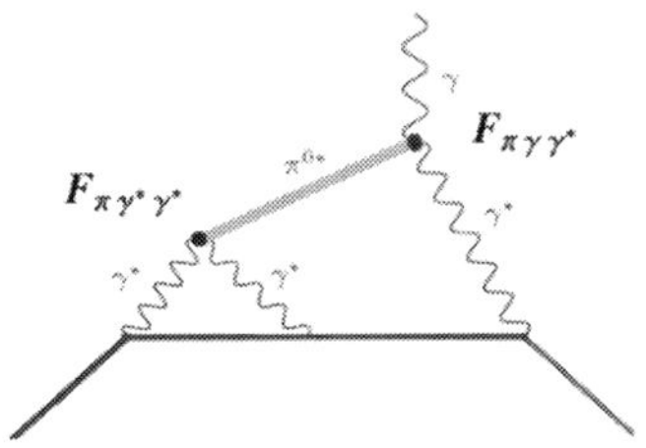

Figure 3. The neutral pion exchange diagram in the hadronic light-by-light scattering contribution to $(g-2)_\mu$

(analytic) expressions for $K(Q_1^2, Q_2^2$ defined by

$$F_{\pi^0\gamma^*\gamma^*}\left(Q_1^2, Q_2^2\right) = -\frac{N_C}{12\pi^2 f_\pi} K(Q_1^2, Q_2^2)\,,$$

can be obtained. For different HQCD models, they were worked out in a series of papers [12–14].

The relevant anomalous 3-point function $\langle AVV\rangle$ is obtained from the 5D Chern-Simons action (6), which contains the following trilinear term

$$S_{CS}^{(\pi VV)} = \frac{N_c}{4\pi^2}\epsilon^{\mu\nu\rho\sigma}\int_0^{z_0} dz\,(\partial_z\alpha(z))\int d^4x\,\pi^a\left(\partial_\rho V_\mu^a\right)\left(\partial_\sigma \hat{V}_\nu\right)\,.$$

Then, using the holographic recipe (1)

$$K(Q_1^2, Q_2^2) = -\int_0^{z_0} \partial_z\alpha(z)\mathcal{J}(Q_1, z)\mathcal{J}(Q_2, z)\,dz, \qquad (7)$$

which contains *(i)* the vector bulk-to-boundary propagator $\mathcal{J}(Q, z)$, which allows to express the 5D action *on-shell*, and *(ii)* the "pion wave function" $\alpha(z)$ coming from the pion pole in the axial bulk-to-boundary propagator.

It should be stressed that any HQCD model provides explicit expressions for $\mathcal{J}(Q, z)$ and $\alpha(z)$, obtained from solutions of second order ordinary differential equations (the 5D equation of motion) in the extra dimension z, and subject to suitable boundary conditions. This leads to explicit expressions for $K(Q_1^2, Q_2^2)$ for any value of the two virtualities. In particular, one can extract the limiting behavior at low momenta and in the deep Euclidean region.

3.1 An Hybrid Strategy: HQCD models as a "Theory Space"

In Ref. [2], the questions we addressed, using HQCD, were the following:

- Which parameters of $F_{\pi^0\gamma^*\gamma^*}$ mostly affect the predicted value of $(g-2)_\mu$?
- What does HQCD predicts for linear and quadratic slopes (low-Q^2 expansion) of $F_{\pi^0\gamma^*\gamma^*}$
- Can we discriminate some HQCD models with respect to others?
- Is Vector Meson Dominance a valid approximation or more resonances give sizable contributions?
- Can we implement QCD low- and high-energy constraints in HQCD models ?

This led us to consider a certain set of HQCD models as a sort of "Theory Space" in which to move trying to answer the questions listed above. Thus, our approach was different from the one followed in [21], were the pion form factor was evaluated in a specific holographic model HW1, *i.e.* HW1, the Hard-Wall model with chiral symmetry breaking induced by a 5D scalar field.

Eventually, our strategy was as follows: from the expressions for $F_{\pi^0\gamma^*\gamma^*}$ obtained from different HQCD models we extracted the corresponding predictions for linear slope $\hat{\alpha}$ at low values of the two photon momenta:

$$K(Q_1^2, Q_2^2) \simeq 1 + \hat{\alpha}\,(Q_1^2 + Q_2^2) + \hat{\beta}\,Q_1^2 Q_2^2 + \hat{\gamma}\,(Q_1^4 + Q_2^4,) \quad (8)$$

Different HQCD models gave different results for the linear slope $\hat{\alpha}$, as shown in Table 2.

Table 2. HQCD predictions for the linear slope. The first three holographic models, HW1, HW2 and SW, give results is in good agreement with the experimental value $\hat{\alpha}_{exp} = -1.76 \pm 0.22$ GeV^{-2}, while this does not hold for the SS model and a simpler version of HW2 with 5D flat space used for further comparison.

Model	$\hat{\alpha}$ (GeV^{-2})
HW1	-1.60
HW2	-1.81
SW	-1.66
SS	-2.04
HW2 (Flat)	-1.37

Somewhat surprisingly, a good agreement with the experimental value of $\hat{\alpha}$ is obtained only for models - HW1, HW2 and SW - which reproduce the partonic log at large-Q^2, thus matching the UV behavior of QCD. Therefore, we restricted our further analysis only to those three models. From those model we also extracted a prediction for the curvatures $\hat{\beta}$ and $\hat{\gamma}$

$$\hat{\beta} = 3.33(32)\,\text{GeV}^{-4}, \quad \hat{\gamma} = 2.84(21)\,\text{GeV}^{-4}. \tag{9}$$

The investigation of (Lowest) Vector Meson Dominance, using the resonance expansion of the bulk-to-boundary propagator

$$\mathcal{J}(Q,z) = \sum_{n=1}^{\infty} \frac{f_n}{Q^2 + m_n^2}\psi_n(z) \tag{10}$$

and consequently of the form factor

$$K(Q_1^2, Q_2^2) = \sum_{k,l=1}^{\infty} \frac{B_{kl}}{(Q_1^2 + m_k^2)(Q_2^2 + m_l^2)}, \tag{11}$$

led to the conclusion that dominant contributions to $\hat{\alpha}$, $\hat{\beta}$ and $\hat{\gamma}$ were obtained from first terms in (10), suggesting that truncating the sums in $K(Q_1^2, Q_2^2)$ to double poles was already a quite good approximation.

The full analytic expressions for $F_{\pi^0\gamma^*\gamma^*}$, which encode also the different spectrum of resonance of each model, would generically lead to complicated integrals in the evaluation of a_μ. More importantly, there is a further QCD high-energy constraint, which we want to impose, but which none of these HQCD models can take into account.

A list of high energy constraints on $F_{\pi^0\gamma^*\gamma^*}$ contains some constraints which are satisfied in HQCD models with asymptotic AdS metric

$$\lim_{Q^2\to\infty} K(Q^2, Q^2) = \frac{8\pi^2 f_\pi^2}{N_c}\frac{1}{Q^2} \tag{12}$$

$$\lim_{Q^2\to\infty} K(0, Q^2) \sim \frac{1}{Q^2}, \tag{13}$$

where the second one is the Brodsky-Lepage scaling [22].

In Ref.[23], the importance a new constraint was stressed

$$\lim_{Q^2\to\infty} F_{\pi_0^*\gamma^*\gamma^*}(Q^2, Q^2, 0) = \frac{f_\pi}{3}\chi_0 + \cdots$$

where f_π is the pion decay constant and χ_0 is a new order parameter of QCD: the quark magnetic susceptibility, which we shall deal with in greater detail in the next Section. In literature

$$0 \le \chi_0 \le 9\ \text{GeV}^{-2}.$$

All these reasons led us to propose an interpolator for the pion form factor (hereafter dubbed DIP and introduced in kaon physics in [24]) to encode both HQCD low-energy predictions and QCD short-distance constraints, and whose expression was also simple enough to allow us to use the formulas of [25] for the evaluation of the one pion exchange to HLbL contribution to a_μ:

$$K^{(DIP)}(q_1^2, q_2^2) = 1 + \lambda\left(\frac{q_1^2}{q_1^2 - m_V^2} + \frac{q_2^2}{q_2^2 - m_V^2}\right) + \eta\,\frac{q_1^2 q_2^2}{(q_1^2 - m_V^2)(q_2^2 - m_V^2)}. \tag{14}$$

Low-energy parameters $\hat{\alpha}$, $\hat{\beta}$ and $\hat{\gamma}$ are easily related to λ, η and m_V and, moreover, it is easy to implement some high- and low-energy constraints and yield predictions for the remaining parameters. For instance, imposing

$$1 + 2\lambda + \eta = 0, \quad \lambda + \eta = -\frac{4\pi^2 f_\pi^2}{3m_V^2} \quad (OPE), \tag{15}$$

$$\frac{\lambda}{m_V^2} = \hat{\alpha} = -1.76 \pm 0.22 \tag{16}$$

one gets

$$m_V = (0.64^{+0.10}_{-0.13})\text{GeV}, \tag{17}$$

$$\chi_0 = \frac{N_c}{4\pi^2 f_\pi^2}(1 + \lambda) = (2.42 \pm 0.17)\,\text{GeV}^{-2}, \tag{18}$$

with the value of m_V in the DIP close to the value of the mass of the first vector resonance, ρ.

3.2 Numerical results

We defer the reader to Ref. [2] for the details of the numerical analysis, which we did in order to extract our prediction for a_μ and estimate the error. Here, we only mention that in the numerical analysis a (mild) generalization of the interpolator was also considered, showing, however, that numerical values were stable and very close to those obtained with the original DIP. Our final number for the pion-exchange HLBL contribution was

$$a_\mu^{\pi^0} = 65.4(2.5) \cdot 10^{-11} \tag{19}$$

with the error mainly driven by the linear slope of $F_{\pi^0\gamma^*\gamma^*}$. We stress that the reported error in (19) has to be interpreted as indicating the allowed range of values of a_μ in the "Theory Space" of HQCD models that we had considered. Another caveat is that, although large values of the magnetic susceptibility χ_0 are disfavored, in the absence of stronger bounds on χ_0, an additional $(10 - 15)\%$ systematic uncertainty on the previous value for $a_\mu^{\pi^0}$ cannot be excluded.

Table 3. A partial list of estimates of $a_\mu^{\pi^0}$ obtained using different models for $F_{\pi^*\gamma^*\gamma^*}$

Models for $F_{\pi^*\gamma^*\gamma^*}$	$a_\mu^{\pi^0} \times 10^{-11}$
Modified ENJL [26–28]	59(9)
VMD/HLS [29–32]	57(4)
VMD+V ($h_2 = 0$) [33]	58(10)
VMD+V ($h_2 = -10\,\mathrm{GeV}^2$) [33]	63(10)
VMD+V (const.$F_{\pi_0\gamma^*\gamma}$) [37]	77(7)
DSE [34, 36]	58(7)
Non local χQM [38]	65(2)
AdS/QCD[21, 39]	69
AdS/QCD/DIP [2]	65.4(2.5)
RχT[40]	65.8(1.2)
CχQM [41]	68(3)

Table 3 shows a partial list of values of $a_\mu^{\pi^0}$ obtained using different models, some of them discussed in other talks at this conference. Our numerical analysis showed also that the values of χ_0 hovered around $2\,\mathrm{GeV}^{-2}$, as in the case shown in (18), with higher values disfavored although not definitely excluded. Higher values, e.g. $\chi_0 \sim 9\,\mathrm{GeV}^{-2}$, would shift the result (19) of at most 15%. Given the precision needed for $(g-2)_\mu$, a better control on χ_0 is compelling. In the next section, we shall discuss the present knowledge about χ_0 and some attempt to pinpoint its value using (extended) HQCD models.

4 Beyond the DIP ?

Can we go beyond the approach we followed in [2] and make a more thorough application of HQCD to the evaluation of $a_\mu^{\pi^0}$?

Some possible options could be the following:

- Given one of the HQCD models, we could use the full analytic expressions for $K(Q_1^2, Q_2^2)$ for the evaluation of $a_\mu^{\pi^0}$ as it was done for HW1 in [21].
- We could try to understand if a given model HQCD effectively comply with all the QCD constraints on the pseudoscalar exchange, included those advocated in [37], which led to an enhancement of the value of $a_\mu^{\pi^0}$, with respect to the other competing models.
- We could try to use the extensions of HQCD models proposed in [42, 43, 63], in which additional 5D tensor fields were added, corresponding in the AdS/CFT dictionary to quark tensor operators, in order to produce an independent prediction on the magnetic susceptibility χ_0.

On the other hand, one would readily led to some problematic issues.

- As we stressed in the introduction, none of the HQCD models that have been proposed is free from problems. That is why we tried the "Theory Space" approach, illustrated in the previous Section. For instance, it would be nice to have a model, with resonances following Regge trajectories, as happens in the SW model, and at the same time recovering the full chiral field for the pion at low momenta, which is the appealing feature of HW2. An attempt in this direction, following some ideas in [45], is currently under study.
- A technical problem, directly related to the evaluation of the HLbL contribution, is that the analytical form of $K(Q_1^2, Q_2^2)$ could not allow the use of the simpler 2-loops integrations as it was the case with DIP. (However, more general expression exists, see for instance the talk of M. Procura in this workshop [46].)
- The extended models of HQCD with 5D Lagrangians containing also to tensor fields, that we mentioned above and that one could try to use in order to get independent determinations of χ_0, seem to have serious problems in reproducing a satisfying QCD phenomenology.

In the following, we shall briefly consider this last issue.

4.1 Some facts about χ_0

In presence of a magnetic background, the tensor current

$$O^{T,a}_{\mu\nu}(x) = \bar{q}(x)t^a\sigma_{\mu\nu}q(x) \tag{20}$$

can acquire a non vanishing VEV

$$\langle\bar{q}t^a\sigma_{\mu\nu}q\rangle = e_q\chi_q\langle\bar{q}q\rangle F^{cl}_{\mu\nu} \tag{21}$$

where $F^{cl}_{\mu\nu} = \partial_\mu A^{cl}_\nu - \partial_\nu A^{cl}_\mu$.

The operator O^T is odd under C but contains both parity even and odd parts and thus can create create both $J^{PC} = 1^{--}$ and 1^{+-} states $\rho^{(n)}, b_1^{(n)}$:

$$\begin{aligned}\langle 0|O^{T,a}_{\mu\nu}(x)|b_1^{(n),c}(k)\rangle &= \frac{i}{\sqrt{2}} f_b^{(n)}\epsilon_{\mu\nu\alpha\beta}\varepsilon^{\perp(n)\alpha}_{(b)}k^\beta\delta^{ac}e^{-ik\cdot x}\\ \langle 0|O^{T,a}_{\mu\nu}(x)|\rho^{(n),c}(k)\rangle &= \frac{i}{\sqrt{2}} f_\rho^{T,(n)}(\varepsilon^{\perp(n)}_{(\rho)\mu}k_\nu - \varepsilon^{\perp(n)}_{(\rho)\nu}k_\mu)\delta^{ac}e^{-ik\cdot x}.\end{aligned}$$

χ_0 can be estimated using QCD sum rules, because it enters in the mixed VT 2-point function

$$\lim_{q^2\to 0}\Pi_{VT}(q^2) = -\chi_0\langle\bar{q}q\rangle \tag{22}$$

Since it first evaluation, done in [47] using QCD sum rules, many different approaches have been followed to calculate χ_0. The comparison of the various results, however, is not immediate, partly because χ_0 depends on the renormalization scale. In comparing the results, it is common to write the magnetic susceptibility in terms of a dimensionless parameter c_χ as

$$\chi_0 = -c_\chi\frac{N_c}{16\pi^2 f_\pi^2} \tag{23}$$

where $f_\pi = 92.4$ is the pion decay constant.

Table 4 shows an incomplete list of values c_χ, obtained using different theoretical approaches.

The first four raws in Table 4 contain the results obtained applying sum rules in different contexts: analysing the of nucleon magnetic moments and $\Delta \to N\gamma$ radiative transitions [47], studying photon distribution amplitudes in QCD [50] or radiative heavy meson decays [51].

 DOI: 10.1051/epjconf/201611801003

Table 4. Values of c_χ obtained with different theoretical methods.

Method (renormalization point)	c_χ
Sum rules (1 GeV) [47]	3.86 ± 0.11
Sum rules (0.5 GeV) [48]	2.56
Sum rules (1 GeV) [49]	1.98 ± 0.18
Sum rules (1 GeV) [50]	1.41 ± 0.13
Sum rules (1 GeV) [51]	1.28 ± 0.22
OPE + Pion Dominance (0.5 GeV) [52]	2
Instanton vacuum (1 GeV) [53]	1.12 ± 0.07
Zero mode of Dirac Op. (1 GeV) [54]	1.58
Lattice (1 GeV) [55]	0.695(1)
Lattice (1 GeV) [56]	1.90 ± 0.08
NJL model (0.63/ GeV) [57]	-1.93
QM model (0.56 GeV) [57]	-2.36
$\langle AVV \rangle$ in HQCD ($\ll$ 1 GeV) [58]	2.15
$\langle AVV \rangle$ in HQCD [59]	2
HQCD + Tensors [60]	0.06
HQCD +Tensors [61]	1–1.5

The result $c_\chi = 2$ was obtained by Vaishtein in [52], studying the QCD $\langle AVV \rangle$ 3-point function, when two virtualities of the external legs are large and one vector current represents the constant external electromagnetic field strength, and then using OPE and pion pole dominance. This result is extremely important, in particular for the comparison with the predictions obtained from HQCD models.

Further results displayed in Table 4 were obtained by using instantons [53], zero mode solutions of quark Dirac equations in QCD [54], lattice [55, 56] and other models in [57].

4.2 HQCD predictions for χ_0

The last four raws in Table 4 show the results of applying (extended) HQCD models to evaluation of the quark magnetic susceptibility.

The first two results were obtained from the study of the anomalous three-point function $\langle AVV \rangle$ in HW1 model [58], and in both HW1 and HW2 (SS) model in [59]. In particular the result of [59] exactly coincides with the one obtained in [52] using QCD OPE, non renormalization theorems and pion pole dominance. The authors of [59] based their results on an anomaly matching for resonances valid in HQCD models that, if true also in real QCD, reproduced Vainshtein result. However, their assumptions were criticised as being actually not true in QCD [62].

The last two raws in Tab 4 show the values of for c_χ obtained in HW1 model extended to include an antisymmetric tensor dual to the QCD operator $O^{T,a}_{\mu\nu}(x)$ in (20).

There is a great discrepancy between the result found in [60] and those found in [58, 59], which were instead very close or coinciding with the one of Vainshtein [52]. The authors of [61], who used the same 5D model HW1 with tensor fields as in [60], in order to make a global fit on low energy phenomenological parameters, raised the question about the real necessity to match the UV behavior of holographic models with the QCD one. They advocated that by relaxing this constraint, a better fit to low energy properties could be obtained. In particular, regarding the quark magnetic susceptibility, they were led to a range of values much closer to the other ones reported in Table 4.

Actually, that was not the first time that a more phenomenological use of the HQCD models was proposed in the literature. In [61], it was argued that the UV matching should be done only in the case of conserved currents, whose scaling dimensions are not modified by the RG flow; this latter property does not hold for the tensor currents $O^{T}_{\mu\nu}(x)$. Indeed, in Ref. [63], using the same model, with holographic correlation functions of the vector and tensor operators reproducing the large-Q^2 behavior expected from QCD, a poor phenomenological agreement of the resulting spectrum of resonances was obtained. Notice that already in [42], we were led to modify the UV boundary conditions, altering the rule in Eq.(2) of the holographic dictionary in the case of the antisymmetric tensor field, in order to get a realistic resonance spectrum.

5 Conclusions

HQCD models allow many analytic calculations of physical quantities which are relevant for non perturbative regime of QCD. In particular, they capture important aspect of QCD at low and intermediate energies, where QCD is at strong coupling, reproducing features like chiral symmetry breaking, vector meson dominance, including whole families of vector and axial vector resonances of increasing masses.

However, there is not yet a model prevailing on the others: any of them has some problematic issue, eventually not shared by the others. This state of affairs led us to consider, in the evaluation of HLBL contribution to $(g-2)_\mu$, the idea of playing in the "Theory Space" of a set of HQCD models, in order to extract generic predictions. The result for the pion exchange contribution to the hadronic ligh-by-light that we obtained is compatible with those obtained in other models.

Our result depends on an additional UV constraint containing the quark magnetic susceptibility χ_0. We have reviewed the attempts that have been made in the literature to get predictions for this quantity, by extending some Hard-Wall models through the introduction of tensor fields dual to QCD tensor operators. These models, however, have not led to an improvement of predictive power. The results for χ_0 can be considered at most preliminary.

We still have to understand how far we can go trying to strike the right balance between the simplicity of the HQCD models and their predictivity.

Acknowledgments

I thank the organizers and the participants of the conference, for the very stimulating atmosphere. I also thank my collaborators Oscar Cata, Giancarlo D'Ambrosio and David Greynat. This work has been supported in part by MIUR under project 2010YJ2NYW and by INFN research initiative PhenoLNF.

References

[1] M. Knecht, *Perspectives on the hadronic contribution to the muon g-2*
[2] L. Cappiello, O. Cata and G. D'Ambrosio, Phys. Rev. D **83**, 093006 (2011) [arXiv:1009.1161 [hep-ph]].
[3] J. M. Maldacena, Adv. Theor. Math. Phys. **2**, 231 (1998) [Int. J. Theor. Phys. **38**, 1113 (1999)];
[4] S. S. Gubser, I. R. Klebanov and A. M. Polyakov, Phys. Lett. B **428**, 105 (1998);
[5] E. Witten, Adv. Theor. Math. Phys. **2**, 253 (1998)
[6] T. Sakai and S. Sugimoto, Prog. Theor. Phys. **113**, 843 (2005);
[7] T. Sakai and S. Sugimoto, Prog. Theor. Phys. **114**, 1083 (2005)
[8] J. Erlich, E. Katz, D. T. Son and M. A. Stephanov, Phys. Rev. Lett. **95**, 261602 (2005)
[9] L. Da Rold and A. Pomarol, Nucl. Phys. B **721**, 79 (2005);
[10] J. Hirn and V. Sanz, JHEP **0512**, 030 (2005);
[11] A. Karch, E. Katz, D. T. Son and M. A. Stephanov, Phys. Rev. D **74**, 015005 (2006)
[12] H. R. Grigoryan and A. V. Radyushkin, Phys. Rev. D **76**, 115007 (2007).
[13] H. R. Grigoryan and A. V. Radyushkin, Phys. Rev. D **77**, 115024 (2008) [arXiv:0803.1143 [hep-ph]].
[14] H. R. Grigoryan and A. V. Radyushkin, Phys. Rev. D **78**, 115008 (2008) [arXiv:0808.1243 [hep-ph]].
[15] D. Hetzog, *Status of muon g-2 experiment at FNAL*
[16] G. W. Bennett *et al.* [Muon g-2 Collaboration], Phys. Rev. D **73**, 072003 (2006) [hep-ex/0602035].
[17] D. Melnikov, *Theory review of the muon (g-2)*
[18] J. Bijnens and J. Prades, Mod. Phys. Lett. A **22**, 767 (2007) [hep-ph/0702170 [HEP-PH]].
[19] F. Jegerlehner and A. Nyffeler, Phys. Rept. **477**, 1 (2009) [arXiv:0902.3360 [hep-ph]].
[20] J. Prades, E. de Rafael and A. Vainshtein, Adv. Ser. Direct. High Energy Phys. **20**, 303 (2009) [arXiv:0901.0306 [hep-ph]].
[21] D. K. Hong and D. Kim, Phys. Lett. B **680**, 480 (2009) [arXiv:0904.4042 [hep-ph]].
[22] G. P. Lepage and S. J. Brodsky, Phys. Rev. D **22**, 2157 (1980).
[23] A. Nyffeler, Phys. Rev. D **79**, 073012 (2009) [arXiv:0901.1172 [hep-ph]].
[24] G. D'Ambrosio, G. Isidori and J. Portoles, Phys. Lett. B **423**, 385 (1998) [arXiv:hep-ph/9708326].
[25] M. Knecht and A. Nyffeler, Phys. Rev. D **65**, 073034 (2002) [arXiv:hep-ph/0111058].
[26] J. Bijnens, E. Pallante and J. Prades, Phys. Rev. Lett. **75**, 1447 (1995) [Phys. Rev. Lett. **75**, 3781 (1995)] [hep-ph/9505251].
[27] J. Bijnens, E. Pallante and J. Prades, Nucl. Phys. B **474**, 379 (1996) [hep-ph/9511388].
[28] J. Bijnens, E. Pallante and J. Prades, Nucl. Phys. B **626**, 410 (2002) [hep-ph/0112255].
[29] M. Hayakawa, T. Kinoshita and A. I. Sanda, Phys. Rev. Lett. **75**, 790 (1995) [hep-ph/9503463].
[30] M. Hayakawa, T. Kinoshita and A. I. Sanda, Phys. Rev. D **54**, 3137 (1996) [hep-ph/9601310].
[31] M. Hayakawa and T. Kinoshita, Phys. Rev. D **57**, 465 (1998) [Phys. Rev. D **66**, 019902 (2002)] [hep-ph/9708227].
[32] M. Hayakawa and T. Kinoshita, hep-ph/0112102.
[33] M. Knecht, A. Nyffeler, M. Perrottet and E. de Rafael, Phys. Rev. Lett. **88**, 071802 (2002) [hep-ph/0111059].
[34] C. S. Fischer, T. Goecke and R. Williams, Eur. Phys. J. A **47**, 28 (2011) [arXiv:1009.5297 [hep-ph]].
[35] T. Goecke, C. S. Fischer and R. Williams, AIP Conf. Proc. **1343**, 539 (2011) [arXiv:1012.0193 [hep-ph]].
[36] T. Goecke, C. S. Fischer and R. Williams, Phys. Rev. D **83**, 094006 (2011) [Phys. Rev. D **86**, 099901 (2012)] [arXiv:1012.3886 [hep-ph]].
[37] K. Melnikov and A. Vainshtein, Phys. Rev. D **70**, 113006 (2004) [hep-ph/0312226].
[38] A. E. Dorokhov and W. Broniowski, Phys. Rev. D **78**, 073011 (2008) [arXiv:0805.0760 [hep-ph]].
[39] D. K. Hong, D. Kim and S. Matsuzaki, Phys. Rev. D **81**, 073005 (2010) [arXiv:0911.0560 [hep-ph]].
[40] K. Kampf and J. Novotny, Phys. Rev. D **84**, 014036 (2011) [arXiv:1104.3137 [hep-ph]].
[41] D. Greynat and E. de Rafael, JHEP **1207**, 020 (2012) [arXiv:1204.3029 [hep-ph]].
[42] L. Cappiello, O. Cata and G. D'Ambrosio, Phys. Rev. D **82**, 095008 (2010) [arXiv:1004.2497 [hep-ph]].
[43] S. K. Domokos, J. A. Harvey and A. B. Royston, JHEP **1105**, 107 (2011) [arXiv:1101.3315 [hep-th]].
[44] R. Alvares, C. Hoyos and A. Karch, Phys. Rev. D **84**, 095020 (2011) [arXiv:1108.1191 [hep-ph]].
[45] L. Cappiello, G. D'Ambrosio and D. Greynat, Eur. Phys. J. C **75**, no. 10, 465 (2015) [arXiv:1505.01000 [hep-ph]].
[46] M. Procura, *Recent developments in the calculation of the hadronic light-by-light contribution to the muon g-2.*
[47] B. L. Ioffe and A. V. Smilga, Nucl. Phys. B **232**, 109 (1984).
[48] V. M. Belyaev and Y. I. Kogan, Yad. Fiz. **40**, 1035 (1984).
[49] I. I. Balitsky, A. V. Kolesnichenko and A. V. Yung, Sov. J. Nucl. Phys. **41**, 178 (1985) [Yad. Fiz. **41**, 282 (1985)].
[50] P. Ball, V. M. Braun and N. Kivel, Nucl. Phys. B **649**, 263 (2003)
[51] J. Rohrwild, JHEP **0709**, 073 (2007)
[52] A. Vainshtein, Phys. Lett. B **569**, 187 (2003).
[53] H. C. Kim, M. Musakhanov and M. Siddikov, Phys. Lett. B **608**, 95 (2005)
[54] B. L. Ioffe, Phys. Lett. B **678**, 512 (2009).

[55] P. V. Buividovich, M. N. Chernodub, E. V. Luschevskaya and M. I. Polikarpov, Nucl. Phys. B **826**, 313 (2010).

[56] V. V. Braguta, P. V. Buividovich, T. Kalaydzhyan, S. V. Kuznetsov and M. I. Polikarpov, Phys. Atom. Nucl. **75**, 488 (2012) [arXiv:1011.3795 [hep-lat]].

[57] M. Frasca and M. Ruggieri, Phys. Rev. D **83**, 094024 (2011) [arXiv:1103.1194 [hep-ph]].

[58] A. Gorsky and A. Krikun, Phys. Rev. D **79**, 086015 (2009).

[59] D. T. Son and N. Yamamoto, arXiv:1010.0718 [hep-ph].

[60] A. Gorsky, P. N. Kopnin, A. Krikun and A. Vainshtein, Phys. Rev. D **85**, 086006 (2012) [arXiv:1201.2039 [hep-ph]].

[61] S. K. Domokos, J. A. Harvey and A. B. Royston, JHEP **1304**, 104 (2013) [arXiv:1210.6351 [hep-th]].

[62] M. Knecht, S. Peris and E. de Rafael, JHEP **1110**, 048 (2011) [arXiv:1101.0706 [hep-ph]].

[63] R. Alvares, C. Hoyos and A. Karch, Phys. Rev. D **84**, 095020 (2011) [arXiv:1108.1191 [hep-ph]].

Status and perspectives of the MEG experiment

Fabrizio Cei[1,a]

[1] *INFN and Universitty of Pisa, Department of Physics, Largo B.Pontecorvo 3, 56127 Pisa (Italy)*
On behalf of the MEG Collaboration

Abstract. We present the status and the perspectives of the MEG experiment which has been searching for the Lepton Flavour Violating decay $\mu^+ \to e^+\gamma$ for several years. In a dataset corresponding to 3.6×10^{14} positive muons stopped on target we didn't find any evidence for this decay and established an upper bound on the $\mu^+ \to e^+\gamma$ branching ratio of 5.7×10^{-13} at 90 % C.L., with a sensitivity of 7.7×10^{-13}, which improves our world best limit by a factor of 4 and the best previous limit by a factor of 20. Our final dataset is about twice as large than the analyzed sample and the expected final sensitivity of the experiment is $\sim 5 \times 10^{-13}$. We also discuss the status of the upgrade of the experiment (MEGII) which aims to improve the sensitivity by a further order of magnitude.

1 Introduction to Lepton Flavour Violation and $\mu^+ \to e^+\gamma$ decay

The Lepton Flavour Violation (LFV) in the charged lepton sector is almost forbidden in the Standard Model (SM), even including neutrino oscillations and mixing [1]. Conversely, the majority of New Physics models (NP) [2], particularly in view of recent measurements of a large θ_{13} at reactor [3] and accelerator [4] experiments, predict measurable branching ratios ($\mathcal{B}$) for LFV reactions such as $\mu^+ \to e^+\gamma$. The discovery of a LFV process would be clear evidence for NP, whereas improvements in the $\mathcal{B}$ upper limits (UL) constitute significant constraints on the parameter space, complementary to those obtainable at high energy colliders. The expected $\mathcal{B}$ for $\mu^+ \to e^+\gamma$ decay ranges from $\sim 10^{-14}$ to $\sim 10^{-12}$; here we show the results and the perspectives of the MEG experiment, which has improved the best world upper limit on this process by a factor of 20 and can reach a sensitivity which covers a large fraction of the allowed parameter space of NP schemes.

2 The MEG Experiment

The MEG experiment [5] at Paul Scherrer Institute (PSI) [6] aims to search for the $\mu^+ \to e^+\gamma$ decay with a sensitivity $\sim 10^{-13}$. The $\mu^+ \to e^+\gamma$ signature for muons decaying at rest is a back-to-back monoenergetic (52.83 MeV each) time-coincident e^+-γ pair. Positron and photon candidates are characterized by their energies (E_γ, E_e), their relative directions ($\theta_{e\gamma}$, $\phi_{e\gamma}$)[1] and emission time ($t_{e\gamma}$). The background has two components: the Radiative Muon Decay $\mu^+ \to e^+\nu_e\bar{\nu}_\mu\gamma$ (RMD), whose rate is proportional to the muon stopping frequency R_μ and the ACCidental Background (ACCB), given by the random coincidence of energetic positrons from the SM Michel decay with photons from RMD, e^+-e^- annihilation-in-flight or bremsstrahlung. The ACCB, whose rate is proportional to ${R_\mu}^2$, is responsible for 93 % of MEG events with $E_\gamma >$ 48 MeV.

The MEG experiment uses the PSI πE5 beam line which is capable of delivering up to 10^8 stopping μ^+/s even if a lower intensity of 3×10^7 μ^+/s is used to control the ACCB. The positive muon beam is stopped in a 205 micron thick plastic target which is slanted $\approx 20°$ with respect to the beam axis. The experiment employs a Liquid Xenon (LXe) detector for the measurement of the photon energy, arrival time and first interaction point as well as a magnetic spectrometer for the measurement of the positron momentum vector and timing. The photon detector is formed by a single volume ($\sim$ 900 ℓ) of LXe, viewed by 846 UV-sensitive photomultiplier tubes (PMTs) submerged in the liquid. The spectrometer is composed by 16 Drift Chambers (DC), each one formed by two staggered layers of sense wires and cathodic foils, and by a double-array of scintillation Timing Counter (TC) located inside a superconducting solenoid (COnstant Bending RAdius, COBRA) with a gradient magnetic field along the beam axis. The magnetic field is arranged to sweep out positrons with small longitudinal momenta and make the positron bending radius almost independent of emission angle. The detector is continuously monitored by a multi-element calibration system [5, 7, 8] including: 1) a Cockroft-Walton accelerator; 2) an array of α sources mounted inside the photon detector; 3) a liquid hydrogen target, used in conjunction with a π^- beam to induce the Charge Exchange (CEX) reaction $\pi^-p \to \pi^0 n$, followed

[a] e-mail: fabrizio.cei@pi.infn.it

[1] $\theta_{e\gamma} = (\pi - \theta_e) - \theta_\gamma$ and $\phi_{e\gamma} = (\pi + \phi_e) - \phi_\gamma$, θ and ϕ being the polar angle and the azimuthal angle respectively, taking the beam-axis as z-axis.

© The Authors, published by EDP Sciences. This is an open access article distributed under the terms of the Creative Commons Attribution License 4.0 (http://creativecommons.org/licenses/by/4.0/).

by the π^0 decay in two photons; 4) a variable monoenergetic positron beam and 5) a neutron generator. The two photons emitted in the CEX calibration are detected in timing and angular coincidence, one in the LXe calorimeter and one in an auxiliary detector, a NaI or a higher resolution BGO array. The relative alignment between the DCs and between the LXe detector and the spectrometer is obtained by looking at straight tracks from cosmic muons and at curved Michel positron tracks and comparing the measurements with an optical survey, improved in 2011 by inserting a laser tracker and prismatic corner cube reflectors mounted on the DC modules. Based on FPGA technology the trigger system is designed to reduce the trigger rate from several MHz to ~ 10 Hz by using fast estimates of photon energy and positron-photon relative timing and direction, with a resulting signal efficiency and live time fraction > 95 % and ≈ 99 % respectively [9]. The readout is performed by a custom-made chip (Domino Ring Sample, DRS), developed at PSI, with a maximum sampling speed of 5 GHz and 12 bit voltage digitization [10].

3 Data sample

The experiment collected data continuously and in stable running conditions from 2009 to 2013. A first UL on $\mathcal{B}$ for $\mu^+ \to e^+\gamma$ decay was published in 2011 [8] based on the 2009 – 2010 data sample. Here we show the results obtained by analyzing a double-size sample formed by data collected from 2009 to 2011 and corresponding to about 3.6×10^{14} positive muons stopped on target.

4 Analysis procedure and results

Our analysis is based on a combination of a blind and a maximum likelihood (ML) approach [8], applied in the analysis region defined by 48 MeV $< E_\gamma <$ 58 MeV, 50 MeV $< E_e <$ 56 MeV, $|t_{e\gamma}| <$ 0.7 ns, $|\theta_{e\gamma}| <$ 50 mrad and $|\phi_{e\gamma}| <$ 50 mrad. We call "time sidebands" the regions defined by 1 ns $< |t_{e\gamma}| <$ 4 ns, "E_γ-sideband" as defined by 40 MeV $< E_\gamma <$ 48 MeV and "angle sidebands" those defined by 50 mrad $< |\phi_{e\gamma}| <$ 150 mrad or 50 mrad $< |\theta_{e\gamma}| <$ 150 mrad. The positron track is reconstructed by combining its measured positions at each DC layer (hit) while longitudinal z-positions are derived from signals induced on the segmented DC cathodes. The positron kinematic variables are extracted by using a Kalman filter track fitting technique [11], with a model for the hits and the track based on the GEANE package [12]. An accurate model for multiple scattering and energy loss in the detector materials along with a detailed map of the magnetic field, measured with a 0.2 % precision, are also included. The fitted positron track is propagated to the TC allowing an iterative refinement of the hits with the positron time measurement. The track fit yields a parameter covariance matrix, resulting in very good agreement with the measured resolutions, extracted by using the two turn method [8]. Consequently, a per-track error is determined which allows us to follow the variable DC performance during the data taking period and is taken into account in the ML analysis. The average hit multiplicity for a track is 10 and only tracks with ≥ 7 hits and ≤ 2 turns in the spectrometer are retained for the analysis. A single positron per event is selected by applying additional track quality cuts. The positron energy resolution, evaluated by fitting the kinematic edge, is well described by the sum of three Gaussian curves with a resolution σ_{E_e} = (300 – 310) keV for the core component (≈ 85 %)[2]. The ϕ_e-resolution has a ϕ_e-dependence with a minimum at $\phi_e = 0$, where it is measured by the two-turn method to be σ_{ϕ_e} = (7 – 8) mrad. Similarly the measured θ_e-resolution is σ_{θ_e} = (10 – 11) mrad. The decay vertex coordinates and the positron direction at the vertex are determined by extrapolating the reconstructed track back to the target. The resolutions on the decay vertex coordinates, also determined by the two-turn method, are described by a Gaussian curve with σ_z = (1.5 – 2.0) mm and in the vertical direction by the sum of two Gaussian curves with σ_y = (1.2 – 1.3) mm for the core component (≈ 85 %).

The LXe detector uses the xenon scintillation light to measure the photon position, first interaction time and total energy release. The 3-D photon interaction point is reconstructed by looking at the pattern of scintillation light detected by the PMTs near the incident position and the photon interaction time by combining the leading edge times of the PMT waveforms. The photon direction is defined as the line connecting the decay vertex to the photon interaction point in the LXe detector. The photon energy reconstruction is based on the sum of the charges collected by all PMTs. Monochromatic 55 MeV photons from π^0 decay are used to determine the absolute energy scale. Pile-up photons, which at $R_\mu = 3 \times 10^7 \mu^+/\mathrm{s}$ affect ≈ 15 % of triggered events, are identified topologically by the PMT light distribution, temporally by the PMT leading edge time distribution and by means of a new algorithm which analyzes waveforms after summing all channels at the end of the full chain of photon reconstruction. This algorithm also allows to re-adjust the charge integration window for the energy estimate, resulting in a better energy reconstruction. The efficiency of photon reconstruction, obtained by comparing CEX data with MC simulations, is ~ 63 %, with a 4 % improvement due to the new algorithm mentioned above. The position reconstruction resolution, evaluated by a MC simulation and validated in CEX runs by placing lead slit collimators in front of the LXe detector, is 5 mm on the photon entrance face and 6 mm along the radial depth w. The timing and energy resolutions are evaluated using 55 and 83 MeV photons from π^0 decay. The LXe timing resolution is σ_{t_γ} = 67 ps at 52.8 MeV. The position-dependent energy resolutions are measured in the CEX data and the average energy resolution, extracted from a Gaussian fit to the high energy side of the spectrum, results 1.7 % and 2.4 % for $w >$ 2 cm and $w <$ 2 cm respectively. The resolutions of the relative directions are obtained by combining the relevant resolutions of positrons and photons discussed above. The results are ≈ 16 mrad for $\theta_{e\gamma}$ and ≈ 9 mrad for

[2] The ranges here and after take into account the small, yearly differences caused by various factors (e.g. variable DC configurations and efficiencies).

DOI: 10.1051/epjconf/201611801004

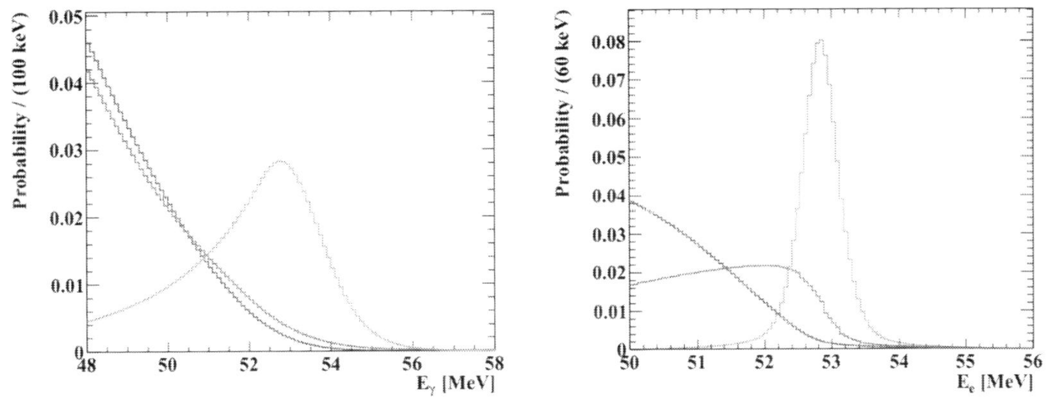

Figure 1. PDF for photon energy (left) and for positron energy (right). In both figures green curves are for signal, red curves for RMD and purple curves for ACCB.

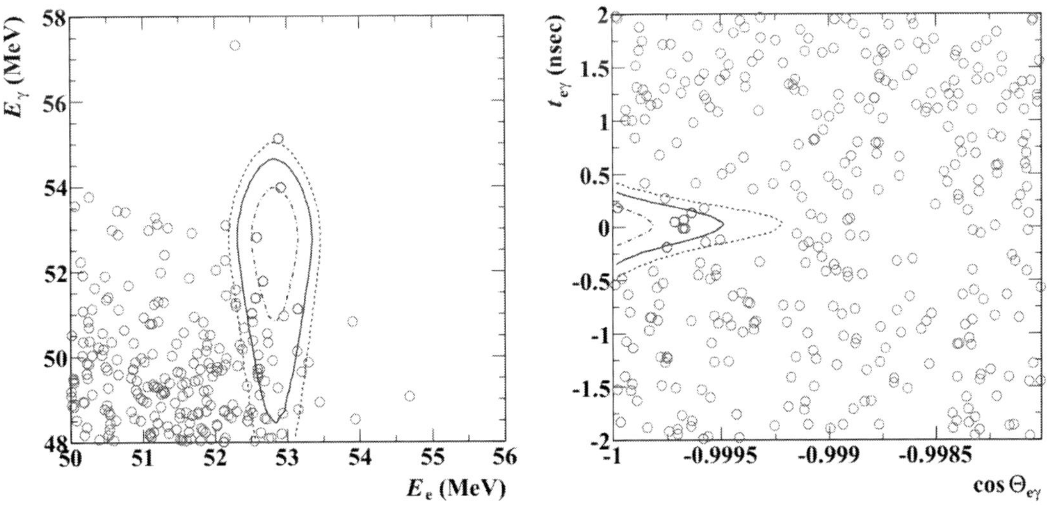

Figure 2. Event distributions for the 2009 – 2011 dataset in the (E_e, E_γ) and $(\cos\Theta_{e\gamma}, t_{e\gamma})$ planes. In the figure on the left (right) a selection of $|t_{e\gamma}| < 0.244$ ns and $\cos\Theta_{e\gamma} < -0.9996$ with 90 % efficiency for each variable (52.4 MeV $< E_e <$ 55 MeV and 51 MeV $< E_\gamma <$ 55.5 MeV with 90 % and 74 % efficiencies for E_e and E_γ, respectively) is applied. The signal PDF contours (1, 1.64 and 2 σ) are also shown.

$\phi_{e\gamma}$. The relative time $t_{e\gamma}$ is derived from the time measurements in the LXe detector and in the TC, after correcting for the particle time-of-flights. The corresponding resolution at 52.8 MeV energy is $\approx$ 130 ps, measured from the RMD peak observed in the E_γ-sideband above the flat ACCB distribution; a small correction takes into account the E_γ-dependence measured in the CEX runs. The position of the RMD-peak ($t_{e\gamma} = 0$), monitored constantly during the data taking period, is stable within 15 ps.

The ML analysis is supplemented by a blind analysis procedure which involves masking a region of 48 MeV $< E_\gamma <$ 58 MeV and $|t_{e\gamma}| < 1$ ns until the Probability Density Functions (PDFs) for the likelihood function are finalized. The time, energy and angle sidebands are used to optimize the analysis, to study ACCB and to extract the corresponding PDFs. Calibration data and measured resolutions are used to build the Signal (S) and RMD events PDFs. Different resolutions and correlations are included in the PDFs on an event-by-event basis. The per-event positron error matrix, estimated by the new Kalman filter, is introduced in the PDFs, together with the dependence on the photon interaction point and the quality of the positron tracking. Examples of PDFs for photon and positron energy are shown in Fig. 1.

All the PDFs are inserted in the global likelihood function [8] which includes also the sideband constraints on the number of RMD and ACCB events. The ML fit is performed to estimate the number of S, RMD and ACCB

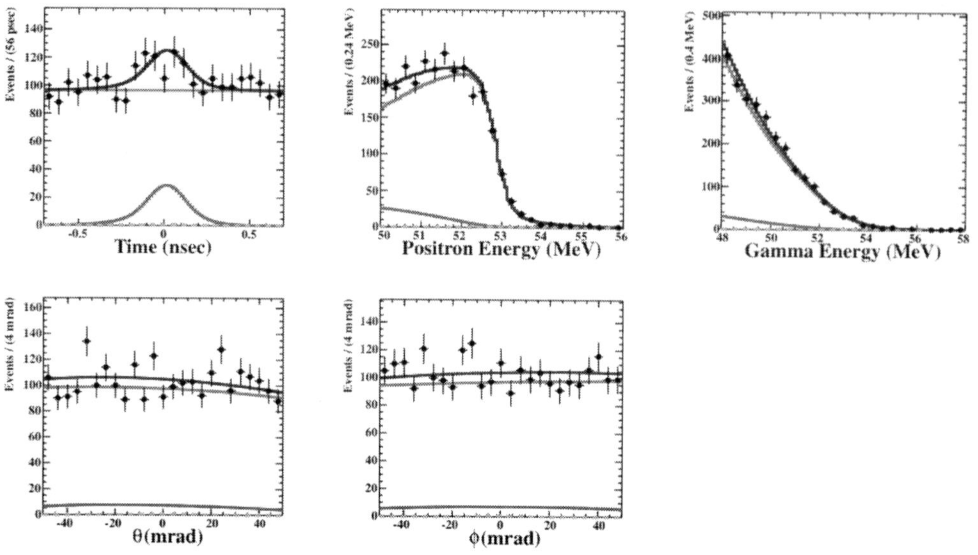

Figure 3. Results of ML fit for the 2009 – 2011 dataset. The black dots are the data and the colour lines represent the contributions extracted from the fit: S (green), RMD (red), ACCB (purple) and total (blue).

events in the analysis region. An analysis with constant PDFs is also performed as a crosscheck and has shown consistent results. The confidence interval for the number of S events is calculated by a frequentist method with a profile likelihood-ratio ordering [8, 13], where the numbers of RMD and ACCB events are treated as nuisance parameters. The experimental sensitivity $\mathcal{S}_{90}$ is evaluated as follows: a large sample of pseudo experiments (toy MCs) with 0 signal events is generated and fit with the ML algorithm. $\mathcal{S}_{90}$ is defined as the median of the distribution of the 90 % C.L. ULs extracted from the toy MC ensemble. Likelihood fits are also performed in fictitious analysis regions in the time and angle sidebands, getting ULs in good agreement with the $\mathcal{S}_{90}$'s. The normalization factor needed to convert a UL on the number of S events into a UL on $\mathcal{B}(\mu^+ \to e^+\gamma)$ is computed by using two independent schemes; either counting the number of Michel positrons selected with a dedicated pre-scaled trigger, or the number of RMD events observed in the muon data. The combined uncertainty on the $\mathcal{B}$ value is 4 %. The systematic uncertainties on the PDF parameters and on the normalization are taken into account in the calculation of the confidence intervals by fluctuating the PDFs by the amount of the uncertainties. The global effect on the observed UL is 1 % with the main contribution coming from the angular PDFs. Fig. 2 shows the event distributions in the (E_e, E_γ) and $(\cos\Theta_{e\gamma}, t_{e\gamma})$ planes for the 2009 – 2011 dataset, where $\Theta_{e\gamma}$ is the opening angle between positron and photon, together with the contours of the averaged signal PDFs. We show in Fig. 3 the results of the ML fit for the dataset 2009 – 2011 where the best fit value for the number of S events is −0.4. The observed profile likelihood ratios as a function of $\mathcal{B}$ are shown in Fig. 4 for three different datasets: 2009 – 2010, 2011 and 2009 – 2011 combined sample. The UL at 90 % C.L.

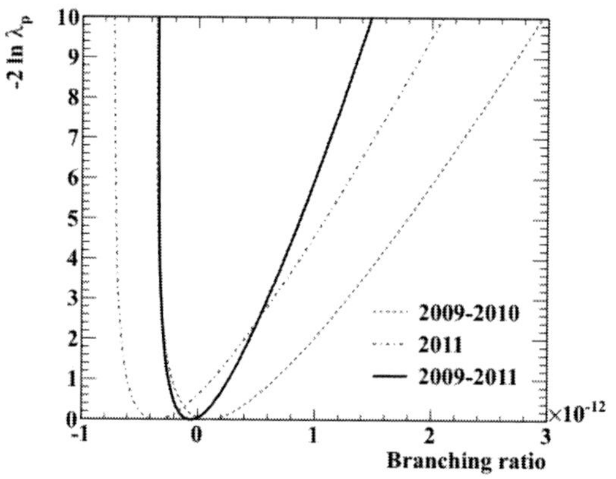

Figure 4. Observed profile likelihood ratios (λ_p) as a function of $\mathcal{B}$ for 3 different data sets: 2009 – 2010, 2011 and the combined 2009 – 2011 sample.

$(\mathcal{B}_{90})$ and $\mathcal{S}_{90}$ for the full 2009 – 2011 dataset are respectively 5.7×10^{-13} and 7.7×10^{-13}. As a quality check the ML fit was repeated omitting the constraint on the number of background events. We obtained $N_{\rm RMD} = 163 \pm 32$ and $N_{\rm ACCB} = 2411 \pm 57$, in good agreement with the expectations estimated from E_γ and time sidebands $\langle N_{\rm RMD}\rangle = 169 \pm 17$ and $\langle N_{\rm ACCB}\rangle = 2415 \pm 25$. The obtained UL is more stringent than $\mathcal{S}_{90}$, but this is not unusual since the probability of having $\mathcal{B}_{90}$ equal or smaller than the observed quantity is calculated to be 24 % with a

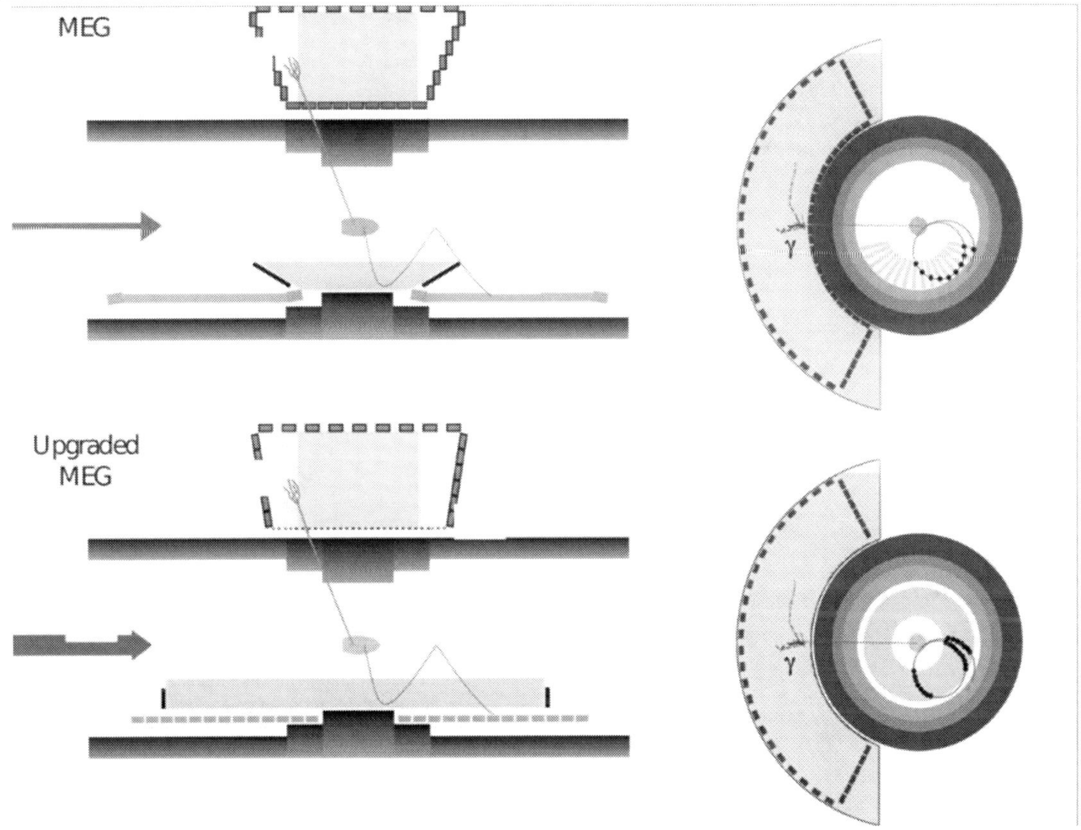

Figure 5. Overview of the MEG upgrade compared with the present MEG configuration.

sample of toy MC experiments. These results were published in 2013 [14] and improve the previous MEG UL [8] by a factor of 4.

5 Expected final sensitivity

The results were obtained by analyzing a data sample corresponding to about half of the total amount of data collected by MEG with the total number of positive muons stopped on target about 7.2×10^{14}. The analysis of the remaining data is in progress and is expected to be completed shortly. The present Upper Limit $\mathcal{B}(\mu^+ \to e^+\gamma) < 5.7 \times 10^{-13}$ at 90% C.L. represents an improvement by a factor of 20 with respect to the pre-MEG limit [15] and a significant constraint for several NP models. When the full statistics will be analyzed we expect to reach a final sensitivity, estimated by using toy MC's, of $\sim 5\times10^{-13}$. It's important to note that, despite the expected doubling of data statistics, the predicted improvement in sensitivity is only of about a factor of 1.5. This is due to the fact that systematic effects dominate the limiting sensitivity because of the presence of an irreducible background. The unique way to obtain a much more significant improvement, at the level of a further order of magnitude in sensitivity, is to reduce the spill-in of background events within the signal region by enhancing the detector resolution. This was the basic motivation for the MEG upgrade project.

6 The MEG upgrade

The MEG upgrade [16] project was based on the long-term experience we gained during MEG phase I. The guideline was the possibility to take advantage of a higher intensity beam (from 3×10^7 to 7×10^7 stopping positive muons per second), thanks to an enlarged detector acceptance and to better resolution sub-detectors, to be obtained with moderate costs and on a relatively short time-scale ($\sim 3-4$ years). An overview of the MEG upgrade is shown in Fig. 5.

The main element of the upgrade is the replacement of the present DC system with a 2 m long unique volume cylindrical drift chamber. With respect to the present MEG configuration, this new device has 1) a higher momentum, 2) a higher angular resolution and 3) a much

Table 1. Resolution (Gaussian σ) and efficiencies for MEG upgrade

PDF parameters	Present MEG	Upgrade scenario
$\sigma_{E_{e^+}}$ (keV)	380	110
e^+ σ_θ (mrad)	11	5
e^+ σ_ϕ (mrad)	9	5
e^+ σ_Z / σ_Y(core) (mm)	2.0/1.2	1.2/0.7
$\frac{\sigma_{E_\gamma}}{E_\gamma}$ (%) w>2 cm	1.7	1.0
γ position at LXe $\sigma_{(u,v)}$-σ_w (mm)	5	2
γ-e^+ timing (ps)	130	80
Efficiency (%)		
trigger	≈ 99	≈ 99
γ reconstruction	63	63
e^+ reconstruction	40	95
event selection	80	85

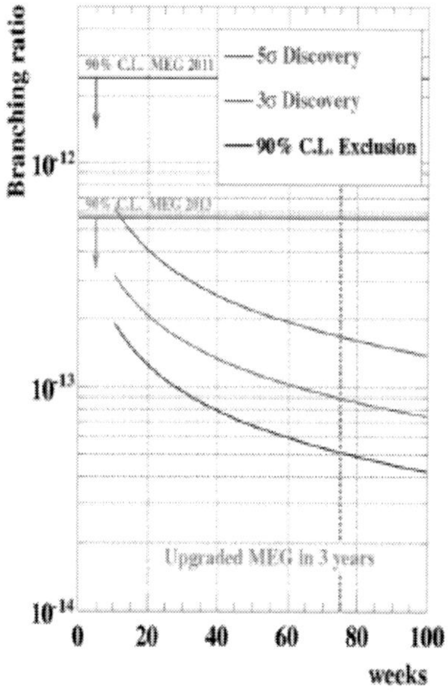

Figure 6. Expected sensitivity of MEG upgrade as a function of DAQ time. The present [14] and previous [8] MEG upper limits are indicated by arrows. The Blue, Red and Black curves show respectively the $\mathcal{B}$ corresponding to a discovery at 5 σ and 3 σ levels and the 90 % C.L. exclusion line.

larger positron efficiency because of the reduction of the material budget traversed by the positron and of the much better coupling with the TC. Moreover, since no cathode foils will be used for the z coordinate reconstruction, there will be no problems of low level signals and of electric discharges due to the formation of charge deposits on the cathodes. The new DC will be equipped with ~ 1400 sense wires and ~ 8000 field and guard wires with a 2π azimuthal coverage and will be operated with a 85:15 He/Isobutane gas mixture. The amount of traversed material corresponds to less than 2×10^{-3} X_0 and the stereo view angle is $(7-8)^\circ$. The single hit resolution is expected to be 120 micron. This value was tested with laboratory measurements on a small and a full length prototype by using cosmic rays and beam tests [17]. The small prototype was also used to check that the expected ageing of the gas mixture during the MEG II DAQ time will produce a maximum gain drop of 40 %, which can be corrected by increasing the HV, and that the z coordinate can be reconstructed with a $\sim$ 10 cm precision by using the charge division on the two sides of signal wires. Other major upgrades involve:

a) a different target configuration, with 140 micron thickness and a slant angle of 15°, to better substain the increased muon stopping rate.

b) The replacement of LXe photon detector inner face PMTs with an array of 12 mm × 12 mm Silicon Photomultipliers (SiPM) sensitive to LXe scintillation light. The increased granularity and active coverage are expected to enhance 1) the light collection efficiency, 1) the uniformity and 3) the pile-up rejection capability with an improvement of a factor of 2 in energy and position resolution that is considerable for shallow events where the photon converts very close to the entering thin window. The photon detection efficiency is also expected to increase by some percent because of the smaller mass to be traversed by the photon entering the LXe detector. About 600 SiPMs were already tested and show a good photon detection efficiency for LXe scintillation light.

c) A redesign of the LXe detector lateral sides and a better suited placement of PMTs on these sides to improve the acceptance and the light collection uniformity.

d) The building of a new pixellated Timing Counter, made by a 2-D array of scintillator tiles, to improve the resolution of positron timing measurement by means of multiple hits in different tiles. The single tile resolution and the dependence of the timing resolution as a function of the number of crossed tiles were measured in beam tests and helped obtain an average positron timing resolution of $\approx$ 35 ps, in agreement with upgrade project needs.

e) The building of a new mixed trigger/digitizer DAQ board (WaveDREAM) in order to fulfill the requests of an increased number of read-out channels and of a higher bandwidth of the analog Front-End.

f) The insertion in the DownStream side of the experiment of a LYSO+plastic scintillator detector to detect low energy positrons emitted in RMD decay and tag the corresponding high-energy photon coming from the same decay. This detector is already under construction.

The figures of merit of the upgraded detector are compared in Table 1 with that of the present detector. The expected sensitivity, the $3-\sigma$ and the $5-\sigma$ discovery curves as a function of the DAQ time are shown in Fig. 6. With 3 years of data taking we expect to reach a sensitivity of $\sim 5\times10^{-14}$ with an improvement of one order of magnitude with respect to the estimated final sensitivity of MEG phase I.

References

[1] F. Böhm & P. Vogel, *Physics of Massive Neutrinos*, Cambridge University Press, (1992), S. T. Petcov, Sov. J. Nucl. Phys. **25** (1977) 340.

[2] R. Barbieri, L. Hall and A. Strumia, *Nucl. Phys. B* **455** (1995) 219, J. Hisano, D. Nomura and T. Yanagida, *Phys. Lett. B* **437** (1998) 351, M. Raidal et al., *Eur. Phys. J. C* **57** (2008) 13, G. Blankenburg et al., *Eur. Phys. J. C* **72** (2012) 2126.

[3] F. P. An et al., *Phys. Rev. Lett.* **108** (2012) 171803, F. P. An et al., *Phys. Rev. Lett.* **112** (2014) 061801, J. K. Ahn et al., *Phys. Rev. Lett.* **108** (2012) 191802, Y. Abe et al., *Phys. Lett. B* **723** (2013) 66.

[4] K. Abe et al., *Phys. Rev. Lett.* **107** (2011) 041801, K. Abe et al., *Phys. Rev. Lett.* **112** (2014) 061802.

[5] J. Adam et al., *Eur. Phys. J. C* **73** (2013) 2365

[6] http://www.psi.ch

[7] J. Adam et al., *Nucl. Instr. Meth. A* **641** (2011) 19, A. M. Baldini et al., *Nucl. Instr. Meth. A* **565** (2006) 589, A. M. Baldini et al., *Nucl. Instr. Meth. A* **545** (2006) 753

[8] J. Adam et al., *Phys. Rev. Lett.* **107** (2011) 171801

[9] L. Galli et al., *JINST* **8** (2013) P01008.

[10] S. Ritt et al., *Nucl. Instr. Meth. A* **623** (2010) 486

[11] P. Billoir, *Nucl. Instr. Meth. A* **225** (1984) 352, R. Frühwirth, *Nucl. Instr. Meth. A* **262** (1987) 444.

[12] V. Innocente and E. Nagy, *Nucl. Instr. Meth. A* **324** (1993) 297, A. Fontana et al., *J. Phys. Conf. Ser.* **119** (2008) 032018.

[13] J. Beringer et al. (Particle Data Group), *Phys. Rev. D* **86** (2012) 010001, G. J. Feldman and R.D. Cousins, *Phys. Rev. D* **57** (1998) 3873.

[14] J. Adam et al., *Phys. Rev. Lett.* **110** (2013) 201801

[15] M. Ahmed et al., *Phys. Rev. D.* **65** (2002) 112002.

[16] A. M. Baldini et al., *MEG Upgrade Proposal*, `physics.ins-det/1301.7225`.

[17] M. Venturini et al., *Nucl. Instr. Meth. A* (2015) doi:10.1016/j.nima.2015.09.063.

DOI: 10.1051/epjconf/201611801005

The muon EDM in the g-2 experiment at Fermilab

Rebecca Chislett[1,a] *on behalf of the* Muon $g-2$ Collaboration[b]

[1] *Department of Physics and Astronomy, UCL, London, WC1E 6BT, UK.*

Abstract. The observation of a muon electric dipole moment would provide an additional source of CP violation which is required to explain the matter anti-matter asymmetry in the universe. The current experimental limit, $|d_\mu| < 1.9 \times 10^{-19} e\cdot$cm, was set by the BNL E821 experiment. This paper discusses how the new experiment at Fermilab, E989 [3], aims to decrease this by two orders of magnitude down to $10^{-21} e\cdot$cm.

1 Introduction

The new g-2 experiment at Fermilab is designed to measure the anomalous part of the muon magnetic dipole moment (MDM) to a precision of 140ppb, an improvement over the 540ppb measurement made at the previous BNL E821 experiment [1]. However, it is also possible to make a measurement or set a limit on a possible muon electric dipole moment (EDM). The new experiment hopes to improve on the E821 limit of $|d_\mu| < 1.9 \times 10^{-19} e\cdot$cm [2] by about two orders of magnitude.

Theoretically, fundamental particles can have an EDM directed along the spin and governed by an equation analogous to the experimentally observed MDM,

$$\vec{d} = \eta \frac{Qe}{2mc} \vec{s}. \quad (1)$$

The Hamiltonian, $\mathcal{H}$, for a spin one-half particle in an electric (E) and magnetic (B) field for a particle with a magnetic moment, μ, is

$$\mathcal{H} = -\vec{\mu} \cdot \vec{B} - \vec{d} \cdot \vec{E}. \quad (2)$$

Considering the transformation properties of the various components, as shown in Table 1, it can be seen that the EDM component, $\vec{d} \cdot \vec{E}$ is a CP odd quantity. The present experimental limits [9–11] are orders of magnitude above the Standard Model predictions, hence any observed EDM would provide an additional source of CP violation. Since the CP violation in the Standard Model is insufficient to explain the matter-antimatter asymmetry in the universe the search for a new source is important. At the current levels of experimental precision the limits on EDMs have started to cut in to the predictions from different beyond the standard model (BSM) theories [5].

The present limit on the electron EDM, $8.7 \times 10^{-29} e\cdot$cm [9], is about 9 orders of magnitude smaller than the muon EDM limit. Assuming a simple mass scaling, $d \propto m$, the electron limit would indicate that the muon EDM would be less than $\sim 10^{-27}$, far below what can be measured currently. However, some BSM models [12, 13] predict non-standard scalings which are quadratic or even cubic in mass. The muon also provides a unique opportunity to search for an EDM in the second generation of particles.

Table 1. The transformation properties of the electric and magnetic fields and dipole moments

	$\vec{E}$	$\vec{B}$	$\vec{\mu}$ or $\vec{d}$
P	-	+	+
C	-	-	-
T	+	-	-

2 Experimental setup

The Fermilab E989 g-2 experiment is based on the same experimental setup as was used in the E821 experiment with some significant improvements to increase the precision [4]. Polarised muons are injected into the storage ring and as they travel around the ring the anomalous magnetic moment, a_μ, causes the spin to precess around faster than the muon momentum. The parity violation inherent in weak decays means that the highest energy positrons are emitted parallel to the muon spin. This allows the spin direction to be tracked through the detection of the decay positrons, done using 24 calorimeter stations around the ring.

The difference between the spin and cyclotron frequencies is given by

$$\vec{\omega}_a = -\frac{Qe}{m_\mu}\left[a_\mu \vec{B} - \left(a - \frac{1}{\gamma^2 - 1}\right)\frac{\vec{\beta} \times \vec{E}}{c}\right] \quad (3)$$

for $\beta \cdot B = \beta \cdot E = 0$. The experiment is run at the "magic" momentum $p_{magic} = m/\sqrt{a_\mu} = 3.09$ GeV/c, such that the equation simplifies to

[a] e-mail: rebecca.chislett.10@ucl.ac.uk

[b] `muon-g-2.fnal.gov/collaboration.html`

© The Authors, published by EDP Sciences. This is an open access article distributed under the terms of the Creative Commons Attribution License 4.0 (http://creativecommons.org/licenses/by/4.0/).

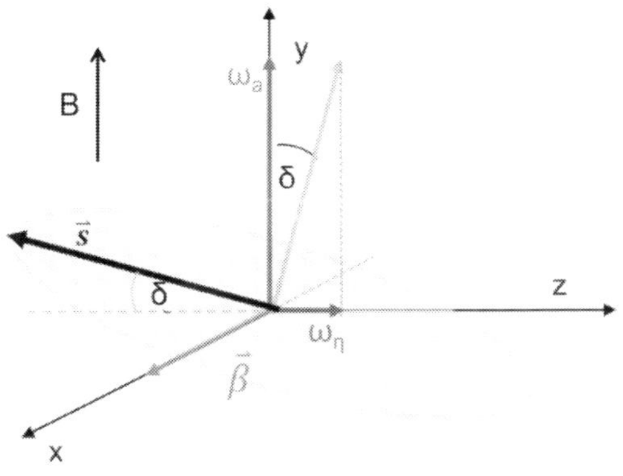

Figure 1. The contributions to the precession frequency from the MDM and the EDM and the resulting spin precession plane.

$$\vec{\omega}_a = -a_\mu \frac{Qe}{m_\mu}\vec{B}. \tag{4}$$

Therefore the anomalous magnetic moment can be determined using the precession frequency extracted from the positron decays and a precise measurement of the magnetic field. However, if the muon has an EDM the equation for the spin precession gains an extra component:

$$\vec{\omega}_\eta = -\eta \frac{Qe}{2m_\mu}\left[\frac{\vec{E}}{c} + \vec{\beta}\times\vec{B}\right]. \tag{5}$$

which add to $\vec{\omega}_a$.

The component of the spin precession from the MDM is directed along the magnetic field, causing the spin to precess in the horizontal plane. The contribution from the EDM is dominated by the motional magnetic field contribution, which is directed towards the centre of the ring, causing a vertical precession of the spin. Adding in the EDM component causes the spin precession plane to tilt, leading to a vertical oscillation $\pi/2$ out of phase with the precession due to the MDM. Assuming that the motional field dominates and the other components can be neglected, the two contributions are perpendicular as shown in Figure 1. The tilt in the precession plane is then given by

$$\delta = \tan^{-1}\left(\frac{\eta\beta}{2a}\right). \tag{6}$$

which produces a milliradian tilt for an EDM of the order of $10^{-19}e\cdot$cm. The additional component from an EDM also slightly increases the magnitude of the precession frequency to:

$$\omega_{a\eta} = \sqrt{\omega_a^2 + \omega_\eta^2}. \tag{7}$$

This leads to two EDM search methods; indirectly, by comparing the measured value of the precession frequency to the Standard Model prediction or directly, by looking for a tilt in the precession plane. Three different direct analyses were performed at the E821 experiment to look for a vertical oscillation, each of which will be discussed in turn along with the possible improvements to the measurements at the new experiment.

3 Measuring the EDM at E821

The E821 EDM measurement was a combination of three different analyses. The first two used scintillator panels on the front of the calorimeter to divide the calorimeter into vertical (and horizontal in some cases) segments. One looked for an oscillation in the average vertical position of the positrons hitting the calorimeter [8] and the other looked for a asymmetry in the phases as a function of vertical position [7]. Both of these measurements were dominated by systematic uncertainties. The third measurement, dominated by statistical uncertainties, used the tracking detector to look for an oscillation in the positron decay angle as a function of time [6].

For a fixed magnetic field and muon momentum the statistical figure of merit is given by NA^2 where N is the total number of positrons and A is the asymmetry. In the case of the MDM this asymmetry is the difference between forward and backward going spins, whereas for the EDM it is the difference between upward and downward going spins. Figure 2 shows this figure of merit as a function of energy for both the MDM and EDM. In the case of the forward-backward asymmetry of the MDM this results in a greater statistical power for positrons at the higher end of the energy spectrum, whereas in the case of the EDM the acceptable energy range is much broader and centred further toward the middle of the energy spectrum. Hence in the EDM search considering a wider range of positron energies is preferable.

3.1 Vertical Position Oscillations

A tilt in the precession plane would cause an oscillation in the vertical angle of the positrons, leading to an oscillation in the average vertical position of the positrons hitting the calorimeter. However, changes in the path length of the positrons hitting the calorimeter cause the vertical width to vary which can feed in to oscillations in the average vertical position if the calorimeter isn't perfectly aligned. To quantify this effect the average width is plotted as a function of time. This distribution is then fit to obtain the different contributions:

$$\begin{aligned} f(t) =& W + S_{g2}sin(\omega t) + C_{g2}cose(\omega t) \\ &+ S_{2g2}sin(2\omega t) + C_{2g2}cos(2\omega t) \\ &+ e^{-t/\tau_{CBO}}\left[S_{CBO}sin\left(\omega_{CBO}(t-t_0)+\Phi_{CBO}\right)\right. \\ &\qquad \left.+C_{CBO}cos\left(\omega_{CBO}(t-t_0)+\Phi_{CBO}\right)\right] \\ &+ L(t) \end{aligned} \tag{8}$$

The first term is the average width. The second set of terms are oscillations in phase with the g-2 oscillation due to the difference in path length between positrons from different spin orientations. Those that travel further to reach the calorimeter will have a greater vertical spread

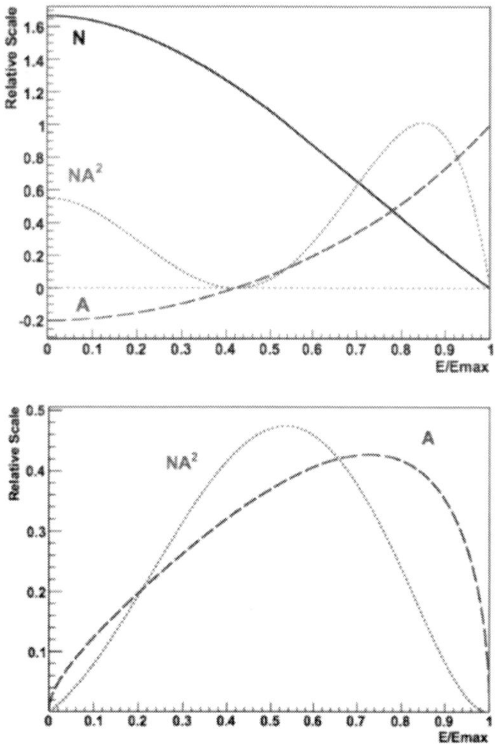

Figure 2. The statistical figures of merit for the MDM measurement (top) and EDM measurement (bottom) as a function of normalised positron energy [2].

by the time they reach the calorimeter, despite being released with the same variation in decay angles. The g-2 frequency is fixed in the fit using the number oscillation results. Lastly, coherent betatron oscillations (CBO) cause variations in the radial position of the beam and hence the distance the positrons travel, again changing the width on the face of the calorimeter. The final term accounts for the calorimeter deadtime which has a greater impact at earlier times.

The CBO frequency, phase and lifetime extracted from the width oscillations are then fixed in the fit to the average vertical position as a function of time, shown in Figure 3. The fit function used is similar to that for the width oscillations,

$$\begin{aligned} f(t) =&K + \left[S_{g2}sin(\omega t) + C_{g2}cos(\omega t)\right] \\ &+ e^{-t/\tau_{CBO}}\left[S_{CBO}sin\left(\omega_{CBO}(t-t_0) + \Phi_{CBO}\right)\right. \\ &\left.+C_{CBO}cos\left(\omega_{CBO}(t-t_0) + \Phi_{CBO}\right)\right] \\ &+ Me^{-t/\tau_M} \end{aligned} \tag{9}$$

where the first term is the average vertical position, the second accounts for variations at the g-2 frequency, the third corresponds to CBO oscillations and the final term is for slow changes in detector response and pileup. As the

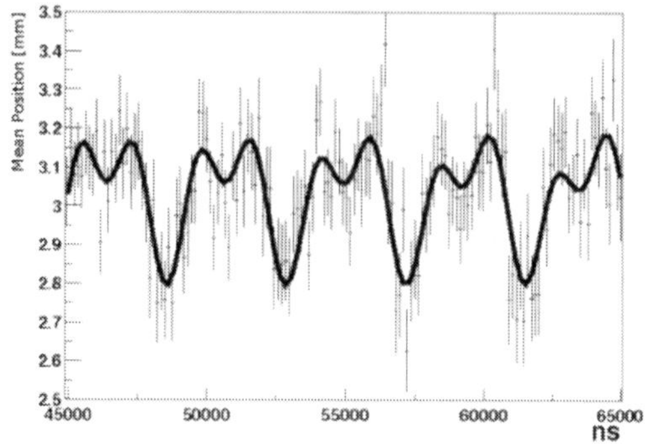

Figure 3. The average vertical position of the positrons hitting the calorimeter as a function of time [2].

Table 2. The uncertainties associated with the vertical position oscillation measurements [2].

Effect	Error(μm)
Detector Tilt	6.1
Vertical Spin	5.1
Quadrupole Tilt	3.9
Timing Offset	3.2
Energy Calibration	2.8
Radial Magnetic Field	2.5
Albedo and Doubles	2.0
Fitting Method	1.0
Total Systematic	10.4
Statistical	5.9
Total Uncertainty	11.9

g-2 number oscillation is aligned with the cosine phase an EDM signal would show up as a non zero value of S_{g2}.

The systematic uncertainties associated with this measurement are summarised in Table 2. The largest systematic comes from the calorimeter potentially being tilted with respect to the beam, such that any horizontal oscillation also appears as a vertical oscillation. An average vertical muon spin component pushes the decay angles off centre and when combined with the varying positron path lengths causes a vertical oscillation. Any radial magnetic field would also have a similar effect due to the deflection of the positrons. It is also important to consider any difference between the top and bottom halves of the calorimeter which could fake an oscillation. Overall, the systematic uncertainties are much larger than the statistical uncertainties and so it is a priority of the E989 experiment to reduce these, as will be detailed in Section 4.

The final measurement of the average vertical position oscillations is $S_{g2} = (1.27 \pm 11.9)\mu$m, which based on simulation translates to an EDM measurement of $d_\mu = (-0.1 \pm 1.4) \times 10^{-19}e$·cm. The resulting limit with 95% confidence level sets the limit as $|d_\mu| < 2.8 \times 10^{-19}e$·cm.

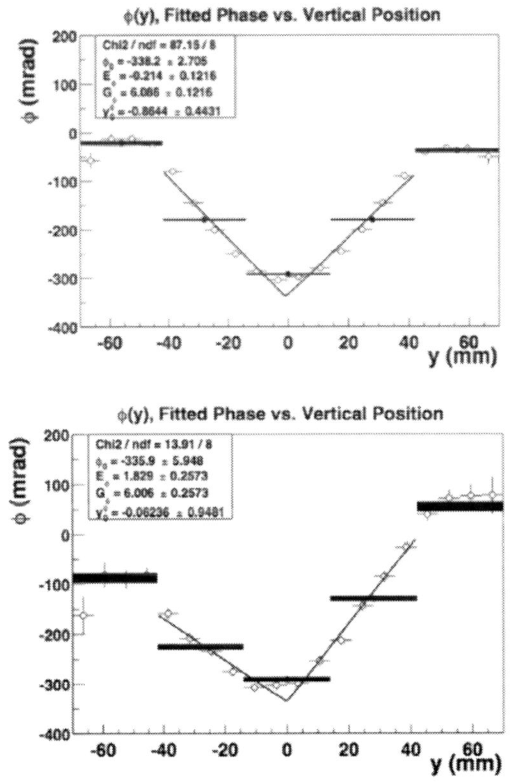

Figure 4. The fitted phase as a function of the vertical position from simulation with no EDM (top) and with an EDM (bottom) [7].

3.2 Vertical phase asymmetry

The positrons emitted in the outward radial direction have a longer path length to the detector and therefore have a larger vertical spread on the calorimeter face. This means that the positrons detected towards the top and bottom are dominated by outward going decays and those in the centre by the inward going positrons leading to a change in the fitted phase with the vertical position. However, the presence of an EDM would tilt the precession plane and lead to more outward going decays in the top half of the calorimeter and more inward going decays at the bottom, suppressing the phase difference in the bottom half of the calorimeter and leading to an asymmetry in the distribution. This effect is shown using simulated events in Figure 4.

The phase as a function of vertical position can be fitted to look for an EDM contribution:

$$\Phi(y) = p_0 + p_1(y - p_2) + |p_3(y - p_2)|. \quad (10)$$

The first term is the value at the centre of the calorimeter, and the last is the symmetric phase shift expected without the presence of an EDM. A non-zero value of the parameter p_1 would be the indication of an asymmetry in the phase and hence of an EDM.

Table 3. The uncertainties associated with the vertical phase asymmetry measurements [2].

Effect	Error(μrad / mm)
Detector Tilt	20
Detector Misalignment	28
Energy Calibration	4.3
Muon Vertical Spin	8.0
Radial Magnetic Field	14.4
Timing	3.4
Total Systematic	38
Statistical	28
Total Uncertainty	47

The systematic uncertainties on this measurement, as shown in Table 3, are similar to those for the vertical position measurement. The dominant uncertainties come from a misalignment or tilt in the calorimeters, where a misalignment can fake an asymmetry by shifting the spectrum slightly off centre and a tilt can cause asymmetric vertical loses again inducing a fake signal. Any non zero average vertical spin or radial magnetic field also has an effect for similar reasons to the discussion in Section 3.1

3.3 Vertical decay angle

The final method used to search for an EDM signal used the tracking detector placed in front of one of the calorimeter stations. This allows for a more direct measurement of the angle of the decaying positrons, but is more statistically limited than the other measurements. Firstly the number oscillation was plotted as a function of time. This was done modulo the precession period to reduce the effects of periodic disturbances at other frequencies. Fitting the distribution provides the phase of the g-2 oscillation:

$$N(t) = e^{-t/\tau_e}(N_0 + Wcos(\omega t + \Phi)). \quad (11)$$

This phase can then be used in the fit to the vertical angle oscillation:

$$\theta(t) = M + A_\mu cos(\omega t + \Phi) + A_{EDM} sin(\omega t + \Phi). \quad (12)$$

as shown in Figure 5. The first term accounts for any offset in the central angle, which could be due to an average vertical muon spin component. The second term is in phase with the number oscillation and the final term is out of phase, where a non zero value of A_{EDM} indicates the presence of an EDM.

The systematic uncertainties associated with this measurement are small compared to the statistical uncertainty, as shown in Table 4. The largest contributions are from any coupling of the CBO oscillations into the vertical decay angle and the acceptance coupling. The acceptance coupling refers to the variation of vertical angle acceptance of the detector with azimuth combined with the correlation between the positron path length and azimuthal position of the decay point. Other contributions that were considered include a radial magnetic field, which would

 DOI: 10.1051/epjconf/201611801005

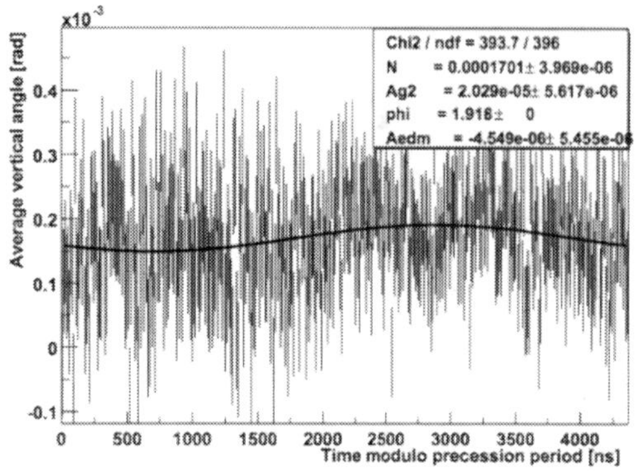

Figure 5. The average vertical decay angle as a function of time modulo the precession period [2].

Table 4. The uncertainties associated with the vertical decay angle measurement [2].

Effect	Error(μrad)
Radial Magnetic Field	0.13
Acceptance Coupling	0.3
Horizontal CBO	0.3
Phase Fit	0.01
Precession Period	0.01
Total Systematic	0.44
Statistical	4.4
Total Uncertainty	4.4

tilt the precession plane, and errors due to the fit. Most of the systematic error estimates are very conservative but despite this remain an order of magnitude below the statistical uncertainties.

In this analysis the measured vertical angle oscillation amplitude was $(-0.1 \pm 4.4) \times 10^{-6}$ rad which produces a limit of $|d_\mu| < 3.2 \times 10^{-19} e\cdot\text{cm}$.

4 Improvements at E989

The new g-2 experiment at Fermilab has many upgrades which will increase the sensitivity to an EDM signal. The easiest measurement to improve upon is the vertical decay angle measurement as the number of positron decays recorded by the trackers will be increased at the new experiment. However, it is worth discussing the changes that might reduce the systematic uncertainties on the calorimeter-based analyses.

Firstly, the calorimeters at E989 will be segmented, using a grid of 6 by 9 cells, removing the need for scintillators to be placed on the front to make the measurements. This segmentation will help with the ability to control pileup and measure both the detector position and detector tilt, both large sources of systematic uncertainty in the E821 EDM measurement. The lower energy acceptance of the new calorimeters is important for EDM measurements as the sensitivity covers a broader energy range than the g-2 number oscillation measurement. In addition a laser calibration system is being introduced to improve the gain calibration and timing information.

Secondly, the electrostatic quadrupole voltage at E989 has been chosen carefully to reduce the amplitude of the CBO by at least a factor of 4. This will impact all three analyses as the CBO was a significant systematic uncertainty. The new E989 experiment will have three straw tracking detectors, compared to the one present at E821 which, although not directly used in the calorimeter analyses, improves the knowledge and monitoring of the beam distribution. The full BMAD and G4Beamline simulations all the way from the production target will provide detailed information on the polarisation and momentum variance of the beam. Lastly, the increased amount of data will not only reduces the statistical uncertainty, but will also decrease the systematic uncertainties that are derived from the data.

All of the improvements mentioned above help to reduce the systematic uncertainties on the calorimeter-based EDM measurement, although the total reduction is yet to be quantified. However, the most promising measurement for the E989 experiment is the vertical decay angle, which will be vastly improved by the increase in the number of muons, the number of trackers and the increased tracker acceptance. Combining these three effects the new experiment is expected to take of the order of 1000 times more statistics than at E821, which should reduce the statistical error by one order of magnitude quickly and further subsequently.

Obviously, once the statistical error is reduced by an order of magnitude the systematic uncertainties start to become important and need to be controlled. Most of the improvements mentioned above also apply to this measurement, most importantly the reduction in the CBO. The fact that the tracker is placed inside the vacuum chamber and has an increased geometrical acceptance makes a significant difference, especially for the acceptance-coupling uncertainty. The trackers will also provide a much improved characterisation of the beam and the alignment of the calorimeters which are critical to achieving the goal of an EDM sensitivity down to $10^{-21} e\cdot\text{cm}$.

5 Conclusions

The new g-2 experiment aims to improve the limit on the muon EDM set at the previous E821 experiment at BNL by approximately two orders of magnitude, down to 10^{-21}e·cm. Three different methods, each with a similar sensitivity, were combined to set the previous limit and the improvements to the new experiment mean that a lower limit will be reached in all of these analyses. However, with the introduction of three new tracking stations, the most promising method searches for an oscillation in the average vertical decay angle of the positrons, which was statistics limited at E821 with low systematic errors. At the new experiment the tracking statistics are expected to increase by about 1000 times, and this coupled with an improved understanding of the key systematic uncertainties will allow and EDM of 10^{-21}e·cm to be probed.

Acknowledgments

The Fermilab $g-2$ experiment is supported in part by the U.S. DOE Offices of High-Energy and Nuclear Physics, the U.S. NSF, the U.K. STFC, the NSFC in China, the XXX in Korea, and by the INFN in Italy. The author is supported by the DOE Office of Nuclear Physics, award DE-FG02-97ER41020.

This work was supported by the STFC grant: ST/L001888/1.

References

[1] E821 Collaboration, G. W. Bennett, et al., Phys. Rev. Lett **92**, 161802 (2004).

[2] E821 Collaboration, G. W. Bennett, et al., Phys. Rev. D **80**, 052008 (2009).

[3] E989 Collaboration, J. Grange et al., FERMILAB-FN-0992-E (2015).

[4] D. Hertzog, this workshop, *Next Generation Muon g-2 Experiments* (2015).

[5] J. Engle, et al., Prog. Part. Nucl. Phys. **71** (2013) 21-74

[6] M. Sossong, *A Search for an Electric Dipole Moment of the Positive Muon)*, University of Illinois PhD thesis(2005).

[7] S. O. Giron, *Measuring the Electric-Dipole Moment of the Muon at BNL E821*, University of Minnesota PhD thesis(2004).

[8] R. S. McNabb, *An Improved Limit on the Electric Dipole Moment of the Muon*, University of Minnesota PhD thesis (2003).

[9] ACME Collaboration, J. Baron, et al., Science **343**, 269 (2014)

[10] W.C. Griffith, et al., Phys. Rev. Lett. **102**, 101601 (2009).

[11] C.A. Baker, et al., Phys. Rev. Lett. **97**, 131801 (2006).

[12] K.S. Babu, Bhaskar Dutta, R.N. Mohapatra, Phys. Rev. Lett. **85** 5064-5067 (2000).

[13] E. O. Iltan, Eur. Phys. J. C **54**, 583 (2008).

DOI: 10.1051/epjconf/201611801006

The Muon g–2 in two-Higgs-doublet models

Eung Jin Chun[1,a]

[1] Korea Institute for Advanced Study, Seoul 130-722, Korea

Abstract. Updating various theoretical and experimental constraints on the four different types of two-Higgs-doublet models (2HDMs), we find that only the "lepton-specific" (or "type X") 2HDM can explain the present muon g−2 anomaly in the parameter region of large $\tan\beta$, a light CP-odd boson, and heavier CP-even and charged bosons which are almost degenerate. The severe constraints on the models come mainly from the consideration of vacuum stability and perturbativity, the electroweak precision data, the b-quark observables like $B_S \to \mu\mu$, the precision measurements of the lepton universality as well as the 125 GeV boson property observed at the LHC.

1 Outline

Since the first measurement of the muon anomalous magnetic moment $a_\mu = (g-2)_\mu/2$ by the E821 experiment at BNL in 2001 [1], much progress has been made in both experimental and theoretical sides to reduce the uncertainties by a factor of two or so establishing a consistent 3σ discrepancy

$$\Delta a_\mu \equiv a_\mu^{\rm EXP} - a_\mu^{\rm SM} = +262\,(85) \times 10^{-11} \qquad (1)$$

which is in a good agreement with the different group's determinations. Followed by the 2001 announcement, there have been quite a few studies in the context of 2HDMs [2–4], however, restricted mainly to the type I and II models out of four different types of 2HDMs ensuring natural flavour conservation. Considering the recent experimental development confirming more precisely the Standard Model (SM) predictions, including the discovery of the 125 GeV Brout-Englert-Higgs boson, it would be timely to revisit the issue of the muon g–2 in 2HDMs.

An additional contribution to the muon g–2 from an extra boson in 2HDMs, shown in Fig. 1, may be the origin of the positive excess in the Δa_μ. This can happen in the type II or X (lepton-specific) 2HDM which allows a light boson having large Yukawa couplings enhanced by $\tan\beta$. While the type II option is completely ruled out by now, the type X model [5] remains an unique option to explain the a_μ anomaly evading all the recent experimental constraints [6–9].

The previous studies on the muon g–2 in the type II 2HDM (2HDM-II) and various experimental constraints were nicely summarised in Ref. [3]:

- The one-loop correction mediated by a light CP-even (CP-odd and charged) boson gives a positive (negative) contribution to Δa_μ, and thus the current 3σ deviation can be explained by a CP-even boson lighter than about 5 GeV. However, such a light boson was already in contradiction to non-observation of radiative Υ decays $\Upsilon \to \gamma + X$.
- Contrary to the one-loop correction, the major two-loop contribution from the Barr-Zee diagram [10] mediated by a light CP odd (even) boson can give a sizable positive (negative) contribution to Δa_μ. Thus, a light CP odd boson A with large $\tan\beta$ can account for the muon g–2 deviation. However, most of the muon g–2 favoured region in the lower and upper M_A part are excluded by the LEP and TEVATRON search on $Z \to b\bar{b}A(b\bar{b})$, respectively, except a small gap of $M_A \approx 25 - 70$ GeV with $\tan\beta \gtrsim 20$.
- However, let us note that such a light A gives a huge contribution to $B_s \to \mu^+\mu^-$ as its rate is proportional to $\tan^4\beta/M_A^4$ [11], and thus the above gap is completely closed now by the LHC measurement which is consistent with the SM prediction [12].

Now the situation can be quite different in the type X 2HDM (2HDM-X) where all the extra boson couplings to quarks (leptons) are proportional to $\cot\beta$ ($\tan\beta$). Due to this, the 2HDM-X becomes hadrophobic in the large $\tan\beta$ limit invalidating most hadron-related constraints applied

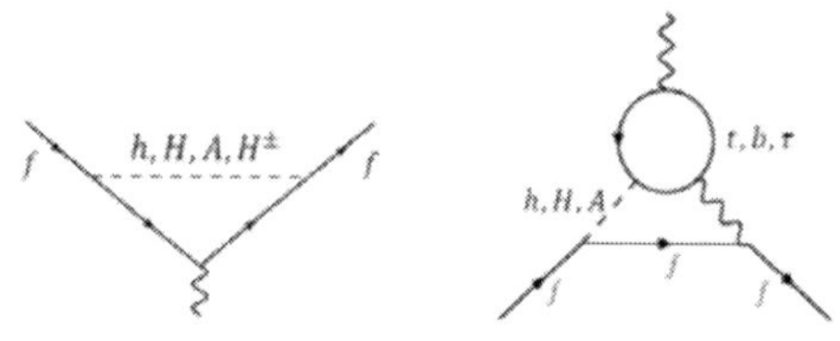

Figure 1. One and two loop diagrams contributing to the muon g−2 in 2HDMs.

[a] e-mail: ejchun@kias.re.kr

© The Authors, published by EDP Sciences. This is an open access article distributed under the terms of the Creative Commons Attribution License 4.0 (http://creativecommons.org/licenses/by/4.0/).

to the type II model. On the other hand, its leptophilic property brings severe constraints from the precision leptonic observables. The key features in confronting the type X model with the muon g–2 anomaly can be summarized as follows [6–9].

- As in the 2HDM-II, the one-loop correction with a light CP even boson H can account for the muon g–2 excess. While the Upsilon decay suppressed by $1/\tan^2\beta$ cannot provide a meaningful constraint, the Belle and LHCb searches for $B \to K\mu\mu$ shut down the muon g–2 favoured region except tiny gaps around $M_H \approx 3$ and 4 GeV [13]. In any case, such a light Higgs boson is excluded by the current measurement of $B_s \to \mu^+\mu^-$ even in the 2HDM-X where $\Gamma(B_s \to \mu^+\mu^-) \propto 1/M_H^4$.
- The Barr-Zee diagram with the tau lepton in the loop can account for the muon g–2 anomaly again in a parameter region of small M_A and large $\tan\beta$ evading all the constraints from the hadron colliders and the b-quark observables [6] except the process $B_s \to \mu^+\mu^-$ which rules out $M_A \lesssim 10$ GeV [7].
- However, the lepton universality test by HFAG [14] combined with the $Z \to \tau\tau$ decay turns out to limit severely the muon g–2 favoured region of the type X model [8] allowing (only at 2σ) a small region below $M_A \approx 80$ GeV and $\tan\beta \approx 60$ [9].
- With such a light A, the exotic decay of the 125 GeV boson $h \to AA$ or $AA^*(\tau\tau)$ becomes generically too large unless a certain cancelation is arranged to suppress the hAA coupling λ_{hAA} which turns out to be possible only in the wrong-sign limit of the lepton Yukawa coupling [7].

2 Four types of 2HDMs

Non-observation of flavour changing neutral currents restricts 2HDMs to four different classes which differ by how the Higgs doublets couple to fermions [16]. They are organized by a discrete symmetry Z_2 under which different Higgs doublets and fermions carry different parities. These models are labeled as type I, II, "lepton-specific" (or X) and "flipped" (or Y). Having two Higgs doublets $\Phi_{1,2}$, the most general Z_2 symmetric scalar potential takes the form:

$$\begin{aligned} V &= m_{11}^2|\Phi_1|^2 + m_{22}^2|\Phi_2|^2 - m_{12}^2(\Phi_1^\dagger\Phi_2 + \Phi_1\Phi_2^\dagger) \\ &+\frac{\lambda_1}{2}|\Phi_1|^4 + \frac{\lambda_2}{2}|\Phi_2|^4 + \lambda_3|\Phi_1|^2|\Phi_2|^2 + \lambda_4|\Phi_1^\dagger\Phi_2|^2 \\ &+\frac{\lambda_5}{2}\left[(\Phi_1^\dagger\Phi_2)^2 + (\Phi_1\Phi_2^\dagger)^2\right], \end{aligned} \tag{2}$$

where a (soft) Z_2 breaking term m_{12}^2 is introduced. Minimization of the scalar potential determines the vacuum expectation values $\langle\Phi_{1,2}^0\rangle \equiv v_{1,2}/\sqrt{2}$ around which the Higgs doublet fields are expanded as

$$\Phi_{1,2} = \left[\eta_{1,2}^+, \frac{1}{\sqrt{2}}\left(v_{1,2} + \rho_{1,2} + i\eta_{1,2}^0\right)\right]. \tag{3}$$

The model contains the five physical fields in mass eigenstates denoted by $H^\pm, A, H$ and h. Assuming negligible CP violation, $H^\pm$ and A are given by

$$H^\pm, A = s_\beta\,\eta_1^{\pm,0} - c_\beta\,\eta_2^{\pm,0} \tag{4}$$

where the angle β is determined from $t_\beta \equiv \tan\beta = v_2/v_1$, and their orthogonal combinations are the corresponding Goldstone modes $G^{\pm,0}$. The neutral CP-even bosons are diagonalized as

$$h = c_\alpha\,\rho_1 - s_\alpha\,\rho_2, \quad H = s_\alpha\,\rho_1 + c_\alpha\,\rho_2 \tag{5}$$

where $h\,(H)$ denotes the lighter (heavier) state.

The gauge couplings of h and H are given schematically by

$$\mathcal{L}_{\rm gauge} = g_V m_V(s_{\beta-\alpha}h + c_{\beta-\alpha}H)VV \tag{6}$$

where $V = W^\pm$ or Z. Taking h as the 125 GeV boson of the SM, the SM limit corresponds to $s_{\beta-\alpha} \to 1$. Indeed, LHC finds, $c_{\beta-\alpha} \ll 1$ in all the 2HDMs confirming the SM-like property of the 125 GeV boson [18].

	y_u^A	y_d^A	y_l^A	y_u^H	y_d^H	y_l^H	y_u^h	y_d^h	y_l^h
Type I	$\cot\beta$	$-\cot\beta$	$-\cot\beta$	$\frac{\sin\alpha}{\sin\beta}$	$\frac{\sin\alpha}{\sin\beta}$	$\frac{\sin\alpha}{\sin\beta}$	$\frac{\cos\alpha}{\sin\beta}$	$\frac{\cos\alpha}{\sin\beta}$	$\frac{\cos\alpha}{\sin\beta}$
Type II	$\cot\beta$	$\tan\beta$	$\tan\beta$	$\frac{\sin\alpha}{\sin\beta}$	$\frac{\cos\alpha}{\cos\beta}$	$\frac{\cos\alpha}{\cos\beta}$	$\frac{\cos\alpha}{\sin\beta}$	$-\frac{\sin\alpha}{\cos\beta}$	$-\frac{\sin\alpha}{\cos\beta}$
Type X	$\cot\beta$	$-\cot\beta$	$\tan\beta$	$\frac{\sin\alpha}{\sin\beta}$	$\frac{\sin\alpha}{\sin\beta}$	$\frac{\cos\alpha}{\cos\beta}$	$\frac{\cos\alpha}{\sin\beta}$	$\frac{\cos\alpha}{\sin\beta}$	$-\frac{\sin\alpha}{\cos\beta}$
Type Y	$\cot\beta$	$\tan\beta$	$-\cot\beta$	$\frac{\sin\alpha}{\sin\beta}$	$\frac{\cos\alpha}{\cos\beta}$	$\frac{\sin\alpha}{\sin\beta}$	$\frac{\cos\alpha}{\sin\beta}$	$-\frac{\sin\alpha}{\cos\beta}$	$\frac{\cos\alpha}{\sin\beta}$

Table 1. The normalized Yukawa couplings for up- and down-type quarks and charged leptons.

Normalizing the Yukawa couplings of the neutral bosons to a fermion f by m_f/v where $v = \sqrt{v_1^2 + v_2^2} = 246$ GeV, we have the following Yukawa terms:

$$-\mathcal{L}_{\rm Yukawa}^{\rm 2HDMs} = \sum_{f=u,d,l} \frac{m_f}{v}\left(y_f^h h\bar{f}f + y_f^H H\bar{f}f - i y_f^A A\bar{f}\gamma_5 f\right) + \left[\sqrt{2}V_{ud}H^+\bar{u}\left(\frac{m_u}{v}y_u^A P_L + \frac{m_d}{v}y_d^A P_R\right)d + \frac{\sqrt{2}m_l}{v}y_l^A H^+\bar{\nu}P_R l + h.c.\right] \tag{7}$$

where the normalized Yukawa couplings $y_f^{h,H,A}$ are summarized in Table I for each of these four types of 2HDMs. Let us now recall that the tau Yukawa coupling $y_\tau \equiv y_l^h$ in Type X (also $y_b \equiv y_d^h$ in Type II) can be expressed as

$$y_\tau = -\frac{s_\alpha}{c_\beta} = s_{\beta-\alpha} - t_\beta c_{\beta-\alpha} \tag{8}$$

which allows us to have the wrong-sign limit $y_\tau \sim -1$ compatible with the LHC data [15] if $c_{\beta-\alpha} \sim 2/t_\beta$ for large $t_\beta \equiv \tan\beta$ favoured by the muon g–2. Later we will see that a cancellation in λ_{hAA} can be arranged only for $y_\tau^h < -1$ to suppress the $h \to AA$ decay.

3 The Muon g–2 from a light CP-odd boson

Considering all the updated SM calculations of the muon g–2, we obtain

$$a_\mu^{\rm SM} = 116591829(57) \times 10^{-11} \tag{9}$$

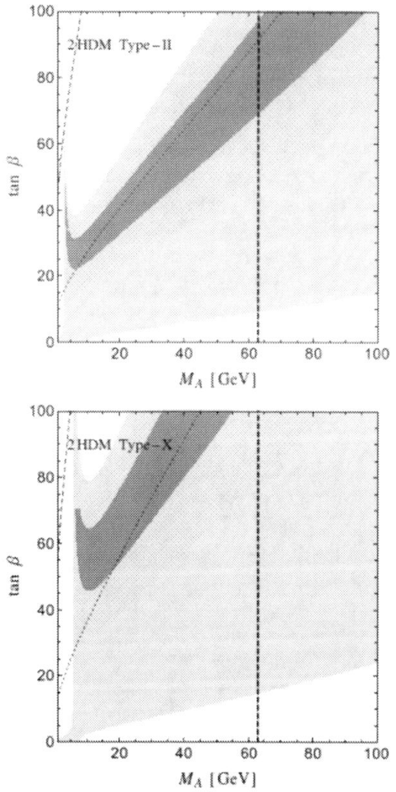

Figure 2. The 1σ, 2σ and 3σ regions allowed by Δa_μ in the M_A-$\tan\beta$ plane taking the limit of $c_{\beta-\alpha} = 0$ and $M_{h(H)} = 125$ (200) GeV in type II (left panel) and type X (right panel) 2HDMs. The regions below the dashed (dotted) lines are allowed at 3σ (1.4σ) by Δa_e. The vertical dashed line corresponds to $M_A = M_h/2$.

comparing it with the experimental value $a_\mu^{\rm EXP} = 116592091\,(63) \times 10^{-11}$, one finds a deviation at 3.1σ: $\Delta a_\mu \equiv a_\mu^{\rm EXP} - a_\mu^{\rm SM} = +262\,(85) \times 10^{-11}$. In the 2HDM, the one-loop contributions to a_μ of the neutral and charged bosons are

$$\Delta a_\mu^{\rm 2HDM}(1\text{loop}) = \frac{G_F\, m_\mu^2}{4\pi^2\sqrt{2}} \sum_j \left(y_\mu^j\right)^2 r_\mu^j\, f_j(r_\mu^j), \tag{10}$$

where $j = \{h, H, A, H^\pm\}$, $r_\mu^j = m_\mu^2/M_j^2$, and

$$f_{h,H}(r) = \int_0^1 dx \frac{x^2(2-x)}{1-x+rx^2}, \tag{11}$$

$$f_A(r) = \int_0^1 dx \frac{-x^3}{1-x+rx^2}, \tag{12}$$

$$f_{H^\pm}(r) = \int_0^1 dx \frac{-x(1-x)}{1-(1-x)r}. \tag{13}$$

These formula show that the one-loop contributions to a_μ are positive for the neutral scalars h and H, and negative for the pseudo-scalar and charged bosons A and $H^\pm$ (for $M_{H^\pm} > m_\mu$). In the limit $r \ll 1$,

$$f_{h,H}(r) = -\ln r - 7/6 + O(r), \tag{14}$$

$$f_A(r) = +\ln r + 11/6 + O(r), \tag{15}$$

$$f_{H^\pm}(r) = -1/6 + O(r), \tag{16}$$

showing that in this limit $f_{H^\pm}(r)$ is suppressed with respect to $f_{h,H,A}(r)$. Now the two-loop Barr-Zee type diagrams with effective $h\gamma\gamma$, $H\gamma\gamma$ or $A\gamma\gamma$ vertices generated by the exchange of heavy fermions gives

$$\Delta a_\mu^{\rm 2HDM}(2\text{loop}-\text{BZ}) = \frac{G_F\, m_\mu^2}{4\pi^2\sqrt{2}} \frac{\alpha_{\rm em}}{\pi} \sum_{i,f} N_f^c\, Q_f^2\, y_\mu^i\, y_f^i\, r_f^i\, g_i(r_f^i), \tag{17}$$

where $i = \{h, H, A\}$, $r_f^i = m_f^2/M_i^2$, and m_f, Q_f and N_f^c are the mass, electric charge and number of color degrees of freedom of the fermion f in the loop. The functions $g_i(r)$ are

$$g_i(r) = \int_0^1 dx \frac{\mathcal{N}_i(x)}{x(1-x)-r} \ln\frac{x(1-x)}{r}, \tag{18}$$

where $\mathcal{N}_{h,H}(x) = 2x(1-x) - 1$ and $\mathcal{N}_A(x) = 1$.

Note the enhancement factor m_f^2/m_μ^2 of the two-loop formula in Eq. (17) relative to the one-loop contribution in Eq. (10), which can overcome the additional loop suppression factor α/π, and makes the two-loop contributions may become larger than the one-loop ones. Moreover, the signs of the two-loop functions $g_{h,H}$ (negative) and g_A (positive) for the CP-even and CP-odd contributions are opposite to those of the functions $f_{h,H}$ (positive) and f_A (negative) at one-loop. As a result, for small M_A and large $\tan\beta$ in Type II and X, the positive two-loop pseudoscalar contribution can generate a dominant contribution which can account for the observed Δa_μ discrepancy. The additional 2HDM contribution $\delta a_\mu^{\rm 2HDM} = \delta a_\mu^{\rm 2HDM}(1\text{loop}) + \delta a_\mu^{\rm 2HDM}(2\text{loop}-\text{BZ})$ obtained adding Eqs. (10) and (17) (without the h contributions) is compared with Δa_μ in Fig. 4.

4 Electroweak Precision Data

Allowing such a light CP-odd boson, there could be a strong limit on the extra boson masses coming from the electroweak precision test. To see this, we compare the theoretical 2HDMs predictions for M_W and $\sin^2\theta_{\rm eff}^{\rm lept}$ with their present experimental values via a combined χ^2 analysis. These quantities can be computed perturbatively by means of the following relations

$$M_W^2 = \frac{M_Z^2}{2}\left[1 + \sqrt{1 - \frac{4\pi\alpha_{\rm em}}{\sqrt{2}G_F M_Z^2}\frac{1}{1-\Delta r}}\right] \tag{19}$$

$$\sin^2\theta_{\rm eff}^{\rm lept} = k_l\left(M_Z^2\right)\sin^2\theta_W\,, \tag{20}$$

where $\sin^2\theta_W = 1 - M_W^2/M_Z^2$, and $k_l(q^2) = 1 + \Delta k_l(q^2)$ is the real part of the vertex form factor $Z \to l\bar{l}$ evaluated at $q^2 = M_Z^2$. We than use the following experimental values:

$$M_W^{\rm EXP} = 80.385 \pm 0.015\ \text{GeV},$$

$$\sin^2\theta_{\rm eff}^{\rm lept,EXP} = 0.23153 \pm 0.00016. \tag{21}$$

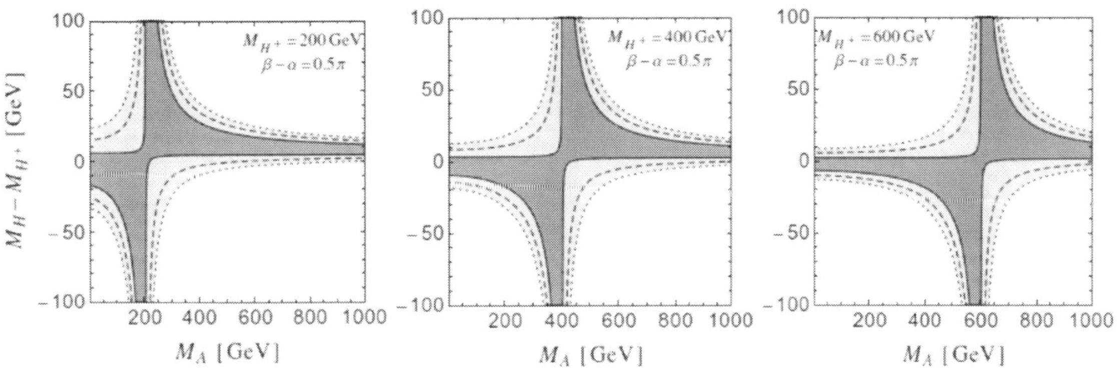

Figure 3. The parameter space allowed in the M_A vs. $\Delta M_H = M_H - M_{H^\pm}$ plane by the electroweak precision constraints. The green, yellow, gray regions satisfy $\Delta\chi^2_{\rm EW}(M_A, \Delta M) < 2.3, 6.2, 11.8$, corresponding to 68.3, 95.4, and 99.7% confidence intervals, respectively.

The results of our analysis are displayed in Fig. 3 confirming a custodial symmetry limit [17] of our interest $M_A \ll M_H \sim M_{H^\pm}$, $M_A \sim M_{H^\pm} \ll M_H$, or $M_A \sim M_H \sim M_{H^\pm}$ although the last two cases are disfavoured.

5 Theoretical Consideration of vaccum stability and perturbativity

While any value of M_A is allowed by the EW precision tests in the limit of $M_H \sim M_{H^\pm}$, there appear upper bounds on $M_{H^\pm,H}$ by theoretical consideration of vacuum stability, global minimum, and perturbativity expressed respectively by

$$\lambda_{1,2} > 0,\ \lambda_3 > -\sqrt{\lambda_1\lambda_2},\ |\lambda_5| < \lambda_3 + \lambda_4 + \sqrt{\lambda_1\lambda_2}, \quad (22)$$

$$m_{12}^2(m_{11}^2 - m_{22}^2\sqrt{\lambda_1/\lambda_2})(\tan\beta - (\lambda_1/\lambda_2)^{1/4}) > 0, \quad (23)$$

$$|\lambda_i| \lesssim |\lambda_{\rm max}| = \sqrt{4\pi}, 2\pi, \text{ or } 4\pi. \quad (24)$$

Taking λ_1 as a free parameter, one can have the following expressions for the other couplings in the large t_β limit:

$$\lambda_2 v^2 \approx s^2_{\beta-\alpha} M_h^2 \quad (25)$$

$$\lambda_3 v^2 \approx 2M^2_{H^\pm} - (s^2_{\beta-\alpha} + s_{\beta-\alpha} y_\tau) M_H^2 + s_{\beta-\alpha} y_\tau M_h^2 \quad (26)$$

$$\lambda_4 v^2 \approx -2M^2_{H^\pm} + s^2_{\beta-\alpha} M_H^2 + M_A^2 \quad (27)$$

$$\lambda_5 v^2 \approx s^2_{\beta-\alpha} M_H^2 - M_A^2 \quad (28)$$

where we have used the relation (8) neglecting the terms of $O(1/t_\beta^2)$.

In the right-sign (RS) limit of the lepton (tau, in particular), $y_\tau s_{\beta-\alpha} \to +1$, one finds a strong upper limit of [6]

$$M_A \ll M_{H^\pm} \sim M_H \lesssim 250\text{GeV} \quad (\text{RS}). \quad (29)$$

On the other hand, in the wrong-sign (WS) limit, $y_\tau s_{\beta-\alpha} \to -1$, the heavy boson masses up to the perturbativity limit,

$$M_A \ll M_{H^\pm} \sim M_H \lesssim \sqrt{4\pi} v \quad (\text{WS}). \quad (30)$$

can be obtained.

Let us finally remark that the hAA coupling is generically order one and thus can lead to a sizeable non-standard decay of $h \to AA$ or $AA^*(\tau\tau)$ if allowed kinematically. Then, one needs to have $|\lambda_{hAA}/v| \ll 1$ to avoid an exotic decay of the SM boson. Noting that $\lambda_{hAA}/v \approx s_{\beta-\alpha}[\lambda_3 + \lambda_4 - \lambda_5]$, one gets

$$\lambda_{hAA} v / s_{\beta-\alpha} \approx -(1 + s_{\beta-\alpha} y_\tau) M_H^2 + s_{\beta-\alpha} y_\tau M_h^2 + 2M_A^2 \quad (31)$$

where we have put $s^2_{\beta-\alpha} = 1$. In the RS or SM limit, the condition $\lambda_{hAA} \approx 0$ can be met for a rather light H with $M_H^2 \approx \frac{1}{2}M_h^2 + M_A^2$ which is disfavoured in the explanation of the muon g-2. On the other hand, one can arrange a cancellation for $\lambda_{hAA} \approx 0$ in the wrong-sign limit for arbitrary value of M_H if the tau Yukawa coupling satisfies

$$y_\tau s_{\beta-\alpha} \approx -\frac{M_H^2 - 2M_A^2}{M_H^2 - M_h^2} < -1. \quad (32)$$

6 Lepton universality tests

In the limit of large $\tan\beta$, the charged boson can generate significant corrections to τ decays at the tree level and furthermore the extra Higgs boson contribution to one-loop corrections can also be significant [19]. The recent study [8] showed that a stringent bound on the charged boson contributions can be obtained from the lepton universality condition obtained by the HFAG collaboration [14]. Given the precision at the level of 0.1 %, the lepton universality data put the strongest bound on the type X 2HDM parameter space in favor of the muon g–2. Thus, let us now make a proper analysis of the HFAG data.

From the measurements of the pure leptonic processes, $\tau \to \mu\nu\nu$, $\tau \to e\nu\nu$ and $\mu \to e\nu\nu$, HFAG obtained the constraints on the three coupling ratios, $(g_\tau/g_\mu) = \sqrt{\Gamma(\tau \to e\nu\nu)/\Gamma(\mu \to e\nu\nu)}$, etc. Defining $\delta_{ll'} \equiv (g_l/g_{l'}) - 1$, let us rewrite the data from the leptonic processes:

$$\delta^l_{\tau\mu} = 0.0011 \pm 0.0015,$$
$$\delta^l_{\tau e} = 0.0029 \pm 0.0015,$$
$$\delta^l_{\mu e} = 0.0018 \pm 0.0014 \quad (33)$$

In addition, combing the semi-hadronic processes $\pi/K \to \mu\nu$, HFAG also provided the averaged constraint on (g_τ/g_μ) which is translated into

$$\delta^{l+\pi+K}_{\tau\mu} = 0.0001 \pm 0.0014. \tag{34}$$

It is important to notice that only two ratios out of the three leptonic measurements are independent and thus the three data (33) are strongly correlated. For a consistent treatment of the data, one combination out of the three has to be projected out. One can indeed check that the direction $\delta^l_{\tau\mu} - \delta^l_{\tau e} + \delta^l_{\mu e}$ has the zero best-fit value and the zero eigenvalue of the covariance matrix, and thus corresponds to the unphysical direction. Furthermore, two orthogonal directions $\delta^l_{\tau\mu}+\delta^l_{\tau e}$ and $-\delta^l_{\tau\mu}+\delta^l_{\tau e}+2\delta^l_{\mu e}$ are found to be uncorrelated in a good approximation. As a result, the 2HDM contribution to $\delta_{ll'}$ are calculated to be

$$\delta^l_{\tau\mu} = \delta_{loop},\ \delta^l_{\tau e} = \delta_{tree} + \delta_{loop},\ \delta^l_{\mu e} = \delta_{tree},\ \delta^{l+\pi+K}_{\tau\mu} = \delta_{loop}. \tag{35}$$

Here δ_{tree} and δ_{loop} are given by [19]:

$$\begin{aligned} \delta_{tree} &= \frac{m_\tau^2 m_\mu^2}{8m_{H^\pm}^4}\tan^4\beta - \frac{m_\mu^2}{m_{H^\pm}^2} t_\beta^2 \frac{g(m_\mu^2/m_\tau^2)}{f(m_\mu^2/m_\tau^2)}, \\ \delta_{loop} &= \frac{G_F m_\tau^2}{8\sqrt{2}\pi^2} t_\beta^2 \left[1+\frac{1}{4}\left(H(x_A) + s^2_{\beta-\alpha}H(x_H) + c^2_{\beta-\alpha}H(x_h)\right)\right]. \end{aligned} \tag{36}$$

where $f(x) \equiv 1-8x+8x^3-x^4-12x^2\ln(x)$, $g(x) \equiv 1+9x-9x^2-x^3+6x(1+x)\ln(x)$, $H(x) \equiv \ln(x)(1+x)/(1-x)$, and $x_\phi = m_\phi^2/m_{H^\pm}^2$. From Eqs. (33), (34) and (35), one obtains the following three independent bounds:

$$\begin{aligned} \frac{1}{\sqrt{2}}\delta_{tree} + \sqrt{2}\delta_{loop} &= 0.0028 \pm 0.0019, \\ \sqrt{\frac{3}{2}}\delta_{tree} &= 0.0022 \pm 0.0017, \\ \delta_{loop} &= 0.0001 \pm 0.0014. \end{aligned} \tag{37}$$

We will use these constraints to put a strong limit on the $(g-2)_\mu$ favoured region in the M_A–$\tan\beta$ plane in the next section. Let us recall that the $Z \to \tau\tau$ data, although less strong than the HFAG data, provides an independent bound [8] which further cuts out some corner of parameter space.

7 Pinning down the whole 2HDM-X parameter space

It is an interesting task to narrow down the allowed region of the type X 2HDM parameter space collecting all the relevant experimental data including those outlined in Section I and the 125 GeV boson data from LHC, in particular. The scan ranges of all the 2HDM-X input parameters are listed in Table 2. For our scan, we adopt the convention $-\pi/2 < \alpha - \beta < \pi/2$ and $0 < \beta < \pi/2$, and use the parameter λ_1 as an input parameter instead of m_{12}^2.

Fig. 4 shows the allowed region in the m_A-$\tan\beta$ plane from the profile-likelihood study taking all the other 2HDM-X parameters as nuisance parameters. To see

2HDM parameter	Range
Scalar boson mass (GeV)	$125 < m_H < 400$
Pseudoscalar boson mass (GeV)	$10 < m_A < 400$
Charged boson mass (GeV)	$94 < m_{H^\pm} < 400$
$c_{\beta-\alpha}$	$0.0 < c_{\beta-\alpha} < 0.1$
$\tan\beta$	$10 < \tan\beta < 150$
λ_1	$0.0 < \lambda_1 < 4\pi$

Table 2. The scan ranges of the 2HDM-X input parameters.

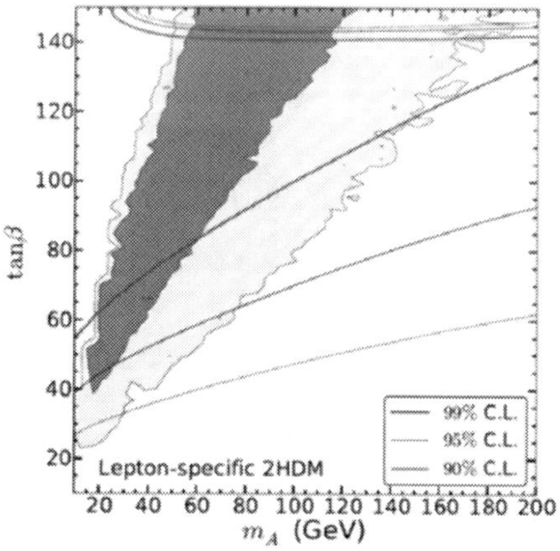

Figure 4. The 68% and 95% allowed regions by Δa_μ and other experimental constraints in the m_A-$\tan\beta$ plane. The contours of the lepton universality likelihood in the 99%, 95% and 90% confidence level are overlayed.

the impact of the lepton universality data by HFAG, we overlay the contour lines of the lepton universality likelihood at the 99%, 95% and 90% confidence level based on the constraints (37). The allowed region opened up for $\tan\beta > 140$ needs a comment. Note that the δ_{loop} is always negative while δ_{tree} becomes positive for larger $\tan\beta/m_{H^\pm}$. Thus, there appears a fine-tuned region around $\tan\beta/m_{H^\pm} \sim 1$ GeV^{-1} where the positive δ_{tree} and the negative δ_{loop} cancel each other to give a good fit. However, such regions are excluded by the $Z \to \tau\tau$ data [8] and thus we are left with the tightly limited region of $M_A \approx 10-80$ GeV and $\tan\beta \approx 25-60$ at the 95% confidence level.

The region allowed in Fig. 4 can be either in the right-sign ($y_\tau \equiv \xi^l_h > 0$) or wrong-sign ($y_\tau \equiv \xi^l_h < 0$) domain as shown in the left and middle panels of Fig. 5. One can see that the rigt-sign limit is tightly constrained to a small region of $m_A \approx 60-80$ GeV while the wrong-sign limit is favoured in a wider range of parameter space. The right panel shows the sizes of the coupling λ_{hAA} restricted by the LHC data on the exotic decay of the 125 GeV boson, putting a generous bound of $\mathrm{Br}(h \to AA^{(*)}) < 40\%$. As explained before, the suppressed value of λ_{hAA} for $m_A \lesssim m_h/2$ is shown to appear only in the wrong-sign domain.

Given the possible existence of a light CP-odd boson explaining the muon g–2 in the type X 2HDM, it would

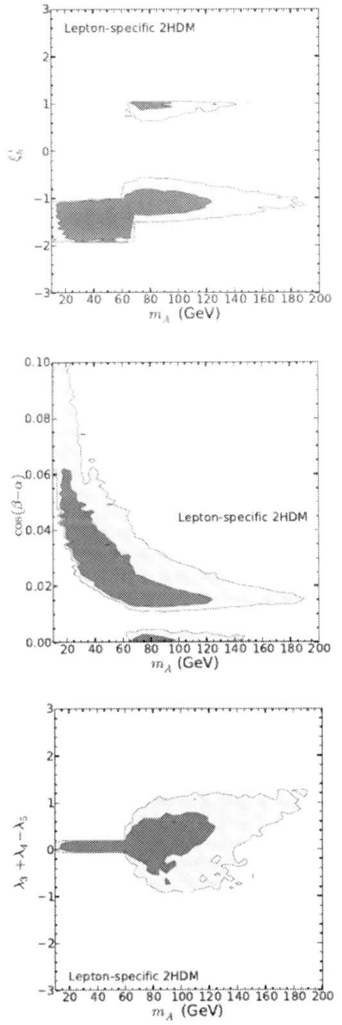

Figure 5. The 68% and 95% allowed regions in the m_A–ξ^l_h ($y_\tau \equiv \xi^l_h$) (left) and m_A–$c_{\beta-\alpha}$ (middle), and m_A–λ_{hAA} ($\lambda_{hAA} \approx \lambda_3+\lambda_4-\lambda_5$) (right) plane.

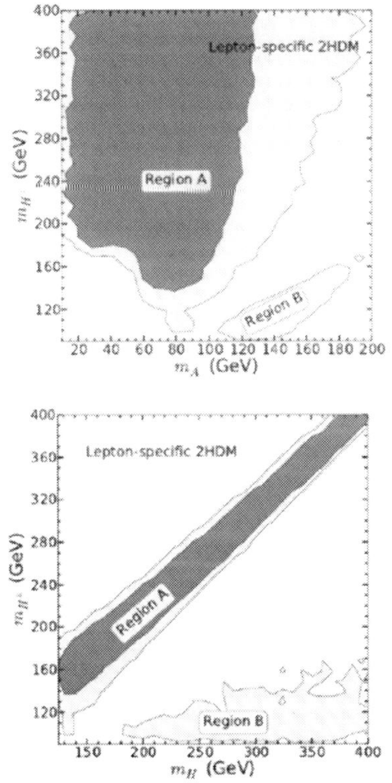

Figure 6. Distribution of the extra boson masses allowed at the 68% and 95% confidence levels.

be important to look for its trails at the LHC. Fig. 6 shows the allowed mass ranges of all the extra bosons. Region A following the pattern of $m_A \lesssim m_H \approx m_{H^\pm}$ is favoured while Region B with $m_A \approx m_{H^\pm} \ll m_H$ is already excluded as discussed before.

8 Tau-rich signatures at the LHC

The bulk parameter space with $m_A \ll m_H \sim m_{H^\pm}$ is a clear prediction of the type X 2HDM as the origin of the muon g–2 anomaly. Since the extra bosons are mainly from the "leptonic" Higgs doublet with large $\tan\beta$, all the three members are expected to dominantly decay into the τ–flavor, leading to τ–rich signatures at the LHC via the following production and ensuing cascade decay chains:

$$\begin{aligned}
pp &\to W^{\pm *} \to H^\pm A \to (\tau^\pm\nu)(\tau^+\tau^-), \qquad (38)\\
pp &\to Z^*/\gamma^* \to HA \to (\tau^+\tau^-)(\tau^+\tau^-),\\
pp &\to W^{\pm *} \to H^\pm H \to (\tau^\pm\nu)(\tau^+\tau^-),\\
pp &\to Z^*/\gamma^* \to H^+H^- \to (\tau^+\nu)(\tau^-\bar{\nu}).
\end{aligned}$$

To probe Region A, we select six benchmark points with different combinations of m_A and m_H presented in Table 3. For each point, we take a simple parametrization of $\tan\beta = 1.25(m_A/\text{GeV}) + 25$ and $m_{H^\pm} = m_H + 15\text{GeV}$. Note that we included the points with $m_A > 80$ GeV for the sake of the LHC study although they are forbidden by the lepton universality tests. In Table 3, we show the production cross-section, the selection cuts and the significance for each benchmark expected for the integrated luminosity of 25/fb at the 14 TeV LHC.

In Fig. 7, we present the exclusion region coming mainly from the chargino-neutralino search at the LHC8, and the expected discovery reaches at LHC14 with the integrated luminosity of 25/fb. A heavy CP-even boson with $m_H > 200$ GeV and a light CP-odd boson with $m_A < 50$ GeV are still allowed, and the LHC14 can explore some of the regions. The sensitivities are weaker for larger m_H just because of smaller cross sections, and for smaller m_A because τs from lighter A become softer and thus the acceptance quickly decreases. Moreover, the

	point A	point B	point C	point D	point E	point F
m_A [GeV]	20	40	100	40	100	180
m_H [GeV]	200	200	200	260	260	260
total $\sigma_{\rm gen}$ [fb]	270.980	241.830	153.580	100.430	71.271	44.163
$n_\ell \geq 3$	6.606	16.681	21.713	7.110	11.962	8.822
$n_\tau \geq 3$	0.894	2.602	4.386	0.888	2.346	1.971
$\not{E}_T$ > 100 GeV	0.201	0.547	1.179	0.209	0.765	0.926
$n_b = n_j = 0$	0.098	0.314	0.857	0.121	0.479	0.631
S/B	0.1	0.5	1.2	0.2	0.7	0.9
$S/\sqrt{B}_{25{\rm fb}^{-1}}$	0.6	1.9	5.2	0.7	2.9	3.8

Table 3. The number of events after applying successive cuts for 14 TeV LHC.

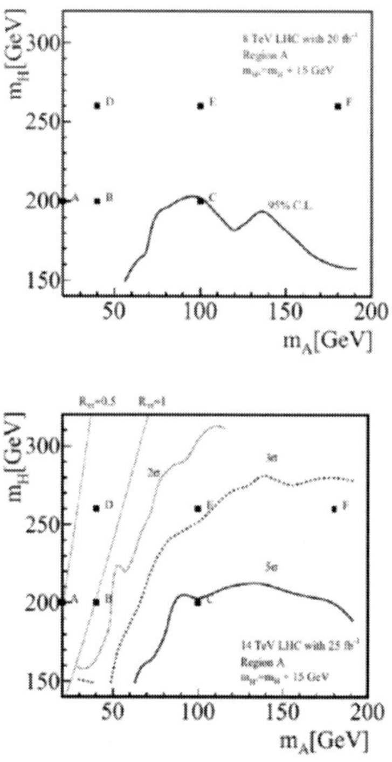

Figure 7. The 95% exclusion contour from LHC8 (left) and the 2σ, 3σ and 5σ discovery reach countours at LHC14 (right) in the m_A–m_H plane.

$H/H^\pm \to AZ/W^\pm$ decay modes also start open to decrease the number of hard τs from direct $H/H^\pm$ decays. In such a region, a light A from heavy H^+/H decay will be boosted, resulting in a collimated τ−pair which becomes difficult to be tagged as two separated τ-jets. It is one of the reasons to have less acceptance for this parameter region. We can estimate the separation $R_{\tau\tau}$ of the τ leptons from A decay:

$$R_{\tau\tau} \sim \frac{2m}{p_T} \sim \frac{4m_A}{m_{H^\pm/H}\sqrt{1-2\frac{m_A^2+m_{W/Z}^2}{m_{H^\pm/H}^2}+\frac{(m_A^2-m_{W/Z}^2)^2}{m_{H^\pm/H}^4}}}. \quad (39)$$

Since the jets are usually defined with $R = 0.5$, the τ−pair starts overlapping. We indicated the region with the overlapping τ problem in red lines in the right panel of Fig. 7. Further studies on how to capture the kinematic features of the boosted $A \to \tau^+\tau^-$ are required to probe such a small m_A region.

9 Summary

The type X 2HDM is still a viable option for the explanation of the muon g–2 in the parameter region with large $\tan\beta$ and a light CP-odd boson A. Being "hadrophobic and leptophilic" in the large $\tan\beta$ limit, it can be easily free from all the hadron-related constraints, particularly, coming from the decay $B_s \to \mu\mu$ which puts only a mild bound of $m_A \gtrsim 10$ GeV. However, such a region is tightly limited by the lepton universality tests from the HFAG and $Z \to \tau\tau$ data. Combining all the current bounds, we find allowed at the 95% confidence level a limited region of $\tan\beta \approx 15$–60 and $m_A \approx 10 - 80$ GeV with $m_H \approx m_{H^\pm} \gg m_A$.

It will be an interesting task to search for such a light CP-odd boson A and the extra heavy bosons $H, H^\pm$ in the next run of the LHC mainly through $pp \to AH, AH^\pm$ followed by the decays $H^\pm \to \tau^\pm\nu$ and $A, H \to \tau^+\tau^-$ which requires further studies to improve the (boosted) tau identification.

References

[1] H. N. Brown *et al.* [Muon g-2 Collaboration], Phys. Rev. Lett. **86** (2001) 2227 [hep-ex/0102017]. G. W. Bennett *et al.* [Muon g-2 Collaboration], Phys. Rev. D **73** (2006) 072003 [hep-ex/0602035].

[2] A. Dedes and H. E. Haber, JHEP **0105** (2001) 006 [hep-ph/0102297]. K. m. Cheung, C. H. Chou and O. C. W. Kong, Phys. Rev. D **64** (2001) 111301 [hep-ph/0103183]. M. Krawczyk, hep-ph/0103223. F. Larios, G. Tavares-Velasco and C. P. Yuan, Phys. Rev. D **64** (2001) 055004 [hep-ph/0103292].

[3] M. Krawczyk, Acta Phys. Polon. B **33** (2002) 2621 [hep-ph/0208076].

[4] K. Cheung and O. C. W. Kong, Phys. Rev. D **68** (2003) 053003 [hep-ph/0302111].

[5] J. Cao, P. Wan, L. Wu and J. M. Yang, Phys. Rev. D **80** (2009) 071701 [arXiv:0909.5148 [hep-ph]].

[6] A. Broggio, E. J. Chun, M. Passera, K. M. Patel and S. K. Vempati, JHEP **1411** (2014) 058 [arXiv:1409.3199 [hep-ph]].

[7] L. Wang and X. F. Han, JHEP **1505** (2015) 039 [arXiv:1412.4874 [hep-ph]].

[8] T. Abe, R. Sato and K. Yagyu, JHEP **1507** (2015) 064 [arXiv:1504.07059 [hep-ph]].

[9] E. J. Chun, Z. Kang, M. Takeuchi and Y.-L. Sming Tsai, arXiv:1507.08067, to apear in JHEP.

[10] S. M. Barr and A. Zee, Phys. Rev. Lett. 65 (1990) 21 [Erratum-ibid. 65 (1990) 2920]. V. Ilisie, arXiv:1502.04199 [hep-ph].

[11] H. E. Logan and U. Nierste, Nucl. Phys. B **586** (2000) 39 [hep-ph/0004139].

[12] CMS and LHCb Collaboration, CMS-PAS-BPH-13-007.

 DOI: 10.1051/epjconf/201611801006

[13] K. Schmidt-Hoberg, F. Staub and M. W. Winkler, Phys. Lett. B **727** (2013) 506 [arXiv:1310.6752 [hep-ph]].

[14] HFAG Collaboration, arXiv:1412.7515 [hep-ph].

[15] P. M. Ferreira, J. F. Gunion, H. E. Haber and R. Santos, Phys. Rev. D **89** (2014) 11, 115003 [arXiv:1403.4736 [hep-ph]]. P. M. Ferreira, R. Guedes, M. O. P. Sampaio and R. San tos, JHEP **1412** (2014) 067 [arXiv:1409.6723 [hep-ph]].

[16] J. F. Gunion and H. E. Haber, Phys. Rev. D **67** (2003) 075019 [hep-ph/0207010]. G. C. Branco, P. M. Ferreira, L. Lavoura, M. N. Rebelo, M. Sher and J. P. Silva, Phys. Rept. **516** (2012) 1 [arXiv:1106.0034 [hep-ph]].

[17] J.-M. Gerard and M. Herquet, Phys. Rev. Lett. **98** (2007) 251802 [hep-ph/0703051 [HEP-PH]].

[18] ATLAS Collaboration, arXiv:1509.00672; D. Chowdhury and O. Eberhardt, arXiv:1503.08216 [hep-ph].

[19] M. Krawczyk and D. Temes, arXiv:hep-ph/0410348.

DOI: 10.1051/epjconf/201611801007

New Physics in B decays

Andreas Crivellin[1] e-mail: andreas.crivellin@cern.ch
[1] CERN Theory Division, CH–1211 Geneva 23, Switzerland

Abstract. While the LHC did not observe direct evidence for physics beyond the standard model, indirect hints for new physics were uncovered in the flavour sector in the decays $B \to K^*\mu^+\mu^-$, $B \to K\mu^+\mu^-/B \to Ke^+e^-$, $B_s \to \phi\mu^+\mu^-$, $B \to D^{(*)}\tau\nu$ and $h \to \tau^\pm\mu^\mp$. Each observable deviates from the SM predictions at the $2-3\,\sigma$ level only, but combining all $b \to s\mu^+\mu^-$ data via a global fit, one finds $4-5\,\sigma$ difference for NP compared to the SM and combining $B \to D^*\tau\nu$ with $B \to D\tau\nu$ one obtains $3.9\,\sigma$.
While $B \to D^{(*)}\tau\nu$ and $h \to \tau\mu$ can be naturally explained by an extended Higgs sector, the $b \to s\mu^+\mu^-$ anomalies point at a Z' gauge boson. However, it is also possible to explain $B \to D^{(*)}\tau\nu$ and $b \to s\mu^+\mu^-$ simultaneously with leptoquarks while their effect in $h \to \tau^\pm\mu^\mp$ is far too small to account for current data. Combining a 2HDM with a gauged $L_\mu - L_\tau$ symmetry allows for explaining the $b \to s\mu^+\mu^-$ anomalies in combination with $h \to \tau^\pm\mu^\mp$, predicting interesting correlations with $\tau \to 3\mu$. In the light of these deviations from the SM we also discuss the possibilities of observing lepton flavour violating B decays (e.g. $B \to K^{(*)}\tau^\pm\mu^\mp$ and $B_s \to \tau^\pm\mu^\mp$).

1 Introduction

The LHC completed the standard model (SM) of particle physics by discovering the Higgs particle while no additional new particles have been observed so far. However, some indirect 'hints' for new physics (NP) in the flavor sector appeared in $B \to K^*\mu^+\mu^-$, $B_s \to \phi\mu^+\mu^-$, $R(K) = B \to K\mu^+\mu^-/B \to Ke^+e^-$, $B \to D^{(*)}\tau\nu$ and $h \to \mu\tau$.

Let us consider the current experimental and theoretical situation is some more detail. Concerning $b \to s\mu^+\mu^-$ transitions, already in 2013 LHCb reported deviations from the SM predictions [1] (mainly in an angular observable called P_5' [2]) in $B \to K^*\mu^+\mu^-$ [3] with a significance of 2–3$\,\sigma$ depending on the assumptions of hadronic uncertainties [4–6] being confirmed in 2015 [7]. Also in the decay $B_s \to \phi\mu^+\mu^-$ [8] LHCb uncovered differences compared to the SM prediction from lattice QCD [9, 10] of $3.5\,\sigma$ [5]. Furthermore, in 2014 LHCb [11] found indications for the violation of lepton flavour universality in

$$R(K) = \frac{\mathrm{Br}[B \to K\mu^+\mu^-]}{\mathrm{Br}[B \to Ke^+e^-]} = 0.745^{+0.090}_{-0.074} \pm 0.036\,, \quad (1)$$

in the range $1\,\mathrm{GeV}^2 < q^2 < 6\,\mathrm{GeV}^2$ which disagrees with the theoretically clean SM prediction $R_K^{\mathrm{SM}} = 1.0003 \pm 0.0001$ [12] by $2.6\,\sigma$. Combining these with other $b \to s$ observables, it is found that NP is preferred compared to the SM by $4-5\,\sigma$ [13, 14]. Symmetry based solutions include NP in $C_9^{\mu\mu}$ (i.e. left-handed $b-s$ current and vectorial muon current) and $C_9^{\mu\mu} = -C_{10}^{\mu\mu}$ (i.e. left-handed $b-s$ and muon current).

Hints for lepton flavour universality violating NP in B decays were observed for the first time by the BaBar collaboration in $B \to D^{(*)}\tau\nu$ [15] in 2012. Recently, these measurements have been confirmed by BELLE [16] and LHCb measured $B \to D^*\tau\nu$ [17]. In summary, these experiments found for the ratios $R(D^{(*)}) \equiv \mathrm{Br}(B \to D^{(*)}\tau\nu)/\mathrm{Br}(B \to D^{(*)}\ell\nu)$ [18]:

$$R(D)_{\mathrm{EXP}} = 0.391 \pm 0.041 \pm 0.028\,, \quad (2)$$

$$R(D^*)_{\mathrm{EXP}} = 0.322 \pm 0.018 \pm 0.012\,. \quad (3)$$

Comparing these measurements to the SM predictions [19] $R_{\mathrm{SM}}(D) = 0.297 \pm 0.017$, $R_{\mathrm{SM}}(D^*) = 0.252 \pm 0.003$, we see that there is a combined discrepancy of $3.9\,\sigma$ [18].

CMS recently also searched for the decay $h \to \tau\mu$ [20] finding a non-zero result of $\mathrm{Br}[h \to \mu\tau] = \left(0.89^{+0.40}_{-0.37}\right)$ which disagrees by about $2.4\,\sigma$ from 0, i.e. from the SM value. This is consistent with the less precise ATLAS measurement [21] giving a combined significance of $2.6\,\sigma$.

In these proceedings, we review NP models which can explain the deviations from the SM discussed above with focus on models with a Z' boson and/or additional Higgs doublets and also briefly mention models with leptoquarks.

2 Tauonic B Decays

Due to the heavy tau lepton in the final state, these decays are sensitive to charged Higgses [24]. A 2HDM of type II (like the MSSM at tree-level) cannot explain the deviations from the SM in tauonic B decays (due to the necessarily destructive interference) without violating bounds from other observables [25] (see left plot in Fig. 1). However, a 2HDM with generic Yukawa coupling (i.e. type III)

© The Authors, published by EDP Sciences. This is an open access article distributed under the terms of the Creative Commons Attribution License 4.0 (http://creativecommons.org/licenses/by/4.0/).

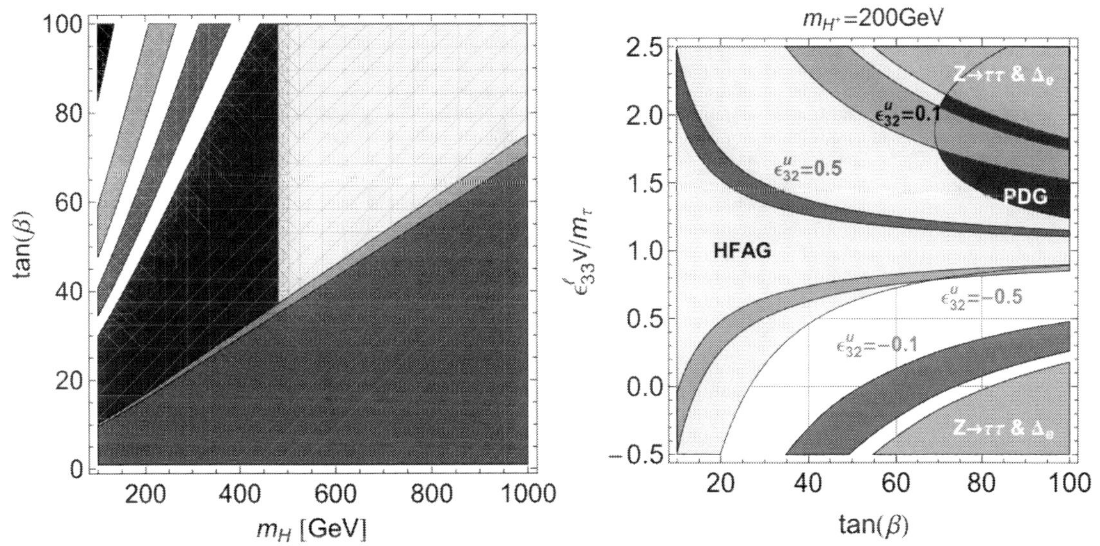

Figure 1. Left: Updated constraints on the 2HDM of type II parameter space. The regions compatible with experiment are shown (the regions are superimposed on each other): $b \to s\gamma$ (yellow) [22], $B \to D\tau\nu$ (green), $B \to \tau\nu$ (red), $B_s \to \mu^+\mu^-$ (orange), $K \to \mu\nu/\pi \to \mu\nu$ (blue) and $B \to D^*\tau\nu$ (black). Note that no region in parameter space is compatible with all processes since explaining $B \to D^*\tau\nu$ would require very small Higgs masses and large values of $\tan\beta$ which is not compatible with the other observables. To obtain this plot, we added the theoretical uncertainty of the SM linearly on the top of the 2σ experimental error. Right: Allowed regions in the $\tan\beta$–$v/m_\tau\epsilon^\ell_{33}$ plane from $\mathcal{R}(D^{(*)})$ and $\tau \to \mu\nu\nu$ at the 2σ level in the perturebed 2HDM of type X [23]. The yellow region is allowed by $\tau \to \mu\nu\nu$ using the HFAG result for $m_H = 30\,\mathrm{GeV}$ and $m_A = 200\,\mathrm{GeV}$, while the (darker) blue one is the allowed region using the PDG result. The red, orange, green, and magenta bands correspond to the allowed regions by $\mathcal{R}(D^{(*)})$ for different values of ϵ^u_{32}. The gray region is excluded by $Z \to \tau\tau$ and $\tau \to e\nu\nu$. For $m_H \simeq m_A$ the allowed regions from $\tau \to \mu\nu\nu$ would be slightly larger.

can account for $B \to D\tau\nu$ and $B \to D^*\tau\nu$ simultaneously, respecting the constraints from all other observables, if the coupling of a right-handed charm to the third generation quark doublet (ϵ^u_{32}) is large [23, 26].

Here, two variants are phenomenologically possible: in the limit of vanishing non-standard couplings, the type III model could reduce either to type II (like the MSSM at tree-level) or type X (leptospecific). While the low energy constraints on the type II model are quite stringent (see left plot in Fig. 1) and it is also challenged by $A \to \tau\tau$ searches [27], the type X model is only weakly constrained (see for example Ref. [28] for a review). Therefore, we will focus on the type X model as a solution [23] which has also the advantage of providing a possible explanation for the anomalous magnetic moment of the muon [29–31] and $\tau \to \mu\nu\nu$. In the right plot in Fig. 1 we show which regions in parameter space can account for the experimental data. As usual, $\tan\beta$ is the ratio of the two vacuum expectation values and m_H (m_A) refer to the additional neutral CP-even (CP-odd) Higgs mass. Interestingly, requiring an explanation for the anomalous magnetic moment of the muon without violating bounds from $\tau \to \mu\nu\nu$ enforces $m_H \ll m_A$ (see left plot of Fig. 2). Together with the large top-charm coupling induced by ϵ^u_{32} (necessary for $R(D^{(*)})$) sizable branching ratios for $t \to Hc$ (see right plot of Fig. 2) are predicted which are well within the reach of the LHC.

Alternative explanations involve leptoquarks [32–36] or R-parity violating SUSY [37].

3 Anomalies in $b \to s\mu^+\mu^-$

A rather large contribution to operator $(\bar s\gamma_\alpha P_L b)(\bar\mu\gamma^\alpha\mu)$, as preferred by the model independent fit [13, 14], can be achieved in models containing a heavy Z' gauge boson (see Refs. [38, 39] for early attempts to explain this anomaly). If one aims at explaining $R(K)$ as well, a contribution to $C_9^{\mu\mu}$ involving muons, but not to C_9^{ee} with electrons is necessary [40–42]. This is naturally the case in models with gauged muon minus tauon number ($L_\mu - L_\tau$) [39, 43, 44][1]. Alternative explanations are again models with leptoquarks [34–36, 54–56].

In Z' models the couplings to quarks can be written generically as

$$L \cup g' \left(\bar d_i \gamma^\mu P_L d_j Z'_\mu \Gamma^{dL}_{ij} + \bar d_i \gamma^\mu P_R d_j Z'_\mu \Gamma^{dR}_{ij}\right) . \qquad (4)$$

where g' is the new $U'(1)$ gauge coupling constant. Unavoidable tree-level contributions to $B_s - \overline{B}_s$ are generated which constrain the coupling to muons to be much larger than the one to $\bar s b$. In the left plot in Fig. 3 the regions in the Γ^L_{sb}–Γ^R_{sb} plane are shown which are in agreement with $B_s - \overline{B}_s$ mixing and $b \to s\mu^+\mu^-$ data within 2σ. Note that in the symmetry limit $\Gamma^R_{sb} = 0$, $B_s - \overline{B}_s$ mixing puts a upper bound on Γ^L_{sb}.

[1] Z' bosons with the desired couplings can also be obtained in other Z' models [45–50]. For an extensive analysis of Z' models prior to the apparence of the anomalies see for example [51–53].

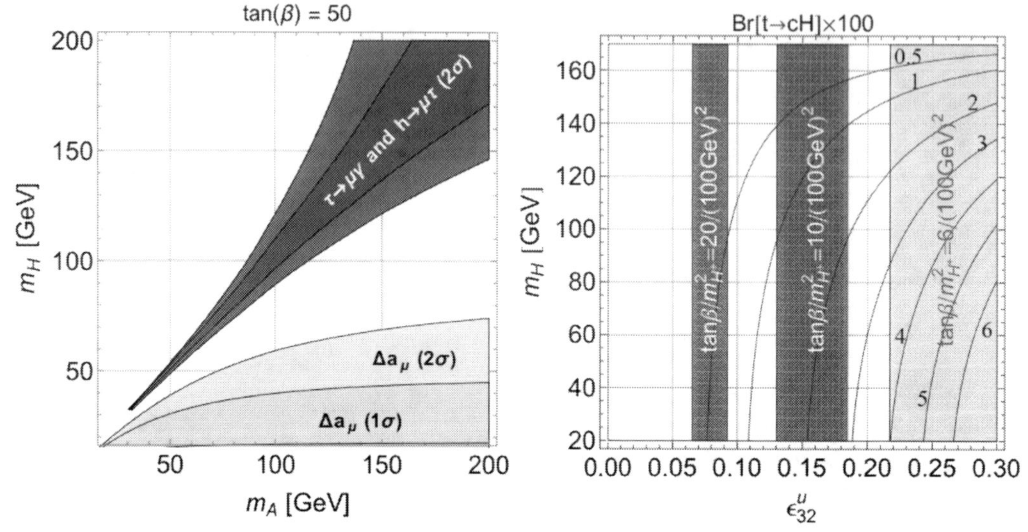

Figure 2. Left: Red, green, and yellow are the allowed regions in the m_A–m_H plane from $(g-2)_\mu$ at the 2σ level for $\tan\beta = 50$, $m_{H^+} = 200\,\text{GeV}$, $\cos(\alpha-\beta) = 0.1$ and different values of ϵ^ℓ_{33} in the perturbed 2HDM X. Blue is the allowed region (again at 2σ) from $\tau \to \mu\gamma$ and $h \to \mu\tau$ for $\epsilon^\ell_{33} = 2m_\tau/v$ and $\cos(\alpha-\beta) = 0.1$, light blue corresponds to $\cos(\alpha-\beta) = 0.2$. The allowed region for Δa_μ covers the three possibilities $\epsilon^\ell_{32} \neq 0$, $\epsilon^\ell_{32} = \epsilon^\ell_{23} \neq 0$ and $\epsilon^\ell_{32} = -\epsilon^\ell_{23} \neq 0$, since the latter ones can give m_τ/m_μ enhanced one-loop contributions. However, the effects turn out to be small, as $\epsilon^\ell_{32,32}$ is stringently constrained from $\tau \to \mu\gamma$. In addition, we checked that the effect of λ_H is very small. The white regions are not compatible with experiment at the 2σ level. Right: The contour lines denote $\text{BR}(t \to Hc) \times 100$ as a function of ϵ^u_{32} and m_H. The colored regions are allowed by $\mathcal{R}(D)$ and $\mathcal{R}(D^*)$ for different values of $\tan\beta/m^2_{H^+}$. Note that H is required to be quite light if one aims at explaining the anomalous magnetic moment of the muon.

3.1 $h \to \tau\mu$

Lepton flavour violating couplings of the SM Higgs are induced by a single operator up to dim-6. Considering only this operator $\text{Br}[h \to \mu\tau]$ can be up to 10% [57, 58]. However, it is in general difficult to get dominant contributions to this operator in a UV complete model, as for example in models with vector-like leptons [59] or leptoquarks [60, 61]. Therefore, among the several attempts to explain this $h \to \mu\tau$ observation, most of them are relying on models with extended Higgs sectors [60, 62]. One particularly elegant solution employs a two-Higgs-doublet model (2HDM) with gauged $L_\mu - L_\tau$ [63].

4 Simultaneous explanation of $b \to s\mu\mu$ and $h \to \tau\mu$ and predictions for $\tau \to 3\mu$

In [43, 44] two models with gauged $L_\mu - L_\tau$ symmetry were presented which can explain $h \to \tau\mu$ simultaneously with the anomalies in $b \to s\mu\mu$ data (including $R(K)$) giving rise to interesting correlated effects in $\tau \to 3\mu$. While in both models the Z' couplings to leptons originate from a gauged $L_\mu - L_\tau$ symmetry, the coupling to quarks is either generated effectively via heavy vector-like quarks charged under $L_\mu - L_\tau$ or directly by assigning horizontal changes to baryons[2].

[2]For pioneering work on horizontal $U(1)$ gauge symmetries see Ref. [64].

4.1 2 Higgs doublets with vector-like quarks

In a 2HDM with a gauged $U(1)_{L_\mu - L_\tau}$ symmetry [63], $L_\mu - L_\tau$ is broken spontaneously by the vacuum expectation value of a scalar Φ (being singlet under the SM gauge group) with $Q^\Phi_{L_\mu - L_\tau} = 1$, leading to the Z' mass $m_{Z'} = \sqrt{2}g'\langle\Phi\rangle \equiv g'v_\Phi$. Two Higgs doublets are introduced which break the electroweak symmetry: Ψ_1 with $Q^{\Psi_1}_{L_\mu - L_\tau} = -2$ and Ψ_2 with $Q^{\Psi_2}_{L_\mu - L_\tau} = 0$. Therefore, Ψ_2 gives masses to quarks and leptons while Ψ_1 couples only off-diagonally to $\tau\mu$:

$$\begin{aligned} \mathcal{L}_Y \supset & -\bar{\ell}_f Y^\ell_i \delta_{fi} \Psi_2 e_i - \xi_{\tau\mu} \bar{\ell}_3 \Psi_1 e_2 \\ & - \bar{Q}_f Y^u_{fi} \tilde{\Psi}_2 u_i - \bar{Q}_f Y^d_{fi} \Psi_2 d_i + \text{h.c.} \, . \end{aligned} \quad (5)$$

Here Q (ℓ) is the left-handed quark (lepton) doublet, u (e) is the right-handed up-quark (charged-lepton) and d the right-handed down quark while i and f label the three generations and the tilde signals charge conjugation. The scalar potential is the one of a $U(1)$-invariant 2HDM [28] with additional couplings to the SM-singlet Φ. We defined as usual $\tan\beta = \langle\Psi_2\rangle/\langle\Psi_1\rangle$ and α is the mixing angle between the neutral CP-even components of Ψ_1 and Ψ_2 (see for example [28]). Therefore, quarks and gauge bosons have standard type-I 2HDM couplings to the scalars. The only deviations from the type I model are in the lepton sector: while the Yukawa couplings $Y^\ell_i \delta_{fi}$ of Ψ_2 are forced to be diagonal by the $L_\mu - L_\tau$ symmetry, $\xi_{\tau\mu}$ gives rise to an off-diagonal entry in the lepton mass matrix:

$$m^{\ell}_{fi} = \frac{v}{\sqrt{2}}\begin{pmatrix} y_e \sin\beta & 0 & 0 \\ 0 & y_\mu \sin\beta & 0 \\ 0 & \xi_{\tau\mu}\cos\beta & y_\tau \sin\beta \end{pmatrix}. \quad (6)$$

It is this τ–μ element that leads to the LFV couplings of h and Z'. The mass basis for the charged leptons is obtained by rotating (μ_R, τ_R) and (μ_L, τ_L) with the angles θ_R and θ_L. A non-vanishing angle θ_R not only gives rise to the LFV decay $h \to \mu\tau$ due to the coupling

$$\frac{m_\tau}{v}\frac{\cos(\alpha-\beta)}{\cos(\beta)\sin(\beta)}\sin(\theta_R)\cos(\theta_R)\bar{\tau}P_R\mu h \equiv \Gamma^h_{\tau\mu}\bar{\tau}P_R\mu h\,, \quad (7)$$

in the Lagrangian, but also leads to off-diagonal Z' couplings to right-handed leptons

$$g' Z'_\nu (\bar{\mu}, \bar{\tau}) \begin{pmatrix} \cos 2\theta_R & \sin 2\theta_R \\ \sin 2\theta_R & -\cos 2\theta_R \end{pmatrix} \gamma^\nu P_R \begin{pmatrix} \mu \\ \tau \end{pmatrix}, \quad (8)$$

while the left-handed couplings are to a good approximation flavour conserving. In order to explain the observed anomalies in the B meson decays, a coupling of the Z' to quarks is required as well, not inherently part of $L_\mu - L_\tau$ models (aside from the kinetic Z–Z' mixing, which is assumed to be small). Following Ref. [39], effective couplings of quarks to the Z' are generated by heavy vector-like quarks [65] charged under $L_\mu - L_\tau$. As a result, the couplings of the Z' to quarks are in principle free parameters. In the limit of decoupled vector-like quarks with the quantum numbers of right-handed quarks, only C_9 is generated, giving a very good fit to data. The results are shown in the right plot of Fig. 3 depicting that for small values of Γ^L_{sb} and θ_R, $b \to s\mu^+\mu^-$ data can be explained without violating bounds from $B_s - \overline{B}_s$ mixing or $\tau \to 3\mu$. In the left plot of Fig. 4 the correlations of $b \to s\mu^+\mu^-$ and $h \to \tau\mu$ with $\tau \to 3\mu$ are shown, depicting that consistency with $\tau \to 3\mu$ requires large values of $\tan\beta$ (not being in conflict with any data as the decoupling limit (i.e. $\xi_{\tau\mu} = 0$) is a type I model) and future searches for $\tau \to 3\mu$ are promising to yield positive results.

4.2 Horizontal charges for quarks

In order to avoid the introduction of vector-like quarks, one can introduce flavour-dependent charges to quarks as well [44]. Here, the first two generations should have the same charges in order to avoid very large effects in K–$\overline{K}$ or D–$\overline{D}$ mixing, generated otherwise unavoidably due to the breaking of the symmetry necessary to generate the measured Cabibbo angle of the CKM matrix. If we require in addition the absence of anomalies, we arrive at the following charge assignment for baryons $Q'(B) = (-a, -a, 2a)$. Here $a \in Q$ is a free model parameter with important phenomenological implications. In this model, at least one additional Higgs doublet which breaks the flavour symmetry in the quark sector is required, and one more is needed if one attempts to explain $h \to \tau\mu$. In case the mixing among the doublets is small, the correlations among $h \to \tau\mu$, $b \to s\mu^+\mu^-$ and $\tau \to 3\mu$ are the same is in the model with vector-like quarks discussed in the last subsection and shown in the left plot of Fig. 4.

The low-energy phenomenology is rather similar to the one of the model with vector like quarks considered in the last section, but the contributions to $B_s - \overline{B}_s$ mixing are directly correlated to $B_d - \overline{B}_d$ and $K - \overline{K}$ mixing as all flavour violation is due to CKM factor. However, concerning direct LHC searches, the implications are very different, as the Z' boson can be directly produced on-shell as a resonance in $p\bar{p}$ collisions since it couples to quarks of the first generation. The resulting strong bounds are shown in right plot of Fig. 4 where they are compared to the allowed regions from $B_s - \overline{B}_s$ mixing and $b \to s\mu^+\mu^-$ data for different values of a.

5 Simultaneous explanation of $b \to s\mu^+\mu^-$ data and $R(D^{(*)})$

5.1 Effective Operators

A scenario with left-handed currents only gives a good fit to $b \to s\mu^+\mu^-$ data [13, 14]. In such a scenario $SU(2)_L$ relations are necessarily present and connect charged to neutral currents. These relations are automatically taken into account once gauge invariant operators are considered. There are only two left-handed 4-fermion operators in the effective Lagrangian

$$\mathcal{L}_{\rm dim6} = \frac{1}{\Lambda^2}\sum O_X C_X\,, \quad (9)$$

where Λ is the scale of NP, which can contribute to $b \to s\ell\ell$ transitions at tree-level [67, 68]:

$$Q^{(1)}_{\ell q} = \left(\bar{L}\gamma^\mu L\right)\left(\bar{Q}\gamma_\mu Q\right), \quad Q^{(3)}_{\ell q} = \left(\bar{L}\gamma^\mu \tau_I L\right)\left(\bar{Q}\gamma_\mu \tau^I Q\right). \quad (10)$$

Here L is the lepton doublet, Q the quark doublet and the flavour indices are not explicitly shown. Writing these operators in terms of their $SU(2)_L$ components (i.e. up-quarks, down-quarks, charged leptons and neutrinos) we find for the terms relevant for the processes discussed in the last section (before EW symmetry breaking)

$$\begin{aligned}\mathcal{L} \supset & \frac{C^{(1)}_{ijkl}}{\Lambda^2}\left(\bar{\ell}_i\gamma^\mu P_L \ell_j \bar{d}_k \gamma_\mu P_L d_l + \bar{\nu}_i \gamma^\mu P_L \nu_j \bar{d}_k \gamma_\mu P_L d_l\right) \\ & + \frac{C^{(3)}_{ijkl}}{\Lambda^2}\left(2\bar{\ell}_i\gamma^\mu P_L \nu_j \bar{u}_k \gamma_\mu P_L d_l - \bar{\nu}_i\gamma^\mu P_L \nu_j \bar{d}_k \gamma_\mu P_L d_l \right. \\ & \left. + \bar{\ell}_i \gamma^\mu P_L \ell_j \bar{d}_k \gamma_\mu P_L d_l\right), \end{aligned} \quad (11)$$

where $C^{(1,3)}_{ijkl}$ are the dimensionless coefficients of the operators of Eq. (10). After EW symmetry breaking the following redefinitions of the fields are performed in order to render the mass matrices diagonal

$$d_L \to D^\dagger d_L,\ u_L \to U^\dagger u_L,\ \ \ell_L \to L^\dagger \ell_L,\ \nu \to L^\dagger \nu\,. \quad (12)$$

Defining

$$\lambda^{(1,3)}\tilde{X}^{(1,3)}_{ij}\tilde{Y}^{(1,3)}_{kl} = L^*_{i'i}L_{j'j}D^*_{k'k}D_{l'l}C^{(1,3)}_{i'j'k'l'}\,, \quad (13)$$

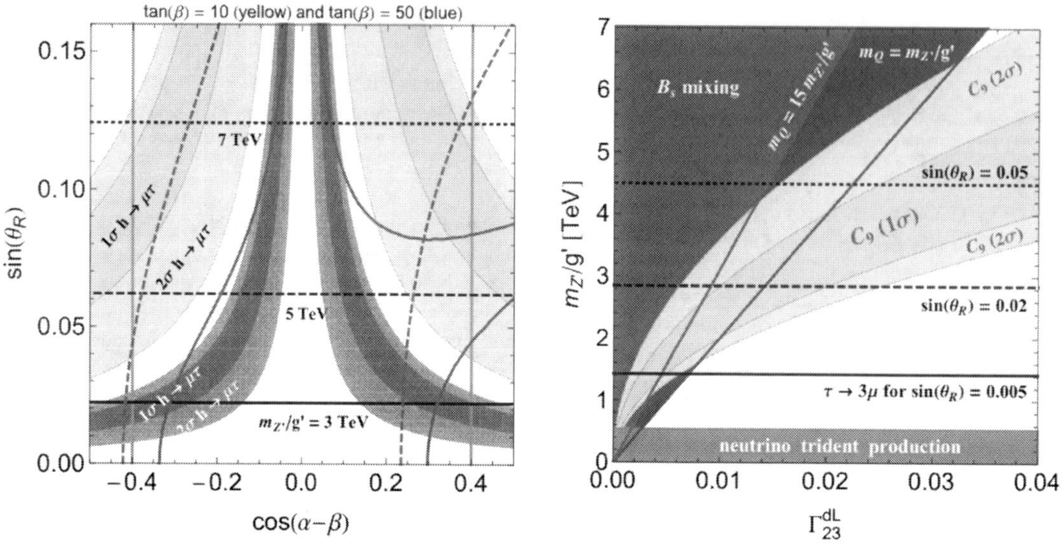

Figure 3. Left: Allowed regions in the $\Gamma^L_{sb}/M_{Z'} - \Gamma^R_{sb}/M_{Z'}$ plane for $g' = 1$ from B_s-$\overline{B}_s$ mixing (blue), and from the $C_9^{\mu\mu} - C_9^{(\prime)\mu\mu}$ fit of Ref. [5] to $b \to s\mu^+\mu^-$ data, with $\Gamma^V_{\mu\mu} = \pm 1$ (red), $\Gamma^V_{\mu\mu} = \pm 0.5$ (orange) and $\Gamma^V_{\mu\mu} = \pm 0.3$ (yellow). Note that the allowed regions with positive (negative) Γ^L_{sb} correspond to positive (negative) $\Gamma^V_{\mu\mu}$. Right: Allowed regions in the Γ^{dL}_{23}–$m_{Z'}/g'$ plane from $b \to s\mu^+\mu^-$ data (yellow) and B_s mixing (blue). For B_s mixing (light) blue corresponds to ($m_Q = 15 m_{Z'}/g'$) $m_Q = m_{Z'}/g'$. The horizontal lines denote the lower bounds on $m_{Z'}/g'$ from $\tau \to 3\mu$ for $\sin(\theta_R) = 0.05,\ 0.02,\ 0.005$. The gray region is excluded by NTP.

where $\lambda^{(1,3)}$ are overall constants, we finally obtain

$$C_9^{ij} = -C_{10}^{ij} = \frac{\pi}{\sqrt{2}\Lambda^2 G_F \alpha V_{tb} V_{ts}^*}\left(\lambda^{(1)}\tilde{X}^{(1)}_{ij}\tilde{Y}^{(1)}_{23} + \lambda^{(3)}\tilde{X}^{(3)}_{ij}\tilde{Y}^{(3)}_{23}\right)$$

$$C_L^{ij} = \frac{\pi}{\sqrt{2}\Lambda^2 G_F \alpha V_{tb} V_{ts}^*}\left(\lambda^{(1)}\tilde{X}^{(1)}_{ij}\tilde{Y}^{(1)}_{23} - \lambda^{(3)}\tilde{X}^{(3)}_{ij}\tilde{Y}^{(3)}_{23}\right),$$

$$C^{cb}_{L\,ij} = -\frac{\lambda^{(3)}}{\sqrt{2}\Lambda^2 G_F}\frac{\tilde{X}^{(3)}_{ij}}{V_{cb}}\sum_k\left(V_{2k}\tilde{Y}^{(3)}_{k3}\right), \quad (14)$$

for the Wilson coefficients relevant for $b \to s\mu^+\mu^-$, $B \to K^{(*)}\nu\bar{\nu}$ and $B \to D^{(*)}\tau\nu$ respectively. Note that in the limit $C^{(1)} = C^{(3)}$ the contribution to $B \to K^{(*)}\nu\bar{\nu}$ vanishes.

We will focus in the following on scenarios with third generation couplings in the EW basis only, which correspond to a general rank 1 matrix in the mass eigenbasis, as suggested in Ref. [69, 70]. In other words we have

$$C^{(1,3)}_{ijkl} = \lambda^{(1,3)}\tilde{X}_{ij}\tilde{Y}_{kl}, \quad (15)$$

$$\tilde{X} = L^\dagger X L,\ \tilde{Y} = D^\dagger Y D, \quad X = Y = \begin{pmatrix} 0 & 0 & 0 \\ 0 & 0 & 0 \\ 0 & 0 & 1 \end{pmatrix}.$$

Taking into account only rotations among the second and third generation one finds

$$\tilde{X} = \begin{pmatrix} 0 & 0 & 0 \\ 0 & \sin^2(\alpha_{\mu\tau}) & -\sin(\alpha_{\mu\tau})\cos(\alpha_{\mu\tau}) \\ 0 & -\sin(\alpha_{\mu\tau})\cos(\alpha_{\mu\tau}) & \cos^2(\alpha_{\mu\tau}) \end{pmatrix}, \quad (16)$$

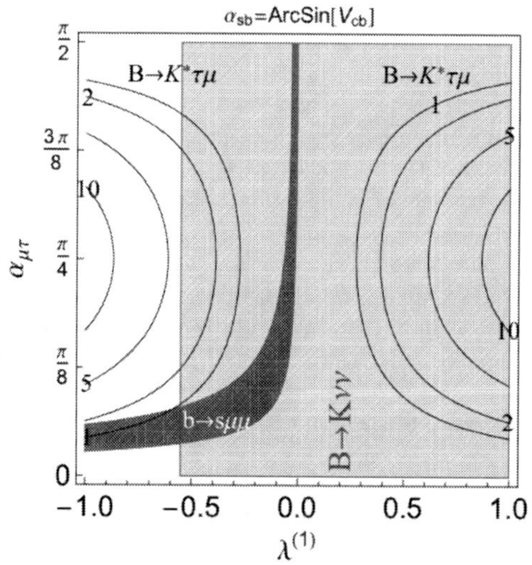

Figure 5. Allowed regions in the $\lambda^{(1)}$–$\alpha_{\mu\tau}$ plane from $b \to s\mu^+\mu^-$ data (blue) and $B \to K\nu\bar{\nu}$ (yellow) for α_{sb} = ArcSin[V_{cb}] and $\Lambda = 1$ TeV. Note that here changing α_{sb} only has the effect of an overall scaling of $\lambda^{(1)}$. The contour lines denote Br[$B \to K^*\tau\mu$] in units of 10^{-6}.

$$\tilde{Y} = \begin{pmatrix} 0 & 0 & 0 \\ 0 & \sin^2(\alpha_{sb}) & -\sin(\alpha_{sb})\cos(\alpha_{sb}) \\ 0 & -\sin(\alpha_{sb})\cos(\alpha_{sb}) & \cos^2(\alpha_{sb}) \end{pmatrix}.$$

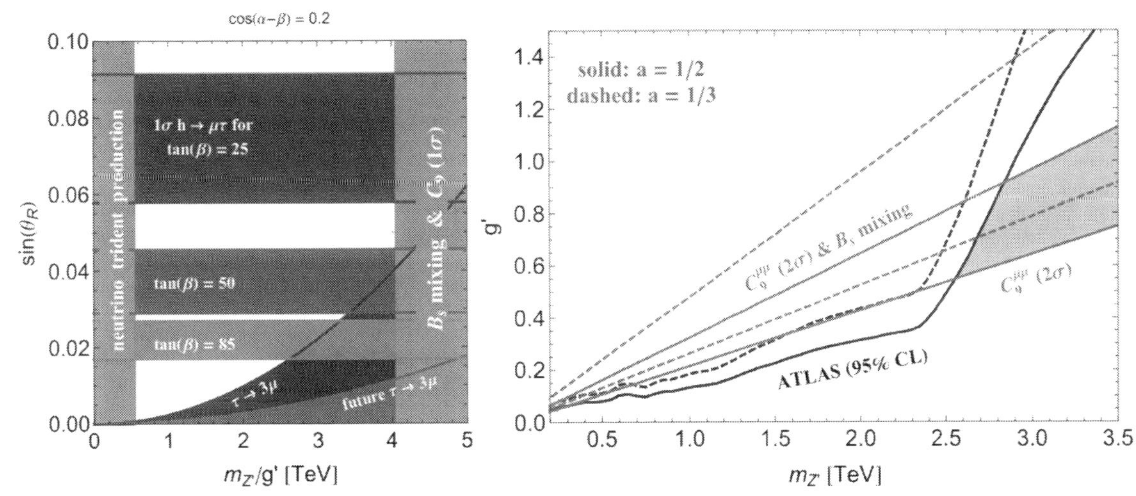

Figure 4. Left: Allowed regions in the $m_{Z'}/g'$–$\sin(\theta_R)$ plane: the horizontal stripes correspond to $h \to \mu\tau$ (1σ) for $\tan\beta = 85, 50, 25$ and $\cos(\alpha-\beta) = 0.2$, (light) blue stands for (future) $\tau \to 3\mu$ limits at 90% C.L. The gray regions are excluded by NTP or B_s–$\bar{B}_s$ mixing in combination with the $1\,\sigma$ range for C_9.
Right: Limits on $q\bar{q} \to Z' \to \mu\bar{\mu}$ from ATLAS [66] (black, allowed region down right) and the 2σ limits on $C_9^{\mu\mu}$ to accommodate $b \to s\mu^+\mu^-$ data (red, allowed regions inside the cone). Solid (dashed) lines are for $a = 1/2$ $(a = 1/3)$. For $a = 1/2$, the green shaded region is allowed (similar for $a = 1/3$ using the dashed bounds).

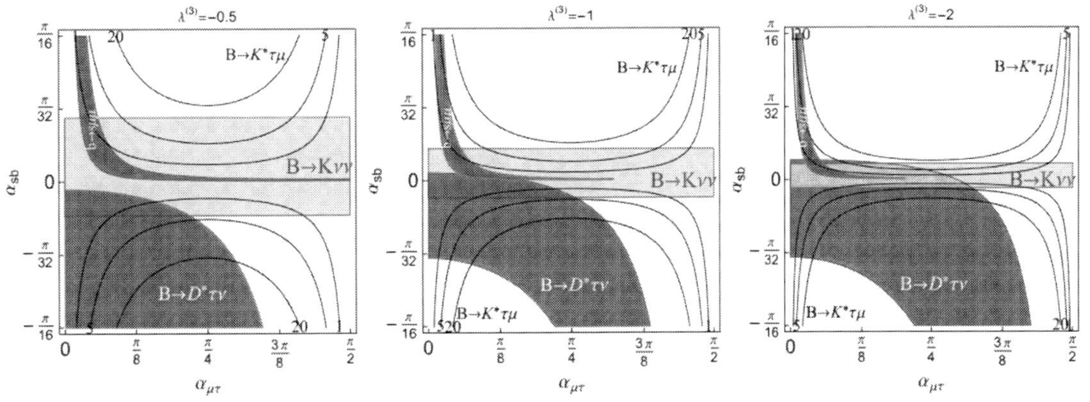

Figure 6. Allowed regions in the $\alpha_{\mu\tau}$–α_{sb} plane from $B \to K\nu\bar{\nu}$ (yellow), $R(D^*)$ (red) and $b \to s\mu^+\mu^-$ (blue) for $\Lambda = 1$ TeV and $\lambda^{(3)} = -0.5$ (left plot), $\lambda^{(3)} = -1$ (middle) and $\lambda^{(3)} = -2$ (right). Note that $\alpha_{sb} = \pi/64$ roughly corresponds to the angle needed to generate V_{cb} and that if $\lambda^{(3)}$ is positive, $R(D^*)$ and $b \to s\mu^+\mu^-$ cannot be explained simultaneously.

Note that a rotation $\sin(\alpha_{sb}) \gg V_{cb}$ would require fine-tuning with the up sector in order to obtain the correct CKM matrix.

5.1.1 $Q_{\ell q}^{(1)}$ operator

In this case we have neutral currents only. As a consequence, there is obviously no effect in $R(D^{(*)})$, but $b \to s\mu^+\mu^-$ is directly correlated to $B \to K^{(*)}\nu\bar{\nu}$ depending on the angle $\alpha_{\mu\tau}$. Note that a change in α_{sb} can be compensated by a change in $\lambda^{(1)}$ and therefore does not affect the correlations among $B \to K^{(*)}\nu\bar{\nu}$ and $b \to s\mu^+\mu^-$ transitions. In Fig. 5 the regions favoured by $b \to s\mu^+\mu^-$ (blue) and allowed by $B \to K\nu\bar{\nu}$ (yellow) are shown together with contour lines for $B \to K^*\tau\mu$ in units of 10^{-6}. Note that $B \to K\nu\bar{\nu}$ rules out branching ratios for $B \to K^*\tau\mu$ above approximately 1×10^{-6} and that the constraint from $B \to K\nu\bar{\nu}$, being inclusive in the neutrino flavours, is independent of $\alpha_{\mu\tau}$.

5.1.2 $Q_{\ell q}^{(3)}$ operator

Here we have also charged currents that are related to the neutral current processes via CKM rotations. In Fig. 6 the regions allowed by $B \to K\nu\bar{\nu}$ (yellow) and giving a good fit to data for $b \to s\mu^+\mu^-$ (blue) and (at the $2\,\sigma$ level) for

$B \to D^*\tau\nu$ (red) are shown for different values of $\lambda^{(3)}$. Note that $b \to s\mu^+\mu^-$ data can be explained simultaneously with $R(D^{(*)})$ for negative $O(1)$ values of $\lambda^{(3)}$ without violating the bounds from $B \to K\nu\bar{\nu}$. Again, in the regions compatible with all experimental constraints, the branching ratios of LFV B decays to $\tau\mu$ final states can only be up to $\approx 10^{-6}$.

5.1.3 $Q_{\ell q}^{(1)}$ and $Q_{\ell q}^{(3)}$ with $\lambda^{(1)} = \lambda^{(3)}$

In this case the phenomenology is then rather similar to the case of $C^{(3)}$ only. The major differences are that, as already mentioned before, the bounds from $B \to K\nu\bar{\nu}$ are evaded and the relative contribution to $b \to s\mu\mu$ compared to $R(D^{(*)})$ is a factor of 2 larger. Again $R(D^{(*)})$ rules out very large branching ratios for lepton flavour violating B decays in the regions compatible with $b \to s\mu^+\mu^-$ data. Note that the MFV-like ansatz [34] with additional flavour rotations phenomenologically only slightly differs from the ansatz with third generation couplings.

5.2 UV completions

Let us briefly discuss UV completions which can give the desired coupling structure[3]. As discussed previously, the 4-Fermi operator $Q_{\ell q}^{(3)}$ is relevant both for $R(K)$ and $R(D^{(*)})$. If $Q_{\ell q}^{(3)}$ is mediated by a single field, then there are only four possibilities: (i) Vector boson (VB) with the SM charges $(SU(3)_c, SU(2)_L, U(1)_Y) = (\mathbf{1}, \mathbf{3}, 0)$, (ii) Scalar leptoquark (SLQ) with $(\mathbf{3},\mathbf{3},-1/3)$, (iii) Vector leptoquark (VLQ) with $(\mathbf{3},\mathbf{1},2/3)$, and (iv) Vector leptoquark with $(\mathbf{3},\mathbf{3},2/3)$. The vector boson $(\mathbf{1},\mathbf{3},0)$ induces only $Q_{\ell q}^{(3)}$. On the other hand, the leptoquark fields result in particular combinations of $Q_{\ell q}^{(1)}$ and $Q_{\ell q}^{(3)}$ [34]. With the assumption of the third generation coupling, the relative size of the effective couplings $\lambda^{(1,3)}$ and the signs are determined as

$$\mathrm{VB}(\mathbf{1},\mathbf{3},0) : \lambda^{(3)} \text{ both positive and negative,} \tag{17}$$

$$\mathrm{SLQ}(\mathbf{3},\mathbf{3},-1/3) : \lambda^{(1)} = 3\lambda^{(3)}, \quad \lambda^{(3)} > 0, \tag{18}$$

$$\mathrm{VLQ}(\mathbf{3},\mathbf{1},2/3) : \lambda^{(1)} = \lambda^{(3)}, \quad \lambda^{(3)} < 0, \tag{19}$$

$$\mathrm{VLQ}(\mathbf{3},\mathbf{3},2/3) : \lambda^{(1)} = -3\lambda^{(3)}, \quad \lambda^{(3)} > 0. \tag{20}$$

The coefficient C_9^{ij} is proportional to $\lambda^{(1)} + \lambda^{(3)}$ and a negative value is favoured by $R(K)$. Therefore, the scalar leptoquark is rejected as a candidate. To explain $R(D^{(*)})$ simultaneously, $\lambda^{(3)}$ itself must also be negative. This condition excludes the triplet vector leptoquark. If the experimental results are explained by the operator $Q_{\ell q}^{(3)}$ under the assumption of third generation coupling only, the possible mediators are the triplet vector boson or the singlet vector leptoquark. According to the analysis of the previous section, a good fit to flavour data requires a mediator mass of $O(1)$ TeV. This opens interesting prospects for the LHC, especially in the case of leptoquarks that can be produced in proton-proton collisions via colour interactions and would decay to one lepton (τ or more interestingly μ) and one jet (possibly a b-jet).

[3]While we focus on leptoquarks here, also massive vectors in the triplet representation of $SU(2)$ are a possible UV completion[71].

6 Lepton flavour violating B decays

As lepton flavour universality is violated in $R(K)$ and $B \to D^{(*)}\tau\nu$, and $h \to \tau\mu$ even violates lepton flavour, it is interesting to examine the possibility of observing lepton flavour violating B decays [69]. Already from the EFT analysis of the last section it is clear that once gauge invariance is emplyed, LFV B decys cannot be very large if one aims at addressing $R(K)$ and $R(D^{(*)})$ simultaneously. In Z' models the additional constraints from $\tau \to 3\mu$ etc. arise. Furthermore, in the UV complete model of Refs. [43, 44] the branching ratios for LFV B decays are tiny, in general these processes are proportional to $\Gamma_{\mu\tau}\Gamma_{sb}$ and can be large in the presence of sizable flavour violation in the quark and in the lepton sector. Here we review $B \to K^{(*)}\tau^\pm\mu^\mp$ and $B_s \to \tau^\pm\mu^\mp$ in Z' models with generic couplings to fermions [72]. Γ_{sb}^L can only be large if there are cancellations originating from Γ_{sb}^R having the same sign but being much smaller. Therefore, the branching ratios for LFV B decays are bounded by fine tuning together with $\tau \to 3\mu$ and $\tau \to \mu\nu\nu$ limiting $\Gamma_{\mu\tau}$. As a result, we find in a scenario in which NP contributions to $C_9^{\ell\ell'}$ only are generated

$$\mathrm{Br}\left[B \to K^{(*)}\tau^\pm\mu^\mp\right] \leq 2.2(4.4) \times 10^{-8}(1 + X_{B_s}), \tag{21}$$

$$\mathrm{Br}\left[B_s \to \tau^\pm\mu^\mp\right] \leq 2.1 \times 10^{-8}(1 + X_{B_s}), \tag{22}$$

where X_{B_s} measures the degree of fine tuning in the B_s system. Note that these limits are obtained for $\Gamma_{\mu\mu} = 0$ (which corresponds to $C_9^{\mu\mu} = 0$) and are even stronger for non-vanishing values of $\Gamma_{\mu\mu}$. For μe final states the possible branching ratios are much smaller due to the stringent constraints from $\mu \to e\gamma$ and $\mu \to e\nu\nu$.

7 Conclusion

In these proceedings we reviewed the impact of the indirect hint for physics beyond the SM in the flavour sector obtained by BABAR, LHCb, CMS and ATLAS on models of NP. We focused on models with Z' bosons and/or additional Higgs doublets but also briefly discussed leptoquarks. While a prime candidate for the explanation of the anomalous $b \to s\mu^+\mu^-$ data is a Z' boson, $h \to \tau\mu$ as well as $B \to D^{(*)}\tau\nu$ can be most naturally explained by an extended scalar sector. Interestingly, models with gauged $L_\mu - L_\tau$ can explain $b \to s\mu^+\mu^-$ data and $h \to \tau\mu$ simultaneously, predicting sizable branching ratios of $\tau \to 3\mu$, potentially observable in future experiments. A simultaneous explanation of $b \to s\mu^+\mu^-$ data and $b \to c\tau\mu$ data is possible in the EFT approach with third generation couplings for which a vector leptoquark could be a UV completion[4].

While the UV complete models [43, 44] predict tiny branching ratios for LFV B decays, these decays can have sizable branching fractions for $\tau\mu$ final states in models with thrid generation couplings and in generic Z' models in the presence of significant fine-tuning in the $B_s - \overline{B}_s$ system.

[4]Rejecting the assumption of third-generation couplings, also a scalar leptoquark would be possible.

Acknowledgments

I thank the organizers, for the invitation and the opportunity to present these results. I am grateful to Christian Gross and Stefan Pokorski for useful comments on the manuscript. This work is supported by a Marie Curie Intra-European Fellowship of the European Community's 7th Framework Programme under contract number (PIEF-GA-2012-326948).

References

[1] U. Egede, T. Hurth, J. Matias, M. Ramon, W. Reece, JHEP **0811**, 032 (2008), 0807.2589
[2] S. Descotes-Genon, T. Hurth, J. Matias, J. Virto, JHEP **1305**, 137 (2013), 1303.5794
[3] R. Aaij et al. (LHCb collaboration), Phys.Rev.Lett. **111**, 191801 (2013), 1308.1707
[4] S. Descotes-Genon, L. Hofer, J. Matias, J. Virto, JHEP **1412**, 125 (2014), 1407.8526
[5] W. Altmannshofer, D.M. Straub (2014), 1411.3161
[6] S. Jäger, J. Martin Camalich (2014), 1412.3183
[7] T.L. Collaboration (LHCb) (2015)
[8] R. Aaij et al. (LHCb), JHEP **09**, 179 (2015), 1506.08777
[9] R.R. Horgan, Z. Liu, S. Meinel, M. Wingate, Phys.Rev.Lett. **112**, 212003 (2014), 1310.3887
[10] R. Horgan, Z. Liu, S. Meinel, M. Wingate (2015), 1501.00367
[11] R. Aaij et al. (LHCb collaboration), Phys.Rev.Lett. **113**, 151601 (2014), 1406.6482
[12] C. Bobeth, G. Hiller, G. Piranishvili, JHEP **0712**, 040 (2007), 0709.4174
[13] W. Altmannshofer, D.M. Straub (2015), 1503.06199
[14] S. Descotes-Genon, L. Hofer, J. Matias, J. Virto (2015), 1510.04239
[15] J. Lees et al. (BaBar), Phys.Rev.Lett. **109**, 101802 (2012), 1205.5442
[16] M. Huschle et al. (Belle), Phys. Rev. **D92**, 072014 (2015), 1507.03233
[17] R. Aaij et al. (LHCb), Phys. Rev. Lett. **115**, 111803 (2015), [Addendum: Phys. Rev. Lett.115,no.15,159901(2015)], 1506.08614
[18] Y. Amhis et al. (Heavy Flavor Averaging Group (HFAG)) (2014), 1412.7515
[19] S. Fajfer, J.F. Kamenik, I. Nisandzic, Phys.Rev. **D85**, 094025 (2012), 1203.2654
[20] CMS (CMS Collaboration) (2014), CMS-PAS-HIG-14-005
[21] G. Aad et al. (ATLAS) (2015), 1508.03372
[22] M. Misiak et al., Phys. Rev. Lett. **114**, 221801 (2015), 1503.01789
[23] A. Crivellin, J. Heeck, P. Stoffer (2015), 1507.07567
[24] P. Krawczyk, S. Pokorski, Phys. Rev. Lett. **60**, 182 (1988)
[25] A. Crivellin, A. Kokulu, C. Greub, Phys.Rev. **D87**, 094031 (2013), 1303.5877
[26] A. Crivellin, C. Greub, A. Kokulu, Phys.Rev. **D86**, 054014 (2012), 1206.2634
[27] C. Collaboration (CMS) (2013)
[28] G. Branco, P. Ferreira, L. Lavoura, M. Rebelo, M. Sher et al., Phys.Rept. **516**, 1 (2012), 1106.0034
[29] A. Broggio, E.J. Chun, M. Passera, K.M. Patel, S.K. Vempati, JHEP **11**, 058 (2014), 1409.3199
[30] L. Wang, X.F. Han, JHEP **05**, 039 (2015), 1412.4874
[31] T. Abe, R. Sato, K. Yagyu, JHEP **07**, 064 (2015), 1504.07059
[32] S. Fajfer, J.F. Kamenik, I. Nisandzic, J. Zupan, Phys.Rev.Lett. **109**, 161801 (2012), 1206.1872
[33] Y. Sakaki, M. Tanaka, A. Tayduganov, R. Watanabe, Phys.Rev. **D88**, 094012 (2013), 1309.0301
[34] R. Alonso, B. Grinstein, J.M. Camalich, JHEP **10**, 184 (2015), 1505.05164
[35] L. Calibbi, A. Crivellin, T. Ota, Phys. Rev. Lett. **115**, 181801 (2015), 1506.02661
[36] M. Bauer, M. Neubert (2015), 1511.01900
[37] N.G. Deshpande, A. Menon, JHEP **01**, 025 (2013), 1208.4134
[38] R. Gauld, F. Goertz, U. Haisch, Phys.Rev. **D89**, 015005 (2014), 1308.1959
[39] W. Altmannshofer, S. Gori, M. Pospelov, I. Yavin, Phys.Rev. **D89**, 095033 (2014), 1403.1269
[40] R. Alonso, B. Grinstein, J. Martin Camalich, Phys.Rev.Lett. **113**, 241802 (2014), 1407.7044
[41] G. Hiller, M. Schmaltz, Phys.Rev. **D90**, 054014 (2014), 1408.1627
[42] D. Ghosh, M. Nardecchia, S. Renner, JHEP **1412**, 131 (2014), 1408.4097
[43] A. Crivellin, G. D'Ambrosio, J. Heeck (2015), 1501.00993
[44] A. Crivellin, G. D'Ambrosio, J. Heeck (2015), 1503.03477
[45] C. Niehoff, P. Stangl, D.M. Straub (2015), 1503.03865
[46] D.A. Sierra, F. Staub, A. Vicente (2015), 1503.06077
[47] A. Celis, J. Fuentes-Martin, M. Jung, H. Serodio (2015), 1505.03079
[48] G. Belanger, C. Delaunay, S. Westhoff, Phys. Rev. **D92**, 055021 (2015), 1507.06660
[49] A. Falkowski, M. Nardecchia, R. Ziegler (2015), 1509.01249
[50] A. Carmona, F. Goertz (2015), 1510.07658
[51] A.J. Buras, F. De Fazio, J. Girrbach, JHEP **1302**, 116 (2013), 1211.1896
[52] A.J. Buras, F. De Fazio, J. Girrbach, JHEP **1402**, 112 (2014), 1311.6729
[53] A.J. Buras, J. Girrbach, JHEP **1312**, 009 (2013), 1309.2466
[54] B. Gripaios, M. Nardecchia, S. Renner, JHEP **1505**, 006 (2015), 1412.1791

[55] D. Becirevic, S. Fajfer, N. Kosnik (2015), 1503.09024

[56] I.d.M. Varzielas, G. Hiller (2015), 1503.01084

[57] S. Davidson, P. Verdier, Phys.Rev. **D86**, 111701 (2012), 1211.1248

[58] J. Kopp, M. Nardecchia, JHEP **1410**, 156 (2014), 1406.5303

[59] A. Falkowski, D.M. Straub, A. Vicente, JHEP **1405**, 092 (2014), 1312.5329

[60] I. Dorner, S. Fajfer, A. Greljo, J.F. Kamenik, N. Konik, I. Niandic, JHEP **06**, 108 (2015), 1502.07784

[61] W. Altmannshofer, S. Gori, A.L. Kagan, L. Silvestrini, J. Zupan (2015), 1507.07927

[62] D. Aristizabal Sierra, A. Vicente, Phys. Rev. **D90**, 115004 (2014), 1409.7690

[63] J. Heeck, M. Holthausen, W. Rodejohann, Y. Shimizu (2014), 1412.3671

[64] H. Terazawa, K. Akama, Y. Chikashige, Phys. Rev. **D15**, 480 (1977)

[65] P. Langacker, Rev.Mod.Phys. **81**, 1199 (2009), 0801.1345

[66] G. Aad et al. (ATLAS), Phys. Rev. **D90**, 052005 (2014), 1405.4123

[67] W. Buchmuller, D. Wyler, Nucl.Phys. **B268**, 621 (1986)

[68] B. Grzadkowski, M. Iskrzynski, M. Misiak, J. Rosiek, JHEP **1010**, 085 (2010), 1008.4884

[69] S.L. Glashow, D. Guadagnoli, K. Lane (2014), 1411.0565

[70] B. Bhattacharya, A. Datta, D. London, S. Shivashankara (2014), 1412.7164

[71] A. Greljo, G. Isidori, D. Marzocca, JHEP **07**, 142 (2015), 1506.01705

[72] A. Crivellin, L. Hofer, J. Matias, U. Nierste, S. Pokorski et al. (2015), 1504.07928

EPJ Web of Conferences **118**, 01008 (2016) DOI: 10.1051/epjconf/201611801008
FCCP2015

Review on Dark Photon

Francesca Curciarello[1,2,3,a]

[1] *Novosibirsk State University, 630090 Novosibirsk, Russia*
[2] *Dipartimento di Fisica e di Scienze della Terra, Università di Messina, Italy*
[3] *INFN Sezione Catania, Italy*

Abstract. e^+e^- collider experiments at the intensity frontier are naturally suited to probe the existence of a force beyond the Standard Model between WIMPs, the most viable dark matter candidates. The mediator of this new force, known as dark photon, should be a new vector gauge boson very weakly coupled to the Standard Model photon. No significant signal has been observed so far. I will report on current limits set on the coupling factor ε^2 between the photon and the dark photon by e^+e^- collider experiments.

1 Introduction

The existence of a low energy dark sector, remained yet undiscovered because of the very small coupling, is predicted by many extensions of the Standard Model (SM) [1–5]. The popular paradigm is to consider a 100 GeV - 10 TeV Weak Interacting Massive Particle (WIMP) as the most viable dark matter candidate and assume that WIMPs are charged under a new weak-scale interaction. In principle, the postulation of a fifth force, under which SM particles are uncharged, is not related to the dark matter puzzle and dates back to the early 1980's [1]. Very recently, its search became a goal for the most important world facilities since it has been advocated as possible explanation of many puzzling astrophysical and terrestrial anomalies [6–12].

The mediator of this new force should be a neutral light vector gauge boson which can kinetically mix with the ordinary photon and thus should be produced in any process in which a virtual or real photon is involved. Coupling to SM photon would occur through loops of heavy dark particles, charged under both electroweak and dark interactions, giving rise to a very weak mixing strength defined as the ratio of dark and fine structure coupling constants ($\varepsilon^2 = \alpha'/\alpha < 10^{-2}$). Moreover, the dark photon (referred to also as U, A' or γ') should get mass by means of a Higgs-like mechanism, suggesting the existence of the equivalent of the SM Higgs boson in the dark sector, the dark Higgs h$'$. Dark photon search is also strongly motivated by particle physics since it should give a positive one loop contribution to the calculated value of the muon magnetic moment anomaly, a_μ, solving the discrepancy for masses of 10 – 100 MeV and coupling of $\varepsilon \sim 10^{-3}$ [13].

Thanks to the available huge data statistics and the good knowledge of backgrounds, high intensity flavour factories are well suited to probe dark forces [14] with a wide and relatively inexpensive physic program. At e^+e^- colliders, dark forces searches are rich with possibilities and can be investigated by exploiting many different processes as radiative meson decays, continuum processes or dark Higgsstrahlung. In the minimal hypothesis of prompt and visible dark photon decays, the KLOE-2 experiment set many constraints by searching for the dark photon in the Dalitz decay $\Phi \to \eta U, U \to e^+e^-$ and in the $e^+e^- \to U\gamma$ continuum process, by considering both $U \to e^+e^-$, $\mu^+\mu^-$ decay channels. Moreover, KLOE-2 investigated also the dark Higgsstrahlung process by assuming the invisible scenario, where $m_U > m_{h'}$ and the dark Higgs escapes detection showing up a missing energy. BaBar and Belle set complementary constraints on dark Higgsstrahlung by considering the visible scenario, $m_U < m_{h'}$, consisting in a multi-lepton or pion final state events. Furthermore, BaBar extracted a stringent constraint by combining limits obtained by using the Initial State Radiation (ISR) processes $e^+e^- \to U\gamma, U \to l^+l^- (l = e, \mu)$ which ruled out most of the remaining $(g-2)_\mu$ favoured region assuming only the visible decay hypothesis.

I will review the status and future prospects of the limits on the mixing strength between the photon and the dark photon set at e^+e^- collider experiments.

2 Dark forces at KLOE

2.1 Φ Dalitz decay

The dark photon is expected to be produced in vector (V) to pseudoscalar (P) meson decays but with a rate ε^2 times suppressed with respect to the ordinary V $\to$ P transitions [15] (see Fig. 1). The branching ratio of the U boson into e^+e^- is non-negligible, thus, V $\to$ P U events are expected to produce a peak in the invariant mass distribution of the electron-positron pair over the continuum Dalitz background.

[a] e-mail: fcurciarello@unime.it

© The Authors, published by EDP Sciences. This is an open access article distributed under the terms of the Creative Commons Attribution License 4.0 (http://creativecommons.org/licenses/by/4.0/).

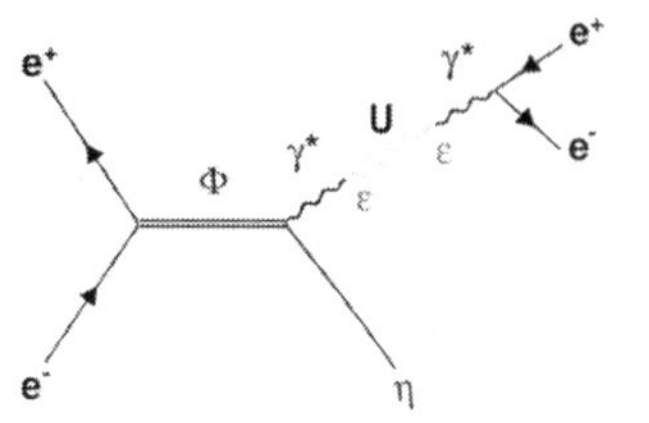

Figure 1. U-boson production through Dalitz ϕ-meson decay.

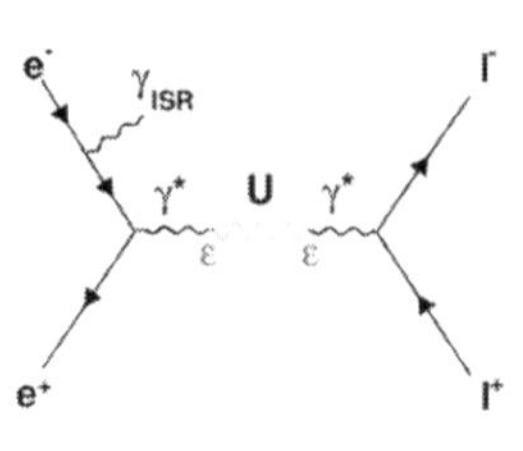

Figure 2. Resonant U-boson production in Uγ events, l = e, μ.

KLOE-2 investigated this hypothesis setting two constraints on the U-boson coupling ε^2, by exploiting the $\phi \to \eta e^+e^-$ decay, where the η meson is tagged by its $\pi^+\pi^-\pi^0$ [16] and $3\pi^0$ decays [17]. The first analysis used a data sample of 1.5 fb^{-1} integrated luminosity with a 2% of background contamination. The limit on the number of U-boson events has been set by using the Confidence Level Signal (CL$_S$) technique [18–20]. This first upper limit (UL) has been then updated, improving sample statistics and background rejection, and combined with a new limit derived by tagging the η meson by its neutral decay into $3\pi^0$ [17]. For this new analysis, 30577 events are selected from a data sample of 1.7 fb^{-1} with 3% of residual background. For each channel, the irreducible background is extracted directly from data after applying a bin-by-bin subtraction of the non-irreducible backgrounds and correcting for the analysis efficiency. The final M_{ee} distribution is then fitted to estimate the expected background, excluding for each U-boson mass hypothesis, the possible signal region used for the upper limit evaluation (5 MeV centred around m_U).

A combined UL on the parameter ε^2 at 90% CL has been derived by using the Vector Meson Dominance expectation for the transition form factor slope ($b_{\phi\eta}$ ~1 GeV^2) resulting in $\varepsilon^2 < 1.7\times10^{-5}$ for $30 < M_U <$ 400 MeV, and for the sub-region $50 < M_U < 210$ MeV, in $\varepsilon^2 < 8.0\times10^{-6}$. The above final combined limit is shown in Fig. 3 and Fig. 6. This limit, published on 2013 [17] was able to rule out a wide range of U-boson parameters as a possible explanation of the a_μ discrepancy in the minimal U-decay hypothesis.

2.2 Uγ events

Radiative U-boson production in $e^+e^- \to U\gamma$, $U \to l^+l^-$, l = e, μ events is considered very promising since it is a clean and simple process and it is independent of the details of the Higgs sector of the dark group (see Fig. 2). The U boson should appear as a resonant peak in the invariant mass distribution of the lepton pair induced by the photon's radiative return mechanism [21]. KLOE investigated both the $U \to \mu^+\mu^-$ and $U \to e^+e^-$ hypotheses. The search for $U \to \mu^+\mu^-$ employed a data sample collected in 2002 at the DAΦNE e^+e^- collider with an integrated luminosity of 239.3 pb^{-1}. Event selection required two tracks of opposite charge with $50° < \theta < 130°$ and an undetected photon whose momentum points at small polar angle ($\theta < 15°$, $> 165°$) [22, 23]. This event selection allowed to reduce the background coming from Final State Radiation and ϕ-resonant processes. Moreover, since the final state γ is emitted with high probability by the initial-state electron or positron, this increases sensitivity on $U \to \mu^+\mu^-$ decay. At the end of analysis selection, the differential cross section $d\sigma_{\mu\mu\gamma}/dM_{\mu\mu}$ has been derived achieving excellent agreement between the measurement and the simulation based on PHOKHARA [24]. No structures are visible in the $M_{\mu\mu}$ spectrum and a limit is extracted on the number of U-boson candidates through the CL$_S$ technique [18–20]. Since the agreement between data and simulation was below 1%, the expected background for the limit extraction has been determined by using the MC PHOKHARA generation. The limit on the number of U-boson events has been converted in terms of the kinetic mixing parameter ε^2 by using the formula reported in Ref. [23]. The limit on ε^2 results in 1.6×10^{-5} and 8.6×10^{-7} in the 520–980 MeV energy range.

The KLOE search for the $U \to e^+e^-$ employed a data sample of an integrated luminosity of 1.54 fb^{-1}. In this analysis the hard ISR photon has been explicitly detected in the calorimeter barrel by requiring a polar angle $50° < \theta < 30°$ for the charged leptons and photon. This large-angle event selection allowed to have sufficient statistics to reach the dielectron mass threshold. At the end of the analysis chain the background contamination is less than 1.5%. Also in this case no resonant U-boson peak was observed and again the CL$_S$ [18–20] technique is applied to estimate U-boson signal events excluded at 90% CL. A limit on the kinetic mixing parameter as a function of m_U was set by using the same equation reported in Refs. [23] and [25]. The resulting exclusion plot is shown in Fig. 3 with all other existing limits in the region 0–1000 MeV [16, 17, 23, 25–31]. The new KLOE limit excludes some of the remaining $g-2$ favoured region (see yellow line in Fig. 3) and has been recently published [25].

2.3 Higgsstrahlung process

If the U boson exists it natural to conceive the breaking of the additional U$_D$(1) symmetry by a Higgs-like mechanism. In this case an additional scalar particle, the dark Higgs h' [32] should exist as well. This suggests to probe the processes shown in Figs. 4 and 7. The Higgsstrahulung process is of particular interest since it is also sensitive to the dark coupling constant $\alpha_D = G_D/4\pi$ (G_D dark gauge coupling) and, in contrast to the above shown processes,

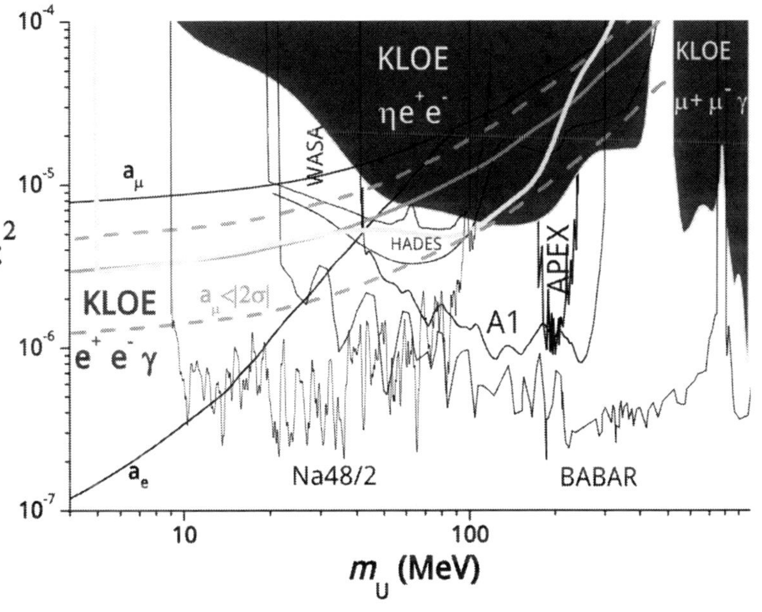

Figure 3. 90% CL exclusion plot for ε^2 as a function of the U-boson mass. The KLOE limits from Φ Dalitz dacay [16, 17], $e^+e^- \to U\gamma \to \mu^+\mu^-(e^+e^-)\gamma$ [23, 25] and from BaBar [26], the limits from the A1 [27] and Apex [28] fixed-target experiments, the WASA [29], HADES [30] and NA48/2 [31] limits are also shown. The solid lines are the limits from the muon and electron anomaly [13], respectively. The gray line shows the U-boson parameters that could explain the observed a_μ discrepancy with a 2σ error band (gray dashed lines) [13].

it is suppressed by only a single factor of ε^2. As for the SM Higgs, the mass of the dark Higgs is not predicted by theory so different scenarios are possible according to U boson and dark Higgs mass hierarchy.

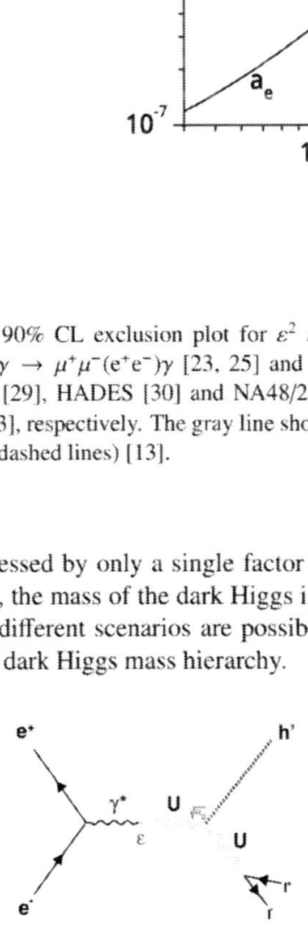

Figure 4. Dark Higgsstrahlung process in the *invisible scenario* with a long lived dark higgs h′ which escapes detection producing a missing energy signal (l = e, μ, π).

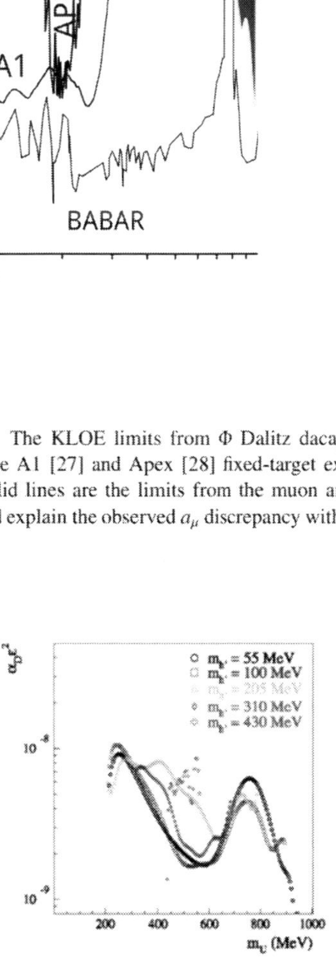

Figure 5. 90% CL combined upper limits in $\alpha_D \times \varepsilon^2$ as a function of m_U for different values of $m_{h'}$.

If the h′ is lighter than the U boson, it turns out to be very long-lived and escapes detection (see Fig. 4). In this case, the expected signal will be a lepton or pion pair from the U-boson decay plus missing energy.

KLOE studied the Higgsstrahlung process in the energy range $2m_\mu < M_U < 1000$ MeV, by considering the process $e^+e^- \to h'U, U \to \mu^+\mu^-$, h′ invisible [33]. The analysis has been performed by using a data sample of 1.65 fb^{-1} collected at center- of-mass energy E_{cm} corresponding to the ϕ peak (~ 1019 MeV) and a data sample of 0.2 fb^{-1} at E_{cm} = 1000 MeV (off-peak sample).

The expected signal would appear in the $M_{\mu\mu} - M_{miss}$ bi-dimensional spectra. No signal signature has been observed and a Bayesian limit on the number of signal events at 90% CL has been evaluated, bin-by-bin, for the on-peak and off-peak sample separately. Results have been translated in terms of $\alpha_D \times \varepsilon^2$ by using the integrated luminosity information, the signal efficiency, the dark Higgsstrahlung cross section and the branching fraction of the $U \to \mu^+\mu^-$

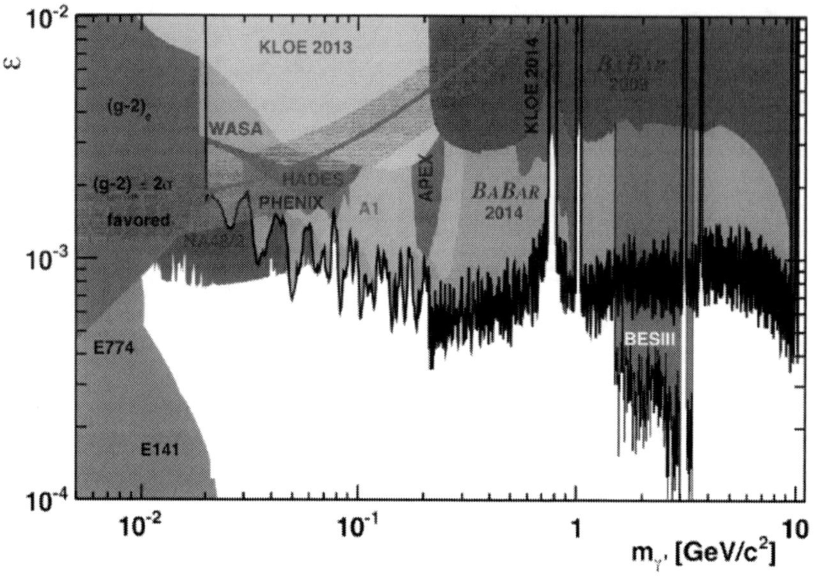

Figure 6. 90% CL currents limits on epsilon (for details see Fig. 3) up to 10 GeV. The preliminary limit from BES-III presented at the CHARM 2015 Conference is also shown [40].

decay [32]. The combined upper limits projected in the $M_{\mu\mu}$ invariant mass, for different dark Higgs mass hypotheses and slightly smoothed, are shown in Fig. 5. Values of the order of $10^{-9} \div 10^{-8}$ in $\alpha_D \times \varepsilon^2$ are excluded at 90% CL for a large range of the dark photon and dark Higgs masses. These limits translate in $\varepsilon \sim 10^{-3} - 10^{-4}$ for $\alpha_D = \alpha$ and are in agreement and complementary with BaBar results [34] as they refer to the same process in a different final state and phase space region.

3 Dark forces at BaBar

3.1 $U\gamma$ events

BaBar also exploited ISR to search for dark photon. The analysis has been performed on a data sample of about 514 fb^{-1} collected mainly at $\sqrt{s} \sim \Upsilon(4s)$ but also at $\Upsilon(2s)$ and $\Upsilon(3s)$ resonances as well as in the regions in between these resonances, exploring the mass range $0.02 < m_U < 10.2$ GeV. Event selection required two oppositely charged tracks plus a photon having a center-of-mass energy greater than 0.2 GeV. For the dielectron channel, the background processes ($e^+e^- \to \gamma\gamma(\gamma)$, $e^+e^- \to e^+e^-\gamma$) have been simulated with BHWIDE [35], while simulation of $e^+e^- \to \mu^+\mu^-\gamma$ events has been performed with KKMC [36]. A very good agreement is achieved between data and MC simulation for both channels. The disagreement below 1 GeV between radiative Bhabha data and simulation is due to MC cut-off and does not affect limit extraction since for both channels the expected background has been derived directly by a fit to data. Uncertainty on expected background modelling is below the statistical uncertainties in the whole investigated range. 90% CL Bayesian upper limits on the $e^+e^- \to U\gamma$ cross-section have been derived (typically $O(1-10)$fb), assuming a flat prior for the cross-section. Resonance regions (±30 MeV around ρ/ω, ± 50 MeV around J/Ψ, $\Psi(2s)$ and $\Upsilon(1s, 2s)$) are excluded from the limit extraction. Results are finally translated into 90% CL upper limits on the mixing strength between the photon and dark photon as a function of the dark photon mass. The final combined BaBar constraint is shown in Fig. 6, set bounds are at the level of 10^{-3}–10^{-4}. This strong limit [26] improves results from many experiments and further constrains the $g-2$ favoured region.

3.2 Higgsstrahlung process

If the dark Higgs has a mass greater than two U-boson masses, then the dark Higgs will decay into a couple of U bosons. In this case, possible final state signatures consist of six particles (inclusive channels) or of four leptons plus a U boson which escapes detection and is detected via missing mass (exclusive channels), see Fig. 7.

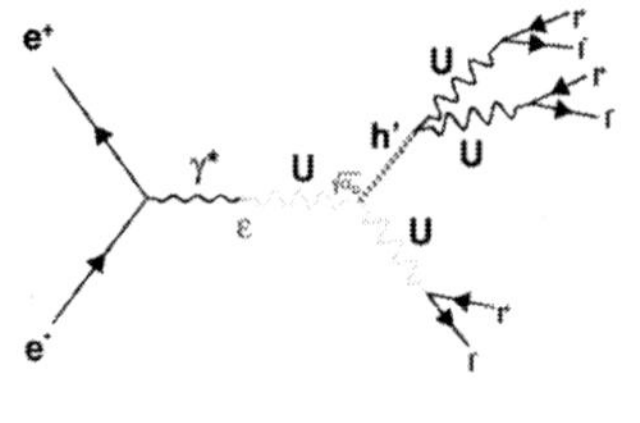

Figure 7. Dark Higgsstrahlung process in the *visible scenario*; since $m_{h'} > 2m_U$ final states signature consists of six-particle events (exclusive channels) or of four-lepton events plus missing energy (inclusive channels), $l = e, \mu, \pi$.

BaBar investigated the Higgsstrahlung process by assuming prompt decays for both dark photon and dark Higgs and by considering $m_{h'} > 2m_U$. Full BaBar statistics, about 521 fb^{-1}, has been exploited to fully reconstruct the $e^+e^- \rightarrow h'U$, $h' \rightarrow UU$ reaction in the $3(l^+l^-)$, $2(l^+l^-) + \pi^+\pi^-$, $2(\pi^+\pi^-) + l^+l^-$ exclusive final states or partially reconstruct it in the $2(\mu^+\mu^-) + X$, $\mu^+\mu^-e^+e^- + X$ inclusive final states, where X denotes any final state other than a pair of pions or leptons while $l = e, \mu$. The $2(e^+e^-) + X$ final state has been excluded because of huge background. Six candidates passed the event selection: $4\pi + 2l$, $4\mu + 2\pi$ and $4\mu + X$. No six-lepton final-state events survived selection criteria and no significant signal is observed in the investigated range ($0.25 < m_U < 3$ GeV, $0.8 < m_{h'} < 10$ GeV). 90% CL upper limits have been set on $\alpha_D \times \varepsilon^2$ at the level of 10^{-8}–10^{-10} (see Fig. 8) which for $\alpha = \alpha_D$ translate in a limit on ε of 10^{-3}–10^{-4}.

4 Dark forces at Belle

4.1 Higgsstrahlung process

Recently Belle also studied the Higgsstrahlung in the same hypotheses of BaBar (*visible scenario* and prompt decay regime, see Fig. 7). By exploiting the full data set (977 fb^{-1}), Belle investigated ten exclusive final states: $3(l^+l^-)$, $2(l^+l^-)+\pi^+\pi^-$, $2(\pi^+\pi^-)+l^+l^-$, and $3(\pi^+\pi^-)$ in the $0.1 < m_U < 3.5$ GeV and $0.2 < m_{h'} < 10.5$ GeV ranges, and three inclusive final states: $2(e^+e^-) + X$, $2(\mu^+\mu^-) + X$, $e^+e^-\mu^+\mu^- + X$ for $m_U > 1.1$ GeV and $2.5 < m_{h'} < 10.5$ GeV. Individual and combined 90% CL Bayesian upper limits on the branching fraction times the Born cross section, $\mathcal{B} \times \sigma_{Born}$, and on the Born cross section, σ_{Born}, have been derived and translated in limits on $\alpha_D \times \varepsilon^2$. These final limits, shown in Fig. 8, improve upon and explore wider mass ranges than the BaBar experiment. The limits from the final states $3(\pi^+\pi^-)$ and $2(e^+e^-) + X$ are the first placed by any experiment [37]. For $\alpha_D = \alpha$, $m_{h'} < 8$ GeV, and $m_U < 1$ GeV, the above limits translate in limits on the mixing parameter, ε of about 8×10^{-4}.

5 Future Prospects

Electron-positron colliders played a leading role in dark force searches and they are expected to play a crucial role in the near future as well, performing new searches and improving above shown limits. The current available data set has not been totally exploited regarding dark forces and new huge amount of data will be soon available thanks to the next data-taking campaigns.

The KLOE-2 run at LNF has started, taking advantage of new sub-detectors and new DAΦNE interaction scheme, with the goal of reaching, within a few years, about 5 fb^{-1}. Moreover, at LNF a planned dedicated fixed-target experiment, PADME [38], will be able to perform a model-independent dark photon search and to probe both visible and invisible decay hypotheses.

The Belle and BaBar experiments in Japan and USA have integrated about 1 ab $^{-1}$ each at $\sqrt{s} \sim 10$ GeV. The aim is to reach ~ 50 ab^{-1} with future generation SuperB factories. With the implementation of a mono-photon trigger Belle-II will be also able to search for an invisibly decaying dark photon [39]. At $\sqrt{s} \sim 3$ GeV, the BES-III detector in Beijing, China, aims to collect an integrated luminosity of ~ 20 fb^{-1}, dark force searches are also in the program. A new preliminary combined limit on $U \rightarrow e^+e^-, \mu^+\mu^-$ decays has already been extracted in the 1.5–3.5 GeV energy range [40], see Fig. 6.

Moreover, it has to be noticed that null results obtained so far on dark photon searches are based on model-dependent searches which assumed only dark photon decays into SM particles. Most of the current set limits are no more valid in a more complex theoretical scenario with a dark photon decaying into light dark matter states. Concerning this non-minimal hypothesis, the $(g-2)_\mu$ favoured region can not be considered completely constrained [39]. In order to rule out dark photon as explanation of the muon anomaly or rule out it completely, a complete series of experiments, exploiting all theoretical possibilities, is needed.

Conclusions

Electron-positron colliders are naturally suited to search for dark forces, however no significant signal has been observed so far. New generation machines with their high statistics datasets would play a leading role by performing model-independent searches which are needed to solve dark force puzzle. Fixed target experiments will be more powerful to probe lower masses and very small couplings for which dark photon decay is no more prompt and its signature can be found by performing displaced vertex searches.

A signal should be observed in the near future, otherwise a change of the WIMP paradigm will be needed.

References

[1] B. Holdom, Phys. Lett. B **166**, 196 (1985)
[2] C. Boehm and P. Fayet, Nucl. Phys. B **683**, 219 (2004)
[3] P. Fayet, Phys. Rev. D **75**, 115017 (2007)
[4] Y. Mambrini, J. Cosmol. Astropart. Phys. **1009**, 022 (2010)
[5] M. Pospelov, A. Ritz and M.B. Voloshin, Phys. Lett. B **662**, 53 (2008)
[6] O. Adriani et al., Nature **458**, 607 (2009)
[7] M. Aguilar et al., Phys. Rev. Lett. **110**, 141102 (2013)
[8] P. Jean et al., Astronomy Astrophysics **407**, L55 (2003)
[9] J. Chang et al., Nature **456**, 362 (2008)
[10] F. Aharonian et al., Phys. Rev. Lett. **101**, 261104 (2008)
[11] A. A. Abdo et al., Phys. Rev. Lett. **102**, 181101 (2009)
[12] R. Barnabei et al., Eur. Phys. J. C **56**, 333 (2008)

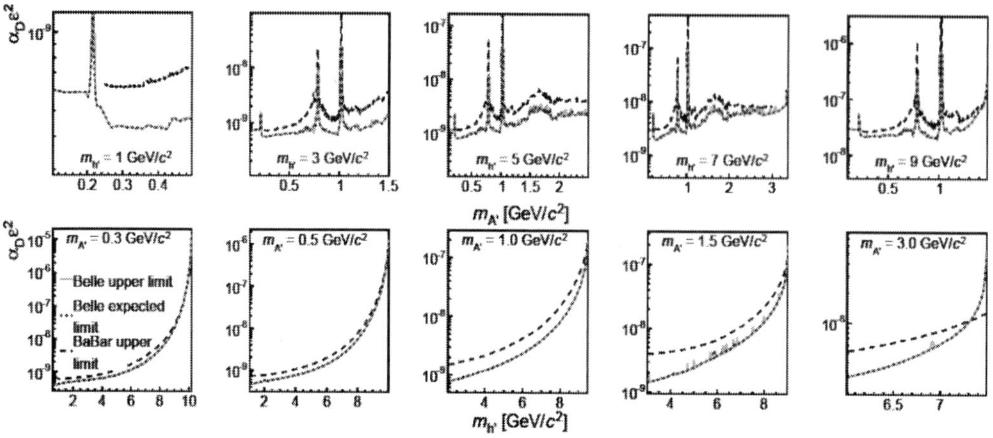

Figure 8. 90% CL upper limit on the product $\alpha_D \times \varepsilon^2$ versus dark photon mass (top row) and dark Higgs boson mass (bottom row) for Belle (solid green curve) and BaBar (dashed black curve [34]). The blue dotted curve shows the expected Belle limit.

[13] M. Pospelov, Phys. Rev. D **80**, 095002 (2009)

[14] R. Essig, P. Schuster and N. Toro, Phys. Rev. textbf80, 015003 (2009)

[15] M. Reece, L.T. Wang, JHEP **07**, 051 (2009)

[16] F. Archilli et al., Phys. Lett. B **706**, 251 (2012)

[17] D. Babusci et al., Phys. Lett. B **720**, 111 (2013)

[18] G. J. Feldman and R. D. Cousins, Phys. Rev. D **57**, 3873 (1998)

[19] T. Junk, Nucl. Instr. Meth. A **434**, 435 (1999)

[20] A. L. Read, J. Phys. G: Nucl. Part. Phys. **28**, 2693 (2002)

[21] L. Barzè et al., Eur. Phys. J. C **71**, 1680 (2011)

[22] D. Babusci et al., Phys. Lett. B **720**, (2013) 336

[23] D. Babusci et al., Phys. Lett. B **736**, (2014) 459

[24] H. Czyż , A. Grzelinska, J.H. Kühn, G. Rodrigo, Eur. Phys. J. C **39**, (2005) 411

[25] A. Anastasi et al., Phys. Lett. B, **750**, 633 (2015)

[26] J. P. Lees et al., Phys. Rev. Lett. **113**, 201801 (2014)

[27] H. Merkel et al., Phys. Rev. Lett. **112**, 221802 (2014)

[28] S. Abrahamyan et al., Phys. Rev. Lett. **107**, 191804 (2011)

[29] P. Adlarson et al., Phys. Lett. B **726**, 187 (2013)

[30] G. Agakishiev et al., Phys. Lett. B **731**, 265 (2014)

[31] J. R. Batley et al., Phys. Lett. B **746**, 178 (2015)

[32] B. Batell et al., Phys. Rev. D **79**, 11508 (2009)

[33] A. Anastatsi et al., Phys. Lett. B **747**, 365 (2015)

[34] J. P. Lees et al., Phys. Rev. Lett. **108**, 211801 (2012)

[35] S. Jadach, W. Placzek and B. F. L. Ward Phys. Lett. B **390**, 298 (1997)

[36] S. Jadach, W. Placzek and Z. Was, Phys. Rev. D **63**, 113009 (2001)

[37] I. Jaegle et al., Phys. ReV. Lett. **114**, 211801 (2015)

[38] M. Raggi and V. Kozhuharov Advances in High Energy Physics **2014**, ID 959802 (2014)

[39] R. Essig et al., arXiv:1309.5084[hep-ph] (2015)

[40] `http://docbes3.ihep.ac.cn/~talks/images/c/cc/CHARM2015-prasad.pdf`

The SHiP facility at CERN

Giovanni De Lellis[1,a]

[1] *Università "Federico II" and INFN, Naples, Italy*

Abstract. Searches for new physics with accelerators are being performed at the LHC, looking for high massive particles coupled to matter with ordinary strength. A new experimental facility meant to search for very weakly coupled particles in the few GeV mass domain has been recently proposed. The existence of such particles, foreseen in different theoretical models beyond the Standard Model, is largely unexplored from the experimental point of view. A beam dump facility, built at CERN in the north area, using 400 GeV protons is a copious factory of charmed hadrons and could be used to probe the existence of such particles. The beam dump is also an ideal source of tau neutrinos, the less known particle in the Standard Model. In particular, tau anti-neutrinos have not been directly observed so far. We report the physics potential of such an experiment and outline the performances of a detector operating at the same facility for the search for the $\tau \to \mu\mu\mu$ decay.

1 Introduction

The discovery of the Higgs boson [1, 2] is certainly a big triumph of the Standard Model. In particular, given its measured mass of 125.5 ± 0.5 GeV, it could well be that the Standard Model is an effective theory working all the way up to the Planck scale. Indeed, the λ coupling of the quartic Higgs potential keeps the same sign making the vacuum stable on the time scale of the Universe [3] and its running value stays finite up to very high energy scales [4]. Nevertheless, there are several facts deserving an explanation that the Standard Model is unable to provide: the existence of dark matter and its nature, the baryonic asymmetry of the Universe and neutrino masses. It is therefore clear the new physics is there and presumably several new particles have still to be discovered.

Searches for new physics with accelerators are being carried out extensively at the LHC, that is especially suited to look for high mass particles with ordinary couplings to matter. Complementary searches for very weakly coupled and therefore long lived particles require a beam dump facility. Such a facility is made of a high density proton target, followed by a hadron stopper and a muon shield. Apart from residual muons, the only remaining particles are electron, muon and tau neutrinos on top of hidden, long lived particles produced either in proton interactions or in secondary particle decays.

A new experiment, Search for Hidden Particles (SHiP), has been proposed [5], designed to operate at a beam dump facility to be built at CERN and to search for weakly coupled particles in the few GeV mass range. Since a high intensity tau neutrino flux is produced by such a facility from D_s decays, the experimental apparatus foresees a neutrino detector to study the tau neutrino cross-section and discover the tau anti-neutrino. The physics case for such an experiment is widely discussed in Ref. [6]. We will review here the physics potential of this experiment for a selection of physics channels and we also give the performances of a dedicated experiment searching for $\tau \to \mu\mu\mu$ decays, operating at the same beam dump Facility at CERN.

2 The SHiP facility and apparatus

The facility will integrate in five years 2×10^{20} 400 GeV protons, produced by the SPS accelerator complex, impinging on a 10 interaction length target made of Molybdenum and Tungsten, followed by an hadron absorber. Hidden particles in the GeV mass range would be produced mostly by the decay of charmed hadrons produced in proton interactions. D_s mesons, copiously produced among charmed hadrons, are a good source of tau neutrinos through their fully leptonic decay. Therefore, the SHiP facility is ideal also to study the physics of tau neutrinos, the less known particle in the Standard Model.

Figure 1 shows the SHiP facility to be placed in the North Area: downstream of the target, the hadron absorber filters out all hadrons, therefore only muons and neutrinos are left. An active muon shield is designed with two sections with opposite polarity to maximize the muon flux reduction: it reduces the muon flux from $\sim 10^{10}$ muons/spill down to about 10^4 per spill. The slow extraction of 4×10^{13} protons is foreseen in each spill, designed to be 1s long to reduce the detector occupancy. A first successful test of the SPS cycle with a 1s long spill was performed in April 2015. The tau neutrino detector is located downstream of the muon shield, followed by the decay vessel and the detector for hidden particles. The Technical Design report of

[a] On behalf of the SHiP Collaboration, e-mail: giovanni.de.lellis@cern.ch

© The Authors, published by EDP Sciences. This is an open access article distributed under the terms of the Creative Commons Attribution License 4.0 (http://creativecommons.org/licenses/by/4.0/).

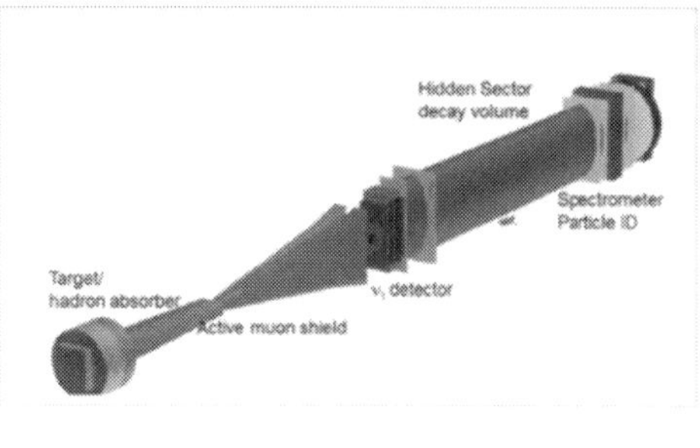

Figure 1. The beam dump facility and the SHiP detector.

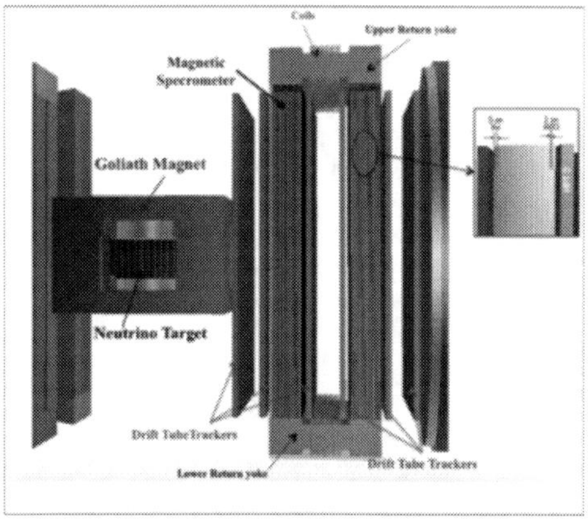

Figure 2. The neutrino detector of the SHiP apparatus.

the SHiP facility is expected to be ready in 2018, the construction and installation to last until the third long shutdown of the LHC and the data taking to start soon after in 2026.

The neutrino detector is made of a magnetised target region, followed by a muon spectrometer, as shown in Fig. 2. The neutrino target is based on the emulsion cloud chamber technology employed by the OPERA experiment [7], with a compact emulsion spectrometer, made of a sequence of very low density material and emulsion films to measure the charge and momentum of hadrons in magnetic field. Indeed, this feature would allow to discriminate between tau neutrinos and anti-neutrinos also in the hadronic decay channels of the tau lepton. The emulsion target is complemented by high resolution tracking chambers to provide the time stamp to the event and connect muon tracks from the target to the muon spectrometer.

The detector for hidden particles, shown in Fig. 3, is located in the downstream part a 60 m long evacuated decay vessel, with an elliptical transverse section of 5×10 m^2, with the longer axis vertically. The hidden particles are supposed to decay within the vessel. The requirement to have less than 1 neutrino interaction in the vessel over the full run sets the pressure to about 10^{-3} mbar. Downstream of the vessel, a magnetic spectrometer is located: it is made of straw tubes with a material budget of 0.5% X_0 per station, achieving a position resolution of

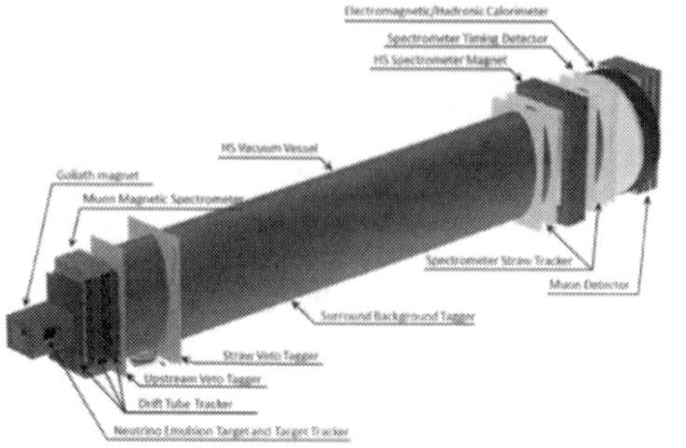

Figure 3. The SHiP detector: the detector for hidden particles is located downstream of the decay vessel.

120 μm per straw, with 8 hits per station on average. This gives a momentum resolution of about 1%. The challenge is to build 5m long straw tubes and the large vacuum vessel. The vessel would be sorrounded by a liquid scintillator layer to tag particles coming from outside. Downstream of the spectrometer, an hadronic and electromagnetic calorimeter and a muon filter are meant for particle identification. A timing detector with a 50ps resolution complements the detector to suppress the combinatorial background.

3 Search for hidden particles

Extensions of the Standard Model in the low mass region foresee the existence of particles as singlets with respect to the Standard Model gauge group. These particles couple to different singlet composite operators (so-called Portals) of the Standard Model. These particles might be fermions, like heavy neutral leptons, scalar particles coupled to the Higgs like the light inflaton, vector particles like the dark photon and eventually pseudoscalar particles like pseudo-Nambu-Goldstone bosons (or "axions"). The SHiP detector located immediately downstream of a beam dump has the potentiality to discover all of them, since it is sensitive to very weakly interacting and long lived particles in a wide unexplored range of their masses and couplings. We describe in the following its sensitivity to dark photons and heavy neutral leptons. For the sensitivity to the other portals, we refer to Ref. [5].

3.1 Search for hidden photons

Dark photons are vectorial particles coupled to the photon with a lagrangian term $L = -kB_{\mu\nu}V^{\mu\nu}/2$. In a beam dump facility, they can be produced both in the proton bremsstrahlung and in the decay of particles like π^0, η, ω and η'. If we assume that no lighter hidden particles exist, the dark photon decays into SM particles with final states like e^+e^-, $\mu^+\mu^-$ or $q\bar{q}$. Recently, data from old experiments have been reanalysed and exclusion plots have been derived together with cosmological limits [8, 9]. Figure 4 shows the exclusion limits at 90% C.L. set by SHiP in case no signal is found. The parameters displayed are

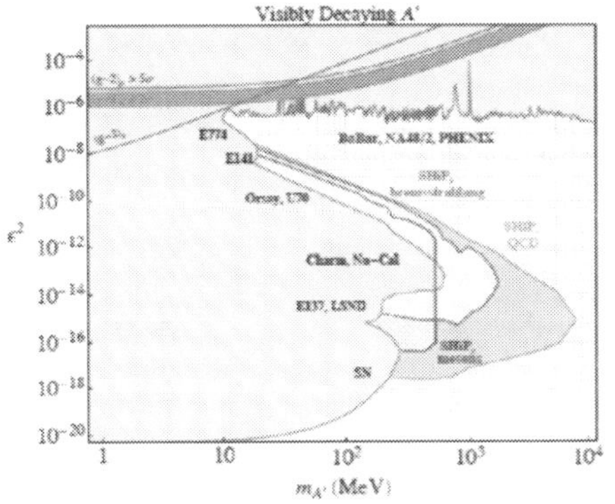

Figure 4. Exclusion limit set by the SHiP experiment in case no signal is found in the plane defined by its mass and ϵ^2 squared coupling. Limits from other experiments are also drawn for comparison.

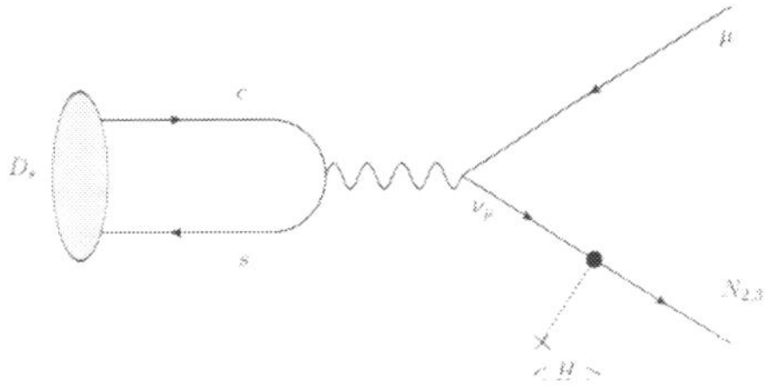

Figure 5. An example of a charmed hadron decay producing $N_{2,3}$ states.

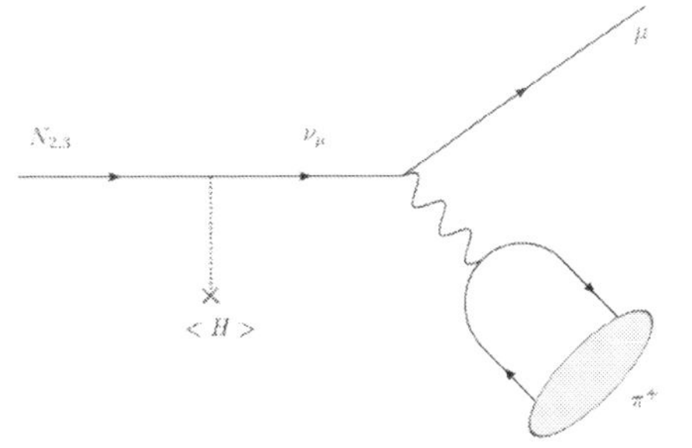

Figure 6. An example of the $N_{2,3}$ decay modes: π and μ are produced in the final state.

the mass and ε^2 squared coupling of the dark photon. The contributions from proton bremsstrahlung and meson decays are indicated: the sensitivity extends up to 10 GeV. It is worth noting that the region investigated by SHiP largely exceeds those explored by previous searches in the region of cosmological interest.

3.2 Search for heavy neutral leptons

Neutrino masses are usually explained by the see-saw mechanism with a lagrangian term given by

$$L = i\bar{N}_I\partial_\mu\gamma^\mu N_I - Y_{I\alpha}\bar{N}_I^C\tilde{H}L_\alpha - M_I\bar{N}_I^C N_I + h.c. \quad (1)$$

where 3 right-handed heavy neutrinos N_I are introduced. The mass of active neutrinos is given by $m_\nu \simeq m_D^2/M_I$ where $m_D \simeq Y_{I\alpha}v$ and $v \simeq 246$ GeV is the Higgs vacuum expectation value. A massive neutrino with M_I of the order of 1 GeV and Yukawa couplings $Y_{I\alpha}$ of the order of 10^{-7} would account for an active neutrino mass $m_\nu \simeq 0.05$ eV, as measured by neutrino oscillation experiments.

If several conditions are satisfied, among which CP is not conserved, these right-handed neutrinos could also explain the leptogenesis and therein the baryogenesis [10, 11]. Such models do not pretend to explain also dark matter: a dark matter candidate could be provided by a fourth right-handed neutrino or an axion. Moreover they do not require two of the three heavy neutrinos to be quasi-degenerate because CP violation does not need to be enhanced by mass degeneracy.

In the attempt of explaining also dark matter with three heavy neutrinos, it was first introduced in 2005 the so-called νMSM [12, 13], later followed by refined versions [14]. The role of the three right-handed neutrinos is different as well as their mass: M_1 could be a dark matter candidate with the mass of several keV while M_2 and M_3 should be states with a mass of about 1 GeV and quasi-degenerate to provide the baryon asymmetry of the Universe. In order to be a dark matter candidate, M_1 has to be so light (several KeV) that its only visible decay channel could be $N_1 \to \gamma\nu$, producing a monochromatic line at $M_I/2$ in the X-ray spectrum. A peak around 3.5 keV was recently reported by two different groups analysing data of astronomical X-ray surveys [15, 16]. The reported peak shows a low statistical significance and it will be studied by next generation space missions with sufficient energy resolution, $\Delta E/E \simeq 10^{-3}$.

The states N_2 and N_3 could be produced in the decay of sufficiently massive particles like charmed hadrons with a diagram shown, as an example, in figure 5. The states produced in this way would be long lived particles that in turn could decay into $\pi\,\mu$ as it is illustrated, as an example, in figure 6. Therefore, by measuring the invariant mass of the $\pi\,\mu$ system, one would expect a peak to show up.

It is worth noticing that there are more cosmological than experimental constraints in the mass region around 1 GeV. Charmed hadrons are the ideal parents of such heavy neutral leptons if their mass is around 1 GeV. Indeed, with kaon decays one is sensitive only to masses lower than 400 MeV, while the cross-section for the production of beauty hadrons is a factor of 20 to 100 lower. Moreover, given that beauty hadrons mostly decay to charmed hadrons, the explored mass range would only extend from 2 to 3 GeV. A proton beam dump facility is the ideal place to copiously produce charmed hadrons and to search for heavy neutral leptons of this mass and lifetime (10^{-5}s) range.

Figure 7 shows the experimental and cosmological bounds on the search for heavy neutral leptons. It also reports superimposed the exclusion limits at 90% C.L. SHiP could set in case no signal is found. The top plot assumes

DOI: 10.1051/epjconf/201611801009

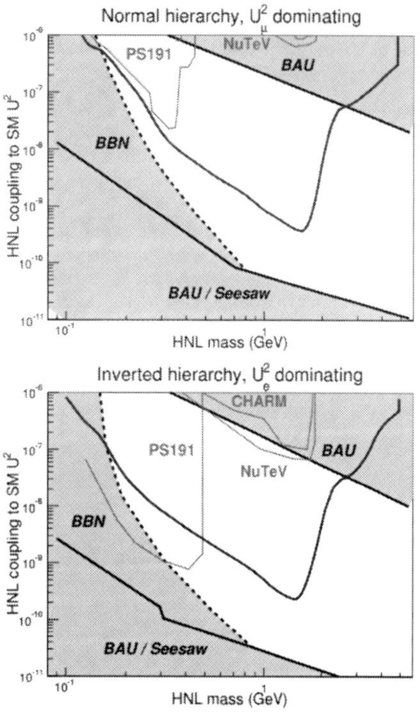

Figure 7. Exclusion limits at 90% C.L. sets by SHiP in case no signal is found. Top: normal hierarchy with U_μ^2 dominating according to Ref. [17]. Bottom: inverted hierarchy with U_e^2 dominating according to Ref. [18].

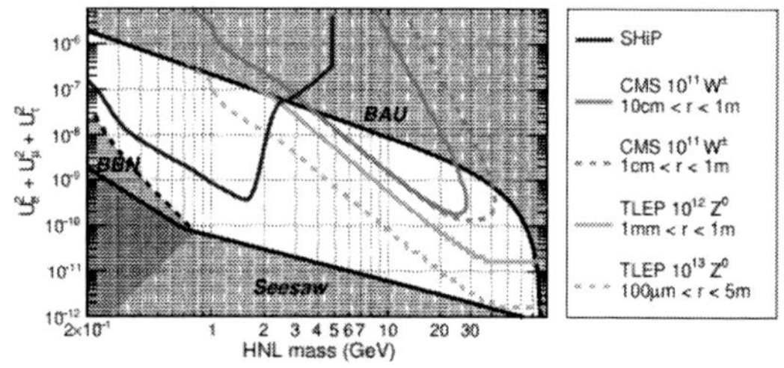

Figure 8. The region explored by the SHiP experiment is compared with cosmological constraints and the projected results of the CMS experiment and TLEP project.

the model reported in Ref. [17] with normal hierarchy and a dominant muon coupling, $U_e^2 : U_\mu^2 : U_\tau^2 = 1 : 16 : 3.8$, while the bottom plot refers to the model in Ref. [18] for inverted hierarchy and a dominant electron coupling, $U_e^2 : U_\mu^2 : U_\tau^2 = 48 : 1 : 1$. The parameter space explored by SHiP extends in the cosmologically relevant region that is experimentally unexplored. Moreover, the region explored by SHiP is complementary to the projected results of LHC projects at the end of their run and with future projects such as TLEP as it is shown in Figure 8.

4 Physics with the neutrino detector

The tau neutrino is the less known particle in the Standard Model. Four candidates were firstly reported in 2001 by the DONUT experiment [19] and the observation of this particle was finally confirmed in 2008 when 9 candidates events were reported with an estimated background of 1.5 [20]. In the same paper they reported, for the first time, the tau neutrino cross-section where the constant term was measured to be $\sigma_{\nu_\tau}^{const} = (0.39 \pm 0.13 \pm 0.13) \times 10^{-38} cm^2 GeV^{-1}$. The large uncertainty is due to the poor statistical sample and to the rather scarce knowledge of the incoming flux. On top of the large uncertainty, DONUT could not separate tau neutrinos from anti-neutrinos. Moreover, the OPERA experiment [7] has detected five tau neutrinos [21–25], discovering the tau neutrino appearance from muon neutrino oscillations. The only leptonic decay observed by OPERA [23] shows negative charge as expected from a ν_τ interaction. Therefore, so far there is no direct evidence for tau anti-neutrinos.

The number of ν_τ and $\bar{\nu}_\tau$ emerging from the molybdenum target can be estimated as:

$$N_{\nu_\tau+\bar{\nu}_\tau} = 4N_p \frac{\sigma_{c\bar{c}}}{\sigma_{pN}} f_{D_s} Br(D_s \to \tau) = 5.7 \times 10^{15} \quad (2)$$

where

- N_p is the number of interacting protons (all incoming ones);
- $\sigma_{c\bar{c}} = 18.1 \pm 1.7\,\mu$barn [26] is the associated charm production per nucleon;
- $\sigma_{pN} = 10.7$ mbarn is the hadronic cross-section per nucleon in a Mo target. The inelastic cross-section pA shows the $A^{0.71}$ dependence [27];
- $f_{D_s} = (7.7 \pm 0.6^{+0.5}_{-0.4})\%$ [28] is the fraction of D_s mesons produced;
- $Br(D_s \to \tau) = (5.54 \pm 0.24)\%$ [29] is the D_s branching ratio into τ;
- the factor 4 accounts for the charm pair production and for the two ν_τ produced per D_s decay.

The SHiP facility is therefore a ν_τ factory, with 5.7×10^{15} tau neutrinos produced, equally divided in neutrinos and anti-neutrinos. The uncertainty on this number is about 14% but the uncertainty on the differential neutrino flux is estimated to be about 20%.

Given the neutrino target mass of 9.6 tons, the neutrino flux, the geometrical acceptance of the detector and the standard model cross-section [30], one expects about 6700 interactions of tau neutrinos and 3400 of tau anti-neutrinos. An uncertainty smaller than 10% affects the cross-section mainly due to uncertainties on scale choices, parton distribution functions and target mass corrections [30]. Therefore the uncertainty on the expected number is dominated by the 20% error on the differential neutrino flux.

The charged-current ν_τ ($\bar{\nu}_\tau$) differential cross-section is given by five structure functions. The contribution to the cross-section of F_4 and F_5 structure functions, introduced by Albright and Jarlskog [31], is negligible in muon

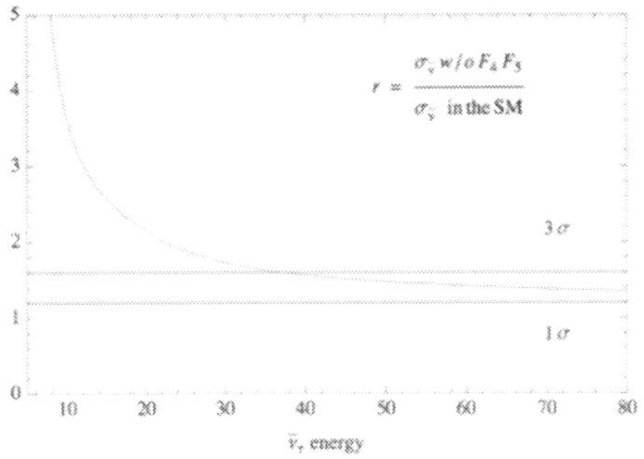

Figure 9. Energy dependence of the ratio between the DIS cross section in the $F_4 = F_5 = 0$ hypothesis and the SM prediction for $\bar{\nu}_\tau$.

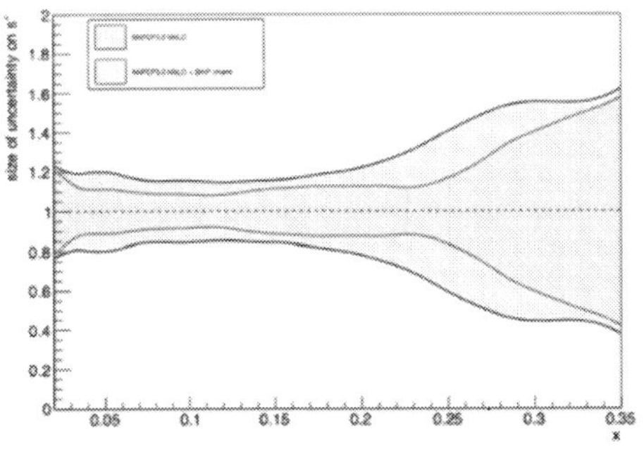

Figure 10. Improvement of the accuracy on s^+ with SHiP (red) compared to the present status (blue) in the $0.02 < x < 0.35$ range.

and electron neutrino interactions due to the charged lepton mass. On the contrary, given the non-negligible mass of the τ lepton, tau neutrino scattering is sensitive to their contribution.

At the leading order, in the limit of massless quarks and target hadrons, the predicted values are $F_4 = 0$ and $2xF_5 = F_2$ [31]. Calculations at NLO show that F_4 gives a contribution to the cross-section of about 1% at the neutrino energy of 10 GeV [30].

SHiP is sensitive to F_4 and F_5. Indeed, the hypothesis of $F_4 = F_5 = 0$ would significantly increase the ν_τ and $\bar{\nu}_\tau$ charged-current deep-inelastic cross sections and the corresponding number of expected ν_τ and $\bar{\nu}_\tau$ interactions. Figure 9 shows the energy dependence of the ratio between the tau anti-neutrino cross-section when F_4 and F_5 vanish and the Standard Model prediction. The region below 40 GeV is highly sensitive to the non-zero values of these structure functions: a 3 σ excess is expected in this region, given that the ratio exceeds 1.6.

4.1 Strange parton distribution

Charmed hadrons are produced in neutrino and anti-neutrino charged-current interactions at the level of about 5%. Experiments based on calorimetric technology identify charmed hadrons only in their muonic decay channel, when two opposite sign muons are produced in the final state. A cut of 5 GeV is applied to muons in order to suppress the background due to punch-through pions. The nuclear emulsion technology, instead, identifies topologically the charmed hadron by detecting its decay vertex. Energy cuts are therefore much looser, thus providing a better sensitivity to the charm quark mass. Moroever, a large statistical gain is provided by the use of hadronic decay modes [32]. Indeed, despite the fact that 1.280.000 ν_μ and 270.000 $\bar{\nu}_\mu$ charged-current interactions were collected by the NuTeV/CCFR Collaboration, only 5102 ν_μ and 1458 $\bar{\nu}_\mu$ events were identified as charm production induced by neutrino charged-current interactions. The largest number of events in an emulsion experiment was collected by CHORUS which integrated 2013 charm candidates from ν_μ and only 32 from $\bar{\nu}_\mu$ [33]. SHiP will integrate about 10^5 charm candidates, more than one order of magnitude larger than the present statistics, with a large ($\sim$ 30%) contribution from anti-neutrinos. Charm production in neutrino scattering is extremely sensitive to the strange quark content of the nucleon, especially with anti-neutrinos where the s-quark is dominant. SHiP will improve significantly the uncertainty on the strange quark distribution in the nucleon as shown in Fig. 10 in terms of $s^+ = s(x) + \bar{s}(x)$ in the $0.02 < x < 0.35$ range.

5 A dedicated experiment to search for $\tau \rightarrow \mu\mu\mu$ decays

The study of charged lepton flavour violation as a probe of new physics is expected to make a big step forward in the next few years with the MEG and MEG-II at PSI, with the proposed search of the $\mu \rightarrow eee$ decay at PSI and with the new searches for μ to e conversion, Mu2e at FNAL and COMET at JPARC. Charged lepton flavour violation in τ decays is illustrated in Fig. 11. So far the best limits were achieved at the B factories where about 10^9 τ leptons were collected. With an almost background free analysis, an upper limit $Br(\tau \rightarrow \mu\mu\mu) < 2.1 \times 10^{-8}$ was set at 90% C.L. [34]. The Belle-II experiment might bring it down to the level of 10^{-9} or even lower. LHCb has collected about 10^{11} τ leptons and performed this search with an irreducible background from the $D_s \rightarrow \eta\mu\nu$ with subsequent $\eta \rightarrow \mu\mu$ decay: they have achieved a similar upper limit, 4.6×10^{-8} [35], with an integrated luminosity of 1.0 fb^{-1} of proton-proton collisions at a centre-of-mass energy of 7 TeV and 2.0 fb^{-1} at 8 TeV. There are prospects for LHCb to improve this limit when 50 fb^{-1} will be integrated but they will hardly compete with B factories.

3×10^{15} τ leptons will be produced in the target of the SHiP Facility. Nevertheless, a thick (10 λ_{int}) target is not suitable for this kind of search because the multiple Coulomb scattering would spoil the momentum resolution of the produced muons, thus spoiling the invariant mass resolution of the 3μ system, essential to identify the τ. The physics case for a $\tau \rightarrow \mu\mu\mu$ dacay experiment at the SHiP Facility has been discussed in Ref. [6]. We revise the main

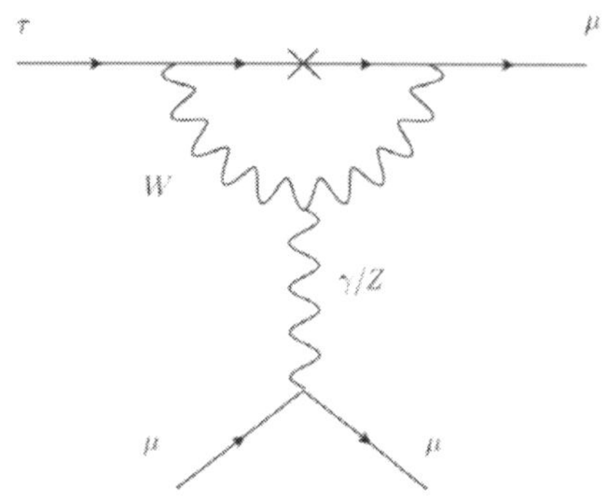

Figure 11. Feynman diagram of the $\tau \to \mu\mu\mu$ decay.

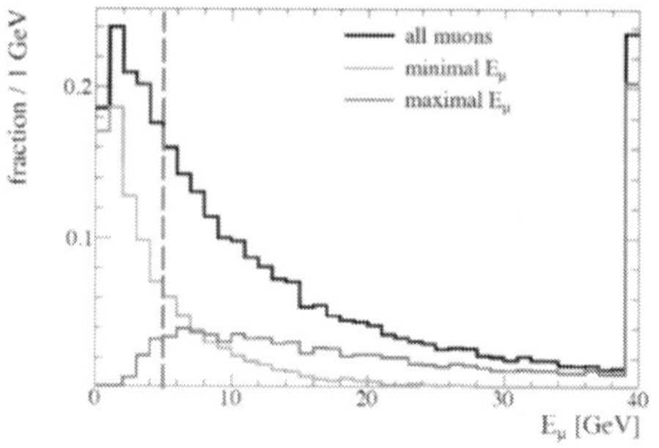

Figure 12. Momentum spectrum of muons produced from the τ decay.

points of the discussion here, together with the coexistence of such a dedicated experiment with the SHiP detector.

We assume a 1 mm (0.01 λ_{int}) tungsten target where only 1% of the protons will interact and 8×10^{12} τs will be produced. About 90% of the τs will decay outside of the target. Downstream of the target, a high precision tracking system will be placed in magnetic field: silicon pixel technology has the required resolution. This tracker, designed with a concept similar to the LHCb VELO [36], would be placed 8 mm far from the beam axis, leaving the remaining 99% of the protons to pass through and reach the SHiP target downstream. This expedient ensures the compatibility with the SHiP experiment. Downstream of the high resolution tracker, a magnetised hadron absorber will act as a muon identifaction system: the magnetic field will ensure that the curvature can be used to improve the track matching with the vertex detector. All three muons will get identified and measured. An optimisation of the material used and their size has to be done to tune the efficiency and mass resolution versus the particle misidentification. A precise timing detector will complete the experimental apparatus to reduce combinatorial background.

The geometrical acceptance of such a detector is $\varepsilon \sim$ 33%, defined by the inner and outer muon chamber efficient in the $3 < \vartheta < 170$ mrad angular range. A cut on the muon momentum above 5 GeV is also applied in order to have muons detected after passing the hadron absorber. Figure 12 shows the momentum distribution of muons originated from τ decays. The main background source is given by the $D_s \to \eta\mu\nu$ decay with subsequent $\eta \to \mu\mu\gamma$, which gives an irreducible background source. If the invariant mass of two muons is required to be larger than 450 MeV, a further reduction of this background is achieved, with about 1.5×10^5 remaining events. Accounting for the expected background and the number of detectable τ decays, a sensitivity to the branching ratio down to 10^{-10} would be achievable.

We therefore conclude that such an experiment could be carried out in parallel to the SHiP experiment and would be competitive in the future landscape of $\tau \to \mu\mu\mu$ searches.

References

[1] ATLAS Collaboration, Phys.Lett. **B716**, 1–29 (2012).
[2] CMS Collaboration, Phys.Lett. **B716**, 30-61 (2012).
[3] G. Degrassi et al., JHEP **1208**, 098 (2012).
[4] S. Heinemeyer, Higgs Physics, arXiv:1405.3781.
[5] M. Anelli et al., CERN-SPSC-2015-016, SPSC-P-350, arXiv:1504.04956.
[6] S. Alekhin et al., CERN-SPSC-2015-017, SPSC-P350-ADD-1, arXiv:1504.04855.
[7] N. Agafonova et al., OPERA Collaboration, JINST **4**, P04018 (2009).
[8] J. Brunner, Phys. Lett. **B731**, 320 (2014).
[9] A. Fradette et al., Phys. Rev. **D90**, 035022 (2014).
[10] M. Drewes, et al., JHEP **1303**, 096 (2013).
[11] M. Drewes, Int. J. Mod. Phys. **E22**, 1330019 (2013).
[12] T. Asaka and M. Shaposhnikov, Phys. Lett. **B620**, 17 (2005).
[13] M. Shaposhnikov, Nucl. Phys. **B763**, 49 (2007).
[14] L. Canetti et al., Phys. Rev. Lett. **110**, 6, 061801 (2013).
[15] E. Bulbul et al., Astrophys. J. **789**, 13 (2014).
[16] A. Boyarsky et al., Phys. Rev. Lett. **113**, 251301 (2014).
[17] D. Gorbunov and M. Shaposhnikov, JHEP **0710**, 015 (2007). Erratum-ibid. **1311**, 101 (2013).
[18] L. Canetti and M. Shaposhnikov, JCAP **1009**, 001 (2010).
[19] K. Kodama et al., DONUT Collaboration, Phys. Lett. **B504**, 218 (2001).
[20] K. Kodama et al., DONUT Collaboration, Phys. Rev. **D78**, 052002 (2008).
[21] N. Agafonova et al., OPERA Collaboration, Phys. Lett. **B691**, 138 (2010).
[22] N. Agafonova et al., OPERA Collaboration, JHEP **1311**, 036 (2013).
[23] N. Agafonova et al., OPERA Collaboration, Phys. Rev. **D89**, 051102 (2014).
[24] N. Agafonova et al., OPERA Collaboration, PTEP **2014**, 101C01 (2014).

[25] N. Agafonova et al., OPERA Collaboration, Phys. Rev. Lett. **115**, 121802 (2015).
[26] C. Lourenco, H.K. Wohri, Phys. Rept. **433**, 127 (2006).
[27] J. Carvalho, Nuclear Physics **A725**, 269 (2003).
[28] H. Abramowicz, ZEUS Collaboration, JHEP **1309**, 058 (2013).
[29] K.A. Olive et al. (Particle Data Group), Chin. Phys. **C38**, 090001 (2014).
[30] M. H. Reno, Phys. Rev. D **74**, 033001 (2006).
[31] C. H. Albright and C. Jarlskog, Nucl. Phys. B **84**, 467 (1975).
[32] G. De Lellis et al., Physics Reports **399**, 227 (2004).
[33] A. Kayis-Topaksu et al., New J. Phys. **13**, 093002 (2011).
[34] Babar and Belle Collaboration, Eur.Phys.J. **C74**, 3026 (2014).
[35] LHCb Collaboraiton, JHEP **1502**, 121 (2015).
[36] C. Parkes et al., Nucl. Instrum. Meth. **A604**, 1 (2009).

EPJ Web of Conferences **118**, 01010 (2016) DOI: 10.1051/epjconf/201611801010
FCCP2015

Mu to electron conversion with the COMET experiment

Peter Dornan

Imperial College London, United Kingdom

Abstract. The COMET experiment at J-PARC is under construction and will search for mu to electron conversion in aluminium at the 3.10^{-17} level, four orders of magnitude better than current limits. The experiment will take place in two phases with a first phase aiming at a two orders of magnitude improvement.

1 Charged lepton flavour violation with COMET

The COMET experiment is currently in preparation at J-PARC. The aim is to investigate charged lepton violation by searching for mu to e transitions from muons stopped in aluminium. The collaboration currently has 177 participants from 33 institutes from Belarus, Canada, China, Czech Republic, France, Georgia, Germany, India, Japan, Malaysia, Russia, Saudi Arabia, South Korea, United Kingdom and Vietnam. First data taking is expected in 2018.

Lepton flavour violation does not feature in the standard model which only describes flavour violations in the charged current interactions of quarks. However neutrino oscillation shows that neutral lepton flavour violation does occur in nature and this, so far, is the one unambiguous observation in particle physics of an interaction beyond the basic standard model. The actual mechanism of lepton flavour violation in the neutrino sector remains to be discovered and so establishing if there is measurable lepton flavour violation in the charged sector is crucial in the search for a wider theory or at least the necessary extension of the standard model to encompass neutrino oscillation.

Flavour is a crucial part of particle physics and whilst it is becoming increasingly well parameterised there is little real appreciation of its role. Consequently it is necessary to acquire as much information as possible if a valid theory is to emerge. An important consequence of flavour changing processes with three families is the mixing matrix with the crucial phase which can yield CP violating processes. Observation of charged lepton flavour processes will almost certainly lead to additional sources of CP violation which could contribute to the unexplained baryon – antibaryon asymmetry.

1.1 Mu to e transitions

A search for a muon changing to an electron is the transition most accessible for experimental investigation of charged lepton violation. Of course it could be that the main transition is from the tau but the lower production rate and the very short lifetime make it much harder to achieve very low rates. Three processes for μ to e transitions are currently being investigated

$$\mu \to e\gamma$$
$$\mu \to eee$$
$$\mu + N \to e + N$$

Present limits are currently: $Br(\mu \to e\gamma) < 5.7.10^{-13}$, $Br(\mu \to eee) < 1.10^{-12}$ and μ to e conversion on gold lower than 6.10^{-13}. Currently $\mu \to e\gamma$ and $\mu \to eee$ are being investigated at PSI. Two experiments are in construction to investigate μ to e conversion, COMET at J-PARC and Mu2e at FNAL.

As a result of neutrino oscillation charged lepton violation must take place at some level This is shown in Figure 1, however the rate from this process is very small, $\sim(m_\nu/m_W)^4$ and so effectively unobservable.

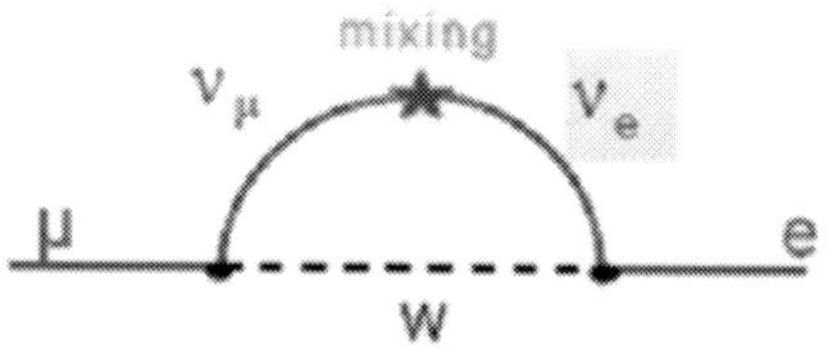

Figure 1. Mu to e transition as a result of neutrino oscillation.

Possible processes for charged lepton violation can be classified as 'Photonic', Figure 2, and Non-Photonic, Figure 3.

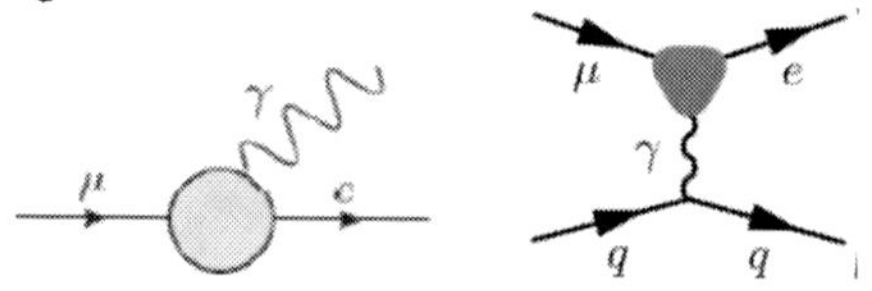

Figure 2. Photonic processes for μ to e Transitions

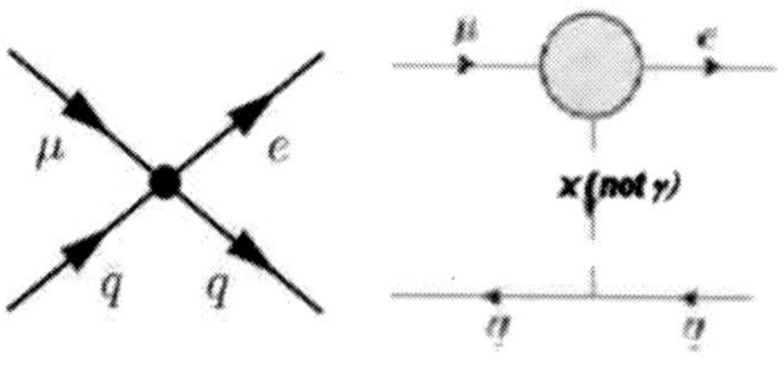

Figure 3. Non-photonic processes for μ to e Transitions

These can be summarised by a Lagrangian[1]

© The Authors, published by EDP Sciences. This is an open access article distributed under the terms of the Creative Commons Attribution License 4.0 (http://creativecommons.org/licenses/by/4.0/).

$$\mathcal{L}_{CLFV} = \frac{m_\mu}{(\kappa+1)\Lambda^2}\bar{\mu}_R\sigma_{\mu\nu}e_L F^{\mu\nu} + h.c.$$
$$\frac{\kappa}{(1+\kappa)\Lambda^2}\bar{\mu}_L\gamma_\mu e_L\left(\bar{u}_L\gamma^\mu u_L + \bar{d}_L\gamma^\mu d_L\right) + h.c.\,.$$

where κ gives the relative strength of photonic and non-photonic processes and Λ is the scale of new physics. For only photonic processes κ = 0 and the μ-e conversion rate is ~1/100 of μ → eγ, however if there are only non-photonic processes, κ → ∞ and then only μ-e conversion and μ →eee will occur.

2 Experimental Aspects of μ to e Conversion

The major problem for experiments seeking to observe and measure very rare events is the elimination of backgrounds. Two aspects of coherent μ to e conversion off a nucleus produce a very characteristic signal which enables high background suppression. These are the energy of the emitted electron and the time following the production of the muon to the electron appearance.

In μ to e conversion the emitted electron is monoenergetic and equal to the muon mass minus the binding energy and nuclear recoil. This comes to 105 MeV for conversion in aluminium. The electron is released after the muon is stopped in the aluminium target and so there is a time delay following the muon production given by the lifetime of a captured muon in aluminium, which is 864 nsec. The signal is therefore an identified electron with 105 MeV energy in a specific time interval following muon production. A pulsed proton beam is used to generate the muons, therefore a sufficient gap between pulses and very high extinction between pulses are essential. The beam timing setup for COMET is shown in Figure 4.

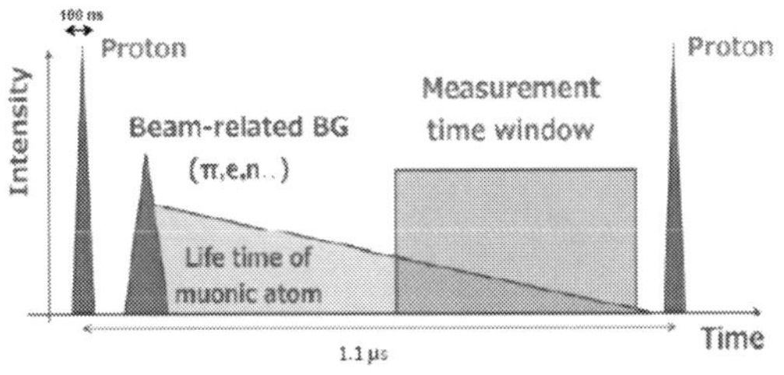

Figure 4. Proton pulses from J-PARC for COMET are separated by 1.1 μsec and the time window for measurement is typically from 0.6 to 1 μsec.

3 The COMET Experiment

The essential features of COMET are shown diagrammatically in Figure 5. An intense 56 kW 8 GeV proton beam is incident on a target in a 5T magnetic field. This forms the pion capture section where soft backward going pions are trapped and directed to the decay and muon transport curved solenoids. The pions quickly decay to muons which are transported and momentum selected by curved solenoids and collimators to be captured in thin aluminium disks. These form the muon stopping target where the mu to electron conversion will take place. Electrons emitted from the conversion process are then transported by further momentum selecting curved solenoids to the electron detector, StrECal. The goal is a single event sensitivity of 3.10^{-17} for μ to e conversion.

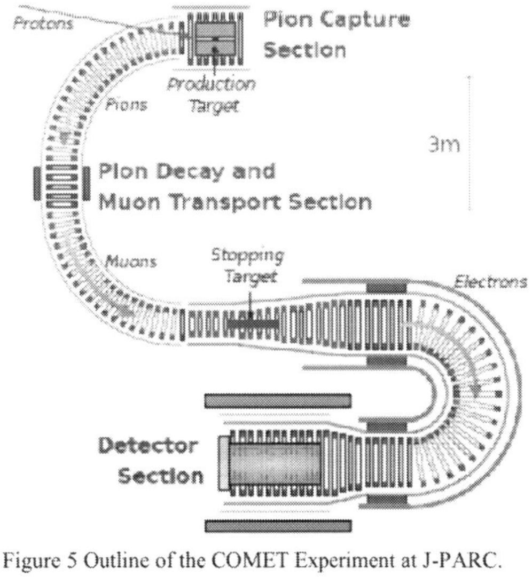

Figure 5 Outline of the COMET Experiment at J-PARC.

3.1 The Curved Solenoids

A critical feature of the experiment is the use of curved solenoids for momentum selection and background minimisation. However in a curved solenoid there is drift in the beam position from the solenoid centre for the nominal momentum as shown in Figure 6 and this requires correction by means of an additional vertical dipole field.

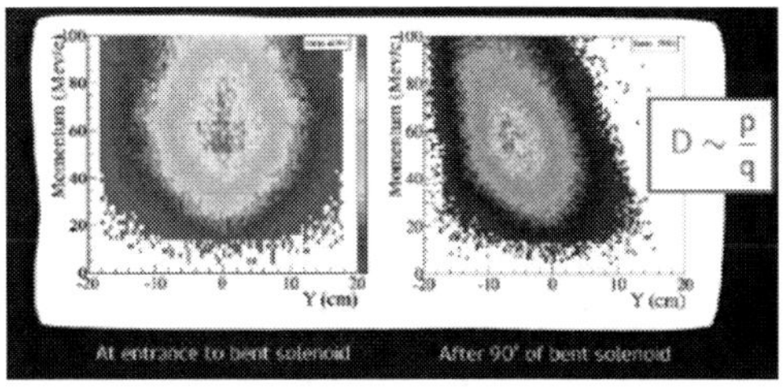

Figure 6. Simulation of the drift of the beam after a 90° transport in the bent solenoid dependent on the momentum, p, and the charge, q.

The first curved solenoid was delivered to J-PARC from Toshiba in March 2015 as shown in Fig. 7

DOI: 10.1051/epjconf/201611801010

Figure 7. Delivery of the first curved solenoid to J-PARC for the COMET experiment.

3.2 The Electron Detector, StrECal

The electron transport and StrECal detector are shown diagrammatically in Figure 8.

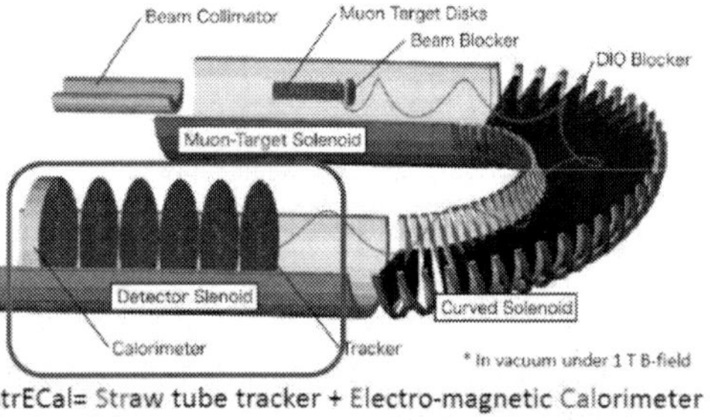

Figure 8. Layout of the electron transport and the elctron detector, StrECal.

The StrECal consists of a straw tube tracker in front of an electromagnetic crystal calorimeter in a 1 T solenoid. The straw tracker will have a minimum of 5 modules each with 4 planes, 2 in x and 2 in y. The exact number is still being optimised. These planes will be installed in vacuum to minimise multiple scattering and provide a spatial resolution < 200μ to give a momentum resolution < 200keV/c for a 105 MeV track. For the ECAL the requirements are an energy resolution better than 5% at 100 MeV and a spatial resolution better than 1 cm. This requires a crystal calorimeter and two crystals GSO and LYSO have been examined in prototypes. The results are shown in Fig 9 and as a result the LYSO crystals have been chosen.

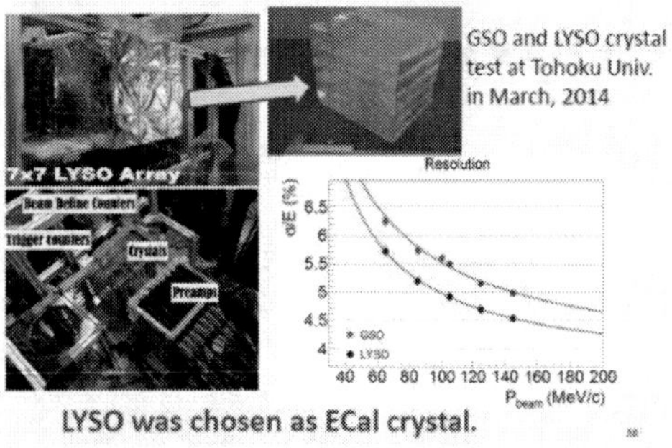

Figure 9. Calorimeter prototyping leading to the choice of LYSO for the COMET ECAL.

4 Backgrounds

Understanding and minimising backgrounds are critical for a very low rate measurement. Some of the most important for COMET are described.

4.1 Muon Decays in Orbit (DIO)

Standard muon decays to evν occur whilst the muon orbits the nucleus. These produce electrons with the same time structure as the signal although with a softer momentum spectrum, emphasising the importance of the momentum-energy resolution of the detector. Figure 10 shows the upper part of the expected DIO spectrum together with a signal corresponding to a μ to e conversion branching ratio of 3.10^{-15}.

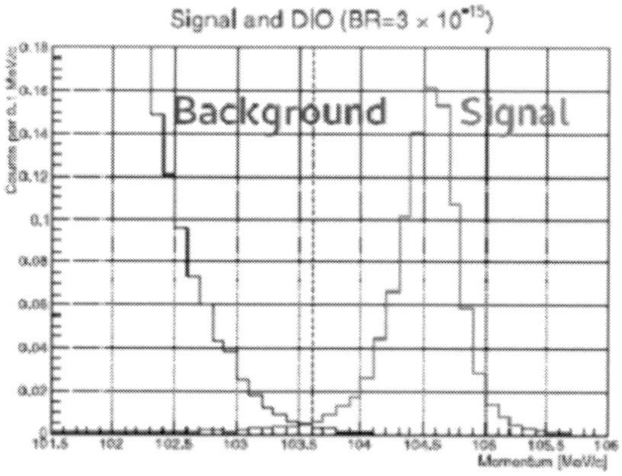

Figure 10. Momentum spectrum expected for the DIO background and a signal of 3.10^{-15}.

4.2 Out of Time Protons on the Target

After the protons hit the target there is a prompt burst of particles which would swamp any signal and this is eliminated with the use of the time window as shown in Figure 4. However should there be additional protons between the bunches these can cause serious backgrounds within the signal time window and so it is crucial that the accelerator has a high extinction level requiring an extinction factor

$$\textit{Extinction Factor} = \frac{\textit{No. of protons between bunches}}{\textit{No of protons in the bunch}} < 10^{-10}$$.

This has now been achieved at J-PARC as shown in Figure 11.

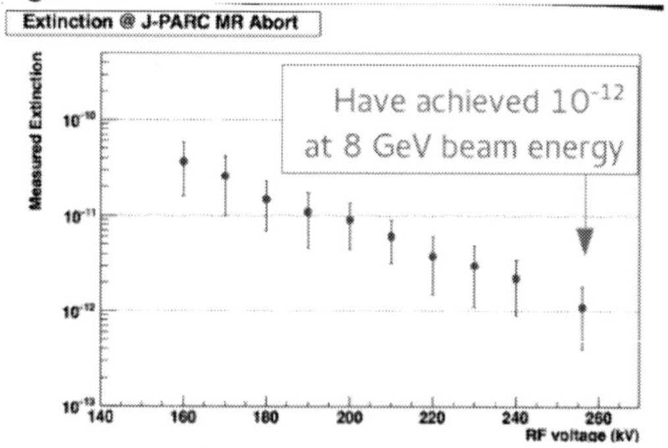

Figure 11. Extinction levels reached in the 8 GeV J-PARC beam

4.3 Other Particles Emitted when a Muon is captured by Aluminium

At the level of accuracy required, predictions are unreliable for the emission of both charged and neutral particles when muons are captured by aluminium. To investigate this a joint COMET-Mu2E dedicated experiment, ALCAP, is taking place at PSI and an example of preliminary results for neutral particles emitted is shown in Figure 12. The indication so far is that current estimates of backgrounds from this source have been made conservatively.

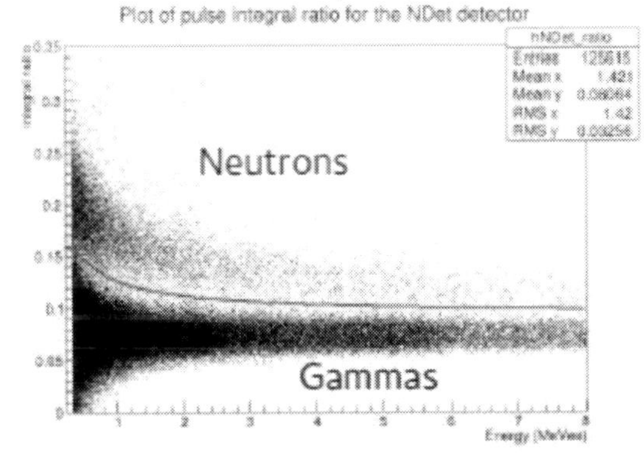

Figure 12. Preliminary results from the ALCAP experiment for neutral particle resulting from muon capture on aluminium

4.4 Summary of expected backgrounds

Current estimates of the expected background rates are summarised in Table 1.

Table 1. Summary of estimated background rates

Beam-related prompt backgrounds	
Radiative pion capture	0.05
Beam electrons	< 0.15
Muon decay in flight	< 0.0002
Pion decay in flight	< 0.0001
Beam-related delayed backgrounds	
Neutron induced	0.021
Delayed pion radiative capture	0.002
Anti-proton induced	0.007
Intrinsic physics backgrounds	
Muon decay in orbit	0.15
Radiative muon capture	<0.001
μ^- Capture with neutral emission	<0.001
μ^- Capture with charged emission	<0.001
Cosmic ray backgrounds	
Cosmic ray muons	0.002
Electrons from cosmic muons	0.002
Total	0.34

5 Overall Performance

The expected running time with the 56 kW beam and predicted sensitivity are given in Table 2

Table 2. Summary of running time and expected sensitivity

Beam Power at 8 GeV	56 kW
Running time	$3x10^{7}$ sec
No. of protons	$8.5x10^{20}$
No. of muon stops	$2x10^{18}$
Est. No. of Background	0.3
Single Event Sensitivity	$2.6x10^{-17}$
Expected limit (90% CL)	$6.0x10^{-17}$

6 Phased Approach

In an attempt to obtain an earlier result and also to examine in detail the assumptions made on beam quality and backgrounds COMET will run in two phases. The full experiment described above is now termed Phase-II and Phase-I will be a simplified version. The differences are shown in Figure 13.

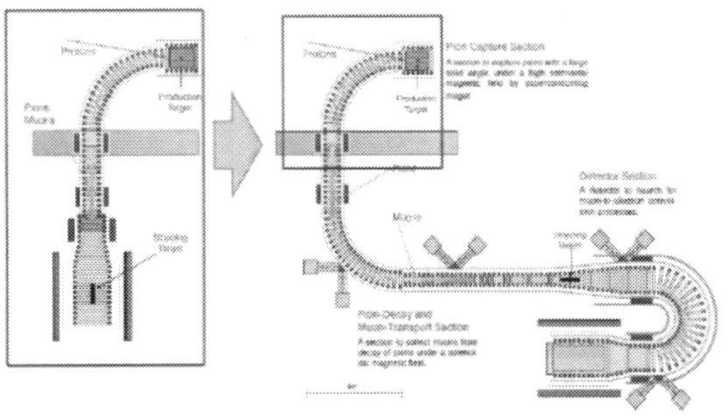

Figure 13. COMET Phase-I and COMET Phase-II

For Phase-I the proton beam power will only be 3.2 kW and the aim will be to make an initial measurement of μ to e conversion with a single event sensitivity of 3.10^{-15}, two order of magnitude better than current limits but two orders of magnitude less than Phase-II. In Phase-I there is no longer a curved solenoid after the muon stopping target and so the electron detection and measurement can no longer be accomplished with an on-axis detector such as the StrECal. Hence for Phase-I the electrons will be detected by a cylindrical drift chamber, CYDET, in a 1T solenoid surrounding the stopping target. The StrECal will be used in Phase-I in special runs in order to monitor the beam and backgrounds in preparation for Phase-II.

6.1 The CYDET

The detector for Phase-I will be a cylindrical drift chamber in a 1 T solenoidal field. It is an all stereo chamber with 18 sense wire layers with stereo angles between 64 and 75 mrad giving a longitudinal resolution around 3 mm. The gas will be helium based to minimise multiple scattering although the final mixture awaits prototyping tests. Inside the inner wall of the chamber hodoscopes will provide the trigger for the experiment and T_{zero} for the drift measurement. A diagram of this detector for Phase-I is shown in Figure 14.

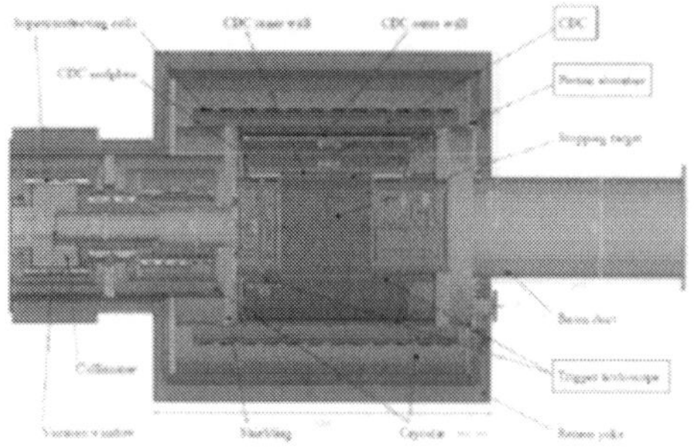

Figure 14. The cylindrical drift chamber and trigger hodoscopes for COMET Phase-I

The CYDET is currently under construction and stringing is in the final stages as can be seen from the photographs in Figure 15.

Figure 15. Construction of the CYDET detector

For Phase-I a prototype StrECal is under construction and will be used to monitor the beam and the backgrounds in preparation for Phase-II.

The predicted performance for Phase-I is given in Table 3

Table 3 Predicted running time and performance of COMET Phase-I

Beam Power at 8 GeV	3.2 kW
Running time	9.5×10^{6} sec
No. of protons	2.3×10^{19}
No. of muon stops	1.2×10^{16}
Est. No. of Background	0.02
Single Event Sensitivity	3.1×10^{-15}
Expected limit (90% CL)	7.0×10^{-15}

7 Summary

Muon to electron conversion provides an excellent chance to observe physics beyond the standard model. The COMET experiment is designed to investigate this with the goal of a four order of magnitude improvement on the present limit. The experiment will adopt a phased approach with a first phase to characterise the beams and backgrounds and also make an initial measurement at the 10^{-15} level.

The beamline and detectors are currently under construction at J-PARC and the collaborating institutes. Data taking for Phase-I will take place in 2018 with Phase-II to follow two to three years later.

References

1. A de Gouvea and P Vogel, arXiv:1303.4097 [hep-ph] (2013)

Muon g-2 and Hadronic Vacuum Polarization: Recent Developments

Simon Eidelman[1,2,a]

[1] *Budker Institute of Nuclear Physics, Novosibirsk 630090, Russia*
[2] *Novosibirsk State University, Novosibirsk 630090, Russia*

Abstract. We discuss various experiments on e^+e^- annihilation into hadrons relevant to the problem of the muon anomalous magnetic moment. They include a status of the ISR measurements of the $e^+e^- \to \pi^+\pi^-$ as well as studies of numerous hadronic final states in experiments with the CMD-3 and SND detectors at the VEPP-2000 e^+e^- collider.

1 Introduction

The anomalous magnetic moment of the muon is one of the most precisely known physical quantities. In 2006 the BNL E821 Collaboration published the final results of their measurement of the muon $a_\mu \equiv (g_\mu - 2)/2$ [1]. Various calculations show that the Standard Model (SM) prediction is about 3.5 standard deviations below the experimental value [2, 3]. At the moment two new measurements of a_μ, each aimed at four-fold increase of accuracy, are planned at Fermilab and J-PARC. If the central value of the experimental result is confirmed, the deviation between experiment and theory will reach 8-10 standard deviations unambiguously pointing to effects of New Physics.

The theoretical prediction accuracy is currently limited by the uncertainties of the hadronic vacuum polarization extracted from the cross sections of e^+e^- annihilation into hadrons measured by a scan method at CMD-2 and SND detectors at the VEPP-2M and initial-state radiation (ISR) at BaBar (Fig. 1), see the review of e^+e^- experiments in Ref. [4].

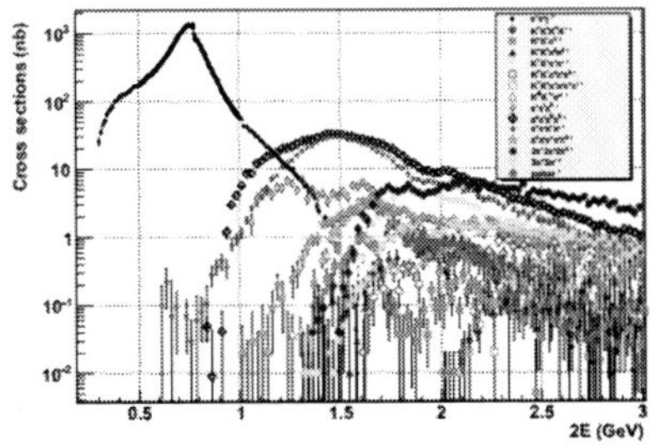

Figure 1. Current status of exclusive measurements [4]

As it is absolutely necessary to improve experimental accuracy, experiments on low-energy e^+e^- annihilation into hadrons are currently in progress in various centers.

In addition to the low-energy measurements, other energy ranges are still of interest for a_μ and particularly for the $\alpha(M_Z^2)$ determination. Recently a new measurement of R was performed between 3.12 and 3.72 GeV in Novosibirsk using the KEDR detector [5]. The achieved systematic uncertainty is 2.1% with a total uncertainty of 3.3%. The results are shown in Fig. 2. Analysis of the KEDR

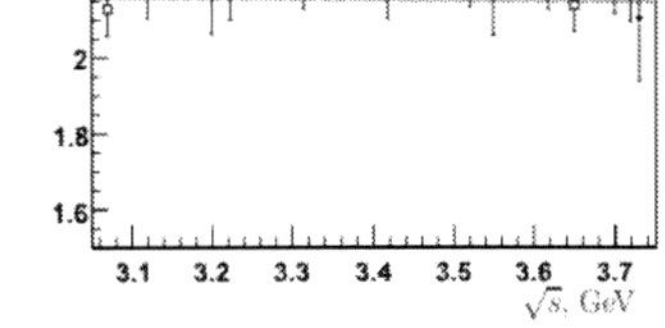

Figure 2. R measurement at KEDR [5]

scan between 1.9 GeV and J/ψ [6] is in progress while R measurements from 2 to 4.6 GeV with BESIII and further studies of the charmonium region with KEDR are also planned. Improvement to about 2% in total can be expected from joint efforts of BESIII and KEDR.

[a] e-mail: simon.eidelman@cern.ch

© The Authors, published by EDP Sciences. This is an open access article distributed under the terms of the Creative Commons Attribution License 4.0 (http://creativecommons.org/licenses/by/4.0/).

DOI: 10.1051/epjconf/201611801011

2 ISR measurements of $e^+e^- \to \pi^+\pi^-$

The process $e^+e^- \to \pi^+\pi^-$ is known to give the largest contribution to the leading-order hadronic term $a_\mu^{\rm LO,had}$, about 73%. Recent ISR measurements of this process substantially improved the accuracy of its cross section. The BaBar Collaboration used a data sample collected at the peak of the $\Upsilon(4S)$ resonance to achieve the record precision of about 0.5% near the ρ meson peak [7, 8].

KLOE was using ISR running at the ϕ meson peak and gradually increased the precision of their measurements to 0.7% [9–11]. However, the results of KLOE and BaBar differ with the discrepancy reaching 5% in some energy regions, far beyond the declared precision.

A very recent precise measurement has been performed by the BES Collaboration running at the $\psi(3770)$ peak [12]. Their ISR measurement reached precision of about 0.9%, see the cross section in Fig. 3.

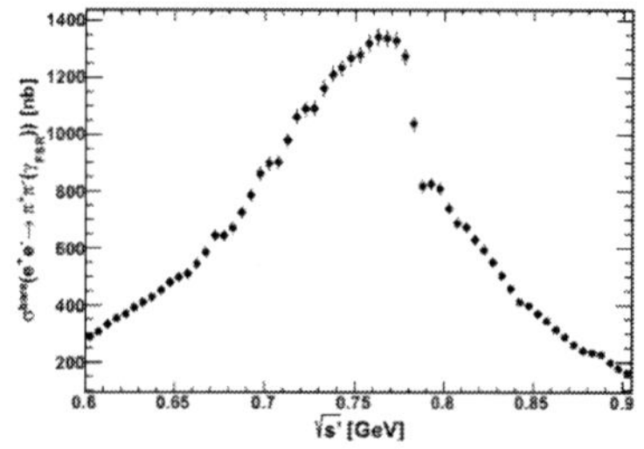

Figure 3. Cross section of $e^+e^- \to \pi^+\pi^-(\gamma)$ at BESIII [12]

Comparison between the results of BESIII and those from other groups – SND [13], CMD-2 [14], BaBar and KLOE shows that there are local discrepancies between the data of BaBar, KLOE and BESIII. This can be seen from Figure 4, which illustrates how the results obtained by ISR look like in terms of the hadronic contribution to muon anomaly after integration. The BESIII result lies between those of BaBar and KLOE being somewhat closer to the latter. Obviously, new measurements with comparable or even better precision are needed.

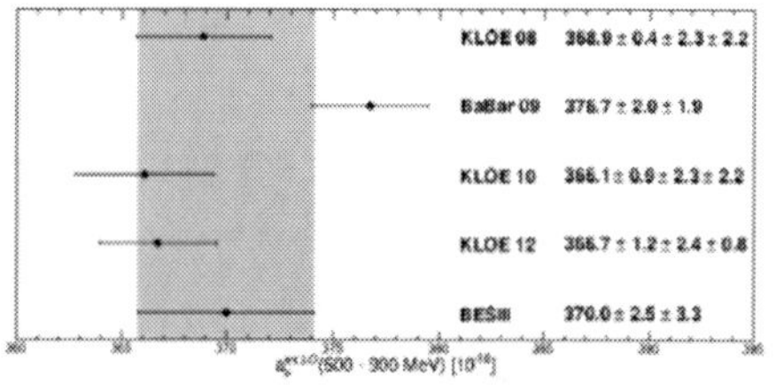

Figure 4. Comparison of BESIII results on a_μ [12]

3 Experiments at VEPP-2000

In 2010 a new low-energy e^+e^- collider, VEPP-2000, was commissioned in Novosibirsk [15]. In 2011 data taking started with two detectors, CMD-3 and SND. An integrated luminosity of $\sim$ 60 fb^{-1} was collected by each in 2011 - 2013 in the center-of-mass (c.m.) energy range 320 - 2000 MeV. Figure 5 illustrates accumulation of luminosity in various energy ranges. Lately analysis has been mainly focused on the c.m. energies above the ϕ meson. In particular, both groups made an attempt to improve the precision of various cross sections important for the muon anomaly problem [16].

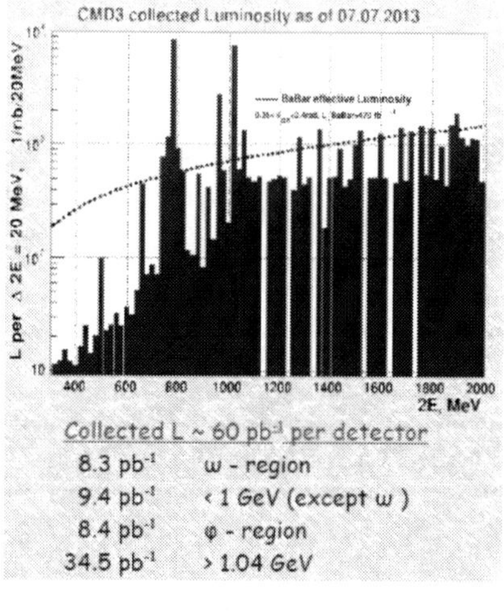

Figure 5. Luminosity collected at VEPP-2000

Measurements of an integrated luminosity at CMD-3 use two processes, $e^+e^- \to e^+e^-$ and $e^+e^- \to \gamma\gamma$, allowing a precision of $\sim$ 1% [17]. At SND, events of large-angle Bhabha scattering are used to determine an integrated luminosity with a systematic accuracy of 2% [18].

CMD-3 declares an aggressive goal of reaching 0.35% accuracy in the measurement of the pion form factor, see their prelimiary results in Fig. 6 [19]. Up to to the c.m. energy of 700 MeV they can separate pions from muons by two independent methods - from the information from the drift chamber and from the electromagnetic calorimeters, the latter being used in the whole energy range.

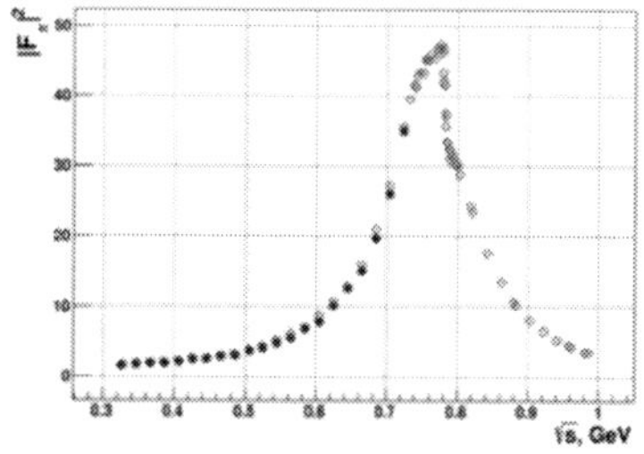

Figure 6. CMD-3 results on the pion form factor

Both detectors have continued analysis of the collected data samples and are reporting measurements of cross sections for various processes with pions and η mesons. SND

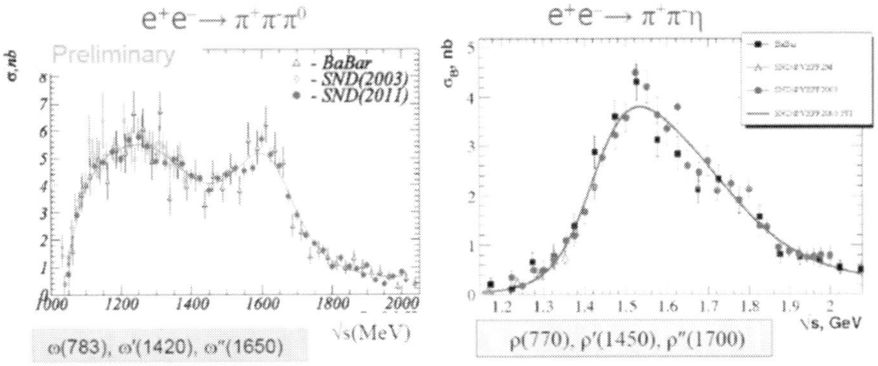

Figure 7. Recent results on hadronic cross sections from SND: (left) $e^+e^- \to \pi^+\pi^-\pi^0$ [20], (right) $e^+e^- \to \pi^+\pi^-\eta$ [18]

published their final results on $e^+e^- \to \pi^+\pi^-\pi^0$ [20], Fig. 7 (left), and $e^+e^- \to \pi^+\pi^-\eta$ [18], Fig. 7 (right).

The cross section of the process $e^+e^- \to \pi^+\pi^-\pi^0$ clearly shows two excitations of the $\omega(783)$ meson – $\omega(1420)$ and $\omega(1650)$. On the contrary, the energy dependence of the process $e^+e^- \to \pi^+\pi^-\eta$ is dominated by the $\rho(770)$ recurrencies – $\rho(1450)$ and $\rho(1700)$. In both cases the SND data from VEPP-2000 are consistent with the results from VEPP-2M and BaBar and have comparable precision. CMD-3 has also measured the cross sections of these processes with comparable precision, not shown here.

CMD-3 and SND studied all three possible charge combinations of the six-pion production. The cross section in Fig. 8. shows the case when all pions are charged with the obvious dip around the $N\bar{N}$ threshold [21].

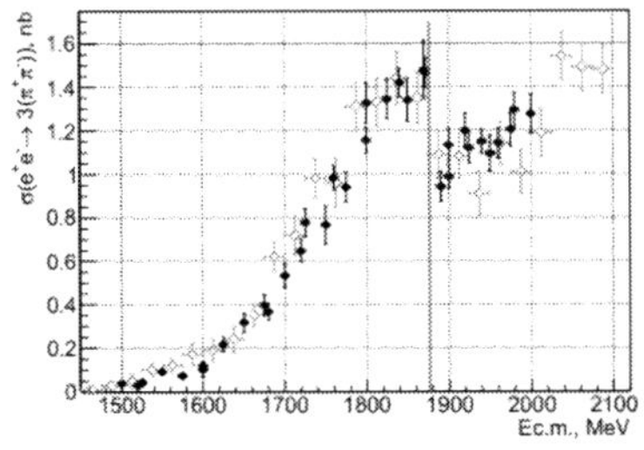

Figure 8. Cross section of $e^+e^- \to 3\pi^+3\pi^-$ at CMD-3 [21]

Such behavior of the cross section has been observed before at BaBar [22] as well as in much earlier e^+e^- and photoproduction measurements, therefore suggesting a probable threshold effect due to the opening of the $N\bar{N}$ channel [23]. Near the $N\bar{N}$ threshold the behavior of the cross section for the $2\pi^+2\pi^-2\pi^0$ final state (see Fig. 9) is also irregular, but somewhat differs from that in the $3\pi^+3\pi^-$ case [24].

Finally, SND has measured for the first time ever the cross section of $e^+e^- \to \pi^+\pi^-4\pi^0$, where the size of the data sample does not allow any conclusions about the $N\bar{N}$ threshold, Fig. 10. These final states have interesting

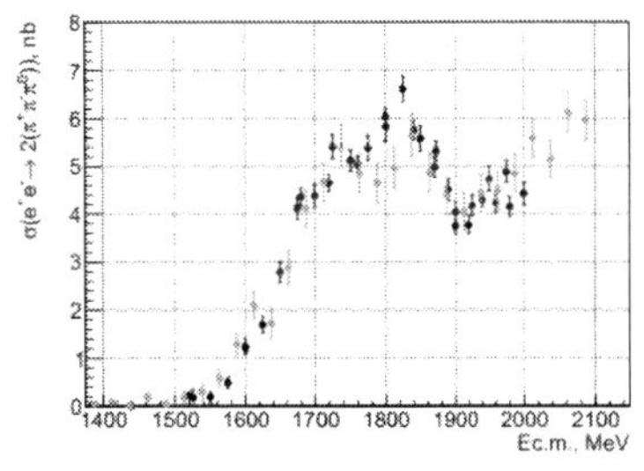

Figure 9. Cross section of $e^+e^- \to 2\pi^+2\pi^-2\pi^0$ at CMD-3

and rich dynamics which analysis should be performed simultaneously for all three final states. For example, the $\rho f_0(1370)$ intermediate mechanism can result in all three possible charge combinations of the six-pion state (the fully neutral $6\pi^0$ state can not be produced in one-photon annihilation because of C-parity violation). There are also mechanisms which do not give all charged final states, e.g., $\omega 3\pi$ or $\eta 3\pi$. An additional problem is that such final states can have negative G-parity whereas normally six-pion states should have positive G-parity.

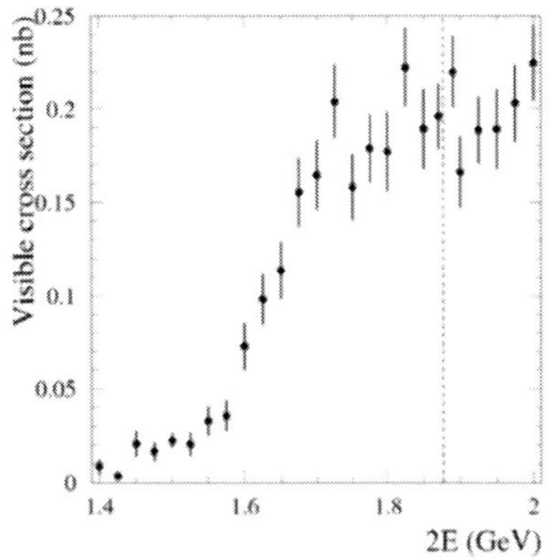

Figure 10. Visible cross section of $e^+e^- \to \pi^+\pi^-4\pi^0$ at SND

DOI: 10.1051/epjconf/201611801011

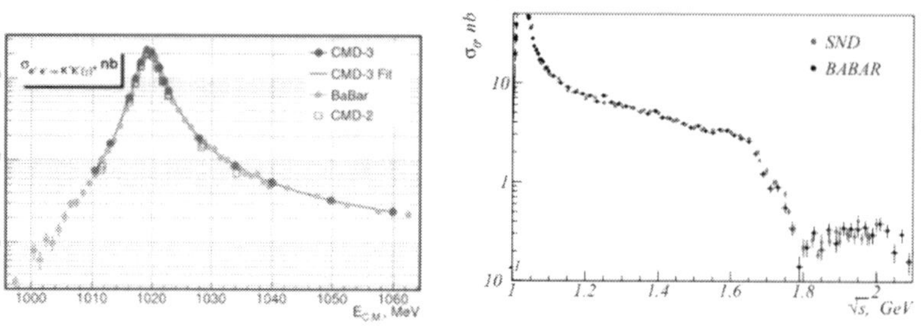

Figure 11. Cross section of $e^+e^- \to K^+K^-$: (left) near the ϕ meson at CMD-3, (right) from 1 to 2 GeV at SND

CMD-3 continued studies of various processes with kaons in the final state using good K/π separation based on measuring dE/dx in the drift chamber. In Fig. 11 we show the cross section of the process $e^+e^- \to K^+K^-$ near the ϕ meson at CMD-3 (left) and in the whole energy range at SND (right) [25].

The CMD-3 group has also reported results on the cross section and dynamics of the $K^+K^-\pi^+\pi^-$ final state, see Fig. 12. While the energy dependence of the cross section is clear and obviously shows the $\phi(1680)$ state, there is still a lot to be done about the dynamics.

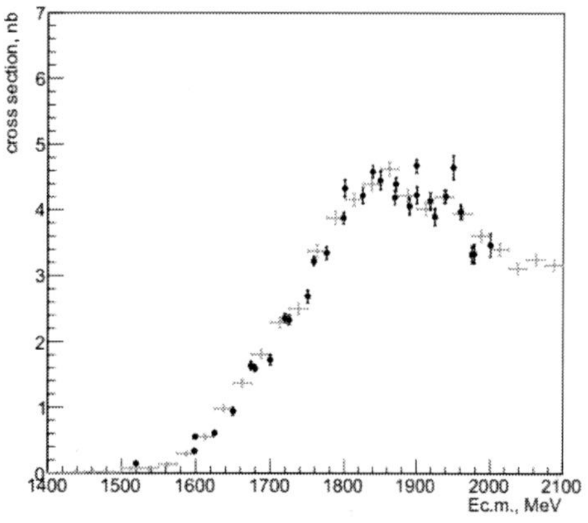

Figure 12. Cross section of $e^+e^- \to K^+K^-\pi^+\pi^-$ at CMD-3 [27].

Using the missing mass method, CMD-3 has also measured the cross section of the process $e^+e^- \to K^+K^-\eta$, Fig. 13. Work is in progress on the process $e^+e^- \to K^+K^-\pi^0$.

SND has already published their measurements of the processes with only neutral particles in the final state – $e^+e^- \to \pi^0\pi^0\gamma$ [26] and $e^+e^- \to \eta\gamma$ [27]. The latter is of particular interest since for the first time events of the process have been found above 1.4 GeV, see Fig. 14.

A study of the nucleon form factors near threshold was continued. SND significantly improved the precision of $\sigma(e^+e^- \to n\bar{n})$ [28] compared to the previous results from FENICE [29], see Fig. 15.

CMD-3 measured the cross section of the process $e^+e^- \to p\bar{p}$ and made an attempt to extract the ratio of

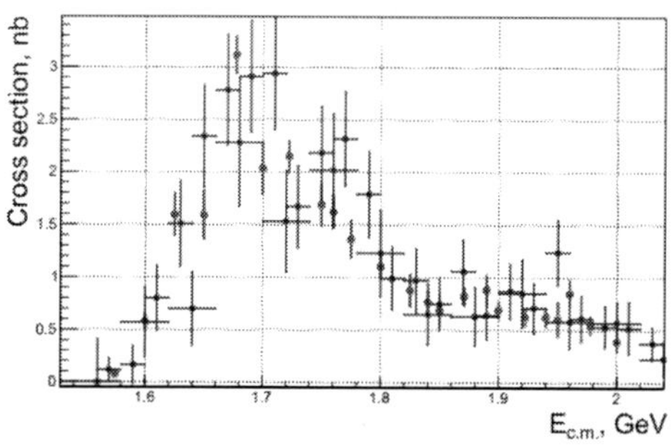

Figure 13. Cross section of $e^+e^- \to K^+K^-\eta$ at CMD-3 [27]

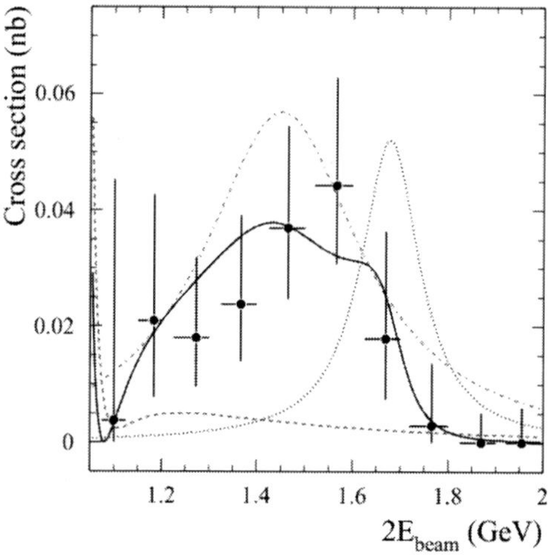

Figure 14. Cross section of $e^+e^- \to \eta\gamma$ at SND [27].

the electric and magnetic form factors based on the angular distribution of the final nucleons [30], see Fig. 16.

Both detectors used an original method of Ref. [31] to measure the partial width of a strongly suppressed $\eta' \to e^+e^-$ decay using the inverse process. CMD-3 reported an upper limit of $\Gamma(\eta' \to e^+e^-) < 0.0024$ eV at 90% CL based on 2.69 pb^{-1} and one mode of η' decay [32]. SND used 2.9 pb^{-1} and five modes of η' decay to improve it to

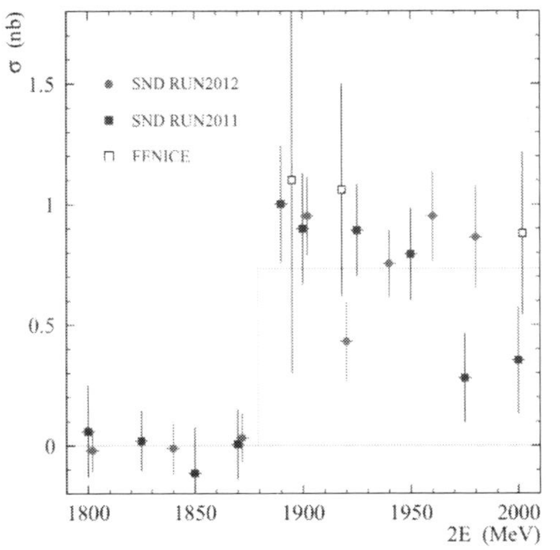

Figure 15. Cross section of the process $e^+e^- \to n\bar{n}$ at SND [28]

< 0.0020 eV. Finally, they combined the data samples of CMD-3 and SND to find $\Gamma(\eta' \to e^+e^-) < 0.0011$ eV at 90% CL [33] which is still about two orders of magnitude below the unitary bound. SND has also performed a feasibility study for a search for $\eta \to e^+e^-$ via $e^+e^- \to \eta$ and concluded that the only promising decay mode for that is $\eta \to 3\pi^0$ [34]. A dedicated two-week run with the luminosity expected at the c.m. energy around the η meson mass will allow to improve the existing limit [35].

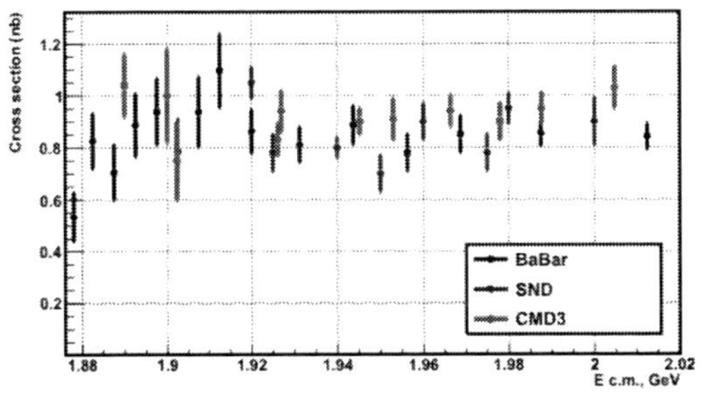

Figure 16. Cross section of the process $e^+e^- \to p\bar{p}$ at CMD-3 [30].

After upgrading the VEPP-2000 collider and commissioning the new injection complex the luminosity of the complex is expected to increase by an order of magnitude. Both detectors will run for another five years with a goal of collecting 1-2 fb^{-1} and increasing significantly the accuracy of all hadronic channels.

Acknowledgments

The author thanks G. D'Ambrosio and other Naples colleagues for their excellent organization of the Conference and very kind hospitality. This work was partially supported by the RFBR grants 13-02-00215, 14-02-91332, 15-02-05674 and the DFG grant HA 1457/9-1.

References

[1] G.W. Bennett et al., Phys. Rev. D **73**, 072003 (2006).
[2] M. Davier et al., Eur. Phys. J. C **71**, 1515 (2011).
[3] K. Hagiwara et al., J. Phys. G **38**, 085003 (2011).
[4] V.P. Druzhinin et al., Rev. Mod. Phys. **83**, 1545 (2011).
[5] V.V. Anashin et al. (KEDR Collaboration), arxiv:1510.02667.
[6] V.V. Anashin et al. (KEDR Collaboration), Phys. Lett. B **703**, 543 (2011).
[7] B. Aubert et al. (BaBar Collaboration), Phys. Rev. Lett. **103**, 231801 (2009).
[8] J.P. Lees et al. (BaBar Collaboration), Phys. Rev. D **86**, 032013 (2012).
[9] F.Ambrosino et al. (KLOE Collaboration), Phys. Lett. B **670**, 285 (2009).
[10] F.Ambrosino et al. (KLOE Collaboration), Phys. Lett. B **700**, 102 (2011).
[11] D. Babusci et al. (KLOE Collaboration), Phys. Lett. B **720**, 336 (2013).
[12] M. Ablikim et al. (BES Collaboration), arXiv:1507.08188.
[13] M.N. Achasov et al. (SND Collaboration), JETP **103**, 380 (2006).
[14] R.R. Akhmetshin et al. (CMD-3 Collaboration), Phys. Lett. B **648**, 28 (2007).
[15] Yu.M. Shatunov et al., Proc. of EPAC 2000, Vienna, Austria, June 2000, p.439.
[16] M.N. Achasov et al. (CMD-3 and SND Collaborations), Int. J. Mod. Phys. Conf. Ser. **35**, 1460388 (2014).
[17] R.R. Akhmetshin et al. (CMD-3 Collaboration), JINST **9**, C09003 (2014).
[18] V.M. Aulchenko et al. (SND Collaboration), Phys. Rev. D **91**, 052013 (2015).
[19] G.V. Fedotovich et al., Phys. Atom. Nucl. **78**, 591 (2015).
[20] V.M. Aulchenko et al. (SND Collaboration), J. Exp. Theor. Phys. **121**, 34 (2015).
[21] R.R. Akhmetshin et al. (CMD-3 Collaboration), Phys. Lett. B **723**, 82 (2013).
[22] B. Aubert et al. (BaBar Collaboration), Phys. Rev. D **73**, 052003 (2006).
[23] J. Haidenbauer et al., Phys. Rev. D **92**, 054032 (2015).
[24] P.A.Lukin et al. (CMD-3 Collaboration), Phys. Atom. Nucl. **78**, 353 (2015).
[25] E.A. Kozyrev et al. (CMD-3 Collaboration), Phys. Atom. Nucl. **78**, 358 (2015).
[26] M.N. Achasov et al. (SND Collaboration), Phys. Rev. D **88**, 054013 (2013).

[27] M.N. Achasov et al. (SND Collaboration), Phys. Rev. D **90**, 032002 (2014).

[28] M.N. Achasov et al. (SND Collaboration), Phys. Rev. D **90**, 112007 (2014).

[29] A. Antonelli et al. (FENICE Collaboration), Nucl. Phys. B **517**, 3 (1998).

[30] R.R. Akhmetshin et al. (CMD-3 Collaboration), Int.J.Mod.Phys.Conf.Ser. **35**, 1460457 (2014).

[31] P. Vorobev et al. (ND Collaboration), Sov. J. Nucl. Phys. **48**, 273 (1988).

[32] R.R. Akhmetshin et al. (CMD-3 Collaboration), Phys. Lett. B **740**, 273 (2015).

[33] M.N. Achasov et al. (SND Collaboration), Phys. Rev. D **91**, 092010 (2015).

[34] M.N. Achasov et al. (SND Collaboration), JETP Lett. **102**, 266 (2015).

[35] G. Agakishiev et al. (HADES Collaboration), Phys. Lett. B **731**, 265 (2014).

Radiative τ leptonic decays and the possibility to determine the τ dipole moments

Matteo Fael[1,a]

[1] *Albert Einstein Center for Fundamental Physics, Institute for Theoretical Physics, University of Bern, CH-3012 Bern, Switzerland*

Abstract. Precise data on radiative leptonic τ decays offer the opportunity to probe the electromagnetic properties of the τ and may allow to determine its anomalous magnetic moment which, in spite of its precise Standard Model prediction, has never been measured. Recently, the branching fractions of the radiative leptonic τ decays ($\tau \to l\nu\bar{\nu}\gamma$, with $l = e, \mu$) were measured by the BaBar collaboration. These precise measurements, with a relative error of about 3%, must be compared with the branching ratios at the next-to-leading order in QED. Indeed the radiative corrections are expected to be of order 10%, for $l = e$, and 3%, for $l = \mu$. Here, we present the prediction of the differential decay rates and branching ratios of the radiative leptonic τ decays in the Standard Model at the next-to-leading order and we compare them with the recent BaBar measurements. Moreover, we report on a dedicated feasibility study for the measurements of the τ anomalous magnetic moment at Belle and Belle II.

1 Introduction

The electric dipole moment (EDM) and the anomalous magnetic moment ($g-2$) of a lepton are physical observables sensitive to quantum corrections induced by virtual particles that populate the vacuum. For this reason they are very well suited to test the Standard Model (SM) of particle physics and to unveil unknown new physics hidden at high energy. The electron and muon $g-2$ have been measured with the wonderful precision of 0.24 ppb [1] and 540 ppb [2], and they agree with the SM predictions at the level of ~ 1.4 and ~ 3.8 standard deviations, respectively [3]. In contrast, the short lifetime of the τ lepton (2.9×10^{-13} s) poses many difficulties for the experimental determination of its dipole moments and indeed it has so far prevented the direct measurement of the $g-2$ by means of the τ spin precession in a magnetic field, like in the electron and muon $g-2$ experiments. In fact, the present bound on the τ $g-2$ is only of $O(10^{-2})$, more than an order of magnitude bigger than the leading contribution $\alpha/(2\pi) \approx 0.001$ – result obtained long ago by Schwinger [4]. Therefore, experiments must attempt the extraction of indirect bounds from τ pair production and decays by comparing sufficiently precise data with the SM predictions.

We propose to measure the electromagnetic dipole moments of the τ lepton via radiative leptonic τ decays ($\tau \to l\gamma\nu\bar{\nu}$ with $l = \mu, e$) by means of an effective Lagrangian approach. Radiative τ leptonic decays can indeed be predicted with very high precision and can be formulated in term of the Bouchiat-Michel-Kinoshita-Sirlin parameters [5–8].

[a] e-mail: fael@itp.unibe.ch

Moreover, the BaBar collaboration performed recently the measurements of the $\tau \to l\gamma\nu\bar{\nu}$ branching fractions for a minimum photon energy $\omega_0 = 10$ MeV in the τ rest frame [9]. The experimental precision of these measurements, around 3%, requires the SM prediction of the branching ratios at next-to-leading order (NLO). Indeed these radiative corrections are expected to be of relative order $(\alpha/\pi)\ln(m_l/m_\tau)\ln(\omega_0/m_\tau)$, corresponding to a large 10% correction for $l = e$, and 3% for $l = \mu$.

After establishing our conventions in sec. 2 and introducing an effective Lagrangian for the study of the τ dipole moments, we briefly review in secs. 3 and 4 the present theoretical and experimental status on the τ $g-2$ and EDM. The SM decay rates and branching ratios at NLO, and their comparison with recent BaBar measurements, are presented in sec. 5. In sec. 6 we report the achievable sensitivity to the τ electromagnetic moments at Belle II experiment. Conclusions are drawn in sec. 7.

2 The τ lepton electromagnetic form factors

The most general vertex function describing the interaction between initial and final states of an on-shell τ lepton, with four-momenta p and p' respectively, and a photon can be written in the form

$$\Gamma^\mu(q^2) = ie\left\{\gamma^\mu F_1(q^2) + \frac{\sigma^{\mu\nu}q_\nu}{2m_f}\left[iF_2(q^2) - F_3(q^2)\gamma_5\right] + \left(\gamma^\mu - \frac{2q^\mu m_f}{q^2}\right)\gamma_5 F_4(q^2)\right\}, \quad (1)$$

© The Authors, published by EDP Sciences. This is an open access article distributed under the terms of the Creative Commons Attribution License 4.0 (http://creativecommons.org/licenses/by/4.0/).

where $e > 0$ is the positron charge, m_τ the mass of the τ, $\sigma_{\mu\nu} = i/2\,[\gamma_\mu, \gamma_\nu]$ and $q = p' - p$ is the ingoing four-momentum of the off-shell photon. The functions $F_2(q^2)$ and $F_3(q^2)$ are related, in the limit $q^2 \to 0$, to the measurable quantities

$$F_2(0) = a_\tau, \qquad F_3(0) = d_\tau \frac{2m_\tau}{e}, \tag{2}$$

where a_τ and d_τ are the anomalous magnetic moment and electric dipole moment of the τ, respectively.

Deviations of the τ dipole moments from the SM values can be analyzed in the framework of dimension-six gauge-invariant operators. Out of the complete set of 59 independent gauge invariant operators in [10, 11], only two of them can directly contribute to the τ $g-2$ and EDM at tree level (i.e., not through loop effects):

$$Q^{33}_{lW} = \left(\bar{l}_\tau \sigma^{\mu\nu} \tau_R\right) \sigma^I \varphi\, W^I_{\mu\nu}, \tag{3}$$

$$Q^{33}_{lB} = \left(\bar{l}_\tau \sigma^{\mu\nu} \tau_R\right) \varphi\, B_{\mu\nu}, \tag{4}$$

where φ and $l_\tau = (\nu_\tau, \tau_L)$ are the Higgs, and the left-handed SU(2) doublets, σ^I the Pauli matrices, $W^I_{\mu\nu}$ and $B_{\mu\nu}$ are the gauge field strength tensors. The leading non-standard effective Lagrangian relevant for our study is therefore given by

$$\mathcal{L}_{\text{eff}} = \frac{1}{\Lambda^2}\left[C^{33}_{lW} Q^{33}_{lW} + C^{33}_{lB} Q^{33}_{lB} + \text{h.c.}\right]. \tag{5}$$

After the electroweak symmetry breaking, the two operators mix and give additional, beyond the SM, contributions to the τ anomalous magnetic moment and EDM:

$$\tilde{a}_\tau = \frac{2m_\tau}{e}\frac{\sqrt{2}v}{\Lambda^2}\,\text{Re}\left[\cos\theta_{\text{W}} C^{33}_{lB} - \sin\theta_{\text{W}} C^{33}_{lW}\right], \tag{6}$$

$$\tilde{d}_\tau = \frac{\sqrt{2}v}{\Lambda^2}\,\text{Im}\left[\cos\theta_{\text{W}} C^{33}_{lB} - \sin\theta_{\text{W}} C^{33}_{lW}\right], \tag{7}$$

where $v = 246$ GeV. Deviations of the τ dipole moments from the SM values could be then determined, possibly down to the level of $O(10^{-3})$, via precise data on τ pair production and τ decays.

3 Status of the τ lepton $g-2$ and EDM

The SM prediction for a_τ is given by the sum of QED, hadronic and electroweak (EW) terms. The QED contribution has been computed up to three loops: $a_\tau^{\text{QED}} = 117\,324\,(2) \times 10^{-8}$ [12], where the uncertainty $\pi^2 \ln^2(m_\tau/m_e)(\alpha/\pi)^4 \sim 2 \times 10^{-8}$ has been assigned for uncalculated four-loop contributions. The errors due to the uncertainties of the $O(\alpha^2)$ and $O(\alpha^3)$ terms, as well as that induced by the uncertainty of α, are negligible. The sum of the one- and two-loop EW contributions is $a_\tau^{\text{EW}} = 47.4\,(5) \times 10^{-8}$ [13–16]. The uncertainty encompasses the estimated errors induced by hadronic loop effects, neglected two-loop bosonic terms and the missing three-loop contribution. It also includes the tiny errors due to the uncertainties in M_{top} and m_τ.

Similarly to the case of the muon $g-2$, the leading-order hadronic contribution to a_τ is obtained via a dispersion integral of the total hadronic cross section of the e^+e^- annihilation (the role of low energies is very important, although not as much as for a_μ). The result of the latest evaluation, using the whole bulk of experimental data below 12 GeV, is $a_\tau^{\text{HLO}} = 337.5\,(3.7) \times 10^{-8}$ [16]. The hadronic higher-order (α^3) contribution a_τ^{HHO} can be divided into two parts: $a_\tau^{\text{HHO}} = a_\tau^{\text{HHO}}(\text{vp}) + a_\tau^{\text{HHO}}(\text{lbl})$. The first one, the $O(\alpha^3)$ contribution of diagrams containing hadronic self-energy insertions in the photon propagators, is $a_\tau^{\text{HHO}}(\text{vp}) = 7.6\,(2) \times 10^{-8}$ [17]. Note that naïvely rescaling the corresponding muon $g-2$ result by a factor m_τ^2/m_μ^2 leads to the incorrect estimate $a_\tau^{\text{HHO}}(\text{vp}) \sim -28 \times 10^{-8}$ (even the sign is wrong!). Estimates of the light-by-light contribution $a_\tau^{\text{HHO}}(\text{lbl})$ obtained rescaling the corresponding one for the muon $g-2$ by a factor m_τ^2/m_μ^2 fall short of what is needed – this scaling is not justified. The parton-level estimate of [16] is $a_\tau^{\text{HHO}}(\text{lbl}) = 5(3) \times 10^{-8}$, a value much lower than those obtained by naïve rescaling. Adding up the above contributions one obtains the SM prediction [16]

$$a_\tau^{\text{SM}} = a_\tau^{\text{QED}} + a_\tau^{\text{EW}} + a_\tau^{\text{HLO}} + a_\tau^{\text{HHO}} = 117\,721\,(5) \times 10^{-8}. \tag{8}$$

Errors were added in quadrature.

The EDM interaction violates the discrete CP symmetry. In the SM, with massless neutrinos, the only source of CP violation is the CKM-phase (and a possible θ-term in QCD sector). It can be shown [18, 19] that all CP-violating amplitudes are proportional to the Jarlskog invariant J defined via the convention-invariant equation [19]

$$\text{Im}\left[V_{ij} V_{kl} V^*_{il} V^*_{kj}\right] = J \sum_{m,n} \varepsilon_{ikm} \varepsilon_{jln}\,. \tag{9}$$

Therefore, a fundamental lepton EDM must arise from virtual quarks linked to the lepton through the W boson and also be sensitive to the imaginary part of the V_{CKM} matrix elements. The leading contribution is naïvely expected at the three-loop level, since two-loop diagram is proportional to $|V_{ij}|^2$. The problem was first analyzed in some detail in [20], but it was subsequently shown that the three-loop diagrams also yield a zero EDM contribution in the absence of gluonic corrections to the quark lines [21]. For this reason, lepton EDMs are predicted to be extremely small in the SM, of the $O(10^{-38} - 10^{-35})\, e\cdot$cm [22], which is far below the current experimental capabilities. Indeed, present experiments can only probe $d_\tau \sim O(10^{-17})\, e\cdot$cm. Also for the electron, $d_e^{\text{exp}} < 0.87 \times 10^{-28} e\cdot$cm [23] while $d_e^{\text{SM}} \sim O(10^{-38})\, e\cdot$cm – it is hard to imagine improvements in the sensitivity by ten orders of magnitude! However, new EDM effects could arise at one or two loop from new physics that violates P and T, and be much larger than the tiny SM prediction even if they come from high mass scales. They generally induce large contributions to lepton and neutron EDMs [24], and although there has been no experimental evidence for an EDM so far, there is considerable hope to gain new insights into the nature of CP violation through this kind of experiments.

4 Experimental determination

The present resolution on the τ anomalous magnetic moment is only of $O(10^{-2})$ [25], more than an order of magnitude larger than its SM prediction in Eq. (8). In fact, while the SM value of a_τ is known with a tiny uncertainty of 5×10^{-8}, the τ short lifetime has so far prevented the determination of a_τ by measuring the τ spin precession in a magnetic field, like in the electron and muon $g-2$ experiments. The present PDG limit on the τ $g-2$ was derived in 2004 by the DELPHI collaboration from $e^+e^- \to e^+e^-\tau^+\tau^-$ total cross section measurements at $\sqrt{s}$ between 183 and 208 GeV at LEP2 (the study of a_τ via this channel was proposed in [26]). The measured values of the cross-sections were used to extract limits on the τ $g-2$ by comparing them to the SM values, assuming that possible deviations were due to non-SM contributions to a_τ. The obtained limit at 95% CL is [25]

$$-0.052 < \tilde{a}_\tau < 0.013, \tag{10}$$

which can be also expressed in the form of central value and error as [25]

$$\tilde{a}_\tau = -0.018\,(17). \tag{11}$$

The present PDG limit on the EDM of the τ lepton at 95% CL is

$$\begin{aligned} -2.2 < \mathrm{Re}(\tilde{d}_\tau) < 4.5 \;\; (10^{-17}\; e\cdot \mathrm{cm}), \\ -2.5 < \mathrm{Im}(\tilde{d}_\tau) < 0.8 \;\; (10^{-17}\; e\cdot \mathrm{cm}); \end{aligned} \tag{12}$$

it was obtained by the Belle collaboration [27] following the analysis of ref. [28] for the impact of an effective operator for the τ EDM in the process $e^+e^- \to \tau^+\tau^-$.

The reanalysis of ref. [29] of various LEP and SLD measurements – mainly of the $e^+e^- \to \tau^+\tau^-$ cross sections – allowed the authors to set the indirect 2σ confidence interval

$$-0.007 < \tilde{a}_\tau < 0.005, \tag{13}$$

a bound stronger than that in Eq. (10). This analysis assumed $\tilde{d}_\tau = 0$.

At the LHC, bounds on the τ dipole moments are expected to be set in τ pair production via Drell-Yan [30, 31] or double photon scattering processes [32]. The best limits achievable in $pp \to \tau^+\tau^- + X$ are estimated to be comparable with present existing ones if one assumes that the total cross section for τ pair production will be measured at the 14% level. Earlier proposals can be found in [33, 34].

Another proposed method to determine the $\tilde{a}_\tau$ would use the channeling of polarized τ leptons in a bent crystal similarly to the suggestion for the measurement of magnetic moments of short-living baryons [35]. This method has been successfully tested by the E761 collaboration at Fermilab, that measured the magnetic moment of the Σ^+ hyperon [36]. However the challenge with this method is to produce a polarized beam of τ leptons; the decay $B^+ \to \tau^+\nu_\tau$ could provide such polarized τ [37] but it has a very tiny branching ratio of $O(10^{-4})$.

The Belle II experiment at the upcoming high-luminosity B factory SuperKEKB [38] will offer new opportunities to improve the determination of the τ electromagnetic properties. The authors of ref. [39] proposed to determine the Pauli form factor $F_{2V}(q^2)$ of the τ via $\tau^+\tau^-$ production in e^+e^- collisions at the Υ resonances ($\Upsilon(1S)$, $\Upsilon(2S)$ and $\Upsilon(3S)$) with a sensitivity of $O(10^{-5})$ or even better. The center-of-mass energy at super B factories is $\sqrt{s} \sim M_{\Upsilon(4S)} \approx 10$ GeV, so that the form factor $F_{2V}(q^2)$ is no longer the anomalous magnetic moment. Furthermore, the contributions to the $e^+e^- \to \tau^+\tau^-$ cross section arise not only from the usual s-channel one-loop vertex corrections, but also from box diagrams, which should be somehow subtracted out. The strategy proposed in [39] to eliminate the contamination from the box diagrams has been to measure the observables on top of the Υ resonances: in this kinematic regime the (non-resonant) box diagrams are numerically negligible and only one-loop corrections to the $\gamma\tau\tau$ vertex are relevant.

However, it is very difficult to resolve the narrow peaks of the $\Upsilon(1S, 2S, 3S)$ in the $\tau^+\tau^-$ decay channel – the $\Upsilon(4S)$ decays almost entirely in $B\bar{B}$ – because of the natural irreducible beam energy spread associated to any e^+e^- synchrotron. If we compare the total width for the Υ resonances ($\Gamma^{\Upsilon}_{\rm tot} \sim 20 - 50$ keV) with the SuperKEKB beam energy spread $\sigma_W = 5.45$ MeV [40], we note that at the Belle II the $\tau^+\tau^-$ events produced with beams at a centre of mass energy $\sqrt{s} \sim M_\Upsilon$ are mostly due to non-resonant interaction. The situation at Belle was similar (the energy spread at KEKB was $\sigma_W = 5.24$ MeV [41]). Eventually, the measurement of the $e^+e^- \to \tau^+\tau^-$ cross section on top of the Υ resonances will not eliminate the contamination of the box diagrams.

5 Radiative τ leptonic decays: theoretical framework

In the next two sections we propose a new method to determine the electromagnetic dipole moments of the τ via measurements of its radiative leptonic decays. The SM prediction, at NLO, for the differential rate of the radiative leptonic decays

$$\tau^\pm \to l^\pm \nu\bar{\nu}\gamma, \tag{14}$$

with $l = e$ or μ, of a polarized $\tau^\pm$ with mass m_τ in its rest frame is

$$\frac{d^6\Gamma^\pm(y_0)}{dx\,dy\,d\Omega_l\,d\Omega_\gamma} = \frac{\alpha\, G_F^2 m_\tau^5}{(4\pi)^6}\, \frac{x\beta_l}{1+\delta_W(m_\mu, m_e)} \times \\ \times \left[G \mp x\beta_l\,\hat{n}\cdot\hat{p}_l\,J \mp y\,\hat{n}\cdot\hat{p}_\gamma\,K + xy\beta_l\,\hat{n}\cdot\left(\hat{p}_l\times\hat{p}_\gamma\right)L\right], \tag{15}$$

where $G_F = 1.166\,378\,7(6) \times 10^{-5}$ GeV^{-2} [42] is the Fermi constant, defined from the muon lifetime, and $\alpha = 1/137.035\,999\,157\,(33)$ is the fine-structure constant [43, 44]. Calling m the mass of the final charged lepton (neutrinos and antineutrinos are considered massless) we define $r = m/m_\tau$ and $r_W = m_\tau/M_W$, where M_W is the W-boson mass; p and $n = (0, \hat{n})$ are the four-momentum and polarization vector of the initial τ, with $n^2 = -1$ and $n\cdot p = 0$. Also, $x = 2E_l/m_\tau$, $y = 2E_\gamma/m_\tau$ and $\beta_l \equiv |\vec{p}_l|/E_l = \sqrt{1-4r^2/x^2}$, where $p_l = (E_l, \vec{p}_l)$ and $p_\gamma = (E_\gamma, \vec{p}_\gamma)$ are the four-momenta of the final charged lepton and photon, respectively. The final charged lepton and photon are emitted at solid angles Ω_l and Ω_γ, with normalized three-momenta $\hat{p}_l$ and $\hat{p}_\gamma$, and c is the cosine of the angle between $\hat{p}_l$ and $\hat{p}_\gamma$. The term $\delta_{\rm W}(m_\mu, m_e) = 1.04 \times 10^{-6}$ is the tree-level correction to muon decay induced by the W-boson propagator [45, 46]. Equation (15) includes the possible emission of an additional soft photon with normalized energy y' lower than the detection threshold $y_0 \ll 1$: $y' < y_0 < y$. The function $G(x, y, c; y_0)$ and, analogously, J and K, are given by

$$G(x, y, c, y_0) = = \frac{4}{3yz^2}\left[g_0(x, y, z) + r_{\rm W}^2\, g_{\rm W}(x, y, z) + \frac{\alpha}{\pi}\, g_{\rm NLO}(x, y, z, y_0)\right], \tag{16}$$

where $z = xy(1 - c\beta_l)/2$; the LO function $g_0(x, y, z)$, computed in [47–50], arises from the pure Fermi V–A interaction, whereas $g_{\rm W}(x, y, z)$ is the leading contributions of the W-boson propagator derived in [46]. The NLO term $g_{\rm NLO}(x, y, z, y_0)$ is the sum of the virtual and soft bremsstrahlung contributions calculated in [51] (see also refs. [52, 53]). The function $L(x, y, z)$, appearing in front of the product $\hat{n} \cdot (\hat{p}_l \times \hat{p}_\gamma)$, does not depend on y_0; it is only induced by the loop corrections and is therefore of $O(\alpha/\pi)$. The (lengthy) explicit expressions of G, J, K and L are provided in [51]. If the initial $\tau^\pm$ are not polarized, eq. (15) simplifies to

$$\frac{d^3\Gamma(y_0)}{dx\,dc\,dy} = \frac{\alpha G_F^2 m_\tau^5}{(4\pi)^6}\,\frac{8\pi^2\, x\beta_l}{1 + \delta_{\rm W}(m_\mu, m_e)}\, G(x, y, c, y_0). \tag{17}$$

At the LO, the analytic integration over the allowed kinematic ranges leads, for a minimum photon energy $y_0 = 2\omega_0/m_\tau$, to [49, 54]

$$\Gamma_{\rm LO}(y_0) = \frac{G_F^2 m_\tau^5}{192\pi^3}\,\frac{\alpha}{3\pi}\, H(y_0), \tag{18}$$

$$\begin{aligned} H(y_0) =\; & 3\,{\rm Li}_2(y_0) - \frac{\pi^2}{2} - \frac{1}{2}\left(6 + \bar{y}_0^3\right)\bar{y}_0 \ln \bar{y}_0 + \\ & + \left(\ln r + \frac{17}{12}\right)\left(6 \ln y_0 + 6\bar{y}_0 + \bar{y}_0^4\right) + \\ & + \frac{1}{48}\left(125 + 45y_0 - 33y_0^2 + 7y_0^3\right)\bar{y}_0, \end{aligned} \tag{19}$$

where $\bar{y}_0 = 1 - y_0$ and the dilogarithm is defined by ${\rm Li}_2(z) = -\int_0^z dt\, \frac{\ln(1-t)}{t}$. Terms depending on the mass ratio r have been neglected in the expression for $H(y_0)$, with the obvious exception of the logarithmic contribution which diverges in the limit $r \to 0$. However, terms in the integrand $G_{\rm LO}(x, y, c)$ proportional to r^2 were not neglected when performing the integral to obtain (19), as they lead to terms of $O(1)$ in the integrated result $H(y_0)$. This feature, first noted in [55], is due to the appearance of right-handed electrons and muons in the final states of (14) even in the limit $r \to 0$, and is a consequence of helicity-flip bremsstrahlung in QED [55–58].

At the NLO, which allows for double photon emission, the branching ratios of the radiative decays (14) can be distinguished in two types due to the double real emission:

- The "inclusive" branching ratio, $\mathcal{B}^{\rm Inc}(y_0)$, where in the final state there is at least one photon with energy higher than y_0.
- The "exclusive" branching ratio, $\mathcal{B}^{\rm Exc}(y_0)$, where in the final state there is one, and only one, photon of energy larger than the detection threshold y_0.

It is clear that at the LO the theoretical predictions for these exclusive and inclusive branching ratios coincide – double bremsstrahlung events are simply not considered.

Exclusive and inclusive branching ratios for the radiative decays (14) were computed, for a threshold $\omega_0 = y_0(m_\tau/2) = 10$ MeV, in ref. [51] and are reported in table 1. Uncertainties were estimated for uncomputed NNLO corrections, numerical errors, and the experimental errors of the lifetimes. The former were estimated to be $\delta\mathcal{B}_{\rm NLO}^{\rm Exc/Inc} \sim (\alpha/\pi) \ln r \ln(\omega_0/m_\tau)\, \mathcal{B}_{\rm NLO}^{\rm Exc/Inc}$. For $\omega_0 = 10$ MeV they are about 10% and 3% for $\tau \to e\bar{\nu}\nu\gamma$ and $\tau \to \mu\bar{\nu}\nu\gamma$, respectively. They appear with the subscript "N" in table 1. Numerical errors, labeled by the subscript "n", are smaller than those induced by missing radiative corrections. These two kinds of uncertainties were combined to provide the theoretical error of the final $\mathcal{B}^{\rm Exc}$ and $\mathcal{B}^{\rm Inc}$ predictions, labeled by the subscript "th". The uncertainty due to the experimental error of the lifetimes is labeled by the subscript "τ".

The recent measurements by the BABAR collaboration of the branching ratios of the radiative decays $\tau \to l\bar{\nu}\nu\gamma$, with $l = e$ and μ, for a minimum photon energy $\omega_0 = 10$ MeV in the τ rest frame, are [9]:

$$\mathcal{B}_{\rm EXP}(\tau \to e\bar{\nu}\nu\gamma) = 1.847\,(15)_{\rm st}(52)_{\rm sy} \times 10^{-2}, \tag{20}$$

$$\mathcal{B}_{\rm EXP}(\tau \to \mu\bar{\nu}\nu\gamma) = 3.69\,(3)_{\rm st}(10)_{\rm sy} \times 10^{-3}, \tag{21}$$

where the first error is statistical and the second is systematic. These results are substantially more precise than the previous measurements of the CLEO collaboration [59]. The experimental values in eqs. (20,21) must be compared with our predictions for the exclusive branching ratios in table 1. For $\tau \to \mu\bar{\nu}\nu\gamma$ decays, the branching ratio measurement and prediction agree within 1.1 standard deviations (1.1σ). On the contrary, the experimental and theoretical values for $\tau \to e\bar{\nu}\nu\gamma$ decays differ by $2.02\,(57) \times 10^{-3}$, i.e. by 3.5$\sigma$. This puzzling discrepancy deserves further researches.

6 τ dipole moments via $\tau \to l\gamma\nu\bar{\nu}$ decays

The effective Lagrangian (5) generates additional non-standard contributions to the differential decay rate in eq. (15). For a $\tau^\pm$ they can be summarised in the shifts

$$G \to G + \tilde{a}_\tau G_a, \qquad J \to J + \tilde{a}_\tau J_a, \tag{22}$$

$$K \to K + \tilde{a}_\tau K_a, \qquad L \to L \mp (m_\tau/e)\, \tilde{d}_\tau L_d. \tag{23}$$

Tiny terms of $O(\tilde{a}_\tau{}^2)$, $O(\tilde{d}_\tau{}^2)$ and $O(\tilde{a}_\tau\tilde{d}_\tau)$ were neglected. Deviations of the τ dipole moments from the SM values can be determined, possibly down to the level of $O(10^{-3})$ comparing the SM prediction for the differential rate in

Table 1. Branching ratios of radiative τ leptonic decays for a minimum photon energy $\omega_0 = 10$ MeV. Inclusive ($\mathcal{B}^{\rm Inc}$) and exclusive ($\mathcal{B}^{\rm Exc}$) predictions are separated into LO contributions ($\mathcal{B}_{\rm LO}$) and NLO corrections ($\mathcal{B}^{\rm Inc/Exc}_{\rm NLO}$). Uncertainties were estimated for uncomputed NNLO corrections (N), numerical errors (n), and the experimental errors of the lifetimes (τ). The first two types of errors were combined to provide the final theoretical uncertainty (th). The last line reports the experimental measurement of ref. [9].

	$\tau \to e\bar{\nu}\nu\gamma$	$\tau \to \mu\bar{\nu}\nu\gamma$
$\mathcal{B}_{\rm LO}$	1.834×10^{-2}	3.663×10^{-3}
$\mathcal{B}^{\rm Inc}_{\rm NLO}$	$-1.06\,(1)_n(10)_N \times 10^{-3}$	$-5.8\,(1)_n(2)_N \times 10^{-5}$
$\mathcal{B}^{\rm Exc}_{\rm NLO}$	$-1.89\,(1)_n(19)_N \times 10^{-3}$	$-9.1\,(1)_n(3)_N \times 10^{-5}$
$\mathcal{B}^{\rm Inc}$	$1.728\,(10)_{\rm th}(3)_\tau \times 10^{-2}$	$3.605\,(2)_{\rm th}(6)_\tau \times 10^{-3}$
$\mathcal{B}^{\rm Exc}$	$1.645\,(19)_{\rm th}(3)_\tau \times 10^{-2}$	$3.572\,(3)_{\rm th}(6)_\tau \times 10^{-3}$
$\mathcal{B}_{\rm EXP}$	$1.847\,(15)_{\rm st}(52)_{\rm sy} \times 10^{-2}$	$3.69\,(3)_{\rm st}(10)_{\rm sy} \times 10^{-3}$

eq. (15), modified by the terms G_a, J_a, K_a and L_d, with sufficiently precise data.

The possibility to set bounds on $\tilde{a}_\tau$ via the radiative leptonic τ decays in (14) was suggested long ago in Ref. [60]. In that article the authors proposed to take advantage of a radiation zero of the LO differential decay rate in (15) which occurs when, in the τ rest frame, the final lepton l and the photon are back-to-back, and l has maximal energy. Since a non-standard contribution to a_τ spoils this radiation zero, precise measurements of this phase-space region could be used to set bounds on its value. However, studies with Belle data show no significant improvement of the existing limits: the $\tilde{a}_\tau$ upper limit (UL) that can be achieved with the whole Belle statistics, about 0.9×10^9 τ pairs, is only UL($\tilde{a}_\tau$) $\simeq 2$ [61].

A more powerful method to extract $\tilde{a}_\tau$ and $\tilde{d}_\tau$ consists in the use of an unbinned maximum likelihood fit of events in the full phase space. In this approach we considered $e^+e^- \to \tau^+\tau^-$ events where both τ leptons decay subsequently into a particular final state: $\tau^\mp$ (signal side) decays to the radiative leptonic mode, the other $\tau^\pm$ (tag side) decays to some well known mode with a large branching fraction. As a tag decay mode we chose $\tau^\pm \to \rho^\pm\nu \to \pi^\pm\pi^0\nu$, which also serves as spin analyser and allows us to be sensitive to the spin dependent part of the differential decay width of the signal decay using effects of spin-spin correlation of the τ leptons [62]. With this technique we analyzed a data sample of $(\ell^\mp\nu\nu\gamma,\ \pi^\pm\pi^0\nu)$ events corresponding to the total amount of data available at Belle and the one planned at the Belle II experiment.

The feasibility study shows that no improvement is expected from Belle data. However the experimental sensitivity on $\tilde{a}_\tau$ at the Belle II experiment, $\sigma_{\tilde{a}} = 0.012$ [61], can already be competitive with Delphi results in (11). On the other hand, the expected sensitivity on the τ EDM, $\sigma_{\tilde{d}} = 6.1 \times 10^{-17} e\cdot$cm [61], is still worse than the most precise measurement of $\tilde{d}_\tau$ done at Belle in τ pair production [27].

7 Conclusions

The magnetic and electric dipole moments of the τ lepton are largely unknown. We proposed to use radiative τ leptonic decays to measure these electromagnetic properties at B factories. Deviations of the τ dipole moments from the SM predictions can be determined via an effective Lagrangian approach.

We studied at the NLO in the SM the differential decay rates and branching ratios of $\tau \to l\gamma\nu\bar{\nu}\,(l = \mu, e)$ decays. Our prediction for $l = \mu$ agrees within 1.1σ with the recent BaBar measurements of $\mathcal{B}(\tau \to \mu\gamma\nu\bar{\nu})$, for a minimum photon energy threshold $\omega_0 = 10$ MeV. On the contrary the measurement of $\mathcal{B}(\tau \to e\gamma\nu\bar{\nu})$ differs from our prediction by 3.5σ. This puzzling discrepancy deserves further researches.

Our feasibility study showed that the measurement of the τ anomalous magnetic moment at Belle II can be already competitive with the current bound from Delphi experiment, while the expected sensitivity to the tau EDM is still worse than the most precise measurement done at Belle.

Acknowledgements

I would like to thank the organizers for the invitation and the opportunity to present these results. I thank S. Eidelman, D. Epifanov and M. Passera for our fruitful and ongoing collaboration. I would like also to thank L. Mercolli for participating in the early stages of this work. This work is supported by the Swiss National Science Foundation.

References

[1] D. Hanneke, S. Fogwell, G. Gabrielse, Phys.Rev.Lett. **100**, 120801 (2008), 0801.1134
[2] G. Bennett et al. (Muon G-2 Collaboration), Phys.Rev. **D73**, 072003 (2006), hep-ex/0602035
[3] F. Jegerlehner (2015), 1511.04473
[4] J. Schwinger, Phys. Rev. **73**, 416 (1948)
[5] L. Michel, Proc. Phys. Soc. **A63**, 514 (1950)
[6] C. Bouchiat, L. Michel, Phys. Rev. **106**, 170 (1957)
[7] T. Kinoshita, A. Sirlin, Phys.Rev. **107**, 593 (1957)
[8] T. Kinoshita, A. Sirlin, Phys. Rev. **108**, 844 (1957)
[9] J.P. Lees et al. (BaBar), Phys. Rev. **D91**, 051103 (2015), 1502.01784
[10] W. Buchmuller, D. Wyler, Nucl.Phys. **B268**, 621 (1986)
[11] B. Grzadkowski, M. Iskrzynski, M. Misiak, J. Rosiek, JHEP **10**, 085 (2010), 1008.4884
[12] M. Passera, Phys.Rev. **D75**, 013002 (2007), hep-ph/0606174
[13] T.V. Kukhto, E.A. Kuraev, Z.K. Silagadze, A. Schiller, Nucl. Phys. **B371**, 567 (1992)
[14] A. Czarnecki, B. Krause, W.J. Marciano, Phys.Rev. **D52**, 2619 (1995), hep-ph/9506256
[15] A. Czarnecki, B. Krause, W.J. Marciano, Phys.Rev.Lett. **76**, 3267 (1996), hep-ph/9512369

[16] S. Eidelman, M. Passera, Mod.Phys.Lett. **A22**, 159 (2007), hep-ph/0701260
[17] B. Krause, Phys.Lett. **B390**, 392 (1997), hep-ph/9607259
[18] C. Jarlskog, Z.Phys. **C29**, 491 (1985)
[19] C. Jarlskog, Phys.Rev.Lett. **55**, 1039 (1985)
[20] F. Hoogeveen, Nucl.Phys. **B341**, 322 (1990)
[21] M. Pospelov, I. Khriplovich, Sov.J.Nucl.Phys. **53**, 638 (1991)
[22] E. Commins, Adv.At.Mol.Opt.Phys. **40**, 1 (1999)
[23] J. Baron et al. (ACME), Science **343**, 269 (2014), 1310.7534
[24] T. Fukuyama, Int.J.Mod.Phys. **A27**, 1230015 (2012), 1201.4252
[25] J. Abdallah et al. (DELPHI Collaboration), Eur.Phys.J. **C35**, 159 (2004), hep-ex/0406010
[26] F. Cornet, J.I. Illana, Phys.Rev. **D53**, 1181 (1996), hep-ph/9503466
[27] K. Inami et al. (Belle Collaboration), Phys.Lett. **B551**, 16 (2003), hep-ex/0210066
[28] W. Bernreuther, O. Nachtmann, P. Overmann, Phys.Rev. **D48**, 78 (1993)
[29] G.A. Gonzalez-Sprinberg, A. Santamaria, J. Vidal, Nucl.Phys. **B582**, 3 (2000), hep-ph/0002203
[30] A. Hayreter, G. Valencia, Phys. Rev. **D88**, 013015 (2013), [Erratum: Phys. Rev.D91,no.9,099902(2015)], 1305.6833
[31] A. Hayreter, G. Valencia, JHEP **07**, 174 (2015), 1505.02176
[32] S. Atag, A.A. Billur, JHEP **11**, 060 (2010), 1005.2841
[33] M.A. Samuel, G. Li, Int.J.Theor.Phys. **33**, 1471 (1994)
[34] F. del Aguila, F. Cornet, J.I. Illana, Phys.Lett. **B271**, 256 (1991)
[35] I.J. Kim, Nucl.Phys. **B229**, 251 (1983)
[36] D. Chen et al. (E761 Collaboration), Phys.Rev.Lett. **69**, 3286 (1992)
[37] M.A. Samuel, G.w. Li, R. Mendel, Phys.Rev.Lett. **67**, 668 (1991)
[38] T. Aushev, W. Bartel, A. Bondar, J. Brodzicka, T. Browder et al. (2010), 1002.5012
[39] J. Bernabeu, G. Gonzalez-Sprinberg, J. Papavassiliou, J. Vidal, Nucl.Phys. **B790**, 160 (2008), 0707.2496
[40] Y. Ohnishi, T. Abe, T. Adachi, K. Akai, Y. Arimoto et al., PTEP **2013**, 03A011 (2013)
[41] H. Koiso, A. Morita, Y. Ohnishi, K. Oide, K. Satoh, Prog. Theor. Exp. Phys. **2013**, 03A009 (2013)
[42] D. Webber et al. (MuLan Collaboration), Phys.Rev.Lett. **106**, 041803 (2011), 1010.0991
[43] T. Aoyama, M. Hayakawa, T. Kinoshita, M. Nio, Phys.Rev.Lett. **109**, 111807 (2012), 1205.5368
[44] T. Aoyama, M. Hayakawa, T. Kinoshita, M. Nio, Phys. Rev. **D91**, 033006 (2015), 1412.8284
[45] A. Ferroglia, C. Greub, A. Sirlin, Z. Zhang, Phys.Rev. **D88**, 033012 (2013), 1307.6900
[46] M. Fael, L. Mercolli, M. Passera, Phys.Rev. **D88**, 093011 (2013), 1310.1081
[47] T. Kinoshita, A. Sirlin, Phys.Rev. **113**, 1652 (1959)
[48] C. Fronsdal, H. Uberall, Phys.Rev. **113**, 654 (1959)
[49] S. Eckstein, R. Pratt, Ann.Phys. **8**, 297 (1959)
[50] Y. Kuno, Y. Okada, Rev.Mod.Phys. **73**, 151 (2001), hep-ph/9909265
[51] M. Fael, L. Mercolli, M. Passera, JHEP **07**, 153 (2015), 1506.03416
[52] A. Fischer, T. Kurosu, F. Savatier, Phys.Rev. **D49**, 3426 (1994)
[53] A. Arbuzov, E. Scherbakova, Phys.Lett. **B597**, 285 (2004), hep-ph/0404094
[54] T. Kinoshita, A. Sirlin, Phys. Rev. Lett. **2**, 177 (1959)
[55] T. Lee, M. Nauenberg, Phys.Rev. **133**, B1549 (1964)
[56] B. Falk, L.M. Sehgal, Phys. Lett. **B325**, 509 (1994)
[57] L.M. Sehgal, Phys. Lett. **B569**, 25 (2003), hep-ph/0306166
[58] V.S. Schulz, L.M. Sehgal, Phys. Lett. **B594**, 153 (2004), hep-ph/0404023
[59] T. Bergfeld et al. (CLEO Collaboration), Phys.Rev.Lett. **84**, 830 (2000), hep-ex/9909050
[60] M. Laursen, M.A. Samuel, A. Sen, Phys.Rev. **D29**, 2652 (1984)
[61] S. Eidelman, D. Epifanov, M. Fael, L. Mercolli, M. Passera (2016), 1601.07987
[62] Y.S. Tsai, Phys.Rev. **D4**, 2821 (1971)

Hadronic contributions to the muon anomaly in the Constituent Chiral Quark Model

David Greynat[1,a]

[1] *Dip. di Scienze Fisiche, Universitá di Napoli "Federico II", Via Cintia, 80126 Napoli, Italia*
INFN-Sezione di Napoli, Via Cintia, 80126 Napoli, Italia

Abstract. The hadronic contributions to the anomalous magnetic moment of the muon which are relevant for the confrontation between theory and experiment at the present level of accuracy, are evaluated within the same framework: the constituent chiral quark model. This includes the contributions from the dominant hadronic vacuum polarization as well as from the next–to–leading order hadronic vacuum polarization, the contributions from the hadronic light-by-light scattering, and the contributions from the electroweak hadronic $Z\gamma\gamma$ vertex. They are all evaluated as a function of only one free parameter: the constituent quark mass. We also comment on the comparison between our results and other phenomenological evaluations.

1 Introduction

This proceeding is a summarized version of our article [1]. Our purpose is to establish a simple *reference model* to evaluate the various hadronic contributions to the anomalous magnetic moment of the muon a_μ within the same framework, and use it as a yardstick to compare with the more detailed evaluations in the literature. Indeed, the experimental world average of the anomalous magnetic moment of the muon a_μ, assuming CPT–invariance, viz. $a_{\mu^+} = a_{\mu^-}$, is (results reported in ref. [2] and references therein)

$$a_\mu^{(\mathrm{exp})} = 116\,592\,080\,(63) \times 10^{-11} \quad (0.54\,\mathrm{ppm})\,, \qquad (1)$$

The prediction of the Standard Model, as a result of contributions from many physicists is (see e.g. refs. [3, 4] and references therein)

$$a_\mu^{(\mathrm{SM})} = 116\,591\,801\,(49) \times 10^{-11}\,, \qquad (2)$$

where the error here is dominated at present by the lowest order hadronic vacuum polarization contribution uncertainty [5] ($\pm 42.0 \times 10^{-11}$), as well as by the contribution from the hadronic light–by–light scattering, which is theoretically estimated to be $(105 \pm 26) \times 10^{-11}$ [6]. The results quoted in (1) and (2) imply a significant 3.6 standard deviation between theory and experiment which deserves attention. It is therefore important to reexamine critically the various theoretical contributions to Eq. (2), in particular the hadronic contributions. Ideally, one would like to do that within the framework of Quantum Chromodynamics(QCD). Unfortunatley, this demands mastering QCD at all scales from short to long distances, something which is not under full analytic control at present. Therefore, one has to resort to experimental information whenever possible, to QCD inspired hadronic models. As a result, all the theoretical evaluations of the hadronic contributions to a_μ have systematic errors which are not easy to pin down rigorously.

The *reference model* which we propose is based on the Constituent Chiral Quark Model (CχQM) [9]. This model emerged as an attempt to reconcile the successes of phenomenological quark models, like the De Rújula-Georgi-Glashow model [10], with QCD. The corresponding Lagrangian proposed by Manohar and Georgi (MG) is an effective field theory which incorporates the interactions of the low–lying pseudoscalar particles of the hadronic spectrum, the Nambu-Goldstone modes of the spontaneously broken chiral symmetry (SχSB), to lowest order in the chiral expansion [11] and in the presence of chirally rotated quark fields. Because of the SχSB, these quark fields appear to be massive.

This model, in the presence of $SU(3)_L \times SU(3)_R$ external sources has been reconsidered by [12]. As emphasized by Weinberg [13], the corresponding effective Lagrangian is renormalizable in the Large–N_c limit; however, the number of the required counterterms depends crucially on the value of the coupling constant g_A in the model and, as shown in [12], it is minimized for $g_A = 1$. With this choice, and a value for the constituent quark mass fixed phenomenologically, the model reproduces rather well the values of several well known low energy constants.

As discussed in ref. [12] the CχQM model has, however, its own limitations. Applications to the evaluation of low–energy observables involving the integration of Green's functions over a full range of euclidean momenta fail, in general, because there is no matching of the model

[a] e-mail: david.greynat@gmail.com

© The Authors, published by EDP Sciences. This is an open access article distributed under the terms of the Creative Commons Attribution License 4.0 (http://creativecommons.org/licenses/by/4.0/).

to the QCD short–distance behaviour. There is, however, an exceptional class of low–energy observables for which the MG–Lagrangian predictions can be rather reliable [12].

We present in this proceeding a summary of the different evaluations with the CχQM Lagrangian which has the advantage of simplicity and can provide a consistency check with the more elaborated phenomenological approaches.

2 Hadronic Vacuum Polarization

Since the dominant contribution to the muon anomaly from the hadronic vacuum polarization shown in Fig. (1) is a well known representation [14] we use it to determine the only free parameter in the CχQM Lagrangian the constituent quark mass M_Q [1].

This contribution

$$a_\mu^{(\mathrm{HVP})} = \int_{4M_Q^2}^{\infty} \frac{dt}{t} K\left(\frac{t}{m_\mu^2}\right) \frac{1}{\pi} \mathrm{Im}\Pi^{(\mathrm{HVP})}(t), \qquad (3)$$

where [2]

$$K\left(\frac{t}{m_\mu^2}\right) = \left(\frac{\alpha}{\pi}\right) \int_0^1 dx \frac{x^2(1-x)}{x^2 + \frac{t}{m_\mu^2}(1-x)}, \qquad (4)$$

and $\frac{1}{\pi}\mathrm{Im}\Pi(t)$ denotes the electromagnetic hadronic spectral function.

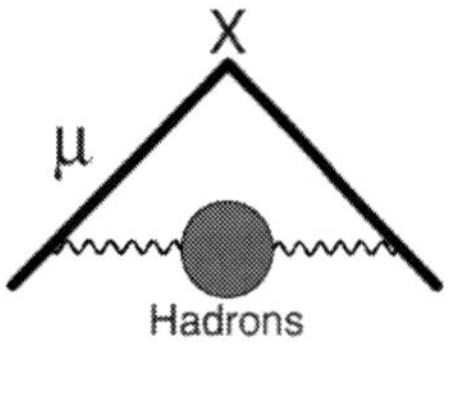

Figure 1. Hadronic Vacuum Polarization contribution to the Muon Anomaly.

The constituent quark fields in the CχQM are assumed to have gluonic interactions, since the Goldstone modes are already in the Lagrangian, the color–SU(3) coupling constant is supposed to be no longer running below a scale $\mu_0 \simeq 2$ GeV where $\alpha_s(\mu_0) \simeq 0.33$ and non-perturbative effects become significant. With inclusion of the leading gluonic corrections in perturbation theory, and to leading order in Large–$\mathrm{N_c}$, the spectral function in Eq. (3) is then

$$\begin{aligned}&\frac{1}{\pi}\mathrm{Im}\Pi^{(\mathrm{HVP})}(t)\\ &= \left(\frac{\alpha}{\pi}\right) N_c \frac{2}{3} \left\{ \delta\left(\frac{1}{2} - \frac{1}{6}\delta^2\right) \right.\\ &\left. + \left[\frac{N_c \alpha_s(\mu_0)}{\pi} \frac{3}{8} \theta(\mu_0^2 - t) + \frac{N_c \alpha_s(\sqrt{t})}{\pi} \frac{3}{8} \theta(t - \mu_0^2) \right] \rho_{\mathrm{KS}}(t) \right\},\end{aligned} \qquad (5)$$

where $\rho_{\mathrm{KS}}(t)$ can be extracted from the early QED calculation of Källen and Sabry [19] (see also ref. [20]):

$$\begin{aligned}&\rho_{\mathrm{KS}}(t)\\ &= \delta\left(\frac{5}{8} - \frac{3}{8}\delta^2 - \left(\frac{1}{2} - \frac{1}{6}\delta^2\right)\log\left[64\frac{\delta^4}{(1-\delta^2)^3}\right]\right)\\ &+ \left(\frac{11}{16} + \frac{11}{24}\delta^2 - \frac{7}{48}\delta^4\right.\\ &\quad + \left(\frac{1}{2} + \frac{1}{3}\delta^2 - \frac{1}{6}\delta^4\right)\log\left[\frac{(1+\delta)^3}{8\delta^2}\right]\right)\log\left[\frac{1+\delta}{1-\delta}\right]\\ &+ 2\left(\frac{1}{2} + \frac{1}{3}\delta^2 - \frac{1}{6}\delta^4\right)\left(2\,\mathrm{Li}_2\left[\frac{1-\delta}{1+\delta}\right] + \mathrm{Li}_2\left[-\frac{1-\delta}{1+\delta}\right]\right).\end{aligned} \qquad (6)$$

Also, at the level of the accuracy expected from the CχQM, it is sufficient to use the one loop expression

$$\frac{\alpha_s(\sqrt{t})}{\pi} \simeq \frac{1}{\left(\frac{11}{6}N_c - \frac{n_f}{3}\right)\log\frac{\sqrt{t}}{\Lambda}}, \qquad (7)$$

with $\Lambda \simeq 250$ MeV and $n_f = 3$. In order to compare the CχQM results for $a_\mu^{(\mathrm{HVP})}$ with the phenomenological determinations which incorporate experimental data, we still have to estimate of the $1/\mathrm{N_c}$–suppressed effects, we then consider the contributions from the $\pi^+\pi^-$ and K^+K^- intermediate states to the spectral function in Eq. (3), as predicted by the CχQM. Notice that in this evaluation, the point like coupling $(-ie)(p^\mu - p'^\mu)$ of scalar QED is replaced by the dressed coupling:

$$(-ie)(p^\mu - p'^\mu) \Rightarrow (-ie)(p^\mu - p'^\mu)\left\{1 + \mathcal{F}(Q^2)\right\}, \qquad (8)$$

with $\mathcal{F}(Q^2)$ the pion (kaon) electromagnetic form factor of the CχQM which, for $g_A = 1$, is given by the expression:

$$\begin{aligned}\mathcal{F}(Q^2) &= \frac{\mathrm{N_c}}{16\pi^2}\frac{Q^2}{f_\pi^2}\left(-\frac{4M_Q^2}{Q^2}\right)\\ &\times\left[1 + \frac{1}{2}\sqrt{1 + \frac{4M_Q^2}{Q^2}}\log\frac{\sqrt{1+\frac{4M_Q^2}{Q^2}} - 1}{\sqrt{1+\frac{4M_Q^2}{Q^2}} + 1}\right].\end{aligned} \qquad (9)$$

The form factor $\mathcal{F}(Q^2)$, however, does not match the QCD behaviour at large–Q^2 values and, therefore, the estimate we propose for the $1/\mathrm{N_c}$–suppresed contributions to the the muon anomaly can only be considered as reasonable

[1] There are many estimates of this contribution, as well as some of the higher order ones, with quark models which can be found in the literature. An earlier reference is [16] and two more recent ones [17] and [18].

[2] The analytic expression of $K\left(t/m_\mu^2\right)$ was first given in ref. [15]; see also ref. [20].

up to values of t in Eq. (3) below $t \sim \mu_0$ where the asymptotic pQCD regime sets in. Contributions beyond $t \sim \mu_0$ have already been taken into account by the second term of the spectral function in Eq. (5).

These considerations provide us with a framework to fix the constituent quark mass M_Q. The prediction of the CχQM, as described above, should be compared to the phenomenological contribution from hadrons formed of u, d and s quarks only, at the level of one–photon exchange. Contributions like for example the one from an intermediate $\pi^0\gamma$ state should therefore be excluded so far (more on that later on), as well as those involving c, b and t quarks. From the numbers quoted in TABLE II of ref. [5], we then find that this restriction reduces the phenomenological determination of the anomaly from hadronic vacuum polarization to a central value

$$a_\mu^{(\mathrm{HVP})}|_{\mathrm{phen.}} \simeq 653 \times 10^{-10} \tag{10}$$

shows that fixing M_Q in the range

$$M_Q = (240 \pm 10)\ \mathrm{MeV}\,, \tag{11}$$

reproduces the phenomenological determination within an error of less than 10%. This determination of the constituent quark mass is the value which we shall systematically use for M_Q when evaluating the predictions for the other hadronic contributions to the muon anomaly. We wish to emphasize, however, that the error of 10 MeV in Eq. (11) only reflects the phenomenological choice that we have made in order to fix M_Q. As discussed in the Introduction the CχQM is only a model of low energy QCD and, as such, there is no a priori way to fix M_Q from first principles. The error in Eq. (11) does not reflect the systematic error due to other plausible ways of fixing M_Q.

3 Hadronic Vacuum Polarization Contributions at Next–to–Leading Order

After having fixed the constituent quark mass, it is then possible to evaluate next-to-leading order contributions to a_μ. For obvious reason, we will present here only the different diagram classification that we used and the results obtained. All details and complete discussion can be found in [1].

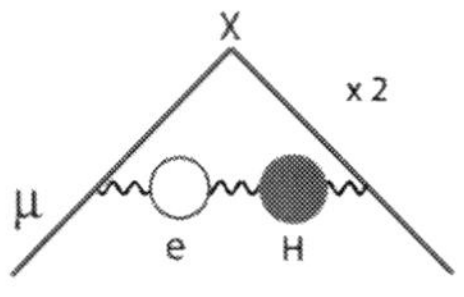

Figure 3. Feynman diagrams corresponding to the Class B contribution to the muon anomaly.

As we can see in table 1, we obtain quite reasonable results except from the contribution in the class D coming from the $\pi^0\gamma$ intermediate state. In order to understand this, we can simplify the phenomenological determination as much as possible to see where the big contribution

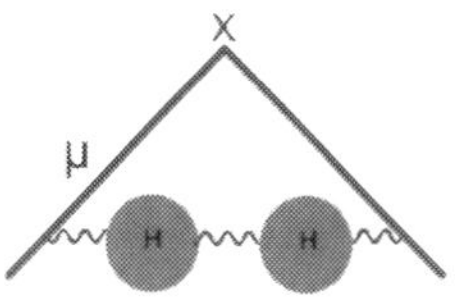

Figure 4. HVP Contributions at $O(\alpha)$ corresponding to the Class C.

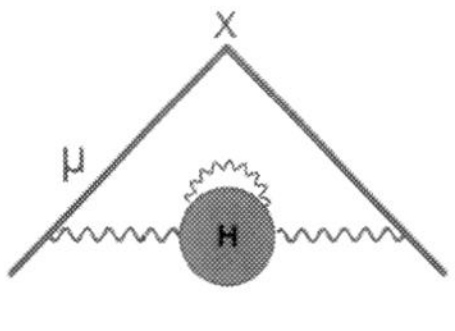

Figure 5. Feynman diagrams corresponding to the Class D contribution to the muon anomaly.

Class	Result $\times 10^{-11}$	Pheno.$\times 10^{-11}$
A	-171 ± 10	$-(207.3 \pm 1.9)$ [24]
B	89 ± 7	106.0 ± 0.9 [24]
C	2.2 ± 0.3	3.4 ± 0.1 [24]
		3.0 ± 0.1 [31]
D($\pi\gamma$)	2.2 ± 0.1	44.2 ± 1.9 [5, 24]
D(Q–loop)	13.5 ± 0.5	
Total	-64 ± 12	$-98.4 \pm 0.6_{\mathrm{exp}} \pm 0.4_{\mathrm{rad}}$ [30]

Table 1. Results for the HVP contributions of $O\left(\frac{\alpha}{\pi}\right)^3$ in the CχQM.

comes from. For that, it will be sufficient to approximate this contribution as follows:

$$a_\mu^{(\pi\gamma)} \sim \left(\frac{\alpha}{\pi}\right)\int_{m_\pi^2}^{\infty}\frac{dt}{t}\frac{1}{3}\frac{m_\mu^2}{t}\frac{1}{\pi}\mathrm{Im}\Pi^{(\pi\gamma)}(t)\,, \tag{12}$$

and use a narrow width expression for the spectral function, which as we shall soon see, is dominated by the ω contribution. This results in the simple formula:

$$a_\mu^{(\pi\gamma)} \sim \frac{1}{3}\frac{m_\mu^2}{M_\omega^2}\frac{4}{\pi}\frac{\Gamma(\omega\to e^+e^-)}{M_\omega}\frac{\Gamma(\omega\to\pi^0\gamma)}{\Gamma_\omega} = 53\times 10^{-11}\,, \tag{13}$$

which reproduces, in order of magnitude, the phenomenological estimates. We can, therefore, see that the *big number* comes from the large experimental value of the branching ratio

$$\frac{\Gamma(\omega\to\pi^0\gamma)}{\Gamma_\omega} \simeq 8\times 10^{-2}\,. \tag{14}$$

Notice that in the case of the ρ contribution the corresponding branching ratio is much smaller:

$$\frac{\Gamma(\rho\to\pi^0\gamma)}{\Gamma_\rho} \simeq 6\times 10^{-4}\,. \tag{15}$$

It is the large branching ratio in Eq. (14) which the CχQM fails to reproduce!

Phenomenologically, the large branching ratio $\frac{\Gamma(\omega\to\pi^0\gamma)}{\Gamma(\rho\to\pi^0\gamma)}$ is due to the ω–ϕ mixing and the fact that the ϕ is an almost pure $s\bar{s}$ state. By construction, the CχQM form factor is $SU(3)$ invariant and, therefore, like any model which is $SU(3)$ invariant, fails to reproduce this phenomenological fact.

4 Hadronic light-by-light contribution

In the CχQM there are two types of contributions: the Constituent Quark Loop (CQL) contribution shown in Fig. (6) and the Goldstone Exchange Contribution shown

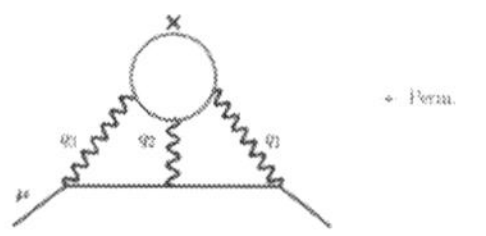

Figure 6. Constituent Quark Loop Contribution to the Muon Anomaly in the CχQM.

in Fig. (7) with constituent quark loops at each vertex. We shall consider these two types of contributions, both leading in the $1/N_c$–expansion, separately.

This contribution can be obtained from the QED analytic calculation of Laporta and Remmidi [33] with the result

$$a_\mu^{(\text{HLbyL})}(\text{CQL}, M_Q = 240\,\text{MeV}) = 82.2\times10^{-11}\,. \quad (16)$$

The second contribution coming from the π^0 exchange lies on [36] the evaluation of the vertex $\pi^0\gamma^*\gamma^*$ form factor in the CχQM,

$$\mathcal{F}^{(\chi\text{QM})}_{\pi^{0*}\gamma^*\gamma^*}\left(q_2^2, q_1^2, q_3^2\right) = -\frac{ie^2N_c}{12\pi^2 f_\pi}\int_0^1 dxx\int_0^1 dy\frac{2M_Q^2}{\mathbb{D}}\,, \quad (17)$$

whith

$$\mathbb{D} = M_Q^2 - x(1-x)(1-y)q_1^2 - x^2y(1-y)q_2^2 - xy(1-x)q_3^2\,. \quad (18)$$

Numerically, one can evaluate this contribution and obtain,

$$a_\mu^{(\text{HLbyL})}(\pi^0)_{\chi\text{QM}} = (68\pm3)\times10^{-11}\,, \quad (19)$$

which does not include the systematic error of the model, agrees well with the phenomenological determinations of this contribution which, according to the most recent update [39] and depending on the underlying phenomenological model for the form factors $\mathcal{F}^{(\chi\text{QM})}_{\pi^{0*}\gamma^*\gamma^*}\left(q_2^2, q_1^2, q_3^2\right)$ vary between

$$a_\mu^{(\text{HLbyL})}(\pi^0)_{\text{phen.}} = (57.4\pm4.6)\times10^{-11} \quad (20)$$

and

$$a_\mu^{(\text{HLbyL})}(\pi^0)_{\text{phen.}} = (80.1\pm4.7)\times10^{-11}\,. \quad (21)$$

It is a fact that asymptotically, for $M_Q\to\infty$, the π^0 contribution largely dominates the Constituent Quark Loop contribution:

$$\left.\frac{a_\mu^{(\text{HLbyL})}(\pi^0)}{a_\mu^{(\text{HLbyL})}(\text{CQL})}\right|_{M_Q\to\infty} \sim \frac{N_cM_Q^2}{16\pi^2 f_\pi^2}\frac{1}{\zeta(3)-\frac{19}{24}}\log^2\frac{M_Q}{m_\pi}\,; \quad (22)$$

however, this asymptotic behaviour is far from being reached at values of M_Q between 230 MeV and 250 MeV, for which the Constituent Quark Loop contribution still dominates over the Goldstone contribution.

For the total hadronic light–by–light contribution in the CχQM, which includes the quark loop contribution as well as the π^0–exchange contributions, we then find

$$148\times10^{-11} \le a_\mu^{(\text{HLbyL})}(\text{C}\chi\text{QM}) \le 153\times10^{-11} \quad (23)$$

for 250 MeV $\ge M_Q \ge$ 230 MeV. This result, which does not include the systematic error of the model, has to be compared with the phenomenological estimate

$$a^{(\text{HLbyL})} = (122\pm18)\times10^{-11}\,, \quad (24)$$

for the total of the hadronic contributions not suppressed in the $1/N_c$–expansion (see e.g. ref. [40] for details). Within the expected systematic uncertainties they compare rather well.

The interesting feature which emerges from this calculation is the observed balance between the Goldstone contribution and the Quark Loop contribution. Indeed, as the constituent quark mass M_Q gets larger and larger, the Goldstone contribution dominates; while for M_Q smaller and smaller it is the Quark Loop contribution which dominates. This is illustrated by the plot of the total $a_\mu^{(\text{HLbyL})}(\text{C}\chi\text{QM})$ versus M_Q shown in Fig. (8). What this plot shows is in flagrant contradiction with the results reported in ref. [41] based in a calculation using a Dyson–Schwinger inspired model. In this model, the authors find a contribution from the π^0–exchange which, within errors, is compatible with the other phenomenological determinations and, in particular, with our CχQM result in Eq. (19); yet their result for the equivalent contribution to the quark loop turns out to be almost twice as large with a total contribution

$$a^{(\text{HLbyL})}(\text{ref. [41]}) = (217\pm91)\times10^{-11}\,. \quad (25)$$

The central value of this result would require a ridicously small value of M_Q in order to be reproduced by the CχQM and, furthermore, for such a small value of M_Q the π^0–exchange contribution would be far too small as compared to all the phenomenological estimates, including the one in ref. [41]. We conclude that a range of values such as

$$170 \le a_\mu^{(\text{HLbyL})}\times10^{11} \le 308\,, \quad (26)$$

allowed by the result quoted in Eq. (25), cannot be digested within the CχQM and in our opinion this casts serious doubts about the compatibility of the model used in ref. [41] with basic QCD features.

5 Hadronic Electroweak Contributions

These are the contributions to the muon anomaly which appear at the two–loop level in the electroweak sector. They are the ones generated by the hadronic $\gamma\gamma Z$ vertex, with one γ and the Z–boson attached to a muon line, as illustrated by the Feynman diagrams in Fig. (9).

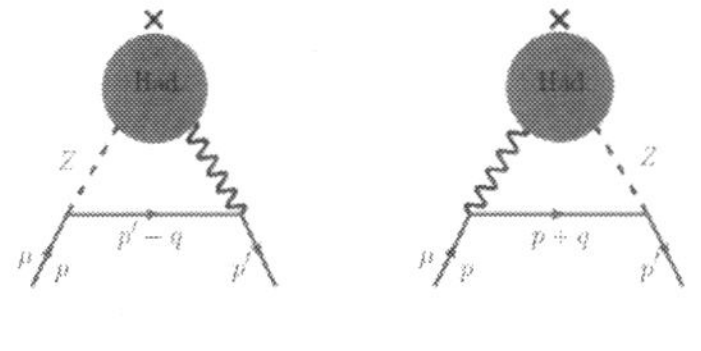

Figure 9. Feynman diagrams with the hadronic $\gamma\gamma Z$ vertex which contributes to the muon anomaly.

These contributions are particularly interesting because, *a priori*, they could be enhanced by a large $\log(M_Z^2/m_\mu^2)$ factor. However, due to the anomaly–free coupling assignments in the Standard Model, there is an important cancellation of UV–scales between the lepton and the quark contributions within a given family [42, 43]. What is left out of this cancellation in the sector of the u, d and s quarks, where the strong interactions play a subtle role at long distances, is governed by the dynamics of spontaneous chiral symmetry breaking [42, 44–47].

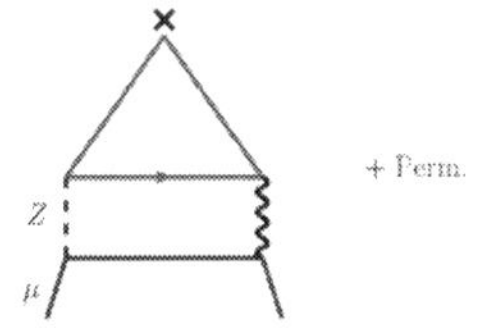

Figure 10. Feynman diagrams in the CχQM of the $\gamma\gamma Z$ vertex type.

The CχQM, where the hadronic blob in Fig. (9) is replaced by a constituent quark loop as illustrated in Fig. (10), offers a simple way to estimate these contributions as, with $g_A = 1$, for $m_c = 1.5$ GeV and $M_Q = 240$ MeV,

$$
\begin{aligned}
& a_\mu^{(\mathrm{HEW})}\Big|_{e,u,d;\mu,s,c}(\mathrm{C}\chi\mathrm{QM}) \\
&= \frac{G_F}{\sqrt{2}}\frac{m_\mu^2}{8\pi^2}\frac{\alpha}{\pi}\left(-\frac{2}{3}\log\frac{M_Q^2}{m_\mu^2} - 4\log\frac{m_c^2}{m_\mu^2} - 5 + \frac{8}{9}\pi^2\right) \\
&= -5.0\times 10^{-11} \qquad (27)
\end{aligned}
$$

with the systematic errors of the model, is compatible with the phenomenological determination [46]:

$$
a_\mu^{(\mathrm{HEW})}\Big|_{e,u,d;\mu,s,c} = (-6.7 \pm 0.5)\times 10^{-11}\,. \qquad (28)
$$

6 Conclusions

From the previous considerations we conclude that the CχQM provides a useful and simple *reference model* to evaluate the hadronic contributions to the anomalous magnetic moment of the muon. The effective Lagrangian of this model is renormalizable in the Large–N$_c$ limit [13] and, as shown in [12], the number of the required counterterms in this limit is minimized for a choice of the axial coupling: $g_A = 1$. The only free parameter of the model is then the mass of the constituent quark mass M_Q which in Section II, from a comparison with the phenomenological determination of the lowest order hadronic vacuum polarization contribution to the muon anomaly, has been fixed to

$$
M_Q = (240 \pm 10)\,\mathrm{MeV}\,, \qquad (29)
$$

This range of values for M_Q reproduces the phenomenological determination within an error of less than 10%. All the other hadronic contributions have then been evaluated for this range of values of M_Q with the results which are summarized in Table 2, despite one exception from the contribution from the $\pi^0\gamma$ intermediate state to hadronic vacuum polarization where the CχQM, because of its $SU(3)$ invariance, fails to reproduce the phenomenological determination which is particularly enhanced because of the large observed branching ratio in Eq. (14).

Class	Result in 10^{-10} units
HVP	$652\,^{+47}_{-42}$
HVP to $O\left(\frac{\alpha}{\pi}\right)^3$	-6.4 ± 1.2
HLbyL	15.0 ± 0.3
HEW	-0.5

Table 2. Summary of results for the hadronic contributions to the muon anomly in the CχQM.

Acknowledgments

The author thanks the organizers, in particular G. D'Ambrosio, for this nice Workshop. The author is also graceful to E. de Rafael for the discussions and suggestions about the presentation and proceeding. This work was supported in part by MIUR under project 2010YJ2NYW and by INFN research initiative PhenoLNF.

References

[1] D. Greynat and E. de Rafael, "Hadronic Contributions to the Muon Anomaly in the Constituent Chiral Quark Model", JHEP **1207**, 020 (2012)

[2] G. Bennett *et al.*, [Muon (g-2) Collaboration], Phys. Rev. **D73** (2006) 072003.

[3] J.P. Miller, E. de Rafael and B.L. Roberts, Rep. Prog. Phys. **70** (2007) 795.

[4] F. Jegerlehner and A. Nyffeler, Phys. Rep. **477** (2009) 1.

[5] M. Davier, A. Hoecker, B. Malaescu and Z. Zhang, Eur. Phys. J. **C71** (2011) 1515.

[6] J. Prades, E. de Rafael and A. Vainshtein, **Lepton Dipole Moments**, Ch. 9 in *Advanced Series on Directions in High Energy Physics–Vol. 20*, Editors: B. Lee Roberts and William J. Marciano
[7] R. Carey *et al*, FERMILAB–PROPOSAL–0989,
[8] T. Mibe, Nucl. Phys. (Proc. Suppl.) **B218** (2011) 242.
[9] A. Manohar and G. Georgi, Nucl. Phys. **B233** (1984) 232.
[10] A. De Rújula, H. Georgi and S. Glashow, Phys.Rev. **D12** (1975) 147
[11] S. Weinberg, Physica **96A** (1984) 327.
[12] E. de Rafael, Phys. Lett. **B703** (2011) 60.
[13] S. Weinberg, Phys.Rev.Lett. **105** (2010) 261601.
[14] C. Bouchiat and L. Michel, J. Phys. Radium **22** (1961) 121.
[15] S.J. Brodsky and E. de Rafael, Phys. Rev. **168** (1968) 1620.
[16] J. Calmet, S. Narison, M. Perrottet and E. de Rafael, Rev. Mod. Phys.**49** (1977) 21.
[17] A.A. Pivovarov, Phys. Atom. Nucl. **66** (2003) 902.
[18] J. Erler and G.T. Sanchez, Phys. Rev. Lett. **97** (2006) 161801.
[19] G. Källen and A. Sabry, Kgl. Danske Videnskab. Selskab, Mat. Fys. Medd, **29** N. 17 (1955).
[20] B.E. Lautrup and E. de Rafael, Phys. Rev. **174** (1968) 1835.
[21] D. Espriu, E. de Rafael and J. Taron, Nucl. Phys. **B 345** (1990) 22.
[22] Xu Feng *et al* (ETMC Collaboration), arXiv:1103.4818v1 [hep-lat].
[23] R. Barbieri and E. Remiddi, Nucl.Phys. **B90** (1975) 273.
[24] K. Hagiwara *et al*, Phys. Rev. **D69** (2004) 093003.
[25] B.E. Lautrup, A. Peterman and E. de Rafael, Phys. Reports **3C** (1972) 193.
[26] E. de Rafael, Phys. Lett. **B322** (1994) 239.
[27] M. Perrottet and E. de Rafael, *unpublished*.
[28] I. Blokland, A. Czarnecki and K. Melnikov, Phys. Rev. Lett. **88** (2002) 07183.
[29] N.N. Achasov *et al.*, SND Collaboration, Eur. Phys. J. **C12** (2000) 25; Phys. Lett. **B559** (2003) 171.
[30] K. Hagiwara *et al*, J. Phys. **G38** (2011) 085003.
[31] F. Jegerlehner, Nucl. Phys. Proc. Suppl. **162** (2006) 22.
[32] J. Aldins, S. Brodsky, A. Dufner and T. Kinoshita, Phys. Rev. Lett. **23** (1969) 441; Phys. Rev. **D1** (1970) 2378.
[33] S. Laporta and E. Remiddi, , Phys. Lett., **B301** (1993) 440 .
[34] R. Boughezal and K. Melnikov, Phys. Lett. **B704** (2011) 114043.
[35] M. Knecht and A. Nyffeler, Phys. Rev. **D65** (2002) 073034.
[36] M. Knecht, A. Nyffeler, M. Perrottet and E. de Rafael, Phys. Rev. Lett. **88** (2002) 071802.
[37] J.-Ph. Aguilar, E. de Rafael and D. Greynat, Phys. Rev. **D77** (2008) 093010.
[38] K. Melnikov and A. Vainshtein, Phys. Rev. **D70** (2004) 113006.
[39] B. Babusci *et al*, arXiv:1109.2461v2 [hep-ph].
[40] J. Prades, E. de Rafael and A. Vainshtein, *Lepton Dipole Moments*, Ch.9, eds. B.L. Roberts and W.J. Marciano, World Scientific (2010).
[41] T. Goecke, C.S. Fisher and R. Williams, Prog. Part. Nucl. Phys. **67** (2012) 563.
[42] S. Peris, M. Perrottet and E. de Rafael, Phys. Lett. **B355** (1995) 523.
[43] A. Czarnecki, B. Krause and W. Marciano, Phys. Rev. **D52** (1995) R2619.
[44] M. Knecht, S. Peris, M. Perrottet and E. de Rafael, JHEP **0211** (2002) 003.
[45] A. Vainshtein, Phys. Lett. **B569** (2003) 187.
[46] A. Czarnecki, W.J. Marciano and A. Vainshtein, Phys. Rev. **D67** (2003) 073006 [Erratum-ibid. **73** (2006) 119901].
[47] M. Knecht, S. Peris, M. Perrottet and E. de Rafael, JHEP **0403** (2004) 035.

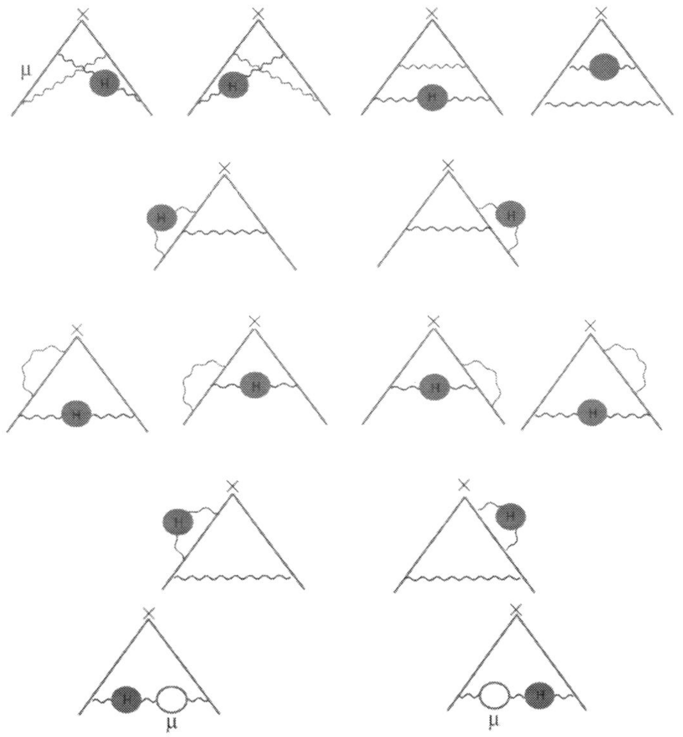

Figure 2. Feynman diagrams corresponding to the Class A contribution to the muon anomaly.

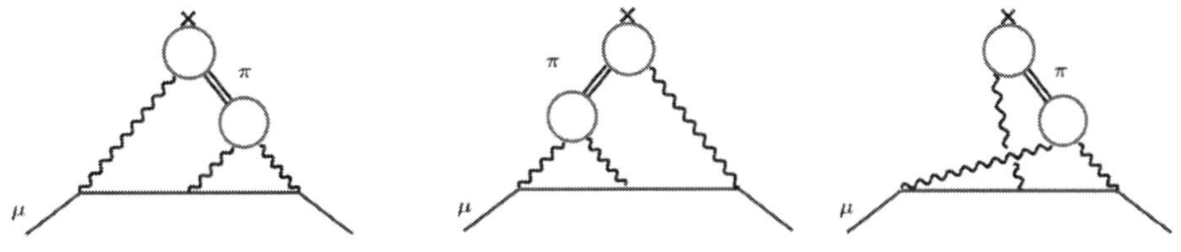

Figure 7. Goldstone Exchange Contribution to the Muon Anomaly in the CχQM.

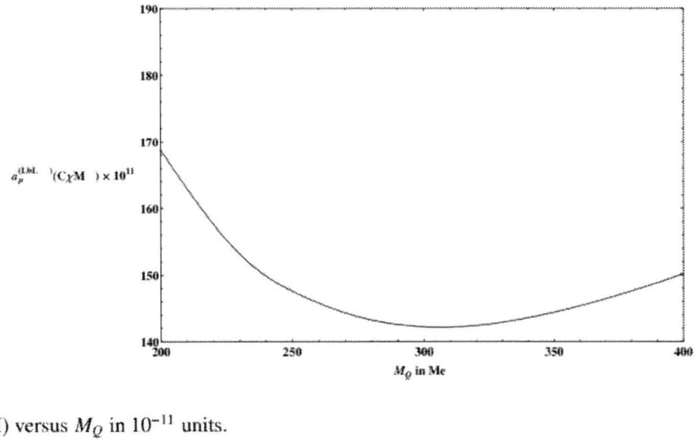

Figure 8. $a_\mu^{(\mathrm{HLbyL})}(\mathrm{C}\chi\mathrm{QM})$ versus M_Q in 10^{-11} units.

EPJ Web of Conferences **118**, 01014 (2016) DOI: 10.1051/epjconf/201611801014
FCCP2015

A two-Higgs-doublet model facing experimental hints

Andreas Crivellin[1,a], Julian Heeck[2,b], and Peter Stoffer[3,c]

[1] *CERN Theory Division, CH–1211 Geneva 23, Switzerland*
[2] *Service de Physique Théorique, Université Libre de Bruxelles, Boulevard du Triomphe, CP225, 1050 Brussels, Belgium*
[3] *Helmholtz-Institut für Strahlen- und Kernphysik (Theory) and Bethe Center for Theoretical Physics, University of Bonn, D–53115 Bonn, Germany*

Abstract. Physics beyond the Standard Model has so far eluded our experimental probes. Nevertheless, a number of interesting anomalies have accumulated that can be taken as hints towards new physics: BaBar, Belle, and LHCb have found deviations of approximately 3.8σ in $B \to D\tau\nu$ and $B \to D^*\tau\nu$; the anomalous magnetic moment of the muon differs by about 3σ from the theoretic prediction; the branching ratio for $\tau \to \mu\nu\nu$ is about 2σ above the Standard Model expectation; and CMS and ATLAS found hints for a non-zero decay rate of $h \to \mu\tau$ at 2.6σ. Here we consider these processes within a lepton-specific two-Higgs doublet model with additional non-standard Yukawa couplings and show how (and which of) these excesses can be accommodated.

1 Introduction

This talk is based on Ref. [1], where a more detailed discussion can be found. Tests of flavor universality or flavor violation serve as a useful tool to search for physics beyond the Standard Model (SM), seeing as the SM predictions are precisely known. Some experiments have reported on deviations from the SM, which we list below.

- Lepton universality in semileptonic B decays, parametrized by the ratio

$$\mathcal{R}(D^{(*)}) \equiv \frac{\text{BR}(B \to D^{(*)}\tau\nu)}{\text{BR}(B \to D^{(*)}\ell\nu)}, \quad \ell = e, \mu, \tag{1}$$

  has been studied by BaBar, Belle and LHCb. Combining their experimental values yields

$$\mathcal{R}(D)_{\text{exp}} = 0.388 \pm 0.047, \quad \mathcal{R}(D^*)_{\text{exp}} = 0.321 \pm 0.021.$$

  Together, these values deviate by more than 3σ from the SM prediction [2–4]

$$\mathcal{R}(D)_{\text{SM}} = 0.297 \pm 0.017, \quad \mathcal{R}(D^*)_{\text{SM}} = 0.252 \pm 0.003.$$

  One possible new-physics explanation comes in the form of a charged scalar [5–7].

- The muon's anomalous magnetic moment $a_\mu \equiv (g-2)_\mu/2$, as measured by the Brookhaven experiment E821 [8], deviates by 2.7σ from its SM value,

$$a_\mu^{\text{exp}} - a_\mu^{\text{SM}} = (236 \pm 87) \times 10^{-11}. \tag{2}$$

  It is not yet clear if this discrepancy is due to new physics or rather underestimated hadronic uncertainties; there are ongoing efforts to reduce the model dependence in the hadronic light-by-light estimate based on dispersion relations [9–12] or lattice QCD [13–16]. A possible interpretation in terms of new physics has been found in two-Higgs-doublet models (2HDM) [17, 18], in particular the lepton-specific 2HDM [19–22].

- For $\tau \to \mu\nu\nu$, the dominant uncertainty in the SM prediction for the branching ratio comes from the τ lifetime. Using the PDG [23] values for τ lifetime and branching ratios

$$B_\mu \equiv \text{BR}(\tau \to \mu\bar{\nu}\nu)_{\text{exp}} = (17.41 \pm 0.04)\%,$$
$$B_e \equiv \text{BR}(\tau \to e\bar{\nu}\nu)_{\text{exp}} = (17.83 \pm 0.04)\%, \tag{3}$$

  we can determine the deviations from the SM prediction $\Delta_\ell \equiv B_\ell/B_\ell^{\text{SM}} - 1$ [24] as

$$\Delta_\mu^{\text{PDG}} = (0.69 \pm 0.29)\%, \quad \Delta_e = (0.28 \pm 0.28)\%. \tag{4}$$

  There is a deviation of about 2.4σ for the muon final state, whereas the electron channel is compatible with the SM prediction, hinting at lepton non-universality. Charged scalars of a 2HDM will modify the rate [24–26].

- Going from lepton non-universality to outright lepton flavor violation, we are drawn to the recent CMS excess of 2.4σ in $h \to \mu\tau$ [27]:

$$\text{BR}(h \to \mu\tau) = \left(0.84^{+0.39}_{-0.37}\right)\%. \tag{5}$$

A similar search at ATLAS [28] finds $\text{BR}(h \to \mu\tau) = (0.77 \pm 0.62)\%$, which slightly bumps the significance to about 2.6σ. Possible explanations naturally require an extended scalar sector [18, 29–35].

[a] e-mail: andreas.crivellin@cern.ch
[b] e-mail: julian.heeck@ulb.ac.be
[c] e-mail: stoffer@hiskp.uni-bonn.de

© The Authors, published by EDP Sciences. This is an open access article distributed under the terms of the Creative Commons Attribution License 4.0 (http://creativecommons.org/licenses/by/4.0/).

Each anomaly individually can be accommodated in SM extensions by scalars, e.g. 2HDMs. The goal of our study is to see if all four anomalies can be explained simultaneously with a fairly minimal model.

2 Modified 2HDM-X

We will study a lepton-specific 2HDM (2HDM-X), defined by the Yukawa couplings in the Lagrangian

$$\mathcal{L}_Y = -\overline{Q}_L Y^u \tilde{\Phi}_2 u_R - \overline{Q}_L Y^d \Phi_2 d_R - \overline{L}_L Y^\ell \Phi_1 e_R + \text{h.c.}, \quad (6)$$

with additional couplings that break the type-X structure

$$\Delta\mathcal{L}_Y = -\overline{Q}_L \xi^u \tilde{\Phi}_1 u_R - \overline{Q}_L \xi^d \Phi_1 d_R - \overline{L}_L \xi^\ell \Phi_2 e_R + \text{h.c.} \quad (7)$$

The scalar interactions with fermions after electroweak breaking can be written as

$$\begin{aligned}\mathcal{L} \supset\ & \bar{\nu}_i \Gamma^{H^+ LR}_{\nu_i \ell_j} P_R \ell_j H^+ \\ & + \bar{u}_i \left(\Gamma^{H^+ RL}_{u_i d_j} P_L + \Gamma^{H^+ LR}_{u_i d_j} P_R\right) d_j H^+ \\ & + \sum_{H_k^0 = h,A,H} \sum_{f=u,d,\ell} \left(\bar{f}_i \Gamma^{H_k^0 LR}_{f_i f_j} P_R f_j H_k^0\right) + \text{h.c.}, \end{aligned} \quad (8)$$

where the couplings – in the limit of large $\tan\beta$ of interest here – are given by

$$\Gamma^{hLR}_{q_i q_j} \simeq -\frac{1}{\sqrt{2}}\left(\frac{m_{q_i}}{v}\delta_{ij}\cos\alpha - \epsilon^q_{ij}\sin\alpha\right), \quad (9)$$

$$\Gamma^{HLR}_{q_i q_j} \simeq -\frac{1}{\sqrt{2}}\left(\frac{m_{q_i}}{v}\delta_{ij}\sin\alpha + \epsilon^q_{ij}\cos\alpha\right), \quad (10)$$

$$\Gamma^{ALR}_{d_i d_j} \simeq -i\frac{1}{\sqrt{2}}\epsilon^d_{ij}, \quad (11)$$

$$\Gamma^{ALR}_{u_i u_j} \simeq i\frac{1}{\sqrt{2}}\epsilon^u_{ij}, \quad (12)$$

$$\Gamma^{H^+ LR}_{u_i d_j} \simeq V_{ij'}\epsilon^d_{j'j}, \quad (13)$$

$$\Gamma^{H^+ RL}_{u_i d_j} \simeq -\epsilon^{u\star}_{j'i} V_{j'j}, \quad (14)$$

$$\Gamma^{hLR}_{\ell_f \ell_i} \simeq \frac{\sin\alpha\tan\beta}{\sqrt{2}}\left(\frac{m_{\ell_i}}{v}\delta_{fi} - \epsilon^\ell_{fi}\right), \quad (15)$$

$$\Gamma^{HLR}_{\ell_f \ell_i} \simeq -\frac{\cos\alpha\tan\beta}{\sqrt{2}}\left(\frac{m_{\ell_i}}{v}\delta_{fi} - \epsilon^\ell_{fi}\right), \quad (16)$$

$$\Gamma^{ALR}_{\ell_f \ell_i} \simeq -i\frac{\tan\beta}{\sqrt{2}}\left(\frac{m_{\ell_i}}{v}\delta_{fi} - \epsilon^\ell_{fi}\right), \quad (17)$$

$$\Gamma^{H^+ LR}_{\nu_f \ell_i} \simeq \tan\beta\left(\frac{m_{\ell_i}}{v}\delta_{fi} - \epsilon^\ell_{fi}\right). \quad (18)$$

V denotes the Cabibbo–Kobayashi–Maskawa mixing matrix, $v \simeq 174\,\text{GeV}$ the vacuum expectation value, and the ϵ matrices parametrize the deviation from the type-X structure (ϵ^f and ξ^f are related by fermion rotations). On phenomenological grounds, we take $\epsilon^d = 0$ and both ϵ^u and ϵ^ℓ of the form

$$\epsilon^{u,\ell} = \begin{pmatrix} 0 & 0 & 0 \\ 0 & 0 & 0 \\ 0 & \times & \times \end{pmatrix}, \quad (19)$$

where $\times$ denotes a non-zero entry, allowing for lepton flavor violation in the μ–τ sector.

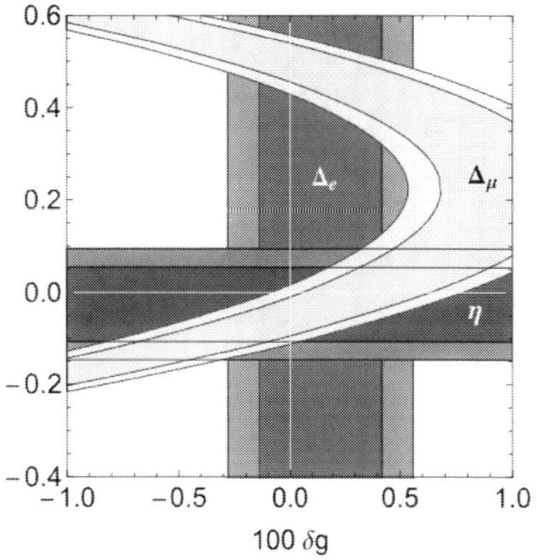

Figure 1. Allowed regions for z and δg from Δ_μ (yellow), Δ_e (blue), and the Michel parameter η (red); see text for definitions. Darker (lighter) regions are at the 2σ (3σ) level, using PDG values [1].

3 Phenomenology

In this section we address the constraints on our model and its potential to resolve the experimental anomalies outlined above.

3.1 Tau decays $\tau \to \ell\nu\nu$

Decays of the tau are modified at tree-level by the charged scalar H^+ (including a change in the Michel parameter η) [24–26] and at loop-level through a modified $W\tau\nu$ coupling $g_{W\tau\nu} \to g_{W\tau\nu}(1 + \delta g)$ [21, 24, 25]:

$$\Delta_e \simeq 2\delta g, \quad (20)$$

$$\Delta_\mu \simeq 2\delta g + \frac{z^2}{4} - 2z\frac{m_\mu}{m_\tau}, \quad (21)$$

$$\eta = -\frac{2z}{4+z^2}, \quad (22)$$

with the 2HDM couplings [21]

$$z \equiv \frac{v^2}{m^2_{H^+}}\Gamma^{LR\,H^+}_{\nu_\tau\tau}\Gamma^{LR\,H^+\star}_{\nu_\mu\mu}, \quad (23)$$

$$\delta g = \frac{\tan^2\beta}{32\pi^2}\left|\frac{m_\tau}{v} - \epsilon^\ell_{33}\right|^2\left[F\left(\frac{m_A^2}{m^2_{H^+}}\right) + F\left(\frac{m_H^2}{m^2_{H^+}}\right)\right]. \quad (24)$$

The experimentally allowed regions are shown in Fig. 1. In the SM we have $\delta g = z = 0$, which is disfavored by $\tau \to \mu\nu\nu$ at more than 2σ (with PDG values). In the 2HDM-X one has $\delta g \leq 0$ and $z > 0$, which makes Δ_μ even worse and puts pressure on the 2HDM-X and 2HDM-II (which has the same lepton couplings). In our modified 2HDM-X, we can however flip the sign of the tau coupling via $\epsilon^\ell_{33} > m_\tau/v$ and obtain $z < 0$, alleviating the tension in $\tau \to \mu\nu\nu$ decays.

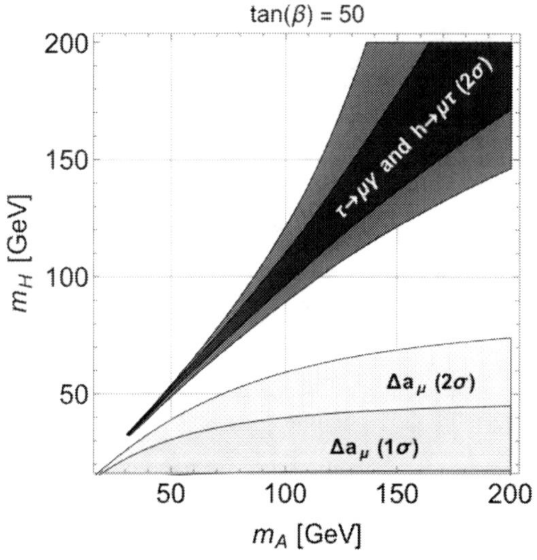

Figure 2. Allowed regions in the m_A-m_H plane from $(g-2)_\mu$, $\tau \to \mu\gamma$ and $h \to \mu\tau$ for $\tan(\beta) = 50$ and $\epsilon_{33}^\ell = 2m_\tau/v$. For $h \to \mu\tau$ blue corresponds to $\cos(\alpha - \beta) = 0.1$ and light blue to $\cos(\alpha - \beta) = 0.2$. The allowed region for Δa_μ is maximal in the sense that we have allowed for the three possibilities $\epsilon_{32}^\ell \neq 0$, $\epsilon_{32}^\ell = \epsilon_{23}^\ell \neq 0$ and $\epsilon_{32}^\ell = -\epsilon_{23}^\ell \neq 0$ as the latter ones can give m_τ/m_μ enhanced one-loop contributions [1].

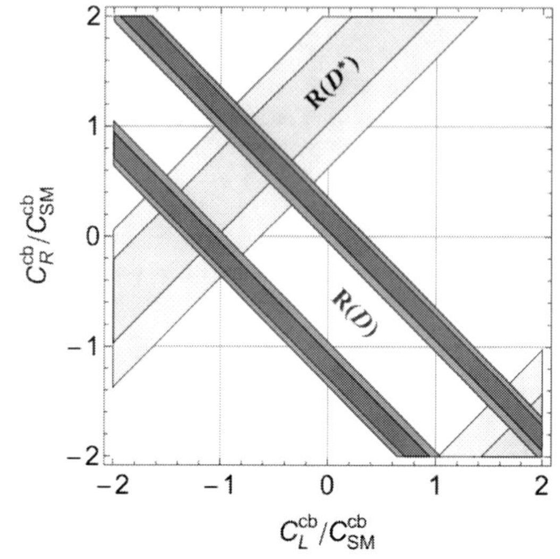

Figure 3. Allowed regions in the C_L^{cb}–C_R^{cb} plane from $\mathcal{R}(D)$ (blue) and $\mathcal{R}(D^*)$ (yellow) for real values of $C_{L,R}^{cb}$. The lighter regions correspond to 2σ experimental uncertainties while the darker regions are correspond to 1σ [1].

3.2 Magnetic moment a_μ and $h \to \mu\tau$

It has been shown that a light pseudoscalar in the 2HDM-X with large $\tan\beta$ can resolve the Δa_μ anomaly via its contribution in a Barr–Zee diagram [36], see for example Refs. [19–21]. As seen above, we need to flip the sign of the tau coupling to the new scalars in order to alleviate the $\tau \to \mu\nu\nu$ discrepancy. Because of this, it is the non-SM-like CP-even scalar that can resolve the Δa_μ anomaly in our model (and not the pseudoscalar). The pseudoscalar then needs to be heavier than the scalar in order not to cancel the contribution to Δa_μ (see Fig 2).

The same Barr–Zee diagrams that give the desired Δa_μ also lead to $\tau \to \mu\gamma$ decays in case $\epsilon_{32}^\ell \neq 0$ (as required for $h \to \mu\tau$). If we want to explain $h \to \mu\tau$, the $\tau \to \mu\gamma$ rate needs to be tuned to small values using $m_H \simeq m_A$, which necessarily also suppresses the contribution to Δa_μ (see Fig 2). As a result, it is possible within our modified 2HDM-X to resolve the anomalies in $\tau \to \mu\nu\nu$ and Δa_μ *or* to resolve $\tau \to \mu\nu\nu$ and $h \to \mu\tau$, but not both at the same time, at least without introducing more parameters.

3.3 Tauonic B decays

The relevant effective Hamiltonian for the semileptonic B decays in our model is

$$\mathcal{H} = C_{\rm SM}^{cb} \bar{c}\gamma_\mu P_L b\, \bar{\tau}\gamma_\mu P_L \nu_\tau + \sum_{X=L,R} C_X^{cb} \bar{c} P_X b\, \bar{\tau} P_L \nu_\tau\,, \quad (25)$$

with SM Wilson coefficient $C_{\rm SM}^{cb} = 4G_F V_{cb}/\sqrt{2}$ and the charged-scalar contribution

$$C_L^{cb} \simeq \frac{\tan\beta}{m_{H^+}^2} \epsilon_{32}^{u*} \left(\frac{m_\tau}{v} - \epsilon_{33}^{\ell*} \right), \qquad C_R^{cb} \simeq 0\,. \quad (26)$$

These Wilson coefficients affect the two ratios $\mathcal{R}(D^{(*)})$ in the following way [4, 37, 38],

$$\frac{\mathcal{R}(D)}{\mathcal{R}(D)_{\rm SM}} = 1 + 1.5\, \Re \left[\frac{C_R^{cb} + C_L^{cb}}{C_{\rm SM}^{cb}} \right] + 1.0 \left| \frac{C_R^{cb} + C_L^{cb}}{C_{\rm SM}^{cb}} \right|^2 , \quad (27)$$

$$\frac{\mathcal{R}(D^*)}{\mathcal{R}(D^*)_{\rm SM}} = 1 + 0.12\, \Re \left[\frac{C_R^{cb} - C_L^{cb}}{C_{\rm SM}^{cb}} \right] + 0.05 \left| \frac{C_R^{cb} - C_L^{cb}}{C_{\rm SM}^{cb}} \right|^2 , \quad (28)$$

leading to the allowed regions of Fig. 3. As can be seen, our new-physics model from Eq. (26) has the right structure to easily resolve the anomaly, e.g. with real $C_L^{cb} \simeq -1.2\,|C_{\rm SM}^{cb}|$.[1]

4 Conclusions

We addressed the measured anomalies in $\mathcal{R}(D^{(*)})$ (3.8σ), a_μ ($\sim 3\sigma$), $\tau \to \mu\nu\nu$ (2.4σ), and $h \to \mu\tau$ (2.6σ) within a simple two-Higgs-doublet model. The Yukawa structure of our model is close to the lepton-specific 2HDM

[1] Efficiency corrections to $\mathcal{R}(D)$ due to the BaBar detector [2] are important in the case of large contributions from $C_{R,L}^{cb}$, i.e. if one wants to explain $\mathcal{R}(D)$ with destructive interference with the SM contribution. As shown in Ref. [39], these corrections can be effectively taken into account by multiplying the quadratic term in $C_{R,L}^{cb}$ of Eq. (27) by an approximate factor of 1.5 (not included in Eq. (27)).

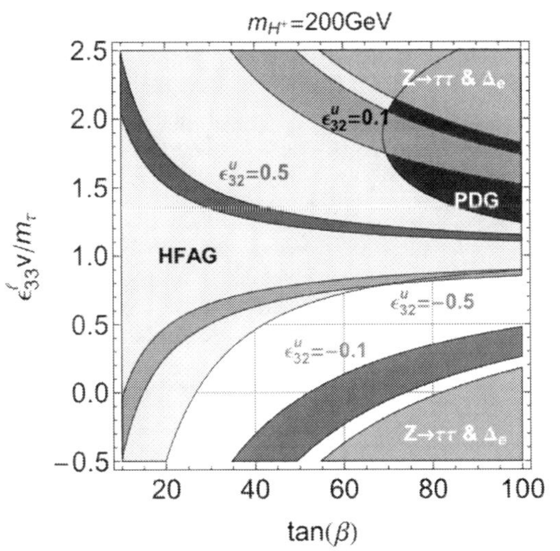

Figure 4. Allowed regions in the $\tan\beta$–$v/m_\tau\epsilon^{\ell}_{33}$ plane from $\mathcal{R}(D^{(*)})$ and $\tau\to\mu\nu\nu$ at the 2σ level. The yellow region is allowed by $\tau\to\mu\nu\nu$ using the HFAG result for $m_H = 30\,\text{GeV}$ and $m_A = 200\,\text{GeV}$, while the (darker) blue one is the allowed region using the PDG result. The red, orange, green, and magenta bands correspond to the allowed regions by $\mathcal{R}(D^{(*)})$ for different values of ϵ^u_{32}. The gray region is excluded by $Z\to\tau\tau$ and $\tau\to e\nu\nu$. For $m_H\simeq m_A$, the allowed regions from $\tau\to\mu\nu\nu$ would be slightly larger [1].

(type X), but with some additional Yukawa couplings involving third-generation fermions that give rise to the b–c (necessary for $\mathcal{R}(D^{(*)})$) and μ–τ transitions (relevant for $h\to\mu\tau$) as well as corrections to $\tau\tau$ couplings (important for $\tau\to\mu\nu\nu$).

The charged scalar H^+ influences $\mathcal{R}(D^{(*)})$ and $\tau\to\mu\nu\nu$ and can resolve both deviations simultaneously (see Fig. 4), as long as the sign of the tau coupling is flipped, meaning $\epsilon^{\ell}_{33} > m_\tau/v$. Because of this, a light scalar (not *pseudo*scalar) is needed to resolve the Δa_μ anomaly, which then induces the top decay $t\to Hc$ [1]. $h\to\mu\tau$ can only be explained if we give up on Δa_μ, otherwise the rate $\tau\to\mu\gamma$ would be too large.

It remains to be seen which of these anomalies stand the test of time and which are simply statistical fluctuations. The fact that some of them can be explained rather naturally within a fairly minimal model gives however hope that we are on the verge of something interesting.

Acknowledgements

We thank the organizers of the FCCP2015 for an interesting workshop. A. Crivellin is supported by a Marie Curie Intra-European Fellowship of the European Community's 7th Framework Programme under contract number PIEF-GA-2012-326948. The work of J. Heeck is funded in part by IISN and by Belgian Science Policy (IAP VII/37). P. Stoffer gratefully acknowledges financial support by the DFG (CRC 16, "Subnuclear Structure of Matter").

References

[1] A. Crivellin, J. Heeck, P. Stoffer (2015), 1507.07567
[2] J. Lees et al. (BABAR Collaboration), Phys.Rev.Lett. **109**, 101802 (2012), 1205.5442
[3] J.F. Kamenik, F. Mescia, Phys.Rev. **D78**, 014003 (2008), 0802.3790
[4] S. Fajfer, J.F. Kamenik, I. Nisandzic, Phys.Rev. **D85**, 094025 (2012), 1203.2654
[5] A. Crivellin, C. Greub, A. Kokulu, Phys.Rev. **D86**, 054014 (2012), 1206.2634
[6] A. Celis, M. Jung, X.Q. Li, A. Pich, JHEP **1301**, 054 (2013), 1210.8443
[7] A. Crivellin, A. Kokulu, C. Greub, Phys.Rev. **D87**, 094031 (2013), 1303.5877
[8] G. Bennett et al. (Muon ($g-2$) Collaboration), Phys.Rev. **D73**, 072003 (2006), hep-ex/0602035
[9] G. Colangelo, M. Hoferichter, M. Procura, P. Stoffer, JHEP **1409**, 091 (2014), 1402.7081
[10] G. Colangelo, M. Hoferichter, B. Kubis, M. Procura, P. Stoffer, Phys. Lett. **B738**, 6 (2014), 1408.2517
[11] V. Pauk, M. Vanderhaeghen, Phys. Rev. **D90**, 113012 (2014), 1409.0819
[12] G. Colangelo, M. Hoferichter, M. Procura, P. Stoffer, JHEP **09**, 074 (2015), 1506.01386
[13] M. Hayakawa, T. Blum, T. Izubuchi, N. Yamada, PoS **LAT2005**, 353 (2006), hep-lat/0509016
[14] T. Blum, M. Hayakawa, T. Izubuchi, PoS **LATTICE2012**, 022 (2012), 1301.2607
[15] T. Blum, S. Chowdhury, M. Hayakawa, T. Izubuchi, Phys.Rev.Lett. **114**, 012001 (2015), 1407.2923
[16] J. Green, O. Gryniuk, G. von Hippel, H.B. Meyer, V. Pascalutsa (2015), 1507.01577
[17] E.O. Iltan, H. Sundu, Acta Phys. Slov. **53**, 17 (2003), hep-ph/0103105
[18] Y. Omura, E. Senaha, K. Tobe, JHEP **1505**, 028 (2015), 1502.07824
[19] A. Broggio, E.J. Chun, M. Passera, K.M. Patel, S.K. Vempati, JHEP **1411**, 058 (2014), 1409.3199
[20] L. Wang, X.F. Han, JHEP **1505**, 039 (2015), 1412.4874
[21] T. Abe, R. Sato, K. Yagyu, JHEP **07**, 064 (2015), 1504.07059
[22] E.J. Chun, Z. Kang, M. Takeuchi, Y.L.S. Tsai (2015), 1507.08067
[23] K. Olive et al. (Particle Data Group), Chin.Phys. **C38**, 090001 (2014)
[24] M. Krawczyk, D. Temes, Eur.Phys.J. **C44**, 435 (2005), hep-ph/0410248
[25] W. Hollik, T. Sack, Phys.Lett. **B284**, 427 (1992)
[26] M. Aoki, S. Kanemura, K. Tsumura, K. Yagyu, Phys.Rev. **D80**, 015017 (2009), 0902.4665
[27] V. Khachatryan et al. (CMS), Phys. Lett. **B749**, 337 (2015), 1502.07400
[28] G. Aad et al. (ATLAS) (2015), 1508.03372

[29] R. Harnik, J. Kopp, J. Zupan, JHEP **03**, 026 (2013), 1209.1397

[30] M.D. Campos, A.E.C. Hernández, H. Päs, E. Schumacher, Phys.Rev. **D91**, 116011 (2015), 1408.1652

[31] D. Aristizabal Sierra, A. Vicente, Phys.Rev. **D90**, 115004 (2014), 1409.7690

[32] J. Heeck, M. Holthausen, W. Rodejohann, Y. Shimizu, Nucl.Phys. **B896**, 281 (2015), 1412.3671

[33] A. Crivellin, G. D'Ambrosio, J. Heeck, Phys.Rev.Lett. **114**, 151801 (2015), 1501.00993

[34] I. Dorsner, S. Fajfer, A. Greljo, J.F. Kamenik, N. Kosnik et al., JHEP **1506**, 108 (2015), 1502.07784

[35] A. Crivellin, G. D'Ambrosio, J. Heeck, Phys.Rev. **D91**, 075006 (2015), 1503.03477

[36] S.M. Barr, A. Zee, Phys. Rev. Lett. **65**, 21 (1990), [Erratum: Phys. Rev. Lett.65,2920(1990)]

[37] A. Akeroyd, S. Recksiegel, J.Phys.G **G29**, 2311 (2003), hep-ph/0306037

[38] Y. Sakaki, H. Tanaka, Phys.Rev. **D87**, 054002 (2013), 1205.4908

[39] S. Fajfer, J.F. Kamenik, I. Nisandzic, J. Zupan, Phys.Rev.Lett. **109**, 161801 (2012), 1206.1872

Next Generation Muon $g-2$ Experiments

David W. Hertzog[1,a]

[1] *Department of Physics, University of Washington, Seattle, WA, 98195, USA*

Abstract. I report on the progress of two new muon anomalous magnetic moment experiments, which are in advanced design and construction phases. The goal of Fermilab E989 is to reduce the experimental uncertainty of a_μ from Brookhaven E821 by a factor of 4; that is, $\delta a_\mu \sim 16 \times 10^{-11}$, a relative uncertainty of 140 ppb. The method follows the same magic-momentum storage ring concept used at BNL, and pioneered previously at CERN, but muon beam preparation, storage ring internal hardware, field measuring equipment, and detector and electronics systems are all new or upgraded significantly. In contrast, J-PARC E34 will employ a novel approach based on injection of an ultra-cold, low-energy, muon beam injected into a small, but highly uniform magnet. Only a small magnetic focusing field is needed to maintain storage, which distinguishes it from CERN, BNL and Fermilab. E34 aims to roughly match the previous BNL precision in their Phase 1 installation.

1 Introduction

At this Workshop, more than 25 presentations were devoted to topics centered on the muon anomalous magnetic moment, $a_\mu \equiv (g-2)/2$. These included a discussion of the two new measurement campaigns – as reported in this paper – and the theoretical issues related to both the Standard Model prediction and various new-physics speculations. In the narrative below, I will focus mainly on the Fermilab E989 experiment, which is in a mature construction state. An important update since this Workshop has been the successful commissioning of the superconducting storage ring magnet to its design field value and start of field-shimming operations. I will also briefly summarize the unique approach to measuring a_μ being taken by J-PARC E34 and point out, where it is important, the key differences between the experiments, [1].

1.1 The $g-2$ test

Muon $g-2$ is a special quantity because it can be both measured and predicted to sub-ppm precision, enabling the so-called $g-2$ test for new physics defined by $a_\mu^{\rm New} \equiv a_\mu^{\rm Exp} - a_\mu^{\rm SM}$. As a flavor- and CP-conserving, chirality-flipping, and loop-induced quantity, a_μ is especially sensitive to new physics contributions [2, 3]. The current $g-2$ test gives:

$$\Delta a_\mu^{\rm New} = [(263-289) \pm 80] \times 10^{-11} \quad (3.3-3.6)\,\sigma.$$

The range here represents different, but standard, evaluations [4, 5] of hadronic vacuum polarization (HVP) loops, and a common averaged value for hadronic light-by-light (HLbL) scattering. In fact, the range is even wider if all efforts to evaluate the SM are considered. These topics received extensive scrutiny at this Workshop. If this range for $\Delta a_\mu^{\rm New}$ is confirmed at a greater significance, the positive sign and relatively large magnitude – several times greater than the electroweak contribution – will provide important clues to the physics it is trying to reveal. The key phrase above is, "at a greater significance." Let's first focus on the combined experimental and theoretical uncertainty of 80×10^{-11} and investigate what contributes to this value and how the combined error might be reduced in the future.

1.2 The Standard Model Inputs

The SM terms are usually listed in five categories:

$$a_\mu^{\rm SM} = a_\mu^{\rm QED} + a_\mu^{\rm Weak} + a_\mu^{\rm HVP} + a_\mu^{\rm Had-HO} + a_\mu^{\rm HLbL}$$

with a quadrature summed total uncertainty of $\sim 50 \times 10^{-11}$.

The QED and Weak term uncertainties are totally negligible. These are impressive calculations. At this Workshop, Melnikov provided an interesting discussion exploring in particular the QED calculations and whether they could be wrong in any way; he strongly argues "no" [6].

The HVP contribution [7] is determined from experiment through a dispersion relation that amounts to an energy-weighted integral of $e^+e^- \to$ *hadron* total cross sections. The uncertainty at 42×10^{-11} is non-negligible and presently dominates the overall δa_μ(SM). This contribution depends on the accuracy of the reported data. Its quoted uncertainty is arrived at from reported cross section errors and, when necessary, errors are expanded to account for independent data sets that differ beyond statistical expectations. A small theoretical uncertainty owing

[a] e-mail: hertzog@uw.edu

© The Authors, published by EDP Sciences. This is an open access article distributed under the terms of the Creative Commons Attribution License 4.0 (http://creativecommons.org/licenses/by/4.0/).

to radiative corrections is also important. New experimental campaigns at BESIII [8] and the upgraded VEPP-2000 facility [9] in Novosibirsk, along with continued analyses of the large and varied BaBar data set [10] can be counted on to reduce the HVP uncertainty.

Higher-order HVP diagrams contribute a value of -98.4×10^{-11} to a_μ(SM), with negligible uncertainty. On the other hand, the higher-order HLbL effect is more problematic. Curiously, it has a total contribution of approximately $+105 \times 10^{-11}$, essentially canceling the HVP-HO term. At present, it can only be estimated using hadronic models, which typically have various strengths, weaknesses, and different limitations. Assigning an uncertainty is almost a guess. We use 26×10^{-11}, which is a consensus value reached by comparing models, but one could as well chose a number 50% larger, which many people do. More troubling is that these models could be badly wrong.

Fortunately intense efforts using lattice QCD have been making rapid progress toward a prediction of HLbL with an uncertainty goal below 10%. Lehner described recent efforts by the RBC and UKQCD Collaborations that aim to give 10% - 20% uncertainty soon on the quark-connected diagrams [11, 12]. Work in progress is focussed on the more difficult disconnected diagram, but the message given is that the end goal is achievable; it will be a matter of statistics more so than method.

An additional HLbL thrust is aimed at developing a data-driven approach using dispersion relations for the photon-photon scattering amplitude. Promising efforts described here by Procura [13] will make use of experimental spectra from BES III and KLOE that are now being harvested.

1.3 The Experimental Inputs

The measurement of a_μ is based on the following principles. When a muon with charge q is circulating in the horizontal plane of a magnetic storage ring, its cyclotron frequency is $\vec{\omega}_c = -q\vec{B}/m\gamma$. The muon spin precesses at frequency $\vec{\omega}_s = -(gq\vec{B}/2m) - [(1-\gamma)q\vec{B}/\gamma m]$, owing to the torque on the magnetic moment and including the Thomas precession effect for the rotating reference frame [14]. The magnitude of ω_s is greater than ω_c for $g \neq 2$. The difference is the anomalous precession frequency defined by

$$\vec{\omega}_a \equiv \vec{\omega}_s - \vec{\omega}_c = -\left(\frac{g-2}{2}\right)\frac{q\vec{B}}{m} = -a_\mu \frac{q\vec{B}}{m}, \tag{1}$$

where we have assumed for now a negligible effect from a non-zero electric dipole moment.

Parity violation in $\mu^+ \to e^+\bar{\nu}_\mu\nu_e$ associates the decay positron energies in the laboratory frame with the average spin direction of the muon at the time of the decay, such that the highest-energy positrons are preferentially emitted when the muon spin is aligned with its momentum and lower-energy positrons are emitted when the spin is reversed. Systems of detectors measure the decay positron times and energies.

To achieve the conditions described above, polarized muon bunches are injected into the magnet, kicked onto a stable storage orbit, and are then observed non-intrusively until they decay. In practice, a > 95% polarized muon beam can be obtained by capturing forward decays of in-flight pions ($\pi \to \mu\nu$) in a beamline lattice made of alternatively focussing then defocussing (FODO) quadrupole magnets. The captured beam will have a small, but finite, fractional transverse momentum component such that, unaided, this beam could not be stored in a simple transverse magnetic field; the muons would escape if not for some kind of vertical containment field. At Fermilab, and earlier at BNL and CERN, this is provided by an electric quadrupole system, which creates a sort of Penning trap.

The motional magnetic field seen by a relativistic muon passing through an electric field $\vec{E}$ contributes an important term to the spin precession rate, represented by

$$\vec{\omega}_a = -\frac{q}{m}\left[a_\mu\vec{B} - \left(a_\mu - \frac{1}{\gamma^2 - 1}\right)\frac{\vec{\beta}\times\vec{E}}{c}\right]. \tag{2}$$

At a muon momentum of $p_\mu = 3.094$ GeV/c, ($\gamma = 29.4$), the 2nd term in Eq. 2 exactly vanishes. The residual effect for muons slightly off the magic momentum, and therefore not centered in the null region of the electric quadrupoles, results in an E-field correction to the measured precession frequency. The beam also executes horizontal and vertical betatron motions at frequencies determined by the weak-focussing index of the storage ring (i.e, the electric field strength). The vertical undulation of the muons means $\vec{p}_\mu$ is not exactly perpendicular to $\vec{B}$, thus a small "pitch" correction is necessary. Combined, these corrections shift a_μ by $86(6) \times 10^{-11}$ [15]; the error was negligible in E821, but will need to be reduced for E989. This will be accomplished by more sophisticated particle tracking and by indirect measurements of the muon beam profile vs. time using an in-vacuum straw tracker system.

This detail is provided to contrast to the J-PARC experiment, which does not need such corrections. They will use a muon beam created by re-accelerating an ultra-cold, stopped muon source to a final injection momentum. The beam is predicted to have a relative transverse momentum of 10^{-5}, a nearly negligible value.

In both experiments, $a_\mu^{\rm Exp}$ is obtained from two independent measurements – the anomalous precession frequency and the magnetic field – plus the evaluation of many systematic error categories. Detectors are used to measure the anomalous precession frequency ω_a and pulsed proton NMR to measure the magnetic field in terms of the proton Larmor precession frequency, ω_p. Both measurements involve frequencies that are measured using highly stable precision oscillators. It is further necessary to know the muon distribution in the storage ring for the muon population that contributes to the ω_a data. This distribution is folded with similarly determined azimuthally averaged magnetic field moments to give the effective magnetic field seen by the muons, $\tilde{\omega}_p$ below. Given these experimentally determined quantities, one obtains a_μ at the precision needed through the relation

$$a_\mu^{\rm Exp} = \frac{g_e}{2}\frac{\omega_a}{\tilde{\omega}_p}\frac{m_\mu}{m_e}\frac{\mu_p}{\mu_e}. \tag{3}$$

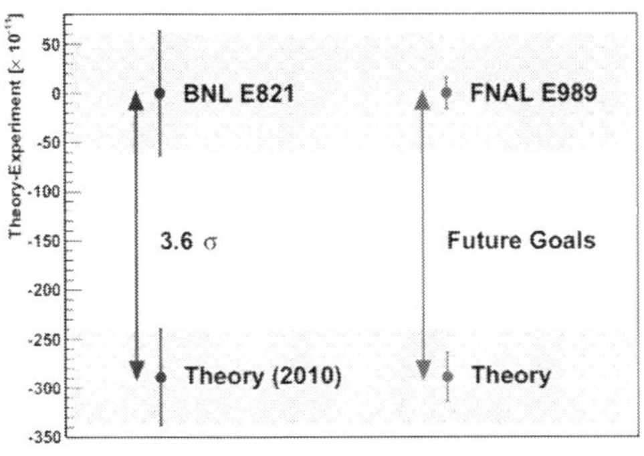

Figure 1. Comparison of Experiment to Theory at present and expected on completion of future Fermilab E989 and factor of ~ 2 improvements in HVP and HLbL calculations.

In this expression, a $g-2$ experiment reports the ratio of the muon precession frequency to the proton precession frequency, $R \equiv \omega_a/\tilde{\omega}_p$, where all systematic errors from the separate uncertainty table entries have been appropriately evaluated and combined in the uncertainty on R. That is the quantity reported by E821 [15]. From different experiments, one obtains the electron g_e factor [16], the muon-to-electron mass ratio, and the proton-to-electron magnetic moment ratio, [17]. Compared to the present 80×10^{-11} combined uncertainty in the $g-2$ test, these quantities are all known quite well. Of course, if any of their values change – e.g., the mass ratio was adjusted after the E821 final paper was published – one can update $a_\mu^{\rm Exp}$ easily.

Table 1 summarizes the latest versions of the absolute and relative uncertainties of the theoretical and experimental quantities used in the $g-2$ test. With the Fermilab E989 goal of $\delta a_\mu \sim 16\times10^{-11}$, the quantities that stand out, suggesting targets for improvement, can be identified. On the experimental side, this will be achieved by a 21-fold increase in statistics and reductions of systematic errors by a factor of 2 to 3. On the theory side, continued improvement in the HVP evaluation – perhaps as much as a factor of 2 eventually? – might be expected. Furthermore, the promise of recent work on the lattice suggests that a reduction of the HLbL uncertainty to $\sim 13\times10^{-11}$ is not an unreasonable goal. If the experiment and theory improvements are met, the uncertainty on the $g-2$ test could reduce from 80 to 30×10^{-11} and the New Physics significance for the range quoted in Eq. 1.1 would approach 9σ. Figure 1 illustrates the $g-2$ test now and in this optimistic future, where the central values are not altered from present values.

Table 1. Uncertainties on the quantities used to determine $a_\mu^{\rm Exp}$ and $a_\mu^{\rm SM}$. Experimental errors from Ref [15]. Theory errors, see Ref [6], except where noted. CODATA ratio uncertainties from the 2014 online update.

Quantity	Uncertainty $\times10^{-11}$	$\delta a_\mu/a_\mu$ (ppb)
Total ω_a Statistical	53	458
Final ω_a Systematic	24	210
Final $\tilde{\omega}_p$ Systematic	20	170
CODATA m_μ/m_e	2.6	22
CODATA μ_p/μ_e	0.35	3
Electron g factor, g_e	0.000035	0.0003
QED	0.08	0.7
Weak	1	8.6
Had-HO	0.7	6
HVP (e.g. Ref [4])	42	360
HLbL	26	223
Net theory	49	420
Final E821	63	540
Goal Fermilab E989	16	140
Goal J-PARC E34	47	400

Table 2. Comparison of various parameters for the Fermilab and J-PARC $g-2$ Experiments. Reproduced from [1].

Parameter	Fermilab E989	J-PARC E24
Statistical goal	100 ppb	400 ppb
Magnetic field	1.45 T	3.0 T
Radius	711 cm	33.3 cm
Cyclotron period	149.1 ns	7.4 ns
Precession frequency, ω_a	1.43 MHz	2.96 MHz
Lifetime, $\gamma\tau_\mu$	64.4 μs	6.6 μs
Typical asymmetry, A	0.4	0.4
Beam polarization	0.97	0.50
Events in final fit	1.5×10^{11}	8.1×10^{11}

2 The Experiments

The classic data display for the anomalous precession frequency is shown in Fig. 2. It represents the arrival time distribution for positrons having energies above threshold E_{th}. This simple method gives a statistical uncertainty on the precession frequency of

$$\delta\omega_a/\omega_a = \frac{1}{\omega_a\gamma\tau_\mu}\sqrt{\frac{2}{NA^2P^2}}. \tag{4}$$

The uncertainty is improved by running at a higher field ($\omega_a \propto B$), higher momentum ($\gamma\tau$), and high polarization of the incoming muon beam (P). The two experiments being mounted have rather different values for several of these parameters. In Table 2 – reproduced from [1] – the key differences are evident, including the precision goals. In both cases, the number of events N and the asymmetry A depend on E_{th}. Optimization of the "figure of merit" NA^2 is achieved for $E_{\rm th}/E_{\rm max} \approx 0.6$, which both experiments assume.

2.1 Fermilab E989

2.1.1 Beamline

Fermilab E989 will obtain high-purity bunches of ~ 97% polarized positive muons from a combination of beamlines that are part of the so-called Muon Campus. The

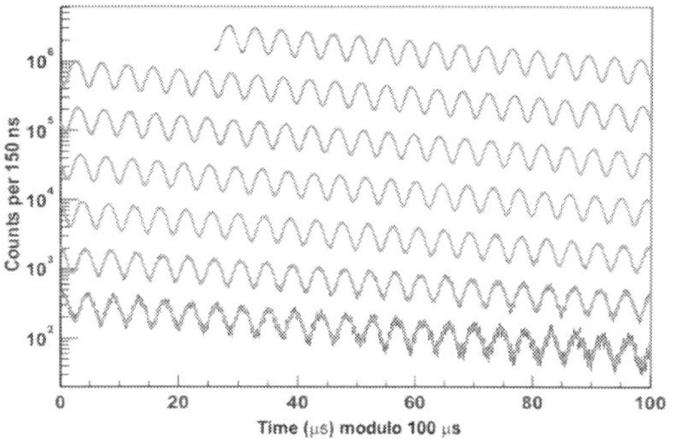

Figure 2. E821 anomalous precession data, including fit. The data is wrapped around every $100\,\mu s$. Figure courtesy E821 collaboration.

sequence is worth describing because it results in an ideal injected beam, having a pion contaminant fraction below 10^{-5}, and no protons. These hadrons could have caused a large hadronic "flash" at injection, probably paralyzing the detector systems and leading to baseline shifts on a slowly decaying background, as was experienced in E821. The clean beam is considered to be a major improvement.

The sequence starts at the Booster, where short batches of 8 GeV protons are delivered to the Recycler Ring. An RF system separates a batch into 4 tighter bunches of $\sim 10^{12}$ protons each. These bunches are extracted one at a time and directed to the former antiproton target station, which is tuned to collect +3.1 GeV/c π^+. Pions are transported along a 270 m FODO lattice and then are injected into the repurposed $\bar{p}$ Delivery Ring (DR), where they make several revolutions before being kicked into a final beamline that terminates at the storage ring entrance. Each bunch is separated by 11 ms or more, and 16 bunches are expected every 1.33-second accelerator cycle leading to an average storage ring fill rate of 12 Hz. The described beam path involves many changes compared to existing equipment. While the target system can be used nearly as is, the long beamlines and the DR needs significant work. A complete end-to-end simulation of the beam, from production to arrival at the storage ring was used to optimize the lattice and to predict the intensity and phase space distribution of muons at injection. We expect up to $7 \times 10^5 \mu^+$ per bunch in a $\Delta p/p \approx 2\%$ momentum bite – which is much greater than the ring acceptance of less than 0.2%.

2.1.2 Storage Ring

The BNL E821 storage ring [18] will be reused at Fermilab. Following a highly publicized move mostly on water from New York to Illinois of the superconducting coils, and a less-publicized move of hundreds of tons of steel yoke and other components by truck, the storage ring has now been fully re-assembled. It resides in the new MC-1 building, which has a strong supportive flooring system and a uniform and stable year-round temperature control.

During summer 2015, the magnet was first powered, followed by minor repairs to a superconducting lead. Full power and full field were achieved in September. No muon would be stored in the magnet without a number of critical storage ring subsystems, many of which are being upgraded significantly for E989. Muons first enter the ring through a bored-out tunnel through the back-leg iron yoke of the C-shaped magnet. They pass through the narrow horizontal constrictions of a superconducting inflector magnet, which creates a nearly field-free corridor. Emerging from the inflector, they traverse a high-gradient field region and cross through the outer electrode Al plate of one of the four electric quadrupole systems before entering the region of uniform magnetic field. One quarter of the way around the ring, a fast magnetic kicker applies an outward transverse 10-11 mrad angular kick during the first turn (only) placing the muon beam on the desired central orbit. Five collimators placed around the ring define the storage volume of 9 cm diameter transverse and circular storage volume.

This sequence is non-trivial to optimize. The incoming beamline Twiss parameters, the inflector angle and field magnitude, and the individual kicker magnitudes and firing sequence can all be adjusted. Modern simulation tools were used to recreate the BNL E821 conditions and consequently led to design improvements for E989. For example, by reducing multiple scattering in the quadrupole plates and high-voltage standoff structures, a significant increase in the stored muon fraction can be obtained. Similarly, if the present close-ended windings on the BNL inflector magnet were to be removed – a new design is in development with exactly this feature – the stored muon fraction will be nearly doubled.

One key device that demanded a major upgrade from the outset is the kicker magnet. For E821, its LCR-based pulse forming network was too slow and the magnitude of the kick too low. These led to an non-optimized storage fraction and large coherent betatron motions of the stored beam – a fact that resulted in a fairly large systematic uncertainty on ω_a. The new kicker will be energized through a Blumlein triaxial transmission line [19]. The pulse rises, flat-tops, and falls, within the 149 ns cyclotron period of the ring – an important fact. With newly shaped kicker plate geometry, the field magnitude can exceed the predicted maximum, such that a tuning sweep can be performed to optimize the storage fraction while also minimizing the coherent betatron amplitude. The aim of these systems is to result in storage of about 18,000 muons per fill immediately after injection, an increase compared to BNL of about 2.5.

2.1.3 Precision Magnetic Field

With the magnet fully powered, the shimming operations designed to smooth out the natural variations in the "as assembled" magnetic field vs. azimuth around the ring have begun. Although the four energizing superconducting coils are continuous, the surrounding steel is built in coarse units. Each 30° yoke section has three pairs of precision pole pieces, 72 adjustable wedge shims, and sets of adjustable edge shims. Placement of these components,

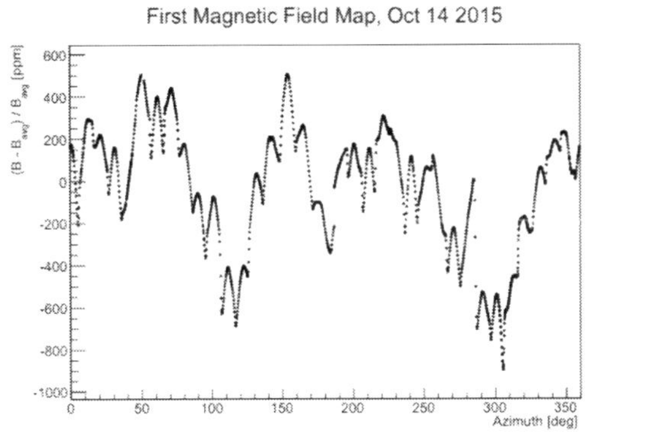

Figure 3. First "as built" raw field map for the reassembled storage ring magnet for E989. The nominal field is 1.4513 T. The lumpy features will be greatly reduced as the shimming process begins (see text).

and the specifics of their dimensions determines the overall field strength in the local region. The pieces are designed to be adjusted to account for the natural variance one starts with owing to tiny (10's of microns) placement errors, and intrinsic magnetic material variances. The task began by establishing the baseline "raw field map", as shown in Fig. 3. The data is obtained using a specialized shimming trolley having 25 NMR probes and Capacitec sensors that precisely measure the pole gap vs azimuth. The location of the shimming trolley is determined reproducibly by a laser alignment system. The field shown is obviously lumpy on the given scale, which is expected. The fine humped features stem from the curvature of each pole surface, which leads to a changing pole-to-pole gap and thus a strongly dependent B field. The longer wavelength features are dominated by three things: the initial conditions of the wedge shims, the top-hat air gap, and the relative pole alignment. Adjustments to flatten these features rely on an iterative process that involves predictive actions based on an `OPERA` simulation, followed by re-measurement.

After coarse shimming is completed, the vacuum chamber system will be installed. It includes the inflector, electric quadrupoles, and the kicker. Further finer adjustments using active current shim coils will then be made. The field in the storage ring volume will be measured by the *in vacuum* trolley with 17 NMR probes, which rides on a set of rails inside the vacuum chamber and records its position using a bar-code reader. It is designed to enter the storage region and map the field exactly where the muons are stored.

All field measurements involve pulsed proton NMR, in which the upgraded data collection procedure will record the free-induction decay (FID) waveforms, rather than simply counting zero crossings. While the E821 system was indeed excellent, the many intervening years degraded much of the inventory. After evaluating the status of the existing systems, a decision was made to build from scratch the 400 fixed probes and the 17 trolley probes. The absolute field probe from E821 has been preserved and will be reused in E989. It was also the same probe used in the muonium hyperfine experiment that established the muon-to-proton magnetic moment ratio [20]. To complement this key tool, new absolute probes are being developed as further cross checks. The electronics to provide the $\pi/2$ pulses and to record the FIDs is mostly new, as is much of the internal controller circuitry inside the trolley.

The uncertainty in the field measurement is entirely systematic. Many of the individual entries in the systematic error table can be reduced naturally by additional, but not radical, efforts. The field systematic error in E821 was already at the 170 ppb level; for E989, the goal is 70 ppb. An underlying improvement that affects many of the systematic issues is the intrinsic field variation in the storage volume – the goal of the intensive shimming program. Better location of the trolley during its measuring cycle, reduced trolley temperature fluctuations, greater field stability from improved experimental hall temperature control, a more complete set of fixed NMR probes, and storage of the FID waveforms which allows field determination in regions of higher gradients, are all important subplots to this unfolding story. As the work has just now begin, we can look forward in about a year to learn how successful the effort has been.

2.1.4 Detector Systems

The detector systems for E989 are entirely new, as is the supporting electronics, and the fast and slow data acquisition systems. The detector systems will include:

- An entrance scintillating paddle and several sets of scintillating fiber hodoscopes to determine the incoming muon beam intensity profile vs. time and space.
- Two sets of rebuilt scintillating fiber hodoscopes placed in the path of the stored muons to measure (somewhat destructively) the stored beam $x - y$ profile in two locations; these detectors are used to optimize the storage and measure the coherent betatron motion. They are rotated out of the way for normal data taking.
- Three sets of in-vacuum straw trackers that reside in the scallop of the vacuum chamber adjacent to the muon storage volume. They provide data of decay positron tracks that can be "traced back" to the point of tangency from the muon that decayed. This provides a transverse stored-muon profile vs. time-in-fill. They also serve to calibrate the calorimeters and they are sensitive to an electric dipole moment of the muon (see [21]).
- Twenty-four stations of electromagnetic calorimeters that are positioned to maximally intercept the higher-energy decay positrons and determine their time of arrival and energy.

The data that appears in the ω_a plot shown in Fig. 2 is extracted from calorimeter signals read out using continuous sampling waveform digitizers. The instantaneous rate at BNL approached several MHz per station. At Fermilab it will exceed that by the 2.5 times higher stored muon rate. Because pileup was a leading systematic error that scales with rate – and must be reduced – a segmented and very fast calorimeter is being designed. Pileup occurs when two decay positrons strike the calorimeter at nearly the same time, and are interpreted as one decay with a larger energy than either individually. If it exceeds E_{th} such that the event is accepted, the phase will be wrong because the two positrons that led to this fake single "high" energy event had each traveled a shorter distance from their parent muons than the single high-energy positron would have that they mimic. That is, the time from muon decay to hit a detector is shorter on average for the low-energy positrons than the higher-energy positron. Mitigating the level of intrinsic pileup is accomplished by segmenting the calorimeter (unlike E821) and by choosing a technology that allows two-pulses to be resolved with a separation of just a few nsec.

Each of the E989 calorimeters will consist of 54 PbF_2 Cherenkov crystals stacked in a 6 high by 9 wide array. The compact placement, see Fig 4, and the proximity to the storage ring field, does not allow conventional PMTs to be attached to the downstream face of the crystals. An ultra-fast readout based on the latest generation of Hamamatsu MPPCs (SiPMs, or silicon photomultipliers) will be used; see inset to Fig 4. Each SiPM has 57,600 $50\,\mu$m pitch pixels, all operated in Geiger mode. The summed current is routed to a bank of custom 800 MHz, 12-bit-depth, waveform digitizers. The SiPMs have excellent gain stability if the temperature and bias are controlled; they are relatively impervious to rate (tested above 10 MHz); and, they are not affected by, nor do they affect, the magnetic field. Continuous digitization of the 54 crystals throughout each fill leads to a raw data rate of 18 GB/s for the 24 calorimeter stations, which is transferred to a bank of GPU processors where the sparsely occurring individual pulse islands are extracted from the data stream and stored.

Paramount to the systematic control is a very high degree of gain calibration and gain stability, both long term and, even more importantly, short term. Over a $700\,\mu$s fill, the instantaneous rate drops by more than four orders of magnitude, a typical recipe in detectors for a gain change. To monitor this precisely, a laser calibration system is being developed [22] that will flash all crystals prior to each fill and, during special calibration runs, sparsely overlay calibration pulses on top of the real data in a pattern that leads to a series of 10^{-4} gain measurements every $5\,\mu$s throughout the fill. The calorimeter system and gain stability were tested at SLAC and the resolution and overall performance was excellent, see [23].

2.1.5 Reduction in Error Summary

The beamline and improved muon storage instrumentation will lead to the needed ×21 in events to reach the desired statistical error of 100 ppb. The improved shimming, upgraded NMR system and analysis, and more frequent trolley maps, will reduce the ω_p systematic uncertainty to a target goal of 70 ppb. The long and pure beamline will largely reduce the muon loss systematic. The improved

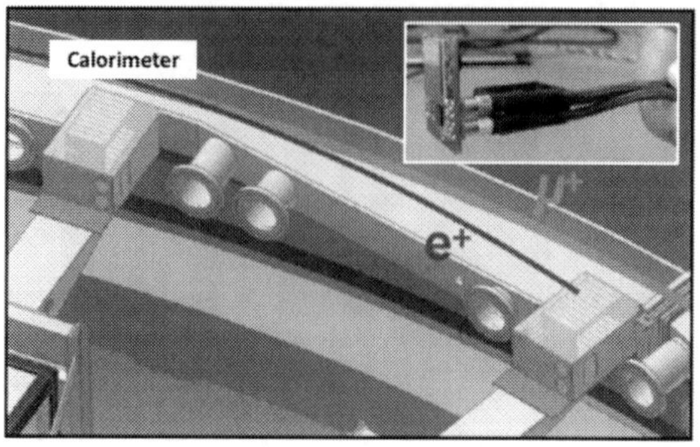

Figure 4. Positioning of two of the segmented calorimeter stations just downstream of the scalloped section of the vacuum chambers. The central muon orbit is sketched in, as is a nominal decay positron. The inset shows the size of the SiPM readout and its custom amplifier and pulse-shaping board.

kicker and adjustment of the weak-focussing tune, will reduce the systematic error related to coherent betatron motions. The E-field and pitch corrections are improved by simulation and measurement. The pileup will be reduced by the segmentation and faster detector response, as well as the many times better resolution and much higher sampling and bit-depth of the digitizers. Finally, the gain stability will be monitored by the new calibration system. Overall, the systematic error goal for ω_a is 70 ppb. The E989 total error budget is the quadrature sum of the above values, giving $\delta a_\mu / a_\mu$ = 140 ppb.

2.2 J-PARC E34

The J-PARC E34 Collaboration is developing a very unique $g-2$ experiment. They start with production of surface muons – those born from pions at rest – which have a momentum of 29 MeV/c. They are stopped in a silica aerogel target where micro-channels have been created using a laser ablation technique [24]. A large fraction of muonium $M \equiv \mu^+ e^-$ atoms formed in the target diffuse out and emerge in a near field-free region in vacuum. The atoms at this point are essentially at rest. The triplet and singlet muonium combinations have relative populations of $\frac{1}{2}, \frac{1}{4}, 0$ and $\frac{1}{4}$, for the (F, M_F) states $(1,1), (1,0), (1,-1)$ and $(0,0)$, respectively.

Two lasers are fired on the M atoms. A λ = 122 nm shot excites the 1S to 2P transition, and a λ = 355 nm shot ionizes the atom, leaving a source of at-rest μ^+. The retained polarization is P_z = 50%, derived exclusively from atoms in the triplet (1,1) state. The (1,0) and (0,0) populations can make transitions at the rate $\nu_{24} \approx 4.5$ GHz, which is so high that no net polarized signal will be left. To date, experiments at TRIUMF have demonstrated [24] a source scaled to J-PARC conditions of about $2 \times 10^5 \mu$/s.

These at-rest, ultra-cold, polarized muons will be accelerated to a momentum of 300 MeV/c using a new muon linac. The beam should have negligible transverse dimensions and transverse momentum. It is, in fact, quite a unique beam that could be used for a $\mu^+ - p$ scattering experiment such as the one proposed by MUSE to help resolve the proton radius puzzle – but, that is another topic (see [1]).

The 10 times lower momentum muons compared to CERN, BNL and Fermilab, implies the need for a much smaller magnet. Figure 5 shows the compact design, which is based on a conventional MRI magnet, operated at 3 T. These magnets have rather uniform magnetic field throughout the inner volume. The muons are injected a steep vertical angle, then they are kicked to the horizontal plane, and finally a small additional magnetic field beyond the solenoid is used to provide vertical containment (it does not appreciably perturb ω_a). Because of the low momentum and high field, the storage radius is just 33 cm and the cyclotron period is only 7.4 ns. The at-most polarization of 50% results in an asymmetry in the precession data of about half compared to Fermilab for the same relative energy threshold. Further, the 10 times shorter lifetime reduces considerably the measuring interval for each fill, which affects the precision.

That said, there are many interesting features here. The magnet will not require nearly the attention to establish its mapped uniformity. The detector system uses a radial arrangement of vanes of silicon strip detectors that provide a uniform acceptance for in-curling positrons from any azimuthal decay position. The rates anticipated are very high, but if the detector system can operate efficiently, it should be relatively free of pileup and perhaps not subject to major gain changes. The EDM measurement, which both experiments will carry out in parallel, is well suited here as it is based on up/down sloping tracks versus time. The E989 experiment will have only a few tracker stations.

The design of E34 is new enough such that even a working systematic error table has not yet been discussed in public, so it is difficult to predict the final uncertainty they can achieve. The Phase 1 statistical goal is roughly 400 ppb, a good comparison to BNL E821, but using a very different method.

3 Summary

These are exciting times. As evidenced at this Workshop by the variety of presentations on specific calculations and measurements, many people are working on $g-2$ in one way or another. The motivation is clear for all. The current discrepancy between measurement and theory is one of the most promising indications of new physics, and the new experimental campaigns and the continued improvement in theory are leading to "discovery level" sensitivity in the $g-2$ test. Fermilab E989 should begin collecting data sometime in 2017. The collaboration will plan for a first result at roughly the precision level of BNL, but the experiment will run continuously to acquire the much larger data set, well into 2018 and possibly beyond. As with most modern precision experiments, the analysis is blinded and the emphasis on systematic uncertainty study completion will rule the timing of any result announcements. We are all looking forward to the next few years of vigorous activity.

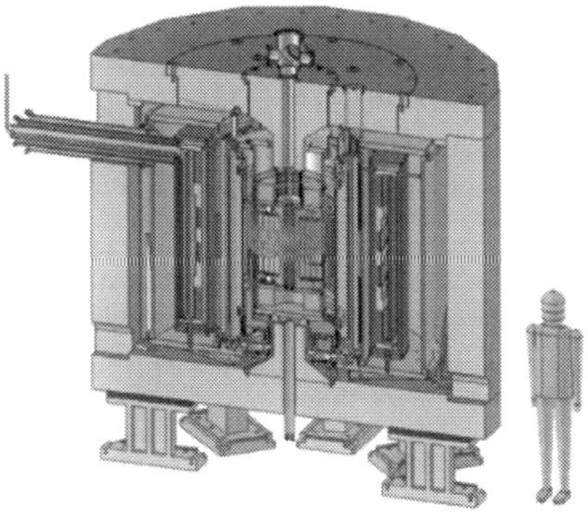

Figure 5. The proposed setup for the J-PARC $g-2$ Experiment. Muons enter at the top left (green trajectory) and spiral into the highly uniform magnetic field region. Their decay positrons curl inward to an array of silicon tracking detectors. Figure courtesy T. Mibe.

Acknowledgments

The author thanks the organizers of this marvelous Workshop, his many colleagues on the Fermilab E989 $g-2$ experiment, and T. Mibe and G. Marshall from J-PARC E34. The author is supported by the DOE Office of Nuclear Physics, award DE-FG02-97ER41020.

References

[1] T.P. Gorringe, D.W. Hertzog, Prog. Part. Nucl. Phys. **84**, 73 (2015), 1506.01465
[2] A. Czarnecki, W.J. Marciano, Phys.Rev. **D64**, 013014 (2001)
[3] D. Stockinger, this workshop, *Recent developments on the MSSM contribution to the muon* $g-2$. (2015)
[4] M. Davier, A. Hoecker, B. Malaescu, Z. Zhang, Eur.Phys.J. **C71**, 1515 (2011)
[5] K. Hagiwara, R. Liao, A.D. Martin, D. Nomura, T. Teubner, J.Phys. **G38**, 085003 (2011)
[6] K. Melnikov, this workshop, *Theory Review of muon* $g-2$. (2015)
[7] R. Chislett, this workshop, *Leading-order hadronic contribution to the electron and muon* $g-2$ (2015)
[8] M. Ablikim et al. (BESIII) (2015), 1507.08188
[9] S. Eidelman, this workshop, *Review on measurement of the R ratio at Novosibirsk and ISR, and radiative corrections.* (2015)
[10] Z. Zhang, this workshop, *Review of recent calculations of the hadronic vacuum polarization contribution to* a_μ. (2015)
[11] C. Lehner, this workshop, *Hadronic light-by-light contribution to* $(g-2)_\mu$ *from lattice QCD.* (2015)
[12] T. Blum, N. Christ, M. Hayakawa, T. Izubuchi, L. Jin, C. Lehner (2015), 1510.07100
[13] M. Procura, *et al*, this workshop, *Dispersion relation for hadronic light-by-light scattering* (2015)

[14] V. Bargmann, L. Michel, V. Telegdi, Phys.Rev.Lett. **2**, 435 (1959)

[15] G. Bennett et al. (Muon G-2 Collaboration), Phys.Rev. **D73**, 072003 (2006)

[16] D. Hanneke, S. Fogwell, G. Gabrielse, Phys. Rev. Lett. **100**, 120801 (2008), 0801.1134

[17] CODATA, *CODATA 2014 Online Recommended Values.* (2014)

[18] G. Danby, L. Addessi, Z. Armoza, J. Benante, H. Brown et al., Nucl.Instrum.Meth. **A457**, 151 (2001)

[19] J. Grange et al. (Muon g-2) (2015), 1501.06858

[20] W. Liu, M.G. Boshier, S. Dhawan, O. van Dyck, P. Egan et al., Phys.Rev.Lett. **82**, 711 (1999)

[21] R. Chislett, this workshop, *The EDM in the $g-2$ experiment* (2015)

[22] A. Anastasi et al. (Muon g-2), Nucl. Instrum. Meth. **A788**, 43 (2015), 1504.00132

[23] A. Fienberg, L. Alonzi, A. Anastasi, R. Bjorkquist, D. Cauz et al., *Studies of an array of PbF2 Cherenkov crystals with large-area SiPM readout* (2015), 1412.5525

[24] G.A. Beer, Y. Fujiwara, S. Hirota, K. Ishida, M. Iwasaki et al., PTEP **2014**, 091C01 (2014)

EPJ Web of Conferences **118**, 01016 (2016) DOI: 10.1051/epjconf/201611801016
FCCP2015

Leading-order hadronic contribution to the electron and muon $g-2$

Fred Jegerlehner[1,2,a]

[1] *Deutsches Elektronen–Synchrotron (DESY), Platanenallee 6, D–15738 Zeuthen, Germany*
[2] *Humboldt–Universität zu Berlin, Institut für Physik, Newtonstrasse 15, D–12489 Berlin, Germany*

Abstract. I present a new data driven update of the hadronic vacuum polarization effects for the muon and the electron $g-2$. For the leading order contributions I find $a_\mu^{\rm had(1)} = (688.57 \pm 4.28)[688.91 \pm 3.52] \times 10^{-10}$ based on e^+e^-data [incl. τ data], $a_\mu^{\rm had(2)} = (-9.92 \pm 0.10) \times 10^{-10}$ (NLO) and $a_\mu^{\rm had(3)} = (1.23 \pm 0.01) \times 10^{-10}$ (NNLO) for the muon, and $a_e^{\rm had(1)} = (185.11 \pm 1.24) \times 10^{-14}$ (LO), $a_e^{\rm had(2)} = (-22.15 \pm 0.16) \times 10^{-14}$ (NLO) and $a_e^{\rm had(3)} = (2.80 \pm 0.02) \times 10^{-14}$ (NNLO) for the electron. A problem with vacuum polarization undressing of cross-sections (time-like region) is addressed. I also add a comment on properly including axial mesons in the hadronic light-by-light scattering contribution. My estimate here reads $a_\mu[a_1, f_1', f_1] \sim (7.51 \pm 2.71) \times 10^{-11}$. With these updates $a_\mu^{\rm exp} - a_\mu^{\rm the} = (31.0 \pm 8.2) \times 10^{-10}$ a 3.8 σ deviation, while $a_e^{\rm exp} - a_e^{\rm the} = (-1.14 \pm 0.82) \times 10^{-12}$ shows no significat deviation.

1 Introduction: hadronic effects in $g-2$.

A well known general problem in electroweak precision physics are the higher order contributions from hadrons (quark loops) at low energy scales. While leptons primarily exhibit the fairly weak electromagnetic interaction, which can be treated in perturbation theory, the quarks are strongly interacting via confined gluons where any perturbative treatment breaks down. Considering the lepton anomalous magnetic moments one distinguishes three types of non-perturbative corrections: **(a)** Hadronic Vacuum Polarization (HVP) of order $O(\alpha^2), O(\alpha^3), O(\alpha^4)$; **(b)** Hadronic Light-by-Light (HLbL) scattering at $O(\alpha^3)$; **(c)** hadronic effects at $O(\alpha G_F m_\mu^2)$ in 2-loop electroweak (EW) corrections, in all cases quark-loops appear as hadronic "blobs". The hadronic contributions are limiting the precision of the predictions.

Evaluation of non-perturbative effects is possible by using experimental data in conjunction with Dispersion Relations (DR), by low energy effective modeling via a Resonance Lagrangian Approach (RLA) (Vector Meson Dominance (VMD) implemented in accord with chiral structure of QCD) [1–3], like the Hidden Local Symmetry (HLS) or the Extended Nambu Jona-Lasinio (ENJL) models, or by lattice QCD. Specifically: **(a)** HVP via a dispersion integral over $e^+e^- \to$ hadrons data (1 independent amplitude to be determined by one specific data channel) (see e.g. [4]) as elaborated below, by the HLS effective Lagrangian approach [5–9], or by lattice QCD [10–15]; **(b)** HLbL via a RLA together with operator product expansion (OPE) methods [16–19], by a dispersive approach using $\gamma\gamma \to$ hadrons data (28 independent amplitudes to be determined by as many independent data sets in principle) [20–22] or by lattice QCD [23, 24]; **(c)** EW quark-triangle diagrams are well under control, because the possible large corrections are related to the Adler-Bell-Jackiw (ABJ) anomaly which is perturbative and non-perturbative at the same time. Since VVV = 0 by the Furry theorem, only VVA (of $\gamma\gamma Z$ -vertex, V=vector, A=axialvector) contributes. In fact leading effects are of short distance type (M_Z mass scale) and cancel against lepton-triangle loops (anomaly cancellation) [25, 26].

2 Leading-order $a_\mu^{\rm had}$ via $\sigma(e^+e^- \to \text{hadrons})$

The leading *non-perturbative* hadronic contribution to $a_\mu^{\rm had}$, represented by the diagram figure 1, can be obtained

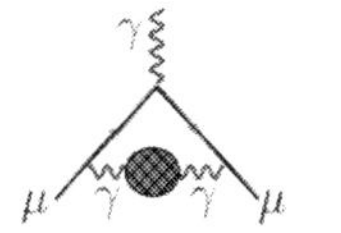

Figure 1. Leading diagram exhibiting a hadronic "blob".

in terms of undressed experimental cross-sections data

$$R_\gamma(s) \equiv \sigma^{(0)}(e^+e^- \to \gamma^* \to \text{hadrons})/\frac{4\pi\alpha^2}{3s}\,, \qquad (1)$$

$s = E_{\rm cm}^2$, $E_{\rm cm}$ the center of mass energy, and the DR:

$$a_\mu^{\rm had} = \left(\frac{\alpha m_\mu}{3\pi}\right)^2 \Bigg(\int\limits_{4m_\pi^2}^{E_{\rm cut}^2} ds \frac{R_\gamma^{\rm data}(s)\,\hat{K}(s)}{s^2} + \int\limits_{E_{\rm cut}^2}^{\infty} ds \frac{R_\gamma^{\rm pQCD}(s)\,\hat{K}(s)}{s^2} \Bigg). \qquad (2)$$

[a] e-mail: fjeger@physik.hu-berlin.de

© The Authors, published by EDP Sciences. This is an open access article distributed under the terms of the Creative Commons Attribution License 4.0 (http://creativecommons.org/licenses/by/4.0/).

 DOI: 10.1051/epjconf/201611801016

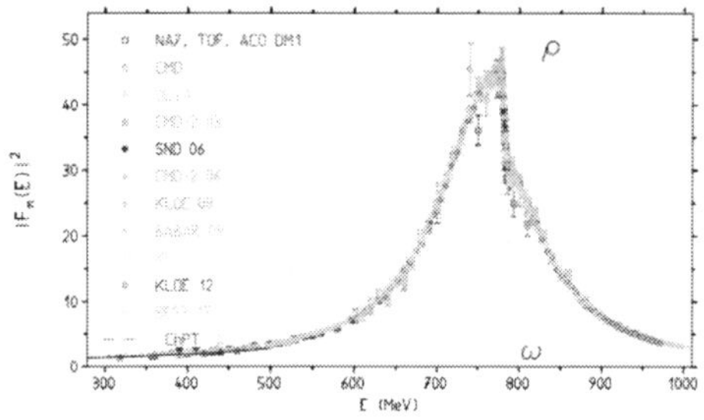

Figure 2. The pion form factor $|F_\pi(s)|^2 = 4R_{\pi\pi}/\beta_\pi^3$ ($\beta_\pi = \sqrt{(1-4m_\pi^2/s)}$) dominated by the ρ resonance peak. Data include measurements from Novosibirsk (NSK) [27–29], Frascati (KLOE) [30–32], SLAC (BaBar) [33] and Beijing (BESIII) [34].

Table 1. Results for $a_\mu^{\mathrm{had}(1)}$ (in units $\times 10^{-10}$).

final state	range (GeV)	$a_\mu^{\mathrm{had}(1)}$ (stat) (syst) [tot]	rel[abs]%
ρ	(0.28, 1.05)	507.55 (0.39) (2.68)[2.71]	0.5 [39.9]
ω	(0.42, 0.81)	35.23 (0.42) (0.95)[1.04]	3.0 [5.9]
ϕ	(1.00, 1.04)	34.31 (0.48) (0.79)[0.92]	2.7 [4.7]
J/ψ		8.94 (0.42) (0.41)[0.59]	6.6 [1.9]
Υ		0.11 (0.00) (0.01)[0.01]	6.8 [0.0]
had	(1.05, 2.00)	60.45 (0.21) (2.80)[2.80]	4.6 [42.9]
had	(2.00, 3.10)	21.63 (0.12) (0.92)[0.93]	4.3 [4.7]
had	(3.10, 3.60)	3.77 (0.03) (0.10)[0.10]	2.8 [0.1]
had	(3.60, 5.20)	7.50 (0.04) (0.01)[0.04]	0.3 [0.0]
pQCD	(5.20, 9.46)	6.27 (0.00) (0.01)[0.01]	0.0 [0.0]
had	(9.46,13.00)	1.28 (0.01) (0.07)[0.07]	5.4 [0.0]
pQCD	(13.0,∞)	1.53 (0.00) (0.00)[0.00]	0.0 [0.0]
data	(0.28,13.00)	680.77 (0.89) (4.19)[4.28]	0.6 [100.]
total		688.57 (0.89) (4.19)[4.28]	0.6 [100.]

The kernel $\hat{K}(s)$ is an analytically known monotonically increasing function, raising from about 0.64 at the two pion threshold $4m_\pi^2$ to 1 as $s \to \infty$. This integral is well defined due to the asymptotic freedom of QCD, which allows for a perturbative QCD (pQCD) evaluation of the high energy contributions. Because of the $1/s^2$ weight, the dominant contribution comes from the lowest lying hadronic resonance, the ρ meson (see figure 2). As low energy contributions are enhanced, about ~ 75% come from the region $2m_\pi < \sqrt{s} < 1$ GeV dominated by the $\pi^+\pi^-$ channel. Experimental errors imply theoretical uncertainties, the main issue for the muon $g-2$. Typically, results are collected from different resonances and regions as presented in table 2. Statistical errors (stat) are summed in quadrature, systematic (syst) ones are taken into account linearly (100% correlated) within the different contributions of the list, and summed quadratically from the different regions and resonances. From 5.2 GeV to 9.46 GeV and above 13 GeV pQCD is used. Relative (rel) and absolute (abs) errors are also shown. The distribution of contributions and errors are illustrated in the pie chart figure 3. As a result we find

$$a_\mu^{\mathrm{had}(1)} = (688.57 \pm 4.28)[688.91 \pm 3.52] \times 10^{-10} \quad (3)$$

based on e^+e^-–data [incl. τ-decay spectra [35]]. In the last 15 years e^+e^- cross-section measurements have dramatically improved, from energy scans [27–29] at Novosibirsk (NSK) and later, using the radiative return mechanism, measurements via initial state radiation (ISR) at meson factories (see figure 4) [30–34]. A third possibility to enhance experimental information useful to improve HVP estimates are τ –decay spectra $\tau \to \bar{\nu}_\tau \pi^0\pi^-, \cdots$, supplied by isospin breaking effects [5–7, 35–40]. In the conserved vector current (CVC) limit τ spectra should be identical to the isovector part $I = 1$ of the e^+e^- spectra, as illustrated in figure 5. Including the $I = 1$ $\tau \to \pi\pi\nu_\tau$ data available from [41–45] in the range [0.63-0.96] GeV one obtains [35]:

$$\begin{aligned} a_\mu^{\mathrm{had}}[ee \to \pi\pi] &= 353.82(0.88)(2.17)[2.34] \times 10^{-10} \\ a_\mu^{\mathrm{had}}[\tau \to \pi\pi\nu] &= 354.25(1.24)(0.61)[1.38] \times 10^{-10} \\ a_\mu^{\mathrm{had}}[\ ee + \tau\ ] &= 354.14(0.82)(0.86)[1.19] \times 10^{-10}, \end{aligned}$$

which improves the LO HVP as given in (3). We briefly summarize recent progress in data collection as follows.

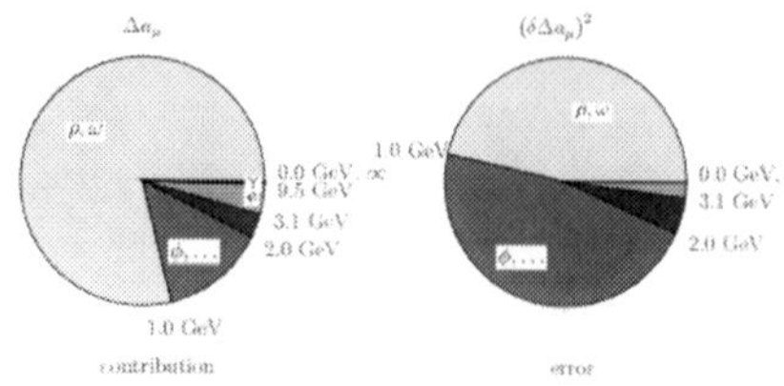

Figure 3. Muon $g-2$: distribution of contributions and error squares from different energy ranges.

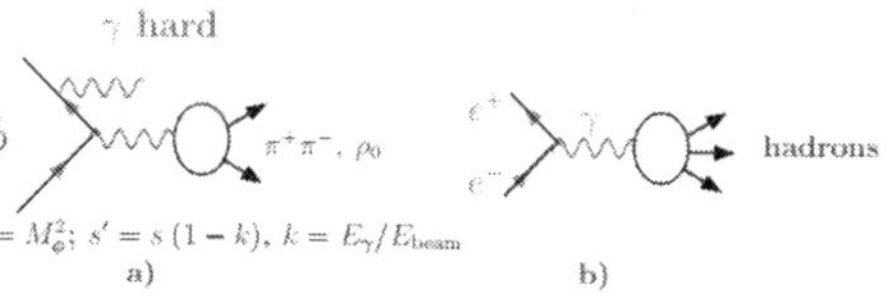

Figure 4. a) Initial state radiation (ISR), b) Standard energy scan.

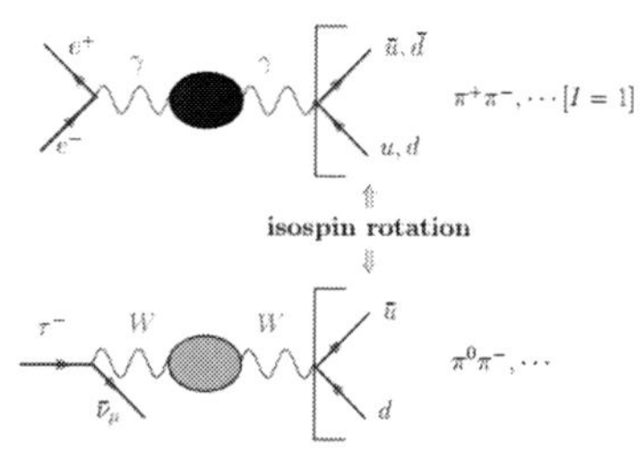

Figure 5. τ-decay data may be combined with I=1 part of e^+e^- annihilation data after isospin rotation $[\pi^-\pi^0] \Leftrightarrow [\pi^-\pi^+]$ and applying isospin breaking (IB) corrections (e.m. effects, phase space, isospin breaking in masses, widths, $\rho^0-\omega$ mixing etc.).

2.1 Data

As I mentioned the most important data are the $\pi\pi$ production data in the range up to 1 GeV. New experimental

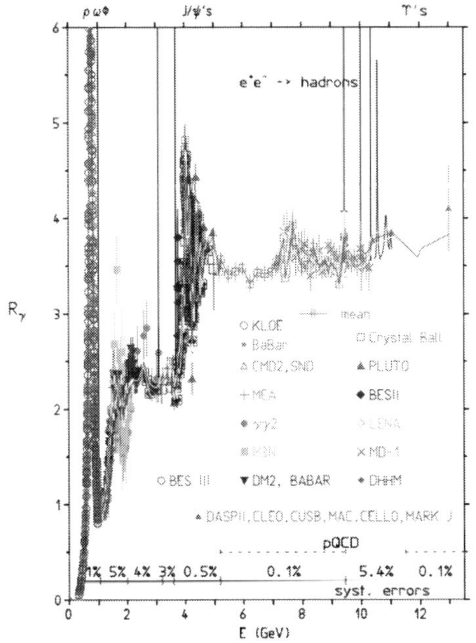

Figure 6. My 2015 compilation of R_γ as a function of energy E.

input for HVP comes from BESIII [34]. Still the most precise ISR measurements from KLOE and BaBar are in conflict, but the new ISR data from BESIII steps to resolve this tension, as it lies in between the two. Other data recently collected, and published up to the end of 2014, include the $e^+e^- \to 3(\pi^+\pi^-)$ data from CMD–3 [46], the $e^+e^- \to \omega\pi^0 \to \pi^0\pi^0\gamma$ from SND [47] and several data sets collected by BaBar in the ISR mode[1] [48–51]. These data samples highly increase the available statistics for the annihilation channels opened above 1 GeV and lead to significant improvements. Recent/preliminary results also included are $e^+e^- \to \pi^+\pi^-\pi^0$ from Belle, $e^+e^- \to K^+K^-$ from CMD-3, $e^+e^- \to K^+K^-$ from SND. The resulting data sample is collected in figure 6, which has indicated the overall precision of the different ranges as well as the pQCD ranges, where data are replaced by pQCD results. Still one of the main issue in HVP is $R_\gamma(s)$ in the region 1.2 to 2.4 GeV, which actually has been improved dramatically by the exclusive channel measurements by BaBar in the last decade. The most important 20 out of more than 30 channels are measured, many known at the 10 to 15% level. The exclusive channel therefore has a much better quality than the very old inclusive data from Frascati (see figure 7).

2.2 NLO and NNLO HVP effects updated

The next-to-leading order (NLO) HVP is represented by diagrams in figure 8. With kernels from [53], the results of an updated evaluation are presented in table 2. The next-to-next leading order (NNLO) contributions have

[1] Including the $p\bar{p}$, K^+K^-, K_LK_S, $K_LK_S\pi^+\pi^-$, $K_SK_S\pi^+\pi^-$, $K_SK_SK^+K^-$ final states.

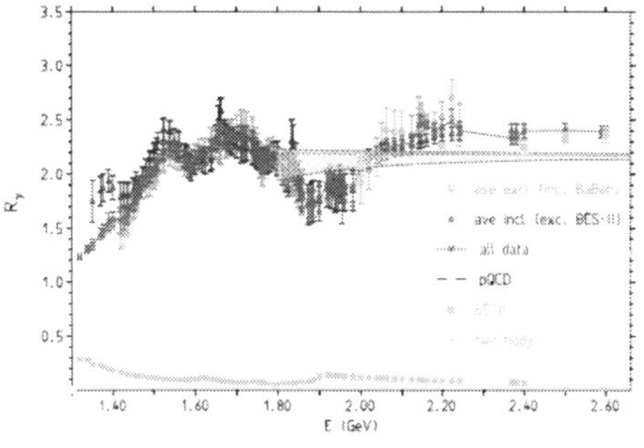

Figure 7. e^+e^- annihilation data in the 1.4 to 2.6 GeV region. Summed up exclusive (excl) channel data are shown together with old inclusive data (incl). Two-body channels represent a small fraction of R_γ only. Above 2 GeV good quality inclusive BES-II data [52] provide a fairly well determined $R_\gamma(s)$.

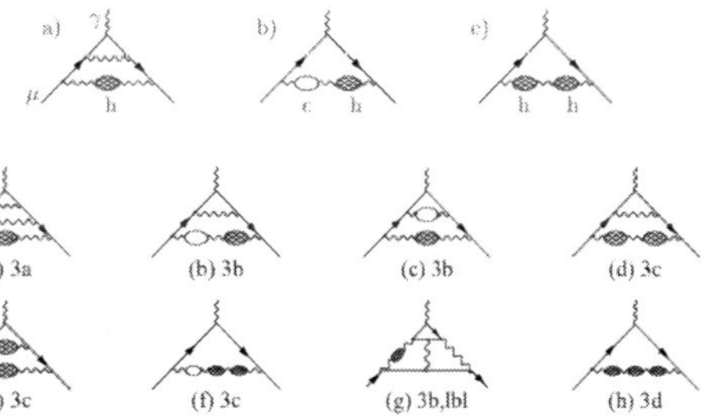

Figure 8. Feynman diagrams with hadronic insertions at NLO (top row) and NNLO.

Table 2. NLO contributions diagrams a) - c) (in units 10^{-11})

$a_\mu^{(2a)}$	$a_\mu^{(2b)}$	$a_\mu^{(2c)}$	$a_\mu^{\mathrm{had}(2)}$
-206.10(1.73)	103.55(0.73)	3.38(0.05)	-99.17 (1.00)

Table 3. NNLO contributions diagrams (a) - (h) (in units 10^{-11})

Class		Kurz et al [54]	my evaluation
$a_\mu^{(3a)}$	=	8.0	7.82(77)
$a_\mu^{(3b)}$	=	−4.1	−4.03(37)
$a_\mu^{(3b,\mathrm{lbl})}$	=	9.1	9.00(77)
$a_\mu^{(3c)}$	=	−0.6	−0.544(7)
$a_\mu^{(3d)}$	=	0.005	0.00522(15)
$a_\mu^{\mathrm{had}(3)}$	=	12.4(1)	12.25(12)

been calculated recently [54, 55]. Diagrams are shown in figure 8 and corresponding contributions evaluated with kernels from [54] are listed in table 3.

The challenge for the future is to keep up with the future experiments [56], which will improve the experimental accuracy from $\delta a_\mu^{\mathrm{exp}} = 63 \times 10^{-11}$ [±0.54 ppm] at present to $\delta a_\mu^{\mathrm{exp}} = 16 \times 10^{-11}$ [±0.14 ppm] the next years. The present results a_μ^{had}[LO VP] $= (6889 \pm 35) \times 10^{-11}$ amount to +59.09 ±0.30 ppm, which poses the major challenge. The subleading results a_μ^{had}[NLO VP] $= (-99.2 \pm$

1.0) $\times 10^{-11}$ and $a_\mu^{\rm had}$[NNLO VP] = $(12.4 \pm 0.1) \times 10^{-11}$ although relevant will be known well enough. These number also compare with the well established weak $a_\mu^{\rm EW}$ = $(154 \pm 1) \times 10^{-11}$ and the problematic HLbL estimated to contribute $a_\mu^{\rm had,LbL}$ = $[(105\div106)\pm(26\div39)]\times10^{-11}$, which is representing a +0.90 ±0.28 ppm effect. Next generation experiments require a factor 4 reduction of the uncertainty optimistically feasible should be a factor 2 we hope.

3 Effective field theory: the Resonance Lagrangian Approach

As we know HVP is dominated by spin 1 resonance physics, therefore we need a low energy effective theory which includes ρ, ω, ϕ mesons. Principles to be implemented are the VMD mechanism, the chiral structure of QCD (chiral perturbation theory), and electromagnetic gauge invariance. A specific realization is the HLS effective Lagrangian [57] (see [8] for a brief account). In our context it has been first applied to HLbL of muon $g-2$ in [1], to HVP in [58]. Largely equivalent is the ENJL model on which the most complete analysis of HLbL in [2] was based. To actually work in practice, the HLS symmetry has to be broken by phenomenologically well known $SU(3)$ and $SU(2)$ flavor breaking and the framework we consider here is the broken HLS (BHLS) model.

We briefly outline the BHLS global fit strategy of a HVP evaluation [9]: one uses data below E_0 = 1.05 GeV (just including the ϕ) to constrain the effective Lagrangian couplings, using 45 different data sets (6 annihilation channels and 10 partial width decays). The effective theory then can be used to predict cross-sections mainly for two-body reactions (besides 3π)

$$\pi^+\pi^-,\ \pi^0\gamma,\ \eta\gamma,\ \eta'\gamma,\ \pi^0\pi^+\pi^-,\ K^+K^-,\ K^0\bar{K}^0\,,$$

while the missing part $4\pi, 5\pi, 6\pi, \eta\pi\pi, \omega\pi$ as well as the regime $E > E_0$ is evaluated using data directly and pQCD for perturbative region and tail. Including self-energy effects is mandatory to properly describe $\rho-\omega-\phi$ mixing and their decays with proper phase space, energy dependent width etc. $\gamma - V$ $(V = \rho, \omega, \phi)$ mixing also turns out to be crucial. The method works in reducing uncertainties by using indirect constraints. It is able to reveal inconsistencies in data, e.g. KLOE vs. BaBar. A goal is to single out a representative effective resonance Lagrangian by a global fit, which is expected to help in improving effective field theory (EFT) calculations of hadronic light-by-light scattering. It has been shown [5–7] that EFT not only helps reducing the uncertainty of HVP, it resolves the τ vs. e^+e^- data puzzle, and it allows us to use, besides the e^+e^- annihilation data, also the τ decay data, as well as other experimental information consistently in a quantum field theory framework. A best fit is obtained for the data configuration NSK+KLOE10+KLOE12+BESIII+τ with a result

$$a_\mu^{\rm had(1)} = (682.40 \pm 3.20) \times 10^{-10}\,, \tag{4}$$

where $(569.04 \pm 1.08) \times 10^{-10}$ results from BHLS predicted channels and $(113.36\pm3.01)\times10^{-10}$ from non-HLS [thereof $(112.02\pm3.01)\times10^{-10}$ from data above 1.05 GeV and $(1.34\pm0.11)\times10^{-10}$ from HLS missing channels below 1.05 GeV]. The global fit including the BABAR sample as well yields $(685.82\pm3.14)\times10^{-10}$. In figure 12 we display the global fit BDDJ15*, the best fit BDDJ15# and BDDJ12 for NSK+τ, which includes scan data only.

An important outcome of effective theory modeling of low energy hadron physics is the observation that a unified treatment of different processes on a Lagrangian level is able to resolve [9, 35] the long standing τ vs. e^+e^- $\pi\pi$ data puzzle [37]. The main effect which distort e^+e^--spectra relative to τ-spectra is $\rho^0-\gamma$ interference in the neutral channel, which is absent in charged channel. A minimal model, which allow us to understand this, is VMD II plus scalar QED [35] for the pion-photon interaction, with effective Lagrangian

$$\mathcal{L} = \mathcal{L}_{\gamma\rho} + \mathcal{L}_\pi \tag{5}$$

where

$$\begin{aligned}\mathcal{L}_{\gamma\rho} &= -\frac{1}{4}F_{\mu\nu}F^{\mu\nu} - \frac{1}{4}\rho_{\mu\nu}\rho^{\mu\nu} + \frac{M_\rho^2}{2}\rho_\mu\rho^\mu + \frac{e}{2g_\rho}\rho_{\mu\nu}F^{\mu\nu}\,,\\ \mathcal{L}_\pi &= D_\mu\pi^+D^{+\mu}\pi^- - m_\pi^2\pi^+\pi^-\ ;\ D_\mu = \partial_\mu - \mathrm{i}\,e\,A_\mu - \mathrm{i}\,g_{\rho\pi\pi}\rho_\mu\,.\end{aligned}$$

Photon and ρ self-energies are then pion-loops, which also implies non-trivial $\gamma-\rho^0$ vacuum polarization (see figure 9). The clue is that the $\rho^0-\gamma$ mixing is uniquely

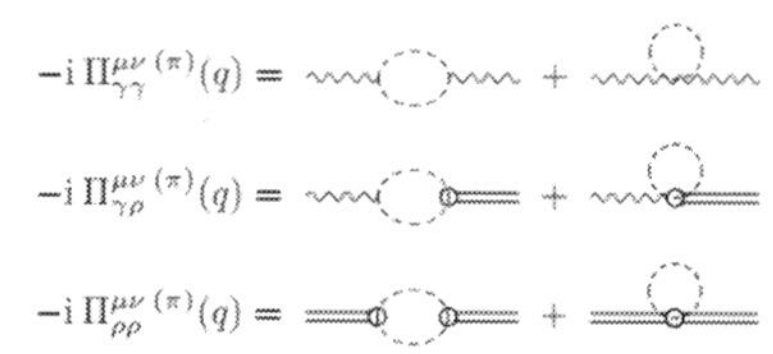

Figure 9. Irreducible self-energy contribution at one-loop.

fixed by the electronic ρ-width $\Gamma_{\rho ee}$. Nothing unknown to be adjusted! Previous calculations à la Gounaris-Sakurai, considered the mixing term to be constant, i.e.

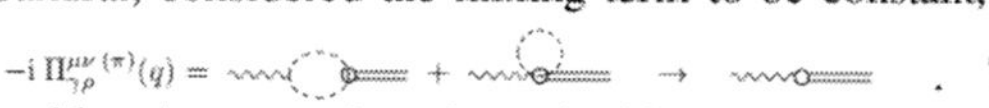

. The problem in comparing charged with neutral channel data turned out to be an inconsistent treatment of quantum loops in the neutral channel Gounaris-Sakurai formula. A consistent calculation requires to consider the 2x2 matrix $\gamma-\rho$ propagator $D_{\alpha\beta}(s)$ $(\alpha \in \gamma,\rho\,,\ \beta \in \gamma,\rho)$ as a starting point. The off diagonal $D_{\gamma\rho}$ element usually has been treated as the well-known VMD $\gamma-\rho$-mixing coupling constant. However, one-loop self-energy effects should be included consistently, especially if it turns out that interference effects are large, in spite of the fact that one of the $g_{\rho\pi\pi}$ couplings in $D_{\rho\rho}$ is replaced by the electromagnetic charge e in $D_{\rho\gamma}$. In our extended VMD model, properly renormalized, the pion form-factor exhibits the four terms shown in figure 10 such that

$$F_\pi(s) \propto e^2\,D_{\gamma\gamma} + e g_{\rho\pi\pi}\,D_{\gamma\rho} - g_{\rho ee}eD_{\rho\gamma} - g_{\rho ee}g_{\rho\pi\pi}\,D_{\rho\rho}\,,$$

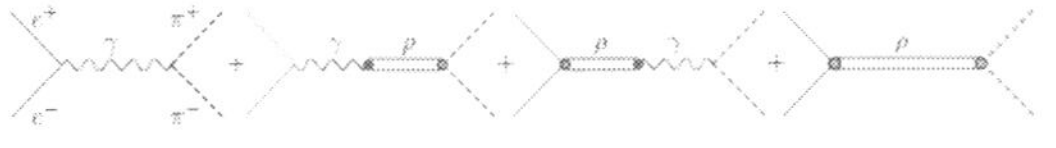

Figure 10. Diagrams contributing to the process $e^+e^- \to \pi^+\pi^-$. Propagators are supposed to be corrected for self-energy effects as displayed in figure 9.

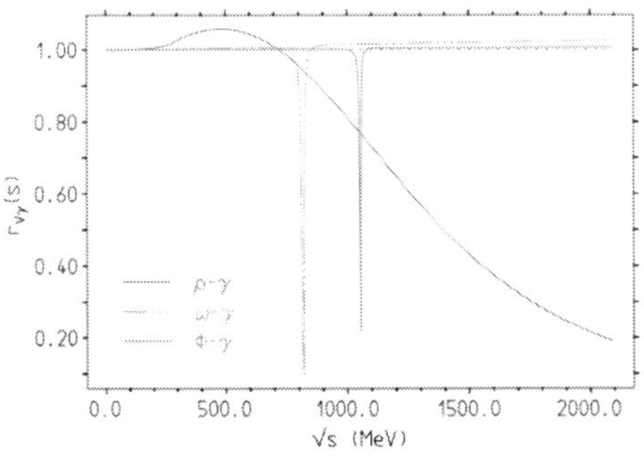

Figure 11. The $\rho^0 - \gamma$ mixing correction to be applied to the τ spectral functions. Corresponding effects for ω, ϕ in $\pi\pi$ off-resonance are tiny (scaled up $\Gamma_V/\Gamma(V \to \pi\pi)$. Caution: the model only applies for energies below about M_ϕ. At the $f_2(1270)$ in $\gamma\gamma \to \pi^+\pi^-, \pi^0\pi^0$ one observes the photons to couple to the quarks rather than to point-like pions.

which replaces the ρ contribution of the GS formula, which usually includes the $\omega - \rho$ mixing and higher ρ contributions $\rho' = \rho(1450)$ and $\rho'' = \rho(1700)$. Properly normalized (VP subtraction: $e^2(s) \to e^2$) we have

$$F_\pi(s) = \left[e^2\, D_{\gamma\gamma} + e\,(g_{\rho\pi\pi} - g_{\rho ee})\, D_{\gamma\rho} - g_{\rho ee} g_{\rho\pi\pi}\, D_{\rho\rho}\right] / \left[e^2\, D_{\gamma\gamma}\right] \quad (6)$$

Typical couplings are $g_{\rho\pi\pi\,\mathrm{bare}} = 5.8935$, $g_{\rho\pi\pi\,\mathrm{ren}} = 6.1559$, $g_{\rho ee} = 0.018149$, $x = g_{\rho\pi\pi}/g_\rho = 1.15128$ [35].

As a result a correction displayed in figure 11 is obtained, a +5 to -10% correction! The proper relationship between e^+e^- and τ spectral functions

$$v_i(s) = \frac{\beta_i^3(s)}{12\pi}\, |F_\pi^i(s)|^2 \quad (i = 0, -)\,, \quad (7)$$

in terms of the pion form-factors F_π and the pion velocities β_i, now reads

$$v_0(s) = r_{\rho\gamma}(s)\, R_{\mathrm{IB}}(s)\, v_-(s)\,, \quad (8)$$

where $R_{\mathrm{IB}}(s)$ is the standard isospin breaking correction (see [35, 38–40]). The τ requires to be corrected for missing $\rho - \gamma$ mixing as well, before being used as I=1 e^+e^- data, because results obtained from e^+e^- data is what goes into the DR (2) (the photon coupled to $\pi^+\pi^-$ not to $\pi^\pm\pi^0$). The correction is large only for the ρ and affects narrower resonances only very near resonance. The effect is part of the experimental data and as ω and ϕ have no charged partners there is nothing to be corrected in these cases. To include further mixing effects, like $\omega - \rho^0$ mixing, one has to extend the Lagrangian and including all possible fields and their possible interactions, which leads to the HLS Lagrangian or a related effective model. For details I refer to [9]. Nevertheless, let me add a few comments on the HLS approach:

Can global fits like our HLS implementation discriminate between incompatible data sets? The problem of inconsistent data is not a problem of whatever model, rather it is a matter of systematics of the measurements. Note that modeling is indispensable for interrelating different data channels. In the HLS global fit τ data play a central role as they are *simple*, i.e. pure I=1, no singlet contribution, no $\gamma - \rho^0 - \omega - \phi$ mixing. In fact, τ -spectra supplemented with PDG isospin breaking, provide a good initial fit for most e^+e^- -data fits, which then are improved and optimized by iteration for a best simultaneous solution.

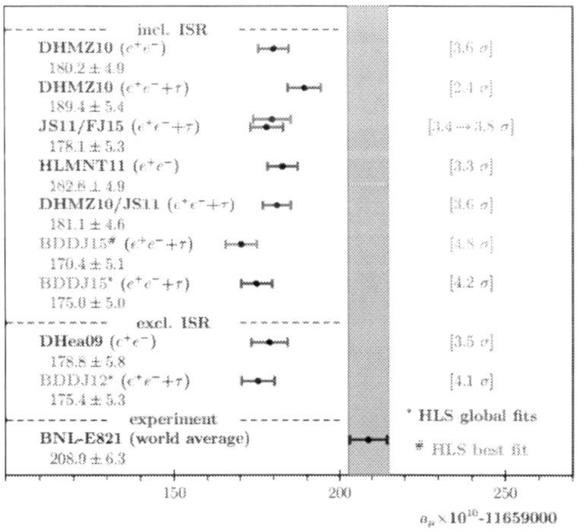

Figure 12. Comparison of recent LO a_μ^{had} evaluations. Note that some results do not include τ data. The HLS best fit BDDJ15$^\#$ (NSK+KLOE10+KLOE12) does not include BaBar $\pi\pi$ data [7], while BDDJ15* does. JS11/FJ15 [35] is updated to include the new BES III data. Further points are BDDJ12 [5], DHMZ10 [59, 60], HLMNT11 [61] and DHea09 [39], (see also [51]).

Why do we get slightly lower results for HVP and with reduced uncertainties? BaBar data according to [59] are in good accord with Belle τ -data, *before* correcting τ -data for the substantial and quite unambiguous $\gamma - \rho^0$ mixing effects! i.e. for the BaBar data alone there seems to be no τ vs. e^+e^- puzzle, while the puzzle exists for all other e^+e^- data sets. This is a problem for the BaBar data. They are disfavored by our global fit! BaBar data rise the HVP estimate quite substantially towards the uncorrected τ data value. In contrast the NSK, KLOE10/12 and the new BESIII data are in very good agreement with the τ + PDG prediction [7], so they dominate the fit and give somewhat lower HVP result[2]! Since, besides the e^+e^- data, additional data constrain the HLS Lagrangian and its parameters, we find a reduced uncertainty and hence an increased significance.

What are the model (using specifically HLS) errors of our estimates? This is hard to say. Best do a corresponding analysis based on different implementations of the res-

[2] We are talking about a 1% shift, which is of the order of the size of the uncertainty.

onance Lagrangian approach. Try to include higher order corrections. However, the fit quality is surprisingly good and we do not expect that one has much flexibility. However, on can improve on photon radiation within a suitably extended HLS approach. Such processes have been implemented recently in the CARLOMAT Monte Carlo [62].

To conclude: our analysis is a first step in a direction which should allow for systematic improvements. A comparison of different estimates and leading uncertainties is shown in figure 12.

4 HVP for the electron anomaly

An up-to-date reevaluation of hadronic VP effects to the electron $g-2$ yields the results given in table 4. The present status is illustrated by the pie chart figure 13.

Table 4. 2015 update of HVP effects contributing to a_e

$$a_e^{\mathrm{had}(1)} = (185.11 \pm 1.24) \times 10^{-14}\ (\mathrm{LO})$$
$$a_e^{\mathrm{had}(2)} = (-22.15 \pm 0.16) \times 10^{-14}\ (\mathrm{NLO})$$
$$a_e^{\mathrm{had}(3)} = (\quad 2.80 \pm 0.02) \times 10^{-14}\ (\mathrm{NNLO})\ [54]$$

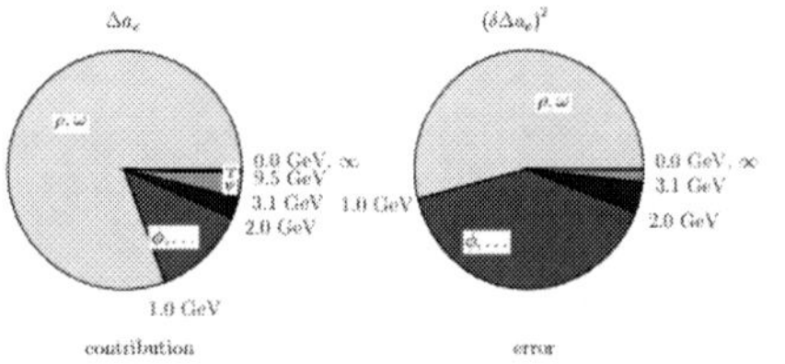

Figure 13. Electron $g-2$: contributions and square errors from different energy ranges.

On the theory side, the by far dominant QED contribution has been calculated to 5-loops [63] with the result

$$\begin{aligned} a_e^{\mathrm{QED}} &= \frac{\alpha}{2\pi} - 0.328\,478\,444\,002\,55(33)\left(\frac{\alpha}{\pi}\right)^2 \\ &\quad +1.181\,234\,016\,816(11)\left(\frac{\alpha}{\pi}\right)^3 \\ &\quad -1.9097(20)\left(\frac{\alpha}{\pi}\right)^4 + 9.16(58)\left(\frac{\alpha}{\pi}\right)^5 . \end{aligned}$$

Together with the hadronic and weak contribution we get the SM prediction (incl. $a_e^{\mathrm{had,LbL}} = (3.7 \pm 0.5) \times 10^{-14}$)

$$a_e^{\mathrm{SM}} = a_e^{\mathrm{QED}} + 1.725(12) \times 10^{-12}\ (\mathrm{hadr\ \&\ weak}) ,$$

which can be used for extracting α_{QED} from a_e at unprecedented precision. Matching the theory prediction with the very precise experimental result of Gabrielse et al. [64]

$$a_e^{\mathrm{exp}} = 0.001\,159\,652\,180\,73(28) ,$$

one extracts

$$\alpha^{-1}(a_e) = 137.0359991685(342)(68)(46)(24)[353] ,$$

which is close [85 → 57] to the value

$$\alpha^{-1}(a_e) = 137.0359991657(342)[0.25\,\mathrm{ppb}] , \tag{9}$$

obtained by [63]. Note that the weak part has been reevaluated as

$$a_e^{\mathrm{weak}} = 0.030 \times 10^{-12} , \tag{10}$$

which is replacing the value 0.039×10^{-12} which has been estimated in [65]. An inconsistency there has been noted by M. Passera [66].

The best test for new physics can be obtained by using α from atomic interferometry [67]. With $\alpha^{-1}(\mathrm{Rb11}) = 137.035999037(91)[0.66\,\mathrm{ppb}]$ as an input one finds

$$a_e^{\mathrm{the}} = 0.001\,159\,652\,181\,87(77) ,$$

such that

$$a_e^{\mathrm{exp}} - a_e^{\mathrm{the}} = -1.14(0.82) \times 10^{-12} , \tag{11}$$

in good agreement. We know that the sensitivity to new physics is reduced by $(m_\mu/m_e)^2 \cdot \delta a_e^{\mathrm{exp}}/\delta a_\mu^{\mathrm{exp}} \simeq 19$ relative to a_μ. Nevertheless, one has to keep in mind that a_e is suffering less form hadronic uncertainties and thus may provide a safer test. Presently, the a_e prediction is limited by the, by a factor $\delta\alpha(\mathrm{Rb11})/\delta\alpha(a_e) \simeq 5.3$ less precise, α available. Combining all uncertainties a_μ is about a factor 43 more sensitive to new physics at present.

5 HVP subtraction of $R_\gamma(s)$: a problem of the DR method?

The full photon propagator is usually obtained by Dyson resummation of the 1pi part (blob) as illustrated by figure 14. As we know this is a geometric series $1 + x + x^2 + \cdots = 1/(1-x)$ which only converges iff $|x| < 1$. Including the external e.m. couplings we have

$$\begin{aligned} \mathrm{i}\,D'_\gamma(q^2) &\equiv \frac{-\mathrm{i}}{q^2} + \frac{-\mathrm{i}}{q^2}(-\mathrm{i}\Pi_\gamma)\frac{-\mathrm{i}}{q^2} + \frac{-\mathrm{i}}{q^2}(-\mathrm{i}\Pi_\gamma)\frac{-\mathrm{i}}{q^2}(-\mathrm{i}\Pi_\gamma)\frac{-\mathrm{i}}{q^2} + \cdots \\ &= \frac{-\mathrm{i}}{q^2}\left\{1 + \left(\frac{-\Pi_\gamma}{q^2}\right) + \left(\frac{-\Pi_\gamma}{q^2}\right)^2 + \cdots\right\} \\ &= \frac{-\mathrm{i}}{q^2}\left\{\frac{1}{1+\frac{\Pi_\gamma}{q^2}}\right\} = \frac{-\mathrm{i}}{q^2 + \Pi_\gamma(q^2)} = \frac{-\mathrm{i}}{q^2}\,\frac{1}{1+\Pi'_\gamma(q^2)} . \end{aligned}$$

Figure 14. The Dyson summation of the photon self-energy.

$$\mathrm{i}\,e^2\,D'_\gamma(q^2) = \frac{-\mathrm{i}}{q^2}\,\frac{e^2}{1+\Pi'_\gamma(q^2)} . \tag{12}$$

The effective charge thus is given by the well-known expression

$$\frac{e^2}{1+\Pi'_\gamma(s)} = \frac{e^2}{1-\Delta\alpha(s)} = e^2(s) . \tag{13}$$

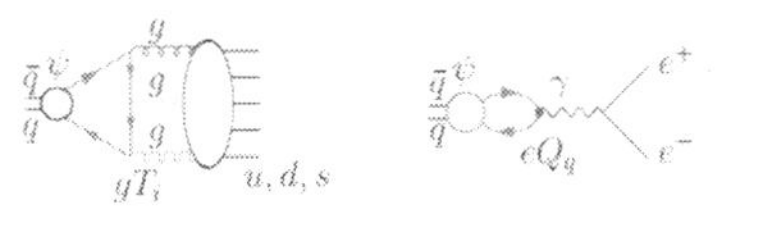

Figure 15. OZI suppressed strong decays let e.m. interaction look to be almost of equal strength.

Usually, $\Delta\alpha(s)$ is a correction i.e $\Delta\alpha(s) \ll 1$ and the Dyson series converges well. Indeed for any type of perturbative effects no problem is encountered (besides possible Landau poles). For non-perturbative strong interaction physics there are exceptions. One would expect that, if there are problems, one would encounter them at low energy, but for the ρ, the ω and the ϕ, in spite of huge resonance enhancements, the hadronic VP contributions to the running charge are small relative to unity, as the effect is suppressed by the e.m. coupling e^2. The exception, surprisingly, we find at pretty high energies, at the narrow OZI suppressed resonances, which are extremely sharp, because they lie below corresponding $q\bar{q}$ -thresholds. While the strong interaction appears heavily suppressed (3 gluons exchange) the electromagnetic channel (1 photon exchange) appears almost as strong as the strong one (see figure 15). Actually, Γ_{ee} is not much smaller than $\Gamma_{\rm QCD}$ (i.e strong decays). This phenomenon shows up for the resonances $J/\psi, \psi_2, \Upsilon_1, \Upsilon_2$ and Υ_3 . The imaginary parts from the narrow resonances read

$$\mathrm{Im}\Pi'_\gamma(\mathrm{s})) = \frac{\alpha}{3}\,\mathrm{R}_\gamma(\mathrm{s}) = \frac{3}{\alpha}\frac{\Gamma_{\rm ee}}{\Gamma} \tag{14}$$

at peak, causing the sharp spikes, which are seen only by appropriate high resolution scans, as we know. Let $\alpha(s)$ denotes the real $\alpha(s) = \alpha/(1 + \mathrm{Re}\Pi'_\gamma(\mathrm{s}))$, we note that,

$$|1 + \Pi'_\gamma(s)|^2 - (\alpha/\alpha(s))^2 = (\mathrm{Im}\Pi'_\gamma(\mathrm{s}))^2$$

and at $\sqrt{s} = M_R$ values for the the different resonances are given by 1.23×10^{-3} (ρ), 2.76×10^{-3} (ω), 1.56×10^{-2} (ϕ), 594.81 (J/ψ), 9.58 (ψ_2), 2.66×10^{-4} (ψ_3), 104.26 (Υ_1), 30.51 (Υ_2) and 55.58 (Υ_3). This shows that near QZI suppressed resonances the Dyson series cannot converge. So we have a problem with the dispersive approach, which requires $R_\gamma(s) \propto \mathrm{Im}\Pi'_\gamma(\mathrm{s})$ as an input. What is measured by an experiment is the full propagator, the summed up Dyson series, $Z = |1/(1-x)|^2$, but we cannot extract x from that since for $|x| \geq 1$ the observable Z has no representation in terms of x. Remember that the object required in the DR is the undressed $R_\gamma(s)$ in (1), which cannot be measured itself, rather we have to extract ($x = -\Pi'_\gamma(s)$)

$$R_\gamma^{\rm bare} = R_\gamma^{\rm phys}\,|1 + \Pi'_\gamma(s)|^2\,. \tag{15}$$

Locally, near OZI suppressed resonances, the usual iterative procedure of getting $R_\gamma^{\rm bare}$ does not converge! The way out usually practiced is to utilize the smooth space-like charge, i.e. $\overline{R}_\gamma^{\rm bare} = R_\gamma^{\rm phys}\,|1 + \Pi'_\gamma(-s)|^2$, expected to do the undressing "in average". This actually does not look too wrong as we see in figure 16. Nevertheless, I see a problem her, not only for the interpretation of resonance data, where one would wish to be able to disentangle electromagnetic form strong interaction effects.

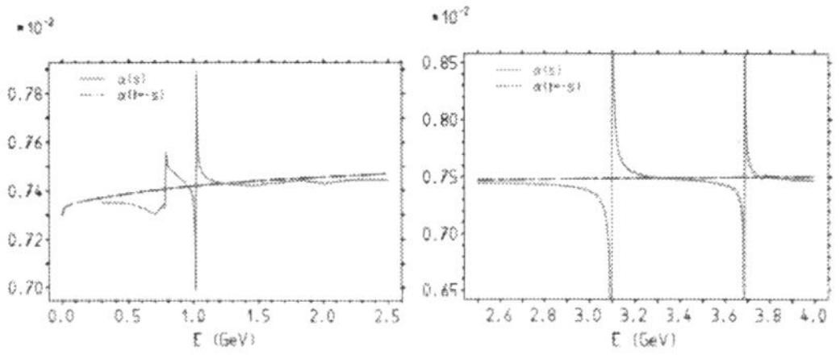

Figure 16. Time-like vs. space-like effective finestructure constant α as a function of the energy E: $\alpha(s)$ in the mean follows $\alpha(t = -s)$ ($s = E^2$). Note that the smooth space-like effective charge agrees rather well with the non-resonant "background" above the ϕ (kind of duality).

For what concerns the proper extraction of the hadronic effects contributing to the running of $\alpha_{\rm QED}$ and to $a_\mu^{\rm had}$, I see no proof that this cannot produce non-negligible shifts!

Fortunately, experimental progress is in sight here: KLOE 2015 [68] has a first direct measurements of the time-like complex running $\alpha_{\rm QED}(s)$! Similar measurements for the J/ψ and other ultra-narrow resonances should be possible with BES III. It is a fundamental problem! An interesting possibility in this respect is a novel approach to determine $a_\mu^{\rm had}$ form a direct space-like measurement of $\alpha(-Q^2)$ as proposed in [69, 70], recently.

6 A comment on axial exchanges in HLbL

The Landau-Yang theorem says that the amplitude $\mathcal{A}($ axial meson $\gamma\gamma)|_{\rm on-shell} = 0$, e.g. $Z^0 \to \gamma\gamma$ is forbidden, while $Z^0 \to \gamma e^+e^-$ is allowed as one of the photons is off-shell. For HLbL such type of contribution has been estimated in [17] to be rather large, which raised the question: Why $a_\mu[a_1, f_1', f_1] \sim 22\times10^{-11}$ is so large? From the data side we know, untagged $\gamma\gamma \to f_1$ shows no signal, while single-tag $\gamma^*\gamma \to f_1$ is strongly peaked when $Q^2 \gg m_{f_1}^2$. The point: the contribution from axial mesons has been calculated assuming a symmetric form-factors under exchange of the two photon momenta. This violates the Landau-Yang theorem, which requires an antisymmetric form-factor. In fact antisymmetrizing the form-factor adopted in [17] reduces the contribution by a factor about 3, and the result agrees with previous findings [1, 2] and with the more recent result [20]. As a result one finds that the estimate $a_\mu^{\rm HLbL,LO} = (116\pm39)\times10^{-11}$ accepted in [65] must be replaced by

$$a_\mu^{\rm HLbL,LO} = (102 \pm 39)\times10^{-11}\,. \tag{16}$$

This also requires a modification of the result advocated in [71]. The evaluation of the axialvector mesons contribution, taking a Landau-Yang modified (i,e, antisymmetrized) Melnikov-Vainshtein form-factors yields [72]

$$a_\mu[a_1, f_1', f_1] \sim (7.51 = [1.89 + 5.04 + 0.58] \pm 2.71)\times10^{-11}\,, \tag{17}$$

where ideal mixing and nonet symmetry results have been averaged. In fact, the sum of the contributions from the f_1 and f_1' depends little on the mixing scheme. The result supersedes $a_\mu[a_1, f_1', f_1] \sim 22(5)\times10^{-11}$ we included in [65].

7 Theory vs. experiment: do we see New Physics?

Table 5. A list of small shifts in theory [in units 10^{-11}]. The error from new entries in the list reduces from the old 5 to 3.6.

New contribution	a_μ	Reference
Old axial exchange HLbL	22±5	[17]
New axial exchange HLbL	7.51±2.71	[20, 72]
NNLO HVP	12.4±0.1	[54]
NLO HLbL	3±2	[73]
Tensor exchange HLbL	1.1±0.1	[20]
Total change	+2.0±3.4	

Here I briefly summarize what is new and where we are. Some new results/evaluations are collected in table 5. We finally compare the SM prediction for a_μ with its experimental value [77] in table 6, which also summarizes the present status of the different contributions to a_μ. A deviation between 3 and 5 σ is persisting and was slightly increasing. Resonance Lagrangian models, like the HLS model, provide clear evidence that there is no τ version HVP which differs from the e^+e^- data result. This consolidates a larger deviation $\Delta a_\mu = a_\mu^{\rm exp} - a_\mu^{\rm the}$. Also the decrease of the axial HLbL contribution goes in this direction, it is compensated however by the new NNLO HVP result. What represents the 4 σ deviation: new physics? Is it a statistical fluctuation? Are we underestimating uncertainties (experimental, theoretical)? Do experiments measure what theoreticians calculate? I refer to [78] for possible interpretations and conclusions.

Table 6. Standard model theory and experiment comparison [in units 10^{-10}].

Contribution	Value	Error	Reference
QED at 5-loops	11 658 471 . 8851	0 . 036	[55, 74, 75]
LO HVP	688 . 91	3 . 52	(3)
NLO HVP	-9 . 917	0 . 100	table 2
NNLO HVP	1 . 225	0 . 012	[54], table 3
HLbL	10 . 6	3 . 9	[65, 72], table 5
EW at 2-loops	15 . 40	0 . 10	[25, 26, 76]
Theory	11 659 178 . 10	5 . 26	–
Experiment	11 659 209 . 1	6 . 3	[77] (updated)
Exp-The 3.8 σ	31 . 0	8 . 2	–

8 Outlook

Although progress is slow, there is evident progress in reducing the hadronic uncertainties, most directly by progress in measuring the relevant hadronic cross-sections. Near future progress we expect from BINP Novosibirsk/Russia and from IHEP Beijing/China. Energy scan as well as ISR measurement of cross-sections in the region from 1.4 to 2.5 GeV are most important to reduce the errors to a level competitive with the factor 4 improvement achievable by the upcoming new muon $g-2$ experiments at Fermilab/USA and at JPAC/Japan [56]. Also BaBar data are still being analyzed and are important for improving the results. Promising is that lattice QCD evaluations come closer to be competitive [15].

Acknowledgements

I thank Z. Zhang for helpful discussions on the isospin breaking corrections and M. Benayoun for close collaboration on HLS driven estimates of $a_\mu^{\rm had}$.

References

[1] M. Hayakawa, T. Kinoshita, A. I. Sanda, Phys. Rev. Lett. **75**, 790 (1995); Phys. Rev. D **54**, 3137 (1996)

[2] J. Bijnens, E. Pallante, J. Prades, Phys. Rev. Lett. **75**, 1447 (1995) [Erratum-ibid. **75**, 3781 (1995)]; Nucl. Phys. B **474**, 379 (1996); [Erratum-ibid. **626**, 410 (2002)];
J. Bijnens, J. Prades, Mod. Phys. Lett. A **22**, 767 (2007)

[3] J. Bijnens, these proceedings

[4] S. Eidelman, F. Jegerlehner, Z. Phys. C **67**, 585 (1995); F. Jegerlehner, Nucl. Phys. (Proc. Suppl.) C **51**, 131 (1996); J. Phys. G **29**, 101 (2003); Nucl. Phys. Proc. Suppl. **126**, 325 (2004)

[5] M. Benayoun, P. David, L. DelBuono, F. Jegerlehner, Eur. Phys. J. C **72**, 1848 (2012)

[6] M. Benayoun, P. David, L. DelBuono, F. Jegerlehner, Eur. Phys. J. C **73**, 2453 (2013)

[7] M. Benayoun, P. David, L. DelBuono, F. Jegerlehner, arXiv:1507.02943 [hep-ph]

[8] F. Jegerlehner, Acta Phys. Polon. B **44**, Vol.11, 2257 (2013)

[9] M. Benayoun, these proceedings.

[10] P. Boyle, L. Del Debbio, E. Kerrane, J. Zanotti, Phys. Rev. D **85** (2012) 074504

[11] X. Feng et al., Phys. Rev. D **88**, 034505 (2013)

[12] C. Aubin, T. Blum, M. Golterman, S. Peris, Phys. Rev. D **88**, 074505 (2013)

[13] A. Francis et al., arXiv:1411.3031 [hep-lat]

[14] R. Malak et al. [Budapest-Marseille-Wuppertal Collab.], PoS LATTICE **2014**, 161 (2015)

[15] M. Petschlies, these proceedings.

[16] M. Knecht, A. Nyffeler, Phys. Rev. D **65**, 073034 (2002)

[17] K. Melnikov, A. Vainshtein, Phys. Rev. D **70**, 113006 (2004)

[18] M. Knecht, Nucl. Part. Phys. Proc. **258-259**, 235 (2015); and these proceedings

[19] A. Nyffeler, these proceedings

[20] V. Pauk, M. Vanderhaeghen, Eur. Phys. J. C **74**, 3008 (2014)

[21] G. Colangelo et al., Phys. Lett. B **738**, 6 (2014)

[22] M. Procura, these proceedings

[23] T. Blum, S. Chowdhury, M. Hayakawa, T. Izubuchi, Phys. Rev. Lett. **114**, 012001 (2015)

[24] Ch. Lehner, these proceedings

[25] M. Knecht, S. Peris, M. Perrottet, E. de Rafael, JHEP **0211**, 003 (2002)

[26] A. Czarnecki, W. J. Marciano, A. Vainshtein, Phys. Rev. D **67**, 073006 (2003) [Erratum-ibid. D **73**, 119901 (2006)]
[27] R. R. Akhmetshin et al. [CMD-2 Collab.], Phys. Lett. B **578**, 285 (2004)
[28] V. M. Aulchenko et al. [CMD-2 Collab.], JETP Lett. **82**, 743 (2005) [Pisma Zh. Eksp. Teor. Fiz. **82**, 841 (2005)]; R. R. Akhmetshin et al., JETP Lett. **84**, 413 (2006) [Pisma Zh. Eksp. Teor. Fiz. **84**, 491 (2006)]; Phys. Lett. B **648**, 28 (2007)
[29] M. N. Achasov et al. [SND Collab.], J. Exp. Theor. Phys. **103**, 380 (2006) [Zh. Eksp. Teor. Fiz. **130**, 437 (2006)]
[30] A. Aloisio et al. [KLOE Collab.], Phys. Lett. B **606**, 12 (2005);
F. Ambrosino et al. [KLOE Collab.], Phys. Lett. B **670**, 285 (2009)
[31] F. Ambrosino et al. [KLOE Collab.], Phys. Lett. B **700**, 102 (2011)
[32] D. Babusci et al. [KLOE Collab.], Phys.Lett. **B720**, 336 (2013)
[33] B. Aubert et al. [BABAR Collab.], Phys. Rev. Lett. **103**, 231801 (2009); J. P. Lees et al., Phys.Rev. **D86**, 032013 (2012)
[34] M. Ablikim et al. [BESIII Collab.], arXiv:1507.08188 [hep-ex]
[35] F. Jegerlehner, R. Szafron, Eur. Phys. J. C **71**, 1632 (2011)
[36] R. Alemany, M. Davier, A. Höcker, Eur. Phys. J. C **2**, 123 (1998)
[37] M. Davier, S. Eidelman, A. Höcker, Z. Zhang, Eur. Phys. J. C **27**, 497 (2003); Eur. Phys. J. C **31**, 503 (2003)
[38] S. Ghozzi, F. Jegerlehner, Phys. Lett. B **583**, 222 (2004)
[39] M. Davier et al., Eur. Phys. J. C **66**, 127 (2010)
[40] Z. Zhang, these proceedings.
[41] R. Barate et al. [ALEPH Collab.], Z. Phys. C **76**, 15 (1997); Eur. Phys. J. C **4**, 409 (1998); S. Schael et al. [ALEPH Collab.], Phys. Rept. **421**, 191 (2005)
[42] M. Davier et al., Eur.Phys.J. **C74**, 2803 (2014)
[43] K. Ackerstaff et al. [OPAL Collab.], Eur. Phys. J. C **7**, 571 (1999)
[44] S. Anderson et al. [CLEO Collab.], Phys. Rev. D **61**, 112002 (2000)
[45] M. Fujikawa et al. [Belle Collab.], Phys. Rev. D **78**, 072006 (2008)
[46] R. Akhmetshin et al. [CMD-3 Collab.], Phys.Lett. **B723**, 82 (2013)
[47] M. Achasov et al. [SND Collab.], Phys.Rev. **D88**, 054013 (2013)
[48] J. Lees et al. [BABAR Collab.], Phys.Rev. **D87**, 092005 (2013)
[49] J. Lees et al. [BABAR Collab.], Phys.Rev. **D88**, 032013 (2013)
[50] J. Lees et al. [BABAR Collab.], Phys.Rev. **D89**, 092002 (2014)
[51] M. Davier, Nucl. Part. Phys. Proc. **260**, 102 (2015)
[52] J. Z. Bai et al. [BES Collab.], Phys. Rev. Lett. **84**, 594 (2000); Phys. Rev. Lett. **88**, 101802 (2002);
M. Ablikim et al., Phys. Lett. B **677**, 239 (2009)
[53] B. Krause, Phys. Lett. B **390** (1997) 392
[54] A. Kurz, T. Liu, P. Marquard, M. Steinhauser, Phys.Lett. **B734**, 144 (2014)
[55] M, Steinhauser, these proceedings
[56] D. Hertzog, these proceedings.
[57] M. Bando, T. Kugo, K. Yamawaki, Phys. Rept. **164**, 217 (1988);
M. Harada, K. Yamawaki, Phys. Rept. **381**, 1 (2003)
[58] M. Benayoun et al., Eur. Phys. J. **C55**, 199 (2008); M. Benayoun, P. David, L. DelBuono, O. Leitner, Eur. Phys. J. **C65**, 211 (2010); Eur. Phys. J. **C68**, 355 (2010)
[59] M. Davier et al., Eur. Phys. J. **C66**, 1 (2009)
[60] M. Davier et al., Eur. Phys. J. **C71**, 1515 (2011)
[61] K. Hagiwara et al., J. Phys. **G38**, 085003 (2011)
[62] K. Kołodziej, Comput. Phys. Commun. **196**, 563 (2015)
[63] T. Aoyama, M. Hayakawa, T. Kinoshita, M. Nio, Phys. Rev. Lett. **109**, 111807 (2012)
[64] G. Gabrielseet al., Phys. Rev. Lett. **97** (2006) 030802 [Erratum-ibid. **99** (2007) 039902]
[65] F. Jegerlehner, A. Nyffeler, Phys. Rept. **477**, 1 (2009)
[66] M. Passera private communication
[67] R. Bouchendira et al., Phys. Rev. Lett. **106** (2011) 080801
[68] The KLOE Collaboration, to be published
[69] C. M. Carloni et al., Phys. Lett. B **746**, 325 (2015)
[70] L. Trentadue, these proceedings
[71] J. Prades, E. de Rafael, A. Vainshtein, Adv. Ser. Direct. High Energy Phys. **20**, 303 (2009)
[72] F. Jegerlehner, Talk at the MITP Workshop "Hadronic contributions to the muon anomalous magnetic moment", 1-5 April 2014, Waldthausen Castle near Mainz, and to be published
[73] G. Colangelo et al., Phys.Lett. **B735**, 90 (2014)
[74] T. Aoyama, M. Hayakawa, T. Kinoshita, M. Nio, Phys. Rev. Lett. **109**, 111808 (2012)
[75] A. Kurz et al., Phys. Rev. D **92**, 073019 (2015)
[76] C. Gnendiger, D. Stöckinger, H. Stöckinger-Kim, Phys. Rev. D **88**, 053005 (2013)
[77] G. W. Bennett et al. [Muon g-2 Collab.], Phys. Rev. D **73** (2006) 072003
[78] D. Stöckinger, these proceedings

Perspectives on the hadronic contribution to the muon g-2

Marc Knecht[1,a]

[1] Centre de Physique Théorique, CNRS/Aix-Marseille Univ./Univ. du Sud Toulon-Var (UMR 7332)
CNRS-Luminy Case 907, 13288 Marseille Cedex 9, France

Abstract. The theoretical aspects of the various hadronic contributions to the anomalous magnetic moment of the muon are reviewed, as well as perspectives of their future evolution.

1 Introduction

The anomalous magnetic moment of the muon a_μ has been measured [1] with an unprecedented relative accuracy of 0.54ppm by the E821 experiment at the Brookhaven AGS. Its value reads [2]

$$a_\mu^{\rm exp} = 11\,659\,209.1(5.4)(3.3) \cdot 10^{-10}. \quad (1)$$

At the theoretical level, the anomalous magnetic moment of the muon receives contributions from all interactions of the standard model. The by far dominant contribution comes from the electromagnetic sector, which accounts for more than 99.99% of its total value. The second most important contribution comes from the strong interactions, and finally the weak interaction provide, as expected, the smallest contribution. The values of these different contributions have been gathered in Table 1 below (in units of 10^{-10}). We will briefly discuss them in turn. Before that, let us consider the two following points.

• If one adds up the various contributions given in Table 1, one obtains a value which differs from the experimental one. The difference ranges from 2.8σ [5, 8]

$$a_\mu^{\rm exp} - a_\mu^{\rm SM} = 23.9(8.6) \cdot 10^{-10}, \quad (2)$$

to 3.5σ [4, 7]

$$a_\mu^{\rm exp} - a_\mu^{\rm SM} = 27.6(8.0) \cdot 10^{-10}, \quad (3)$$

depending on the input values used. Although this discrepancy is with us for quite some time now, it is nevertheless too small to provide conclusive evidence for the existence of physical degrees of freedom beyond those described by the standard model. For an overview of the possibilities concerning the latter, see e.g. Ref. [11], as well as Refs. [12–14] for more specific scenarios.

• Two new experiments are being planed, one at FNAL [15, 16] and the other one at J-PARC [17], with the aim of reducing the experimental uncertainty by a factor of 4 [18]. If all other values remain unchanged, i.e. those shown in Table 1 as well as the central value in Eq. (1), then this reduction of the uncertainties would by itself lead to roughly a 5σ discrepancy between theory and experiment. Things need, however, not develop as nicely as that, and in order to avoid ending up in a similar inconclusive situation as the present one regarding the existence of new physics, these experimental efforts need to be met by a comparable improvement of the theoretical accuracy.

Table 1. The contributions to a_μ (in units of 10^{-10}) using the latest published values

QED	+11 658 471.9	[3]
HVP-LO	+692.3(4.2)	[4]
	+694.9(4.3)	[5]
HVP-NLO	−9.84(7)	[5]
HVP-NNLO	+1.24(1)	[6]
HLxL	+10.5(2.6)	[7]
	+11.6(3.9)	[8]
EW 1 loop	+19.48(1)	[9]
EW 2 loops	−4.12(10)	[10]

2 QED and weak contributions

The contributions to a_μ coming from quantum electrodynamics and from the weak interactions belong to the realm of perturbation theory, which makes their evaluation straightforward, although quite tedious when it comes to including higher orders.

The value coming from pure (multi-flavour) QED includes contributions up to order $\mathcal{O}(\alpha^5)$, i.e five loops. The one-loop contribution was computed by J. Schwinger [19] long time ago. Analytical results are also available for the two- and three-loop corrections (for a survey and references to the original publications, see [20]). Finally, the corrections at four and five loops were obtained more recently [3, 21], through numerical integration of Feynman-parameterised loop integrals. There are no uncertainties attached to the QED value at the level of precision shown here, which is sufficient to match the present experimental

[a] e-mail: marc.knecht@cpt.univ-mrs.fr

© The Authors, published by EDP Sciences. This is an open access article distributed under the terms of the Creative Commons Attribution License 4.0 (http://creativecommons.org/licenses/by/4.0/).

error on a_μ, but also the precision that forthcoming experiments expect to reach. The contributions at order $O(\alpha^4)$ that dominate the coefficient $C_\mu^{(8)}$ of the perturbative series

$$a_\mu^{\text{QED}} = \sum_n C_\mu^{(2n)} \left(\frac{\alpha}{\pi}\right)^n, \tag{4}$$

come from Feynman graphs with internal electron loops, which are enhanced because they produce powers of $\ln(m_\mu/m_e)$. These contributions have been evaluated recently [22, 23], using asymptotic expansion techniques in the small quantity m_e/m_μ, thus providing an independent and welcome check (at a level of precision admittedly lower than the direct numerical evaluation [3, 21], but still higher than the one forthcoming experiments are aiming at). Looking more closely at the structure of the coefficients $C_\mu^{(2n)}$ of the perturbative series in Eq. (4), one notices that they increase drastically starting from the third order,

$$\begin{aligned} &C_\mu^{(2)} = 1/2 \quad C_\mu^{(4)} = 0.765\,857\,425(17) \\ &C_\mu^{(6)} = 24.050\,509\,96(32) \\ &C_\mu^{(8)} = 130.877\,3(61) \quad C_\mu^{(10)} = 751.92(93). \end{aligned} \tag{5}$$

This naturally raises the question whether a large coefficient $C_\mu^{(12)}$ could at least partly explain the $\sim 3\sigma$ discrepancy between theory and experiment mentioned above. Looking for diagrams that are enhanced by the presence of a maximal number of $\ln(m_\mu/m_e)$ factors leads to the estimate [3]

$$C_\mu^{(12)} \left(\frac{\alpha}{\pi}\right)^6 \sim 0.6 \cdot 10^4 \cdot \left(\frac{\alpha}{\pi}\right)^6 \sim 1 \cdot 10^{-12}. \tag{6}$$

This is way too small to explain even part of the discrepancy.

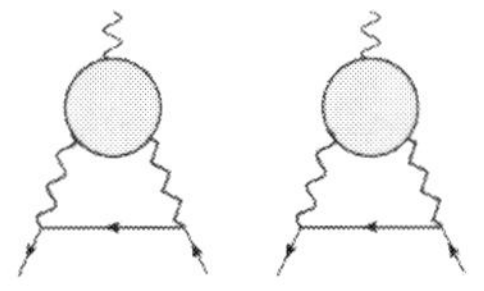

Figure 1. The hadronic contribution to the two-loop electroweak corrections to a_μ. The shaded blob represents the $\langle VVA\rangle$ QCD three-point function. It is connected to the muon line by a photon line and by the neutral Z^0, through the axial coupling of the latter to quarks and leptons.

The evaluation of the contributions at one loop coming from the electroweak sector [9] go back to the first loop calculations done within Weinberg's model for leptons [24]. The two-loop corrections were addressed somewhat later only [25–29]. Table 1 shows the value obtained by the recent reevaluation from Ref. [10]. The main change is a reduction of the uncertainty, due to the fact that the mass of the electroweak scalar boson is now known. These two-loop corrections also involve hadronic contributions, in particular those from the diagrams shown in Fig. 1, which are enhanced by a $\ln(M_Z/m_\mu)$ factor (which is eventually canceled by the corresponding contribution involving lepton loops instead of the blob). Although this contribution provides a substantial fraction of the uncertainty of the total two-loop electroweak correction to a_μ, there is no need to increase its accuracy for the time being. Notice that the discrepancy between theory and experiment represents about two times the total (one and two loops) correction from the electroweak sector. Moreover, the two-loop correction shown in Table 1 also contains an estimate of the leading-log three loop corrections [29]. There is therefore no reason to expect that higher-order electroweak contributions could account even for part of the discrepancy between theory and experiment.

Let us then leave the realm of perturbation theory and turn towards the remaining corrections, where the strong interactions are predominant.

3 Hadronic vacuum polarization

The contribution known as hadronic vacuum polarization (HVP) occurs first at order $O(\alpha^2)$, cf. Fig. 2. It provides to date the largest hadronic correction to a_μ, but also the largest contribution to the theoretical uncertainty (see Table 1). The leading order HVP correction can be expressed in the following way [30–32]

$$a_\ell^{\text{HVP-LO}} = \frac{1}{3}\left(\frac{\alpha}{\pi}\right)^2 \int_{4M_\pi^2}^\infty \frac{ds}{s} K(s) R^{\text{had}}(s), \tag{7}$$

with

$$K(s) = \int_0^1 dx \frac{x^2(1-x)}{x^2 + (1-x)\frac{s}{m_\ell^2}}, \tag{8}$$

and $R^{\text{had}}(s)$ represents the R-ratio of the cross section for $e^+e^- \to$ hadrons. This contribution can therefore be evaluated using available experimental input. The two most recent evaluations from Refs. [4] and [5] are shown in Table 1 (for more details concerning these two evaluations, see Ref. [33]). They are in good agreement (being based on the same data sets, this should not be a surprise) and give a relative precision of 0.6%, despite some tension between, for instance, the high-precision data collected by both BaBar and KLOE in the region of the ρ resonance, as shown in Table 2. Results from data collected in this region by the BESIII experiment have recently become

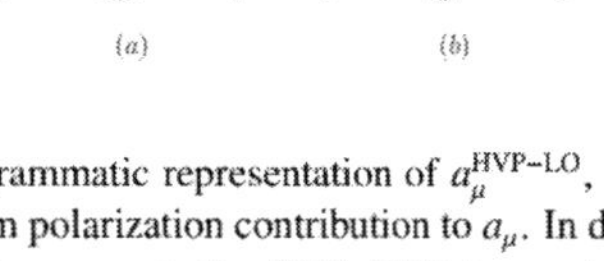

Figure 2. Diagrammatic representation of $a_\mu^{\text{HVP-LO}}$, the leading hadronic vacuum polarization contribution to a_μ. In diagram (a), the shaded blob represents the $\langle VV\rangle$ QCD two-point function. Also shown is a next-to-leading contribution, diagram (b), actually included in $a_\mu^{\text{HVP-LO}}$ (see text).

Table 2. The contribution to $a_\mu^{\text{HVP-LO}}$ (in units of 10^{-10}) coming from the measurement of the $e^+e^- \to \pi^+\pi^-$ cross section in the region between 600 and 900 MeV. Values taken from Fig. 7 in Ref. [34].

Experiment	$a_\mu^{\text{HVP-LO}\,2\pi}(600-900\text{ MeV})$
BaBar	376.7(2.0)(1.9)
KLOE 08	368.9(0.4)(2.3)(2.2)
KLOE 10	365.3(0.9)(2.3)(2.2)
KLOE 12	366.7(1.2)(2.4)(0.8)
BESIII	368.2(2.5)(3.3)

available [34] and are also shown in this Table. Ref. [35] provides an updated evaluation of $a_\mu^{\text{HVP-LO}}$ including these and other data more recent than those available at the time of Refs. [4, 5]. One may notice that the discrepancy between theory and experiment given in Eqs. (2) and (3) represents about 3% or 4% of the value of $a_\mu^{\text{HVL-LO}}$, which, as already mentioned, is itself known with a relative accuracy of about 0.6%. T he evaluation of $a_\mu^{\text{HVP-LO}}$ would thus have to be wrong by several standard deviations in order to explain this discrepancy.

At present, the contribution from hadronic vacuum polarization is the main source of the total theoretical error. In order to match the experimental precision future experiments plan to achieve, the uncertainty on this quantity should be reduced below the 0.5% level. Forthcoming high-precision data from BaBar and KLOE-2, but also from BESIII, Belle II, the experiments at VEPP-2000 in Novosibirsk [36], or perhaps still other sources [37], will hopefully help to reach this goal.

In recent years, a lot of efforts have been devoted to explore the possibilities to obtain reliable determinations of $a_\mu^{\text{HVP-LO}}$ from QCD simulations on the lattice. Several strategies are being considered in order to overcome the difficulties that this computation presents, and I refer to a dedicated presentation [38] at this workshop for more details. I will only make a remark at this stage. What is usually understood under $a_\mu^{\text{HVP-LO}}$ is not only the contribution shown on the diagram on the left in Fig. 2, but also comprises the contribution shown on the diagram on the right of this same Figure. The reason for this situation is rather easy to understand. From an experimental point of view, it is not possible to correct the data for contributions arising from the exchanges of virtual photons between the final-state hadrons, or for the emission of soft photons with energies below the detection threshold of the experiment. Furthermore, some radiative modes, like $e^+e^- \to \pi^0\gamma$, are explicitly included in the evaluation of $a_\mu^{\text{HVP-LO}}$. This last final state, for instance, measured in the energy range between 600 MeV and 1030 MeV, produces by itself a contribution of $4.4(1.9)\cdot 10^{-10}$ to $a_\mu^{\text{HVP-LO}}$. Should the future lattice evaluations of HVP reach the sub-percent level, then comparison with the data-based determinations will be meaningful only if the second contribution in Fig. 2 and radiative corrections are evaluated separately, or if the numerical simulations include both QCD and QED field configurations, which might eventually turn out to be the most expedient way to address this issue.

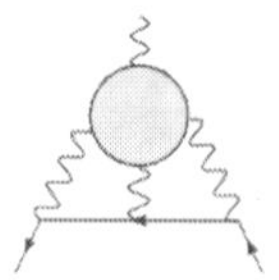

Figure 3. The hadronic light-by-light contribution to a_μ. The blob represents the hadronic four-point function (see text).

4 Hadronic light-by-light

Let us consider the second hadronic correction of importance as far as the contribution to the total theoretical uncertainty is concerned, namely hadronic light-by-light scattering (HLxL), represented on Fig. 3. The vertex function it depicts is expressed as follows, with p (resp. p') the four-momentum of the ingoing (resp. outgoing) muon, and $k = p' - p$,

$$\begin{aligned}
&\bar{\mathrm{u}}(p')\Gamma_\rho^{\text{HLxL}}(p',p)\mathrm{u}(p) = \\
&= \bar{\mathrm{u}}(p')\left[\gamma_\rho F_1^{\text{HLxL}}(k^2) + \frac{i}{2m}\sigma_{\rho\tau}k^\tau F_2^{\text{HLxL}}(k^2)\right]\mathrm{u}(p) \\
&= -ie^6\int\frac{d^4q_1}{(2\pi)^4}\int\frac{d^4q_2}{(2\pi)^4}\frac{1}{q_1^2\,q_2^2\,(q_1+q_2-k)^2} \\
&\times\frac{1}{(p'-q_1)^2-m^2}\,\frac{1}{(p'-q_1-q_2)^2-m^2} \\
&\times\bar{\mathrm{u}}(p')\gamma^\mu(\not p'-\not q_1+m)\gamma^\nu(\not p'-\not q_1-\not q_2+m)\gamma^\lambda\mathrm{u}(p) \\
&\times k^\sigma\frac{\partial}{\partial k^\rho}\,\Pi_{\mu\nu\lambda\sigma}(q_1,q_2,k-q_1-q_2,-k) \qquad (9)
\end{aligned}$$

in terms of the QCD correlator $\Pi_{\mu\nu\lambda\sigma}(q_1,q_2,q_3,q_4)$, $q_1+q_2+q_3+q_4=0$, the Fourier transform of the vacuum expectation value of the time ordered product of four light-quark electromagnetic currents. To extract a_μ^{HLxL}, the corresponding contribution to the anomalous magnetic moment, from this vertex function, one needs to project on the Pauli form factor $F_2^{\text{HLxL}}(k^2)$, and then let the momentum k^μ go to $(0,0,0,0)$. In contrast to a_μ^{HVP}, there is no direct and simple connection of a_μ^{HLxL} with an experimental observable, i.e. something that would look like Eq. (7). In order to evaluate this contribution, one either has to rely on models that aim at reproducing the dynamics of the strong interactions in the whole ranges of integration over the loop

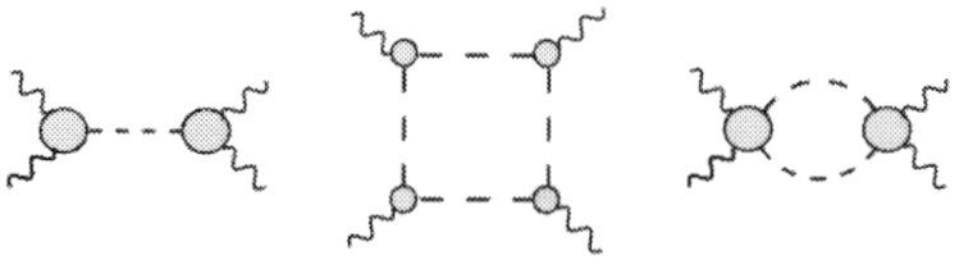

Figure 4. Various contributions to the four-point function $\Pi_{\mu\nu\rho\sigma}(q_1,q_2,q_3,q_4)$: poles due to single-meson intermediate states (left), charged meson loops (middle), two-meson intermediate states (right). The shaded blobs indicate the corresponding form factors.

momenta q_1 and q_2 in Eq. (9), or proceed through numerical simulation of QCD on a space-time lattice. Concerning the latter, see Ref. [39]. In order to implement the first approach, one needs some guiding principles. At our disposal, we have the chiral low-energy expansion and the $1/N_c$ expansion [40]. The second one turns out to be more useful from a practical point of view: the integration over the two momenta q_1 and q_2 in Eq. (9) does not only receive important contributions from the region where the chiral expansion stays relevant. From the point of view of the large-N_c expansion, one may identify individual contributions to the four-point function $\Pi_{\mu\nu\lambda\sigma}(q_1, q_2, q_3, q_4)$, like single-meson poles, or meson loops and two-particle intermediate states, and so on, as illustrated on Fig. 4,

$$\Pi_{\mu\nu\rho\sigma}(q_1, q_2, q_3, q_4) = \Pi_{\mu\nu\rho\sigma}^{\pi^0,\eta,\eta' \text{ poles}}(q_1, q_2, q_3, q_4) + \Pi_{\mu\nu\rho\sigma}^{\pi^\pm, K^\pm \text{ loops}}(q_1, q_2, q_3, q_4) + \Pi_{\mu\nu\rho\sigma}^{\pi\pi, K\bar{K}}(q_1, q_2, q_3, q_4) + \cdots . \tag{10}$$

Only two full calculations, Refs. [42, 43] and [44–46], have to date attempted to provide a description as complete as possible of the various contributions to the four-point function $\Pi_{\mu\nu\rho\sigma}(q_1, q_2, q_3, q_4)$. Although there are differences if specific contributions are compared on an individual basis, both groups obtained similar results for their final numbers for $a_\mu^{\rm HLxL}$, with later updates correcting for the sign of the pion-pole contribution, which was found to be wrong in the original publications, as first pointed out in Ref. [47]. In order to describe the QCD dynamics in the crucial region of intermediate energies, models like the extended Nambu–Jona-Lasinio model [48, 49] and/or the hidden local symmetry model [50, 51], which couples the lowest-lying vector resonance to the pseudoscalars, were used. The final result turns out to be numerically quite close to the contribution due to the exchange of a single neutral pseudoscalar between pairs of photons (see the first contribution on Fig. 4), the remaining contributions canceling each others. For a critical review, as well as updates, see Ref. [52]. Subsequent developments have focused on specific aspects, like: i) the reevaluation of individual contributions coming from the exchanges of single mesons, the pseudoscalars [8, 47, 53], the axial [35, 54], scalar or tensor mesons [54], ii) the large-N_c expansion, iii) short-distance constraints, and iv) dispersive approaches.

Concerning the large-N_c expansion, the general structure of a_μ^{HLxL} reads [7, 55, 56]

$$a_\mu^{\rm HLxL} = N_c \left(\frac{\alpha}{\pi}\right)^3 \frac{N_c}{F_\pi^2} \frac{m_\mu^2}{48\pi^2} \left[\ln^2 \frac{M_\rho}{M_\pi} + c_\chi \ln \frac{M_\rho}{M_\pi} + \kappa\right] + O(N_c^0). \tag{11}$$

The coefficient of the $\ln^2$ constitutes an exact result [47, 55], whereas the constant c_χ is related to the amplitude for the decay $\pi^0 \to e^+e^-$ [55, 56] and can in principle be obtained from a sufficiently accurate measurement of the corresponding branching fraction. Nothing can be said from first principles on the constant κ and on the subleading $O(N_c^0)$ corrections, and here one is redirected to the existing model calculations, as provided by Refs. [42, 43] and [44–46].

An important aspect for the theoretical description of the four-point function $\Pi_{\mu\nu\rho\sigma}(q_1, q_2, q_3, q_4)$ are its properties at short-distances. Various possibilities have to be considered. First, since $\Pi_{\mu\nu\rho\sigma}(q_1, q_2, q_3, q_4)$ is not an order parameter, one has to match the low- and medium-energy part, that is described in terms of resonances, to the perturbative regime of QCD when all momenta reach high (euclidian) virtualities. Another quite interesting situation is provided by the limit, first considered in Ref. [57] (for q_1 in the euclidian region)

$$\lim_{\lambda\to\infty} \Pi_{\mu\nu\rho\sigma}(\lambda q_1, q_2 - \lambda q_1, q_3, q_4) = = \frac{1}{\lambda}\frac{2i}{3}\frac{q_1^\gamma}{q_1^2}\epsilon_{\mu\nu\gamma}{}^{\tau} W_{\rho\sigma\tau}(q_3, q_4, q_2) + O(1/\lambda^2), \tag{12}$$

with $q_2 + q_3 + q_4 = 0$. In this expression, $W_{\mu\nu\rho}(q_1, q_2)$ denotes the Fourier transform of the the three-point QCD correlation function involving two hadronic electromagnetic currents and the flavour-neutral axial currents (including the flavour-singlet one). Finally, one might in principle also consider the behaviour in the limit

$$\lim_{\lambda\to\infty} \Pi_{\mu\nu\rho\sigma}(\lambda q_1, \lambda q_2, q_3 - \lambda q_1 - \lambda q_2, -q_3), \tag{13}$$

which doesn't seem to have been discussed so far. The limit where all virtualities become large (in the euclidian region) starts with a loop of massless quarks, supplemented with corrections from perturbative QCD,

$$\lim_{\lambda\to\infty} \Pi_{\mu\nu\rho\sigma}(\lambda q_1, \lambda q_2, \lambda q_3, \lambda q_4) = \Pi_{\mu\nu\rho\sigma}^{\rm pQCD}(q_1, q_2, q_3, q_4) + \cdots, \tag{14}$$

with $q_1 + q_2 + q_3 + q_4 = 0$. In the approaches mentioned above, this condition is met through the presence of a constituent quark loop, which reproduces the correct limit at leading order, while the sum of the other contributions has then to vanish in this limit. In contrast to the previous situation, the constraint of Eq. (12) appears to be more "democratic", in the sense that all terms in the decomposition (10) can in principle contribute. This means that, in the limit considered in Eq. (12), they have to combine in order to build up, in the limit considered, the three-point correlation function on the right-hand side.

Recent (non lattice) developments concerning a_μ^{HLxL} have come from the side of dispersion relations. One approach [58–60] uses a decomposition of $\Pi_{\mu\nu\rho\sigma}(q_1, q_2, q_3, q_4)$ into a set of invariant functions, for which dispersion relations are written. In the corresponding absorptive parts one then isolates the one-particle (poles), two-particle, and so on, contributions, as illustrated on Fig. 4. The large-N_c expansion, together with the suppression of contributions involving higher invariant masses, provide the justification for considering contributions with only a low number of intermediate states. Notice, however, that, for instance, the two-pion states also contains resonant contributions (scalars or tensors) that, like the poles from the pseudoscalar mesons, occur already at leading order in the large-N_c limit. The idea is then to evaluate these dispersion relations in terms of experimental input for, for instance, various form factors they involve [61]. Whether experiment will be able to provide the necessary input data for all form factors involved at the required accuracy remains an open question [62]. A second approach [63, 64] writes a dispersion relation directly for the form factor $F_2(k^2)$ itself, using the representation in Eq. (9). The task consists then again to identify and evaluate the various discontinuities. So far, it has been shown [63] that one can recover the contribution from the pion pole through this method, using a vector-meson-dominance representation for the corresponding form factor. What is so far missing in these dispersive approaches, is how the short-distance properties will be implemented, that is how the various individual contributions will be made to talk to each others in the relevant asymptotic regimes.

The present discrepancy between theory and experiment represents about three times the values of $a_\mu^{\rm HLxL}$ reported in Table 1. Although the present determinations of a_μ^{HLxL} are still highly model dependent, they seem to incorporate the relevant physics in an appropriate manner. Other determinations relying on simple models, like, for instance, the constituent chiral quark model [65, 66] or holographic QCD [67, 68], do not show significant deviations from the results given in Table 1, although the uncertainties coming from the models themselves are of course difficult to estimate. Finally, higher-order corrections to $a_\mu^{\rm HLxL}$ have also been estimated [69], and found to be quite small.

5 Conclusion

There is a persistent discrepancy between the theory prediction of the anomalous magnetic moment of the muon within the standard model and its measured value, as shown in Eqs. (2) and (3). From the theoretical point of view, we see no hint of any need of higher order corrections at the level of the QED and weak contributions, even with the improvement in the experimental precision that future experiments plan to achieve. An independent cross-check of some of the tenth-order QED corrections enhanced by powers of $\ln(m_\mu/m_e)$, along the lines developed in Refs. [22, 23] for the eighth-order terms, if feasible, would be welcome, though. As far as the contributions of hadronic vacuum polarization is concerned, further improvement should be possible, since more precise data should become available in the future [36]. Actually, inclusion of more recent data on $e^+e^- \to$ hadrons in the determination of $a_\mu^{\rm HVP}$ even enhances the difference between theory and experiment [35]. Finally, there remains the contribution from hadronic light-by-light. There is no sign that important physics is missing in the existing evaluations. The issue here will rather consist in improving on the present precision and in obtaining more reliable estimate of the uncertainties.

Acknowledgements

I would like to thank the organizers of the Workshop "Flavour changing and conserving processes 2015" for the kind invitation to attend this very stimulating meeting. This work has been partially supported by the OCEVU Labex (ANR-11-LABX-0060) and the A*MIDEX project (ANR-11-IDEX-0001-02) funded by the "Investissements d'Avenir" French government program managed by the ANR.

References

[1] G. W. Bennett et al., Phys. Rev D **73**, 072003 (2006).

[2] K. A. Olive et al. [Particle Data Group], Chin. Phys. C **38**, 090001 (2014).

[3] T. Aoyama, M. Hayakawa, T. Kinoshita and M. Nio, Phys. Rev. Lett. **109**, 111808 (2012) [arXiv:1205.5370 [hep-ph]].

[4] M. Davier, A. Hoecker, B. Malaescu and Z. Zhang, Eur. Phys. J. C **71**, 1515 (2011) [Eur. Phys. J. C **72**, 1874 (2012)] [arXiv:1010.4180 [hep-ph]].

[5] K. Hagiwara, R. Liao, A. D. Martin, D. Nomura and T. Teubner, J. Phys. G **38**, 085003 (2011) [arXiv:1105.3149 [hep-ph]].

[6] A. Kurz, T. Liu, P. Marquard and M. Steinhauser, Phys. Lett. B **734**, 144 (2014) [arXiv:1403.6400 [hep-ph]].

[7] J. Prades, E. de Rafael and A. Vainshtein, chapter 9 in *Lepton Dipole Moments*, Advanced Series on Directions in High Energy Physics – Vol. 20, B. Lee Roberts and William J. Marciano Eds, World Scientific Co. Pte. Ltd. (2010).

[8] F. Jegerlehner and A. Nyffeler, Phys. Rept. **477**, 1 (2009) [arXiv:0902.3360 [hep-ph]].

[9] R. Jackiw, S. Weinberg, Phys. Rev. D **5**, 2396 (1972); I. Bars, M. Yoshimura, Phys. Rev. D **6**, 374 (1972); K. Fujikawa, B.W. Lee, A.I. Sanda, Phys. Rev. D **6**, 2923 (1972); G. Altarelli, N. Cabbibo, L. Maiani, Phys. Lett. B **40**, 415 (1972); W. A. Bardeen, R. Gastmans, B.E. Lautrup, Nucl. Phys. B **46**, 315 (1972).

[10] C. Gnendiger, D. Stöckinger and H. Stöckinger-Kim, Phys. Rev. D **88**, 053005 (2013) [arXiv:1306.5546 [hep-ph]].

[11] D. Stöckinger, chapter 12 in *Lepton Dipole Moments*, quoted in [7].

[12] See the contribution of D. Stöckinger to this workshop.

[13] See the contribution of E. J. Chun to this workshop.

[14] See the contribution of K. M. Patel to this workshop.

[15] R. M. Carey et al.. The New g-2 experiment - 2009. Fermilab. Proposal 0989.

[16] B. Lee Roberts, Chin. Phys. C **34**, 741 (2010).

[17] T. Mibe [J-PARC g-2 Collaboration], Chin. Phys. C **34**, 745 (2010).

[18] See the contribution of D. Hertzog to this workshop.

[19] J. Schwinger, Phys. Rev. **73**, 416L (1948).

[20] T. Kinoshita (chapter 3), S. Laporta and E. Remiddi (chapter 4) in *Lepton Dipole Moments*, quoted in [7].

[21] T. Kinoshita, M. Nio, Phys. Rev. D **73**, 053007 (2006). T. Aoyama et al., Phys. Rev. D **78**, 053005 (2008); Phys. Rev. D **78**, 113006 (2008); Phys. Rev. D **81**, 053009 (2010); Phys. Rev. D **82**, 113004 (2010); Phys. Rev. D **83**, 053002 (2011); Phys. Rev. D **83**,

053003 (2011); Phys. Rev. D **84**, 053003 (2011); Phys. Rev. D **85**, 033007 (2012).
[22] See the contribution of M. Steinhauser to this workshop.
[23] A. Kurz, T. Liu, P. Marquard, A. V. Smirnov, V. A. Smirnov and M. Steinhauser, Phys. Rev. D **92**, 073019 (2015) [arXiv:1508.00901 [hep-ph]].
[24] S. Weinberg, Phys. Rev. Lett. **19**, 1264 (1967).
[25] T. V. Kukhto, E. A. Kuraev, Z. K. Silagadze and A. Schiller, Nucl. Phys. B **371**, 567 (1992).
[26] S. Peris, M. Perrottet and E. de Rafael, Phys. Lett. B **355**, 523 (1995) [hep-ph/9505405].
[27] A. Czarnecki, B. Krause and W. J. Marciano, Phys. Rev. D **52**, 2619 (1995) [hep-ph/9506256]; Phys. Rev. Lett. **76**, 3267 (1996) [hep-ph/9512369].
[28] M. Knecht, S. Peris, M. Perrottet and E. De Rafael, JHEP **0211**, 003 (2002) [hep-ph/0205102].
[29] A. Czarnecki, W. J. Marciano and A. Vainshtein, Phys. Rev. D **67**, 073006 (2003) [Phys. Rev. D **73**, 119901 (2006)] [hep-ph/0212229].
[30] C. Bouchiat and L. Michel, J. Phys. Radium 22, 121 (1961).
[31] L. Durand, Phys. Rev. **128**, 441 (1962); Err.-ibid. **129**, 2835 (1963).
[32] M. Gourdin and E. de Rafael, Nucl. Phys. B **10**, 667 (1969).
[33] See the contribution of Z. Zhang to this workshop.
[34] M. Ablikim *et al.* [BESIII Collaboration], arXiv:1507.08188 [hep-ex].
[35] See the contribution of F. Jegerlehner to this workshop.
[36] See the contribution of S. Eidelman to this workshop.
[37] See the contribution of L. Trentadue to this workshop.
[38] See the contribution of M. Petschlies to this workshop.
[39] See the contribution of C. Lehner to this workshop.
[40] E. de Rafael, Phys. Lett. B **322**, 239 (1994) [hep-ph/9311316].
[41] J. Bijnens, E. Pallante and J. Prades, Phys. Rev. Lett. **75**, 1447 (1995) [Phys. Rev. Lett. **75**, 3781 (1995)] [hep-ph/9505251].
[42] J. Bijnens, E. Pallante and J. Prades, Nucl. Phys. B **474**, 379 (1996) [hep-ph/9511388].
[43] J. Bijnens, E. Pallante and J. Prades, Nucl. Phys. B **626**, 410 (2002) [hep-ph/0112255].
[44] M. Hayakawa, T. Kinoshita and A. I. Sanda, Phys. Rev. Lett. **75**, 790 (1995) [hep-ph/9503463].
[45] M. Hayakawa, T. Kinoshita and A. I. Sanda, Phys. Rev. D **54**, 3137 (1996) [hep-ph/9601310].
[46] M. Hayakawa and T. Kinoshita, Phys. Rev. D **57**, 465 (1998) [Phys. Rev. D **66**, 019902 (2002)] [hep-ph/9708227].
[47] M. Knecht and A. Nyffeler, Phys. Rev. D **65**, 073034 (2002) [hep-ph/0111058].
[48] J. Bijnens, C. Bruno and E. de Rafael, Nucl. Phys. B **390**, 501 (1993) [hep-ph/9206236].
[49] J. Bijnens, Phys. Rept. **265**, 369 (1996) [hep-ph/9502335].
[50] M. Bando, T. Kugo and K. Yamawaki, Nucl. Phys. B **259**, 493 (1985).
[51] M. Bando, T. Kugo and K. Yamawaki, Phys. Rept. **164**, 217 (1988).
[52] See the contribution of J. Bijnens to this workshop.
[53] A. Nyffeler, Phys. Rev. D **79**, 073012 (2009) [arXiv:0901.1172 [hep-ph]].
[54] V. Pauk and M. Vanderhaeghen, Eur. Phys. J. C **74**, 3008 (2014) [arXiv:1401.0832 [hep-ph]].
[55] M. Knecht, A. Nyffeler, M. Perrottet and E. de Rafael, Phys. Rev. Lett. **88**, 071802 (2002) [hep-ph/0111059].
[56] M. J. Ramsey-Musolf and M. B. Wise, Phys. Rev. Lett. **89**, 041601 (2002) [hep-ph/0201297].
[57] K. Melnikov and A. Vainshtein, Phys. Rev. D **70**, 113006 (2004)
[58] G. Colangelo, M. Hoferichter, M. Procura and P. Stoffer, JHEP **1409**, 091 (2014) [arXiv:1402.7081 [hep-ph]].
[59] G. Colangelo, M. Hoferichter, M. Procura and P. Stoffer, JHEP **1509**, 074 (2015) [arXiv:1506.01386 [hep-ph]].
[60] See the contribution of M. Procura to this workshop.
[61] G. Colangelo, M. Hoferichter, B. Kubis, M. Procura and P. Stoffer, Phys. Lett. B **738**, 6 (2014) [arXiv:1408.2517 [hep-ph]].
[62] See the contribution of A. Nyffeler to this workshop.
[63] V. Pauk and M. Vanderhaeghen, Phys. Rev. D **90**, 113012 (2014) [arXiv:1409.0819 [hep-ph]].
[64] See the contribution of M. Vanderhaeghen to this workshop.
[65] D. Greynat and E. de Rafael, JHEP **1207**, 020 (2012) [arXiv:1204.3029 [hep-ph]].
[66] See the contribution of D. Greynat to this workshop.
[67] L. Cappiello, O. Cata and G. D'Ambrosio, Phys. Rev. D **83**, 093006 (2011) [arXiv:1009.1161 [hep-ph]].
[68] See the contribution of L. Cappiello to this workshop.
[69] G. Colangelo, M. Hoferichter, A. Nyffeler, M. Passera and P. Stoffer, Phys. Lett. B **735**, 90 (2014) [arXiv:1403.7512 [hep-ph]].

EPJ Web of Conferences **118**, 01018 (2016) DOI: 10.1051/epjconf/201611801018
FCCP2015

Lepton Universality and Lepton Flavour Violation tests at the B-factories

Alberto Lusiani[1,2,a], on behalf of the BABAR Collaboration

[1] *Scuola Normale Superiore, Pisa, Italy*
[2] *INFN sezione di Pisa, Pisa, Italy*

Abstract. We review the experimental status of the lepton universality tests and lepton flavour violation searches after the completion of the data-taking and most of the data analysis of the B-factories BABAR and Belle. The universality of the Standard Model charged weak couplings has been confirmed and moderately improved in precision by the B-factories results. Lepton Flavour violation in the τ lepton decays has been searched in several decay modes and no evidence has been found. We set 90% confidence level upper limits on the Lepton Flavour violating branching ratios in the 10^{-7}–10^{-8} range, greatly extending the previous limits set mainly by CLEO in the 10^{-6} range.

1 Introduction

In the Standard Model (SM) the lepton couplings are the same for each lepton generation. The universality of the charged weak coupling for charged leptons can be tested by combining the measurements of decay widths with leptons in the initial and/or final state, which can be obtained by measuring the relevant branching fractions, the lepton masses and the leptons lifetimes.

In the SM with zero neutrino masses, the lepton flavour is exactly (accidentally) conserved. When we take into account the experimentally measured neutrino mass squared differences and neutrino mixing, lepton-flavour-violating (LFV) branching fractions, such as for instance $\tau \to \mu\gamma$, $\mu \to e\gamma$, $\tau \to 3\mu$, are predicted to be non-zero but suppressed by a factor of order $(m_\nu/m_W)^4$ to values several orders of magnitude smaller than the current experimental reach.

Models beyond the SM like Supersymmetry, Grand Unification or Extra Dimensions typically include additional lepton-flavour-violating processes, which can produce observable LFV τ lepton decay rates, even when the model parameters are constrained with the available experimental data [1–9].

Since about year 2000, the B-factories BABAR and Belle have collected unprecedented samples of clean e^+e^- collisions at and around the $\Upsilon(4S)$ peak, which have improved the experimental test of the SM lepton universality. The B-factories have ceased data-taking in 2008 (BABAR) and 2010 (Belle) and have now completed and published most of the data-analyses, increasing the experimental knowledge on lepton universality and lepton flavour violation.

Before the B-factories BABAR and Belle, experimental measurements have permitted testing the lepton universality to a precision around 0.2%. The most important contributions have been the τ lifetime measurements by the LEP experiments, the τ leptonic branching fraction measurements by the LEP experiments and in particular ALEPH, and the BES τ mass measurement. We report in the following the present status after the B-factories results.

BABAR and Belle have provided effective conditions to search for LFV τ lepton decay branching fractions with clean and unambiguous experimental signatures [10], like a final state with invariant mass matching the τ mass and with total energy matching half the event energy in the center-of-mass frame. We report in the following significant advancements that have been obtained with respect to the pre-existing limits set mainly by CLEO around 10^{-6}.

2 Lepton Universality Tests

In the Standard Model, the partial width of a heavier lepton λ decaying to a lighter lepton ρ are, neglecting neutrino masses and including radiative corrections [11],

$$\Gamma[\lambda \to \nu_\lambda \rho \bar{\nu}_\rho(\gamma)] = \frac{G_\lambda G_\rho m_\lambda^5}{192\pi^3} f\left(\frac{m_\rho^2}{m_\lambda^2}\right) r_W^\lambda r_\gamma^\lambda ,$$

where

$$G_\rho = \frac{g_\rho^2}{4\sqrt{2}M_W^2} \qquad f(x) = 1 - 8x + 8x^3 - x^4 - 12x^2 \ln x$$

$$r_W^\lambda = 1 + \frac{3}{5}\frac{m_\lambda^2}{M_W^2} \qquad r_\gamma^\lambda = 1 + \frac{\alpha(m_\lambda)}{2\pi}\left(\frac{25}{4} - \pi^2\right) .$$

Computing the proper ratios of the above partial widths using the HFAG 2014 τ world averages [12], the Review of Particle Physics (PDG) 2014 [13] and CODATA 2010 [14]

[a] e-mail: alberto.lusiani@pi.infn.it

© The Authors, published by EDP Sciences. This is an open access article distributed under the terms of the Creative Commons Attribution License 4.0 (http://creativecommons.org/licenses/by/4.0/).

results we obtain the following charged weak current coupling ratios:

$$\left(\frac{g_\tau}{g_\mu}\right) = \sqrt{\frac{B_{\tau e}}{B_{\mu e}}\frac{\tau_\mu m_\mu^5 f_{\mu e} r_W^\mu r_\gamma^\mu}{\tau_\tau m_\tau^5 f_{\tau e} r_W^\tau r_\gamma^\tau}} = 1.0011 \pm 0.0015$$

$$\left(\frac{g_\tau}{g_e}\right) = \sqrt{\frac{B_{\tau\mu}}{B_{\mu e}}\frac{\tau_\mu m_\mu^5 f_{\mu e} r_W^\mu r_\gamma^\mu}{\tau_\tau m_\tau^5 f_{\tau\mu} r_W^\tau r_\gamma^\tau}} = 1.0029 \pm 0.0015$$

$$\left(\frac{g_\mu}{g_e}\right) = \sqrt{\frac{B_{\tau\mu}}{B_{\tau e}}\frac{f_{\tau e}}{f_{\tau\mu}}} = 1.0018 \pm 0.0014$$

The data are consistent with lepton universality, and the precision of the tests improved from 0.20–0.23% before the B-Factories to 0.14–0.15%, thanks primarily to the Belle tau lifetime measurement [15]. The BABAR measurement of the ratio between the muon and electron branching fraction [16] and the BABAR and Belle τ mass measurements [17, 18] provide negligible contributions. The precision of the tests is now mainly limited by the uncertainties on the τ lifetime (0.09%) and on the τ leptonic branching fractions (0.11–0.12%). The overall consistency of the data with the SM is summarized in figure 1, which reports the experimental measurements of the τ lifetime and of the "universality improved" $\tau^- \to e^-\bar{\nu}_e\nu_\tau$ branching fraction [19], together with the constraint on them from the SM, which relies on the measurements of the τ lifetime and the muon mass and lifetime.

Partial widths of τ decays to hadrons compared to the same hadron decay to muons measure the tau-muon universality of charged weak couplings as follows:

$$\left(\frac{g_\tau}{g_\mu}\right)^2 = \frac{B(\tau \to h\nu_\tau)}{B(h \to \mu\bar{\nu}_\mu)}\frac{2m_h m_\mu^2 \tau_h}{(1+\delta_h)m_\tau^3\tau_\tau}\left(\frac{1-m_\mu^2/m_h^2}{1-m_h^2/m_\tau^2}\right)^2 ,$$

where $h = \pi$ or K and the radiative corrections are $\delta_\pi = (0.16\pm0.14)\%$ and $\delta_K = (0.90\pm0.22)\%$ [20], we measure:

$$\left(\frac{g_\tau}{g_\mu}\right)_\pi = 0.9963 \pm 0.0027 , \quad \left(\frac{g_\tau}{g_\mu}\right)_K = 0.9858 \pm 0.0071 .$$

The BABAR measurement of the ratio between the kaon and electron branching fractions [16] improved these tests. Similar tests could be performed with decays to electrons, however they are less precise because the hadron two body decays to electrons are helicity-suppressed. Averaging the three g_τ/g_μ ratios we obtain

$$\left(\frac{g_\tau}{g_\mu}\right)_{\tau+\pi+K} = 1.0001 \pm 0.0014 ,$$

accounting for statistical correlations.

Additional electron-muon universality tests can be obtained comparing pion and kaon leptonic branching fractions. The B-factories did not improve these tests.

3 Searches for Lepton Flavour Violation

The B-factories have searched for an extensive range of LFV τ decay modes: no evidence has been found and 90% confidence level (CL) upper limits have been set. The most recent published result are the searches for $\tau \to$ 3 charged leptons [21, 22], $\tau \to \ell K_S^0$ and $\tau \to \ell K_S^0 K_S^0$ [23], $\tau \to \ell V^0$ [24], $\tau \to \ell hh'$ [25]. Recently, the LHCb collaboration has published the result of the search for $\tau \to \mu\mu\mu$ [26].

The most recent published result by the B-factories Belle and BABAR and by LHCb are the searches for $\tau \to$ 3 leptons by Belle [21],

$B(\tau \to e^-e^+e^-) < 2.7 \cdot 10^{-8}$,
$B(\tau \to \mu^-e^+e^-) < 1.8 \cdot 10^{-8}$,
$B(\tau \to e^-\mu^+\mu^-) < 2.7 \cdot 10^{-8}$,
$B(\tau \to \mu^-\mu^+\mu^-) < 2.1 \cdot 10^{-8}$,
$B(\tau \to e^-\mu^+e^-) < 1.5 \cdot 10^{-8}$,
$B(\tau \to \mu^-e^+\mu^-) < 1.7 \cdot 10^{-8}$,

by BABAR [22]

$B(\tau \to e^-e^+e^-) < 2.9 \cdot 10^{-8}$,
$B(\tau \to \mu^-e^+e^-) < 2.2 \cdot 10^{-8}$,
$B(\tau \to e^-\mu^+\mu^-) < 3.2 \cdot 10^{-8}$,
$B(\tau \to \mu^-\mu^+\mu^-) < 3.3 \cdot 10^{-8}$,
$B(\tau \to e^-\mu^+e^-) < 1.8 \cdot 10^{-8}$,
$B(\tau \to \mu^-e^+\mu^-) < 2.6 \cdot 10^{-8}$,

and by LHCb [26]

$B(\tau \to \mu^-\mu^+\mu^-) < 4.6 \cdot 10^{-8}$,

the searches for $\tau \to \ell K_S^0$ and $\tau \to \ell K_S^0 K_S^0$ by Belle [23]

$B(\tau \to e^-K_S^0) < 2.6 \cdot 10^{-8}$,
$B(\tau \to \mu^-K_S^0) < 2.3 \cdot 10^{-8}$,
$B(\tau \to e^-K_S^0K_S^0) < 7.1 \cdot 10^{-8}$,
$B(\tau \to \mu^-K_S^0K_S^0) < 8.1 \cdot 10^{-8}$,

the searches for $\tau \to \ell V^0$ by Belle [24]

$B(\tau \to e^-\rho^0) < 1.8 \cdot 10^{-8}$,
$B(\tau \to \mu^-\rho^0) < 1.2 \cdot 10^{-8}$,
$B(\tau \to e^-K^*(892)^0) < 3.2 \cdot 10^{-8}$,
$B(\tau \to \mu^-K^*(892)^0) < 7.2 \cdot 10^{-8}$,
$B(\tau \to e^-\bar{K}^*(892)^0) < 3.4 \cdot 10^{-8}$,
$B(\tau \to \mu^-\bar{K}^*(892)^0) < 7.0 \cdot 10^{-8}$,
$B(\tau \to e^-\phi) < 3.1 \cdot 10^{-8}$,
$B(\tau \to \mu^-\phi) < 8.4 \cdot 10^{-8}$,
$B(\tau \to e^-\omega) < 4.8 \cdot 10^{-8}$,
$B(\tau \to \mu^-\omega) < 4.7 \cdot 10^{-8}$,

and the searches for $\tau \to \ell hh'$ by Belle [25]

$B(\tau \to e^-\pi^+\pi^-) < 2.3 \cdot 10^{-8}$,
$B(\tau \to \mu^-\pi^+\pi^-) < 2.1 \cdot 10^{-8}$,
$B(\tau \to e^-\pi^+K^-) < 3.7 \cdot 10^{-8}$,
$B(\tau \to \mu^-\pi^+K^-) < 8.6 \cdot 10^{-8}$,
$B(\tau \to e^-K^+\pi^-) < 3.1 \cdot 10^{-8}$,
$B(\tau \to \mu^-K^+\pi^-) < 4.5 \cdot 10^{-8}$,
$B(\tau \to e^-K^+K^-) < 4.4 \cdot 10^{-8}$,
$B(\tau \to \mu^-K^+K^-) < 7.1 \cdot 10^{-8}$,

The Heavy Flavour Averaging Group has collected all the upper limits on the LFV τ branching fractions [12] and we report here just the summary plot in figure 2. In addition, we report in figure 3 the upper limits combinations computed in the HFAG 2014 report. The significant progress made with the B-factories results is evident.

4 Conclusions and prospects

The B-factories helped improving the precision of the lepton universality tests, which continue to confirm the

 DOI: 10.1051/epjconf/201611801018

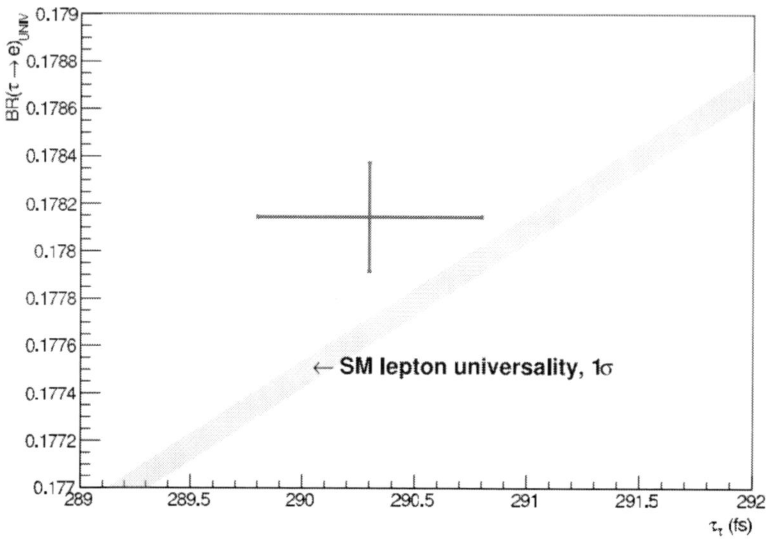

Figure 1. Experimental measurements of the τ lifetime and of the "universality improved" $\tau \to e$ branching fraction compared with the Standard Model prediction (yellow band). The width of the band is dominated by the uncertainty on the τ lifetime.

SM predictions with a precision that increased from 0.20–0.23% to 0.14–0.15%. The relevant experimental measurements are systematically limited, and there are only modest prospects of further improvements at the planned future high luminosity B-factory Belle-II. A high luminosity LEP-like e^+e^- collider at the Z^0 peak would most probably provide better conditions for precision measurements of the τ branching fractions.

The B-factories have experimentally probed the LFV τ branching fractions from the CLEO limits around 10^{-6} up to 10^{-7}–10^{-8} without finding any evidence of a signal. Future high luminosity B-factories like Belle II offer good prospects to advance the experimental reach close to 10^{-7} for decay modes such as $\tau \to \mu\gamma$ and close to 10^{-8} for the cleanest decay modes such as $\tau \to 3$ charged leptons.

References

[1] J.R. Ellis, M.E. Gomez, G.K. Leontaris, S. Lola, D.V. Nanopoulos, Eur. Phys. J. **C14**, 319 (2000), hep-ph/9911459
[2] J.R. Ellis, J. Hisano, M. Raidal, Y. Shimizu, Phys. Rev. **D66**, 115013 (2002), hep-ph/0206110
[3] A. Masiero, S.K. Vempati, O. Vives, Nucl. Phys. **B649**, 189 (2003), hep-ph/0209303
[4] T. Fukuyama, T. Kikuchi, N. Okada, Phys. Rev. **D68**, 033012 (2003), hep-ph/0304190
[5] G. Cvetic, C. Dib, C.S. Kim, J.D. Kim, Phys. Rev. **D66**, 034008 (2002), hep-ph/0202212
[6] C.x. Yue, Y.m. Zhang, L.j. Liu, Phys. Lett. **B547**, 252 (2002), hep-ph/0209291
[7] A. Dedes, J.R. Ellis, M. Raidal, Phys. Lett. **B549**, 159 (2002), hep-ph/0209207
[8] A. Brignole, A. Rossi, Phys. Lett. **B566**, 217 (2003), hep-ph/0304081
[9] E. Arganda, M.J. Herrero, Phys. Rev. **D73**, 055003 (2006), hep-ph/0510405
[10] A. Lusiani, PoS **HQL2010**, 054 (2010), 1012.3733
[11] W.J. Marciano, A. Sirlin, Phys. Rev. Lett. **61**, 1815 (1988)
[12] Y. Amhis et al. (Heavy Flavor Averaging Group (HFAG)) (2014), 1412.7515
[13] K.A. Olive et al. (Particle Data Group), Chin. Phys. **C38**, 090001 (2014)
[14] P.J. Mohr, B.N. Taylor, D.B. Newell, Rev. Mod. Phys. **84**, 1527 (2012), 1203.5425
[15] K. Belous et al. (Belle), Phys. Rev. Lett. **112**, 031801 (2014), 1310.8503
[16] B. Aubert et al. (BaBar), Phys. Rev. Lett. **105**, 051602 (2010), 0912.0242
[17] B. Aubert et al. (BaBar), Phys. Rev. **D80**, 092005 (2009), 0909.3562
[18] K. Abe et al. (Belle), Phys. Rev. Lett. **99**, 011801 (2007), hep-ex/0608046
[19] M. Davier, A. Hocker, Z. Zhang, Rev. Mod. Phys. **78**, 1043 (2006), hep-ph/0507078
[20] R. Decker, M. Finkemeier, Phys. Lett. **B334**, 199 (1994)
[21] K. Hayasaka et al. (Belle), Phys. Lett. **B687**, 139 (2010), 1001.3221
[22] J.P. Lees et al. (BaBar), Phys. Rev. **D81**, 111101 (2010), 1002.4550
[23] Y. Miyazaki et al. (Belle), Phys. Lett. **B692**, 4 (2010), 1003.1183
[24] Y. Miyazaki (Belle), Phys. Lett. **B699**, 251 (2011), 1101.0755

DOI: 10.1051/epjconf/201611801018

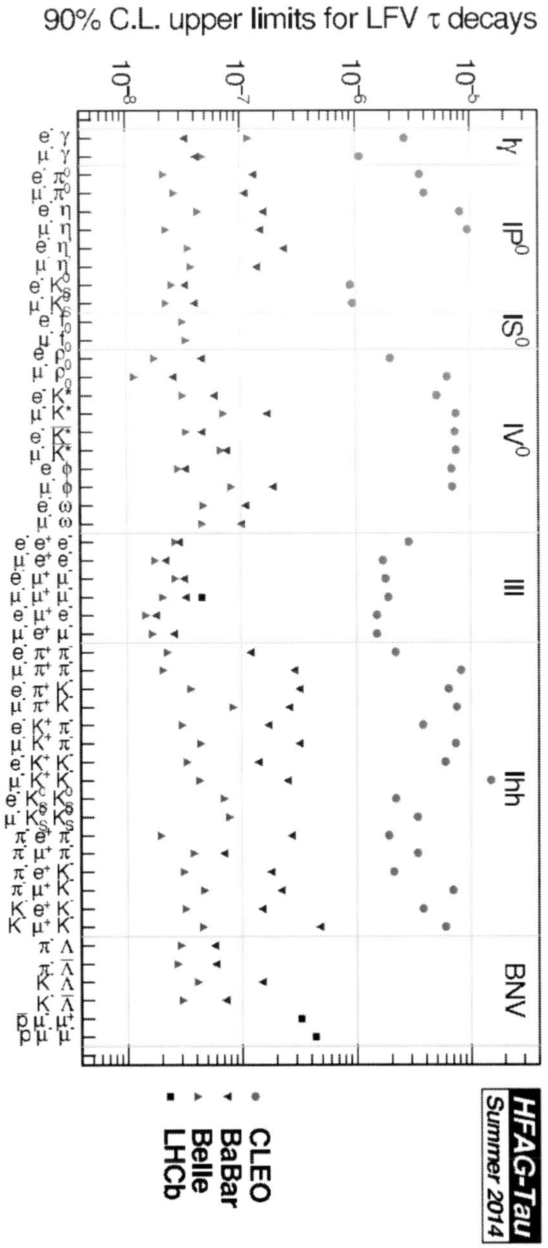

Figure 2. τ lepton-flavour-violating branching fraction upper limits summary plot from the HFAG 2014 report.

[25] Y. Miyazaki et al. (Belle), Phys. Lett. **B719**, 346 (2013), 1206.5595

[26] R. Aaij et al. (LHCb), JHEP **02**, 121 (2015), 1409.8548

DOI: 10.1051/epjconf/201611801018

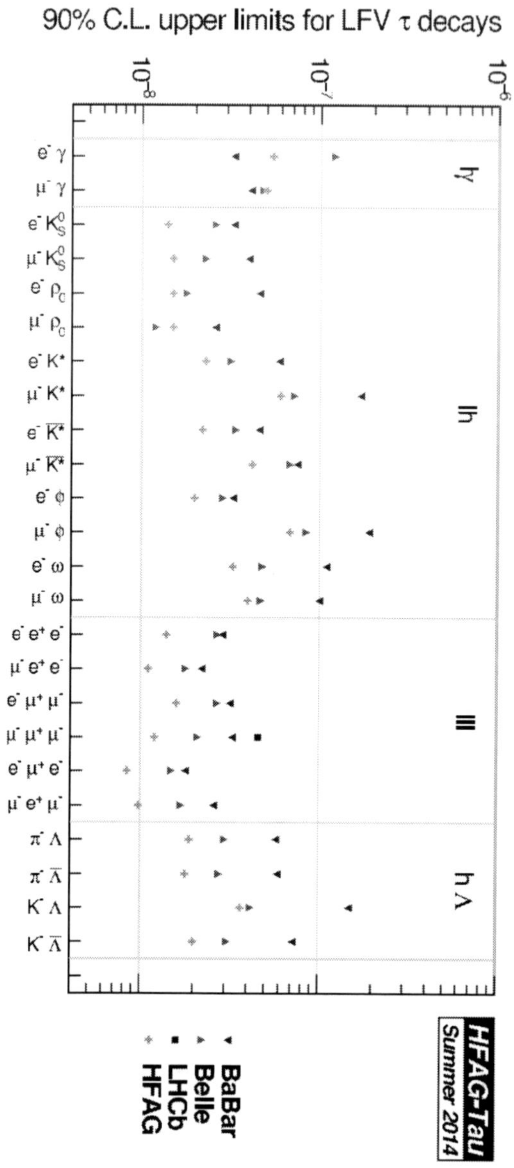

Figure 3. Combinations of the upper limits published by *BABAR*, Belle and LHCb on a list of searches for τ LFV branching fractions, from the HFAG 2014 report.

Pseudoscalar transition form factors: $(g-2)$ of the muon, pseudoscalar decays into lepton pairs, and the $\eta-\eta'$ mixing

Pere Masjuan[1,a] and Pablo Sanchez-Puertas[1,b]

[1] *PRISMA Cluster of Excellence, Institut für Kernphysik, Johannes Gutenberg-Universität, D-55099 Mainz, Germany*

Abstract. We present our model-independent and data-driven method to describe pseudoscalar meson transition form factors in the space- and (low-energy) time-like regions. The method is general and conforms a toolkit applicable to any other form factor, of one and two variables, with the potential to include both high- and low-energy QCD constraints altogether. The method makes use of analyticity and unitary properties of form factors, it is simple, systematic and can be improved upon by including new data. In the present discussion, the method is used to show the impact of experimental data for precision calculations in the low-energy sector of the Standard Model. In particular, due to its relevance for New Physics searches, we have considered the hadronic light-by-light scattering contribution to the anomalous magnetic moment of the muon (the pseudoscalar exchange contribution), the pseudoscalar decays into lepton pairs, and the determination of the mixing parameters of the η and η' system. For all of them we provide the most updated results in a data-driven manner.

1 Introduction

The pseudoscalar meson transition form factors (TFF) describe the effect of the strong interaction on the $\gamma^*\gamma^* - P$ transition (where $P = \pi^0, \eta, \eta' \cdots$) and is represented by a function $F_{P\gamma^*\gamma^*}(q_1^2, q_2^2)$ of the photon virtualities q_1^2, and q_2^2.

From the experimental point of view, one can study the TFF from both space- and time-like energy regimes. The time-like TFF can be accessed from a single Dalitz decay $P \to l^+l^-\gamma$ process which contains an off-shell photon with the momentum transfer q_1^2, a $F_{P\gamma^*\gamma}(q_1^2, 0)$, covering the $4m_l^2 < q^2 < m_P^2$ region. The space-like TFF can be accessed in e^+e^- colliders by the two-photon-fusion reaction $e^+e^- \to e^+e^-P$, see Fig. 1. The common practice is to extract the TFF when one of the outgoing leptons is tagged and the other is not, that is, the single-tag method. The tagged lepton emits a highly off-shell photon with the momentum transfer $q_1^2 \equiv -Q^2$ and is detected, while the other, untagged, is scattered at a small angle and its momentum transfer q_2^2 is almost zero, i.e., $F_{P\gamma^*\gamma}(Q^2) \equiv F_{P\gamma^*\gamma^*}(-Q^2, 0)$.

Experimental information on its doubly virtuality can be obtained through the double-tag method which demands the tagging of both final-state leptons. Due to the decrease of the cross section for both virtual photons, and the difficulties of the detection of all the particles entering into the process, data for the doubly virtuality are not yet available for the lowest pseudoscalars.

Theoretically, the limits $Q^2 = 0$ and $Q^2 \to \infty$ are well known in terms of the axial anomaly in the chiral-limit of QCD [2] and perturbative QCD [3], respectively. The TFF is then calculated as a convolution of a perturbative hard-scattering amplitude and a gauge-invariant meson distribution amplitude (DA) which incorporates the non-perturbative dynamics of the QCD bound-state [3]. At that point, a model needs to be used either for the DA or the TFF itself [3]. The discrepancy among different approaches reflects the model-dependency of that procedure. On top, when the desired object should be parameterized for the whole space-like range, and available experimental data are going to be used, the asymptotic scale of QCD should be fixed and a contact with the low-energy description performed. Theoretical efforts have not yet succeeded on performing such an endeavor, even worse for ascribing a theoretical errors to the procedure. A different procedure might be, then, desirable.

Let us remark that form factors are not interesting by themselves as represent the knowledge of QCD in a nutshell, but also for their important role on precision calculations of low-energy Standard Model observables such as the anomalous magnetic moment of the muon, the pseudoscalar decays into lepton pairs, the radius of the proton, etc, where deviations between the theoretical prediction and the experimental measure point towards an indirect search of New Physics phenomena.

The experimental information on the TFFs together with the theoretical knowledge on their kinematic limits yield the opportunity for a nice synergy between experiment and theory in a simple, easy, systematic, and user-friendly method. This synergy is the purpose of our work.

In this respect, our proposal can be summarized taking the following considerations:

[a] e-mail: masjuan@kph.uni-mainz.de
[b] e-mail: sanchezp@kph.uni-mainz.de

© The Authors, published by EDP Sciences. This is an open access article distributed under the terms of the Creative Commons Attribution License 4.0 (http://creativecommons.org/licenses/by/4.0/).

- We want a method, not a *model*. Our proposal is actually a TOOLKIT.
- It is a method based on the analyticity and unitary of the TFF, important ingredients when heading towards errors at the 10%.
- Should be a method as simple as possible, maximally transparent.
- If possible, the method should not use any assumption, only approaches. That means, a method improvable without *ad hoc* statements.
- We shall provide a systematic method in two different senses: easy to update whenever new experimental data or new theoretical calculations are available; capable of provide a purely theoretical error from the approaches performed.
- Finally, should be predictive and checkable.

The answer to this catalogue of wishes can be found within the Theory of Padé approximants. The connexion with the mathematical problem is given by the well-defined *general rational Hermite interpolation problem*. This problem corresponds with the situation where a function should be approximated but previous information about it is scarce and spread over certain information on a given set of points together with a set of derivatives. Our goal is to make a contact with this problem from our needs and provide a model-independent and data-driven parameterization of the TFFs useful for the problems we have at the precision low-energy frontier of the Standard Model.

1.1 Why Padé Theory?

In front of other more sophisticated models and methods, it may seem curious to appeal now to an old tool such as Padé Theory to perform precise calculations within the low-energy frontier of the Standard Model. In this respect, two considerations must be taken into account. First, Padé Theory is an active field of research in applied mathematics. Few of the results considered here have been the subject of research in the last years only. Second, the corpus of Padé Theory defines precisely the problem we have at hand (as we already announced, the general rational Hermite interpolation problem) and provides with a solution in the form of convergence theorems and tools.

From a theoretical point of view, one should notice that for precise Standard Model calculations at low energies considered here, what is needed is not directly the TFF but an integral over the TFF with a particular weight. (The exception to this point discussed in this talk is the extraction of the $\eta - \eta'$ mixing parameters from the corresponding TFFs.)

Another interesting assertion is the fact that parameterizations such as the Vector Meson Dominance, Lowest Meson Dominance (and extensions), models from holographic QCD [4] are already a certain kind of Padé approximant (the so-called Padé-Type approxiamnts [5] where the poles of the Padé are given in advanced). These parameterizations should take advantage of the theory of Padé approximants if a robust calculation is to be claimed. For example, the uncertainties due to the truncation of the PA sequence which are never accounted for when using these parameterizations, do not need to be small to be neglected [6][1]. On top, it is proven that Padé Type approximants converge slower than PAs, specially when considering integrals up to infinity (cf. second Ref. in [7]). In this respect, heading towards 10% errors for the pseudoscalar contributions to the HLBL (the typical size of the foreseen experimental uncertainty in the new experiment at FermiLab), all these considerations are a must.

The PAs feature an interesting relation with dispersion relations. The state-of-the-art dispersive formalism (DR) accounting for the HLBL (see Ref. [8] and specially the discussion in Ref. [9]) has to face the lack of high-energy constraints coming from QCD (cf. as well [4]). In this respect, PAs can provide a natural tool to extend the DR results and interpolate up to the high-energy region, including the well-known pQCD behavior into the game as it should. Actually, the PA themselves can be used by the DR as a parameterization where to impose the dispersive constraints, obtain the PA coefficients, and then extrapolate far away in the space-like region. From another point of view, the complementarity between PA and DR may come from the interplay between TL and SL regions. In this case, DR after accounting for the TL data (a region not accessible by the PA built from the SL), can provide the LECs to be used for reconstructing the PA in the SL. In this situation, PA would represent an extension of the DR into the SL. In an interactive procedure, SL experimental data can be used, as well, to constrain the PA parameters for later on being used back in the DR and so on.

Last but not least, as a user-friendly and simple tool, PA can be used in the analysis of experimental results as a corpus of fitting functions. In this situation, instead of using a VMD to fit the experimental data, the highest PA would yield the most accurate theoretical result including systematic errors without the need to use a particular model.

2 Transition form factors from rational approximants

The toolkit we have introduced is a powerful method to deal with functions of one and two variables for which experimental data points are known and theoretical constraints available. The method is, of course, general. We developed it focusing on the need to properly introduce experimental data in a systematic manner in observables that manifest a discrepancy with the Standard Model calculation. For the HLBL and pseudoscalar decays into lepton pairs, the object are meson TFFs. In this section we collect our main results for π^0, η and η'.

We proposed in Refs. [11, 12] to use a sequence of rational approximants $P^N_M(Q^2) = \frac{\mathcal{P}_N(Q^2)}{\mathcal{P}_M(Q^2)}$, called Padé approximants (PA) [7], to fit the space-like data [1, 14–16].

[1] In Ref. [6] we showed how the correct treatment of these parameterizations *á la Padé* may yield results up to 20% higher, an effect that goes beyond any attempt of a theoretical quoted error so far.

DOI: 10.1051/epjconf/201611801019

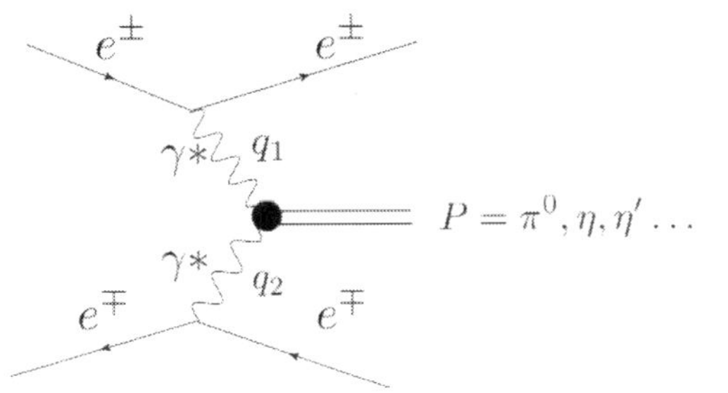

Figure 1. $e^+e^- \to e^+e^-P$ space-like process. q_1 and q_2 represent the photon virtualities and the black blob in the interaction point represents the meson transition form factor $F_{P\gamma^*\gamma^*}(q_1^2, q_2^2)$.

Table 1. π^0, η, and η' slope b_P, curvature c_P, and asymptotic limit ($Q^2 \to \infty$) from Refs. [11–13].

	b_P	c_P	$\lim_{Q^2\to\infty} Q^2 F_{P\gamma^*\gamma}(Q^2)$
π^0	0.0324(22)	$1.06(27)\cdot 10^{-3}$	$2f_\pi$
η	0.60(7)	0.37(12)	0.160(24)GeV
η'	1.30(17)	1.72(58)	0.255(4)GeV

Since PAs are constructed from the Taylor expansion of the $F_{P\gamma^*\gamma}(Q^2)$, from the fits we can obtain the derivatives of the $F_{P\gamma^*\gamma}(Q^2)$ at the origin of energies in a simple, systematic and model-independent way [11, 12]. The idea behind is that the TFF can be expanded as:

$$F_{P\gamma*\gamma}(Q^2, 0) = a_0^P(1 + b_P\frac{Q^2}{m_P^2} + c_P\frac{Q^4}{m_P^4} + \ldots), \quad (1)$$

where a_0^P is related to the decay of the pseudoscalar into two photons through

$$|F_{\eta'\gamma\gamma}(0)|^2 = \frac{64\pi}{(4\pi\alpha)^2}\frac{\Gamma(\eta' \to \gamma\gamma)}{m_{m_{\eta'}}^3}. \quad (2)$$

b_P and c_P in Eq. (3) are the slope and curvature of the TFF respectively. The low-energy parameters are fundamental quantities for constructing the PAs and any hadronic model with appropriate low-energy properties to be used to evaluate hadronic contributions to the observables we are discussing.

Since the analytic properties of TFFs are not known in general although the time-like region at low energies show the well-known unitary $\pi\pi$ cut, the kind of PA sequence to be used is not determined in advance. We consider two different sequences and the comparison among them should reassess our results. The first one is a $P_1^L(Q^2)$ sequence inspired by the success of the simple vector meson dominance ansatz [10], and the second one is a $P_N^N(Q^2)$ sequence which satisfy the pQCD constrains $Q^2F_{P\gamma\gamma^*}(Q^2) \sim$ constant. After combining both sequence's results, slope and curvature results from Refs. [11–13] are shown in Table 1, where $\lim_{Q^2\to\infty} Q^2F_{P\gamma^*\gamma}(Q^2)$ from the $P_N^N(Q^2)$ is also shown.

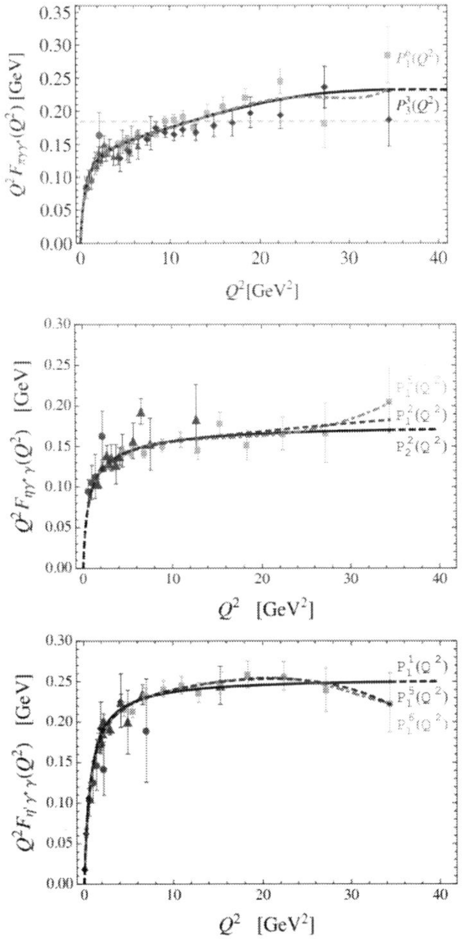

Figure 2. π^0 (upper panel), η (middle panel), and η' (lower panel) TFFs. Green-dot-dashed lines show our best $P_1^L(Q^2)$ fit, and black-solid lines show our best $P_N^N(Q^2)$ fit. Black-dashed lines display the extrapolation of the $P_N^N(Q^2)$ at $Q^2 = 0$ and $Q^2 \to \infty$. Experimental data are from CELLO (red circles), CLEO (purple triangles), and *BABAR* (orange squares) Colls. [14]. The π^0 figure contains also data from BELLE (blue diamonds) [15]; and the η' figure data from L3 (blue diamonds) [16]. Figures form Refs. [11, 12].

3 Applications of the transition form factors

3.1 Hadronic light-by-light scattering contribution to the muon $(g-2)$

The HLBL cannot be directly related to any measurable cross section and requires the knowledge of QCD contributions at all energy scales. Since this is not known yet, one needs to rely on hadronic models to compute it. Such models introduce systematic errors which are difficult to quantify. The large-N_c limit of QCD [17] provides a very useful framework to approach this problem and has been used to perform what nowadays are the reference numbers for the HLBL [4]. This, however, has a shortcoming. Cal-

Table 2. Preliminary results for the pseudoscalar pole contributions to $a_\mu^{HLBL;\pi^0,\eta,\eta'}$ for different values in the chosen $a_{1,1}$ range discussed in Ref. [23].

Units of 10^{-10}	π^0	η	η'	Total
$a_{1,1} = 2b_P^2$ [*OPE*]	6.64(33)	1.69(6)	1.61(21)	$9.94(40)_{stat}(50)_{sys}$
$a_{1,1} = b_P^2$ [Fact]	5.53(27)	1.30(5)	1.21(12)	$8.04(30)_{stat}(40)_{sys}$
$a_{1,1} = 0$	5.10(23)	1.16(7)	1.07(15)	$7.33(28)_{stat}(37)_{sys}$

culations carried out in the large-N_c limit demand an infinite set of resonances. As such sum is not known in practice, one ends up truncating the spectral function in a resonance saturation scheme, the so-called Minimal Hadronic Approximation [18]. The resonance masses used in each calculation are then taken as the physical ones from PDG [19] instead of the corresponding (but unknown) masses in the large-N_c limit. Both problems might lead to large systematic errors not included so far [5, 11, 20].

It was pointed out in Ref. [5] that, in the large-N_c framework, the Minimal Hadronic Approximation can be understood from the mathematical theory of PAs to meromorphic functions. Obeying the rules from this mathematical framework, one can compute the desired quantities in a model-independent way and even be able to ascribe a systematic error to the approach [21].

Our proposal is, then, to use, instead of a hadronic model for the TFF, a sequence of PAs built up from the low-energy expansion obtained in the previous section. The form factor needed in this calculation is the one of double virtuality. For this reason we invoke and extend the PA method to the Canterbury approximants (CA) [22]. Now, the Taylor expansion of the doubly-virtual TFF reads:

$$F_{P\gamma*\gamma*}(Q_1^2,Q_2^2) = a_0^P\left(1 + b_P\frac{Q_1^2+Q_2^2}{m_P^2} + a_{1,1}\frac{Q_1^2Q_2^2}{m_P^4} + \dots\right),$$

where $a_{1,1}$ corresponds to the slope of double virtuality and we have made use of the Bose symmetry to simplify our expression. Having experimental data in some energy region, even without high precision (30% or even 50% statistical error), one can attempt the reconstruction in a systematical way by a sequence of doubly virtual approximants fitted to such data. CAs can accommodate the high-energy constraints from QCD as well [22]:

$$C_1^0(Q_1^2,Q_2^2) = \frac{a_0}{1 + \frac{b_P}{m_P^2}(Q_1^2+Q_2^2) + \frac{2b_P^2-a_{1,1}}{m_P^4}Q_1^2Q_2^2}. \quad (3)$$

Knowing the Taylor expansion of the $F(Q_1^2,Q_2^2)$, Eq. (3) would be unique: a_0 is determined from the $\Gamma(\pi^0\to\gamma\gamma)$ as before, b_p is the slope of the single virtual TFF, and $a_{1,1}$ corresponds to the doubly-virtual slope.

Experimental data for $F(Q_1^2,Q_2^2)$ is not available yet and we cannot extract $a_{1,1}$ from them. The OPE tells us that $\lim_{Q^2\to\infty} F(Q^2,Q^2)\sim Q^{-2}$ and implies $a_{1,1}=2b_p^2$. On the other hand, at low energies, the factorization approach $F_{P\gamma^*\gamma^*}(Q_1^2,Q_2^2) = F_{P\gamma^*\gamma}(Q_1^2,0)\times F_{P\gamma\gamma^*}(0,Q_2^2)$ works well as it is already seen in the construction of the approximant (3) where the parameter $a_{1,1}$ enters as a correction to the $2b_p^2$. Taking the factorization into account, then $a_{1,1}=b_p^2$. To be on a conservative side then, we consider for the numerical analysis $b_p^2\le a_{1,1}\le 2b_p^2$. This range should effectively take into account our ignorance on $a_{1,1}$. As soon as experimental data will be available, this range will be immediately shrunk. Our results for the pseudoscalar pole contribution to the HLBL for π^0,η and η' are collected in Table 2. Let us remark that this is the most robust calculation of the η and η' contribution to the HLBL up to now.

On top, the TFF is considered to be off-shell. To match the large momentum behavior with short-distance constraints from QCD, calculable using the OPE, we consider the relations obtained in Ref. [4]. In practice this will amount to extend the CAs we are using to match the set of high-energy OPE constraints. These results are still preliminary and will be given elsewhere [24].

The HLBL is dominate for Q^2 ranging from 0 to 2 GeV^2, in particular above around 0.5 GeV^2. Therefore a good description of TFF in such region is very important. Such data are not yet available, but any model should reproduce the available one. That is why in [11, 12], in contrast to other approaches, we did not used data directly but the low-energy parameters of the Taylor expansion for the TFF and reconstructed it *via* PAs. The LECs certainly know about all the data at all energies and as such incorporates all our experimental knowledge at once. This procedure implies a model-independent result together with a well-defined way to ascribe a systematic error. In other words, it is the first procedure that can be considered an *approximation*, in contrast to the *assumptions* considered in other approaches.

Let us emphasize the role of experimental data by using the LECs in Table 1 together with the $\pi^0\to\gamma\gamma$ to match the free parameters of the LMD+V model introduced in [25]. We would obtain $a_\mu^{\mathrm{HLBL};\pi^0} = 7.5\times10^{-10}$ which contrasts with the original $a_\mu^{\mathrm{HLBL};\pi^0} = 6.3\times10^{-10}$ [25]. The impact of the new experimental data is then clear, inducing a 20% effect. On top, since the LMD+V is not a PA but a well-educated model, it is difficult to ascribe a systematic error due to the large-N_c approach. PAs in turn, already account for such corrections which in Refs. [11, 12] were found to be of the order of 5% − 10% less dramatic than the naive 30% from the N_c counting, but still required to be taken into account.

3.2 pseudoscalar decays into lepton paris

Pseudoscalar decays into lepton pairs provide a unique environment for testing our knowledge of QCD. As such de-

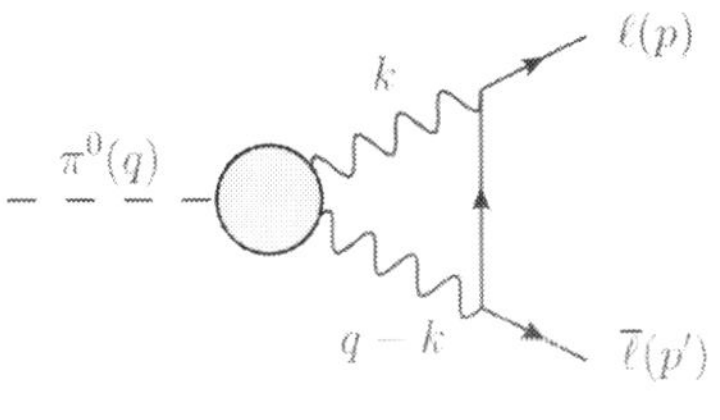

Figure 3. Feynman Diagram for $\pi^0 \to e^+e^-$ process.

cays are driven by a loop process, encode, at once, low and high energies. For the π^0 decay, the process (neglecting electroweak corrections) proceeds in two steps as shown in Fig. 3. The loop does not diverge due to the presence of the pseudoscalar transition form factor on the $\pi^0 \to \gamma^*\gamma^*$ anomalous vertex [2], the $F_{P\gamma^*\gamma^*}(k^2,(q-k)^2)$ with $k^2,(q-k)^2$ space-like photon virtualities. For η and η' intermediate $\pi\pi$ cuts must be taken into account as well [36].

The branching ratio (BR) for this decay may be expressed in terms of the two photon decay width as

$$\frac{\mathrm{BR}(P\to\bar{\ell}\ell)}{\mathrm{BR}(P\to\gamma\gamma)} = 2\left(\frac{\alpha m_\ell}{\pi m_P}\right)^2 \beta_\ell \left|\mathcal{A}(m_P^2)\right|^2, \quad (4)$$

where $\mathcal{A}(q^2)$ represents the loop amplitude (see Ref. [22] and references therein for details)

$$\mathcal{A}(q^2) = \frac{2i}{\pi^2 q^2}\int d^4k \frac{[q^2k^2-(q\cdot k)^2]\tilde{F}_{P\gamma^*\gamma^*}(k^2,(q-k)^2)}{k^2(q-k)^2((p-k)^2-m_\ell^2)}, \quad (5)$$

which is unknown as far as the normalized TFF ($\tilde{F}_{P\gamma\gamma}(0,0)=1$) is unspecified. The role of the doubly virtual TFF is actually rather important as the given integral, similarly to the HLBL case, is UV-divergent. Remarkably, for the π^0 case, it is possible to go further without a single clue on the TFF. Being the lightest hadron, it is not possible to find any additional intermediate hadronic state which may be on-shell, and contribute therefore to the imaginary part. Consequently, its imaginary part is solely given by the intermediate $\gamma\gamma$ state, the well-known unitary bound discussed by Drell [28], which gives

$$\mathrm{Im}(\mathcal{A}(m_{\pi^0}^2)) = \frac{\pi}{2\beta_\ell} ln\left(\frac{1-\beta_\ell}{1+\beta_\ell}\right) \quad (\beta_\ell = \sqrt{1-4m_\ell^2/m_{\pi^0}^2}), \quad (6)$$

Due to the presence of the photon propagators, the kernel of the integral is peaked at around the electron mass. Then, one can expand that kernel in terms of m_e/m_P but also m_e/Λ, being Λ the cut off of the loop integral, or the hadronic scale driven by the TFF. [33] resummed such power corrections and found them negligible [34]. Then, using a Vector Meson Dominance for the TFF they found $\mathrm{BR}(\pi^0\to e^+e^-) = (6.23\pm0.09)\times10^{-8}$ [34], 3σ off the KTeV result.

Radiative corrections were recently reconsidered in [35] and yield the new KTeV result as $\mathrm{BR}^{\text{"KTeV"}}(\pi^0\to e^+e^-) = (6.87\pm0.36)\times10^{-8}$.

In [22], we investigated the role of the TFF of doubly virtuality using the CAs. Beyond considering the eventual effects of different New Physics scenarios (which are negligible), we consider for the numerical analysis that $0 \le a_{1,1} \le 2b_P^2$, and obtained the new Standard Model value. Extending our calculations to the η and η' decays now including as well the $\mu\mu$ mode, the preliminary results are found in Table 3.The error comes from $P\to\gamma\gamma$ and b_π together with the evaluation of the systematic error from our approximation [22], and the two main numbers from the ranging of $a_{1,1}$. To shrink the window here provided, experimental data would then be very welcome. For the π^0, this final number represents still a deviation of the measured BR between $2.6-1.4\sigma$.

Forcing our approximant (3) to reproduce the KTeV result and then used for the π^0 contribution to the HLBL [4], we obtain $a_\mu^{HLBL;\pi^0} = 2.9\times10^{-10}$, a deviation of $2-3\times10^{-10}$ with respect to the standard result. Taking into account that the global theoretical SM error for the muon $(g-2)$ is 6×10^{-10} [4], the role of the $\pi^0\to e^+e^-$ is certainly remarkable, and never been considered so far. Similar effect is also found for the $\eta\to\mu^+\mu^-$ decay, indicating that the current precision of the SM error on the muon $(g-2)$ may be underestimated.

3.3 $\eta-\eta'$ mixing parameters

The physical η and η' mesons are an admixture of the $SU(3)$ Lagrangian eigenstates [26]. Deriving the parameters governing the mixing is a challenging task. Usually, these are determined through the use of $\eta(')\to\gamma\gamma$ decays as well as vector radiative decays into $\eta(')$ together with $\Gamma(J/\Psi\to\eta'\gamma)/\Gamma(J/\Psi\to\eta\gamma)$ [26]. However, since pQCD predicts that the asymptotic limit of the TFF for the $\eta(')$ is essentially given in terms of these mixing parameters, we use our TFF parametrization to estimate the asymptotic limit and further constrain the mixing parameters with compatible results compared to standard (but more sophisticated) determinations [12].

In this section we reanalyze the $\eta-\eta'$ mixing as we did in Ref. [12, 13]. In these works, we took advantage of the flavor basis, where the η and η' pseudoscalar decay constants defined in terms of the axial current, $J^a_{5\mu} \equiv \bar{q}\gamma_\mu\gamma_5\frac{\lambda^a}{2}q$, as $\langle 0|J^a_{5\mu}|P\rangle \equiv ip_\mu F^a_P$ (where $a=q,s$ refers to light and strange quarks, resp.) can be expressed as

$$(F_P^a) \equiv \begin{pmatrix} F_\eta^q & F_\eta^s \\ F_{\eta'}^q & F_{\eta'}^s \end{pmatrix} = \begin{pmatrix} F_q\cos\phi_q & -F_s\sin\phi_s \\ F_q\sin\phi_q & F_s\cos\phi_s \end{pmatrix}, \quad (7)$$

with $\lambda^q = \mathrm{diag}(1,1,0)$ and $\lambda^s = \mathrm{diag}(0,0,\sqrt{2})$. This basis has become popular since large-N_c chiral perturbation theory NLO predictions yield [37, 38]

$$F_qF_s\sin(\phi_q-\phi_s) = \frac{\sqrt{2}}{3}F_\pi^2\Lambda_1, \quad (8)$$

where F_π is the pion decay constant, and Λ_1 is an OZI-violating parameter expected to be small. Assuming Λ_1 to be neglected, Eq. (8) implies $\phi_q=\phi_s\equiv\phi$, an approximation which has been successful in phenomenological

Process	BR(th)	BR(exp)
$\pi^0 \to e^+e^-$	$(6.20 \div 6.35)(5) \times 10^{-8}$	$7.48(38) \times 10^{-8}$ [27]
$\eta \to e^+e^-$	$(5.31 \div 5.44)(5) \times 10^{-9}$	$\leq 2.3 \times 10^{-6}$ [29]
$\eta \to \mu^+\mu^-$	$(4.72 \div 4.52)(8) \times 10^{-6}$	$5.8(8) \times 10^{-6}$ [30]
$\eta' \to e^+e^-$	$(1.82 \div 1.86)(19) \times 10^{-10}$	$\leq 5.6 \times 10^{-9}$ [31, 32]
$\eta' \to \mu^+\mu^-$	$(1.37 \div 1.49)(33) \times 10^{-7}$	—

Table 3. Preliminary results for the range $a_{1,1} \in (b_P^2 \div 2b_P^2)$, i.e. $(OPE \div Fact)$. The errors refers to the statistical error for $BR(P \to \gamma\gamma)$, the error from b_P and the systematic, combined in quadrature.

applications. Large-N_c ChPT predicts also that

$$F_q^2 = F_\pi^2 + \frac{2}{3}F_\pi^2\Lambda_1 . \quad (9)$$

Here, phenomenological studies [12, 13, 37, 38] do not support $\Lambda_1 = 0$ since they find $F_q > F_\pi$. Therefore, to be consistent we will consider the most general case $\phi_q \neq \phi_s$ and work in the so-called octet-singlet basis, where the decay constants are defined as

$$F_P^{80} \equiv \begin{pmatrix} F_\eta^8 & F_\eta^0 \\ F_{\eta'}^8 & F_{\eta'}^0 \end{pmatrix} = \begin{pmatrix} F_8\cos\theta_8 & -F_0\sin\theta_0 \\ F_8\sin\theta_8 & F_0\cos\theta_0 \end{pmatrix}. \quad (10)$$

In this basis, Large-$N_c\chi PT$ at NLO predicts [37, 38]

$$F_8^2 = \frac{4F_K^2 - F_\pi^2}{3}, \qquad F_0^2 = \frac{2F_K^2 + F_\pi^2}{3} + F_\pi^2\Lambda_1, \quad (11)$$

$$(\theta_8 - \theta_0) = -\frac{4\sqrt{2}}{3}\left(\frac{F_K}{F_\pi} - 1\right), \quad \theta_8 + \theta_0 = 2\theta_P, \quad (12)$$

where F_K is the kaon decay constant and θ_P the mixing angle at NLO. To derive (11) and (12) we made use of the relation between the low-energy constant L_5 appearing in the Large-$N_c\chi PT$ Lagrangian at NLO [39] and the ratio F_K/F_π. F_0 is renormalization group dependent. This is connected to the $J^0_{5\mu}$ anomalous dimension implying [26, 40]

$$\mu\frac{dF_0}{d\mu} = -N_F\left(\frac{\alpha_s(\mu)}{\pi}\right)^2 F_0, \quad (13)$$

where N_F is the number of active flavors. Solving this equation at order α_s,

$$F_0(\mu) = F_0(\mu_0)\left(1 + \frac{2N_F}{\beta_0}\left(\frac{\alpha_s(\mu)}{\pi} - \frac{\alpha_s(\mu_0)}{\pi}\right)\right) \equiv F_0(1+\delta),$$

with $\beta_0 = 11N_c/3 - 2N_F/3$. In the singlet-octet basis, the different limiting behaviors of the TFF, $F_{P\gamma\gamma} \equiv F_{P\gamma^*\gamma}(0)$, and $P_\infty \equiv \lim_{Q^2\to\infty} Q^2 F_{P\gamma^*\gamma}(Q^2)$ take the simple form

$$F_{\eta\gamma\gamma} = \frac{1}{4\pi^2}\frac{\hat{c}_8 F_{\eta'}^0 - \hat{c}_0(1+\Lambda_3)F_{\eta'}^8}{F_{\eta'}^0 F_\eta^8 - F_{\eta'}^8 F_\eta^0}, \quad (14)$$

$$F_{\eta'\gamma\gamma} = \frac{1}{4\pi^2}\frac{-\hat{c}_8 F_\eta^0 + \hat{c}_0(1+\Lambda_3)F_\eta^8}{F_{\eta'}^0 F_\eta^8 - F_{\eta'}^8 F_\eta^0}, \quad (15)$$

$$\eta_\infty = 2(\hat{c}_8 F_\eta^8 + \hat{c}_0(1+\delta)F_\eta^0), \quad (16)$$

$$\eta'_\infty = 2(\hat{c}_8 F_{\eta'}^8 + \hat{c}_0(1+\delta)F_{\eta'}^0), \quad (17)$$

where $\hat{c}_8 = 1/\sqrt{3}$ and $\hat{c}_0 = \sqrt{8/3}$ are charge factors and $\delta = -0.17$ [13] accounts for the F_0 running up to $Q^2 \to \infty$ [40]. In addition, we have included the OZI-violating parameter Λ_3, which has been neglected in our previous studies since it enters at the same level as Λ_1.

The set of Eqs. (14-17) form a system of 4 equations with 5 unknowns ($F_{\eta^{(\prime)}}^{(8,0)}, \Lambda_3$). Then it could seem that, at least taking $\Lambda_3 = 0$, we may solve the system. However, as explained in [13], such a system is not linear independent as there is the relation

$$\eta_\infty F_{\eta\gamma\gamma} + \eta'_\infty F_{\eta'\gamma\gamma} = \frac{3}{2\pi^2}\left(1 + \frac{8}{9}(\delta + \Lambda_3 + \delta\Lambda_3)\right) \quad (18)$$

which is free of mixing parameters. Indeed, Eq. (18) determines Λ_3 once its left-hand-side is (experimentally) known but, we still have to face the fact that our system is linear dependent. In order to overcome this problem, we notice that at NLO in large-$N_c\chi PT$, Eqs. (11,12), provides a clean prediction both for F_8 and $(\theta_8 - \theta_0)$ in terms of the well-known value for F_K/F_π [19]. Taking either F_8 or $(\theta_8 - \theta_0)$ constraint, would add an additional equation to the previous system, which would provide a unique solution. Taking both, would lead to an overdetermined system, which in general has no solution. For this reason, we adopt a democratic procedure in which we perform a fit including both, F_8 and $(\theta_8 - \theta_0)$ constraints, together with Eqs. (14-17). In addition, we include the theoretical uncertainties from Large-$N_c\chi PT$ predictions, Eqs. (11,12) by noticing that F_K/F_π typically receives 5% corrections from the NNLO. Consequently, we add this error in quadrature on top of the one from [19] for our fitting procedure.

The solutions are collected in Tab. 4, first row. The value for χ^2/DOF is excellent which indicates a good agreement with large-N_c χPT but predicts non-negligible NNLO corrections accounted here as a 5%. Without this 5%, the χ^2/DOF would grow up to 1.5. In addition we can use Eqs. (11),(12), to predict the values for the mixing angle θ_P and the OZI-violating parameter Λ_1:

$$\theta_P = -15(1)^\circ, \qquad \Lambda_1 = 0.21(5). \quad (19)$$

In Fig. 4 we collect our results (orange crosses) and compare them to different predictions in the literature [26, 37, 38, 41] (red dots), together with our previous results [13] in blue-empty squares. We see that the main difference with respect to our previous work [13], where we did not use the η' TFF asymptotic value and assumed $\phi_q = \phi_s$, appears in F_0. This is to be expected as the inclusion of Λ_1 and Λ_3 affects the singlet part exclusively. In addition, we have reduced our errors thanks to the constraints

F_8/F_π	F_0/F_π	θ_8	θ_0	Λ_3	$R_{J/\Psi}$	χ^2/DOF
1.32(7)	1.25(3)	−22.8(1.1)	−7.6(2.2)	0.05(3)	5.6(7)	1.0

Table 4. Predictions for the mixing parameters in the singlet-octet basis. $F_\pi = 92.2$MeV and $\theta_{8,0}$ in degrees.

from Large-$N_c\chi PT$ with respect to our previous work. Our prediction for Λ_3 may be compared with the one in [41], $\Lambda_3 = -0.03(2)$. Both of them point towards a small value for this parameter, though they differ in sign. We find that Λ_3 actually plays an important role not only in fulfilling the degeneracy condition, Eq. (18), but in the $\eta(\eta') \to \gamma\gamma$ decays as well. In addition, the Λ_1 term is rather important and affects specially the η' results, where deviations of order 10% appear if this is omitted. Finally, we stress that the use of the RG equation for F_0 is fundamental, increases η_∞ and diminishes η'_∞, bringing in agreement experiment and theory. We encourage then the future analysis to account for it.

In addition, we can predict the ratio $R_{J/\psi} \equiv BR(J/\psi \to \eta'\gamma)/BR(J/\psi \to \eta\gamma)$, which is given in terms of ϕ_q alone [37] as $R_{J/\psi} = 5.6(7)$, just at 1.2σ from the experimental value $R_{J/\psi} = 4.7(2)$ [19].

4 Conclusions and Outlook

Pseudoscalar meson transition form factors are a good laboratory to study the properties of pseudoscalar mesons. Their interest goes, however, much beyond the mesons themselves as they play a key role on precision calculations of Standard Model observables at low energies where hadronic contributions are the cornerstone of the error evaluation. We propose the method of Padé approximants as a toolkit to analyze them. The method is easy, systematic, user-friendly, and can be improved upon by including new data. Provides, as well, information about the underlying structure of the TFF and can be used to extrapolate experimental information to extract the low-energy parameters of the form factor together with their asymptotic limits. The former makes contact with the experimental measurement of the pseudoscalar decay into two photons as well as calculations of the slope and curvature, and the later provides insights into the perturbative QCD regime.

Once LEPs and asymptotic values are known, the TFFs allow us to study the $\eta - \eta'$ mixing in a framework consistent with chiral perturbation theory within the large-N_c limit at NLO including OZI-rule-violating parameters.

The most relevant feature of the method here described is their excellent performance as an interpolation tool. As such, it is the most compelling method to provide an accurate parameterization for the TFF in the whole space-like region. This is of primordial importance for the correct assessment, in a data-driven approach, of the pseudoscalar contribution to the HLBL [6, 23]. But this does not stop here. Since the approximants can as well penetrate into the time-like region below the first resonance, precise experimental data can be easily incorporated. Beyond the knowledge the approximants bring with respect to the features of the TFFs at time like, with this extension pseudoscalar decays into lepton pairs can be studied. In this regards, our method provides the most accurate data-driven and model-independent result consistent not only with the well-known QCD features at high and low energies, but as well with a well with a mathematical theory. This excellent performance is here exemplified by calculating the pseudoscalar pole contributions to the HLBL and pseudoscalar decays into leptons pairs (for π^0, η and η') resulting in the most updated and data-driven results up to now.

Acknowledgments

Work partially supported by the Deutsche Forschungsgemeinschaft DFG through the Collaborative Research Center "The Low-Energy Frontier of the Standard Model" (SFB 1044). P. M. wants to thank the organizers of the FCCP2015 conference for encouragement and support.

References

[1] B. Aubert *et al.* [BaBar Collaboration], Phys. Rev. D **80** (2009) 052002.
[2] S. L. Adler, Phys. Rev. **177** (1969) 2426; J. S. Bell and R. Jackiw, Nuovo Cim. A **60** (1969) 47.
[3] G. P. Lepage and S. J. Brodsky, Phys. Rev. D **22** (1980) 2157; T. Feldmann and P. Kroll, Phys. Rev. D **58** (1998) 057501 [hep-ph/9805294].
[4] See, for example, the talks by H. Bijnens, M. Knecht, A. Nyffeler, D. Greynat, and L. Cappiello in this conference. See as well F. Jegerlehner and A. Nyffeler, Phys. Rept. **477** (2009) 1; J. Prades, E. de Rafael and A. Vainshtein, [arXiv:0901.0306 [hep-ph]]; P. Masjuan and M. Vanderhaeghen, arXiv:1212.0357 [hep-ph].
[5] P. Masjuan and S. Peris, JHEP **0705** (2007) 040; Phys. Lett. B **663** (2008) 61.
[6] P. Masjuan, Nucl. Part. Phys. Proc. **260** (2015) 111 [arXiv:1411.6397 [hep-ph]].
[7] G.A.Baker and P. Graves-Morris, Encyclopedia of Mathematics and its Applications, Cambridge Univ. Press, 1996; P. Masjuan Queralt, arXiv:1005.5683 [hep-ph].
[8] See the discussion by M. Proccura in this conference.
[9] See the discussion by A. Vainshtein in this conference.
[10] P. Masjuan, S. Peris and J. J. Sanz-Cillero, Phys. Rev. D **78** (2008) 074028;
[11] P. Masjuan, Phys. Rev. D **86** (2012) 094021.
[12] R. Escribano, P. Masjuan and P. Sanchez-Puertas, Phys. Rev. D **89** (2014) 034014 [hep-ph/1307.2061].
[13] R. Escribano, P. Masjuan and P. Sanchez-Puertas, Eur. Phys. J. C **75** (2015) 9, 414 [arXiv:1504.07742 [hep-ph]].

 DOI: 10.1051/epjconf/201611801019

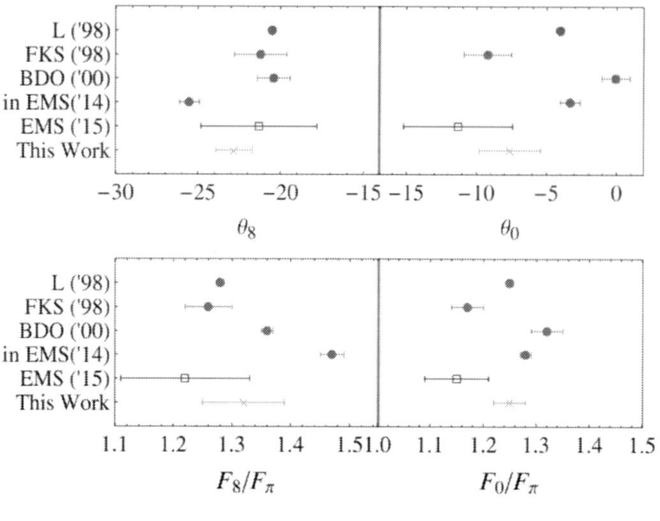

Figure 4. $\eta - \eta'$ mixing parameters in the octet-singlet basis from L [26], FKS [38], BDO [41], EF [37], EMS(14) [12], EMS(15) [13].

[14] H. J. Behrend *et al.* [CELLO], Z. Phys. C **49** (1991) 401; J. Gronberg *et al.* [CLEO], Phys. Rev. D **57** (1998) 33; P. del Amo Sanchez *et al.* [BaBar], Phys. Rev. D **84** (2011) 052001.

[15] S. Uehara *et al.* [Belle Coll.], Phys. Rev. D **86** (2012) 092007;

[16] M. Acciarri *et al.* [L3 Coll.], Phys. Lett. B **418** (1998) 399.

[17] G. 't Hooft, Nucl. Phys. B **72** (1974) 461; E. Witten, Nucl. Phys. B **160** (1979) 57.

[18] S. Peris, M. Perrottet and E. de Rafael, JHEP **9805** (1998) 011.

[19] K. A. Olive *et al.* [Particle Data Group Collaboration], Chin. Phys. C **38** (2014) 090001.

[20] P. Masjuan, E. Ruiz Arriola and W. Broniowski, Phys. Rev. D **85** (2012) 094006; Phys. Rev. D **87** (2013) 014005.

[21] P. Masjuan and S. Peris, Phys. Lett. B **686** (2010) 307; P. Masjuan and J. J. Sanz-Cillero, Eur. Phys. J. C **73** (2013) 2594.

[22] P. Masjuan and P. Sanchez-Puertas, arXiv:1504.07001 [hep-ph].

[23] P. Sanchez-Puertas and P. Masjuan, arXiv:1510.05607 [hep-ph].

[24] P. Sanchez-Puertas and P. Masjuan, in preparation.

[25] M. Knecht and A. Nyffeler, Phys. Rev. D **65** (2002) 073034 [hep-ph/0111058].

[26] H. Leutwyler, Nucl. Phys. Proc. Suppl. **64** (1998) 223; T. Feldmann, P. Kroll and B. Stech, Phys. Rev. D **58** (1998) 114006; R. Escribano and J. -M. Frere, JHEP **0506** (2005) 029.

[27] E. Abouzaid *et al.* [KTeV Collaboration], Phys. Rev. D **75** (2007) 012004 [hep-ex/0610072].

[28] S. D. Drell, Nov. Cim., XI, 5 (1959)

[29] G. Agakishiev *et al.* [HADES Collaboration], Phys. Lett. B **731** (2014) 265 [arXiv:1311.0216 [hep-ex]].

[30] R. Abegg *et al.*, Phys. Rev. D **50** (1994) 92.

[31] V. M. Aulchenko *et al.* [SND Collaboration], Phys. Rev. D **91** (2015) 5, 052013 [arXiv:1412.1971 [hep-ex]].

[32] R. R. Akhmetshin *et al.* [CMD-3 Collaboration], Phys. Lett. B **740** (2015) 273 [arXiv:1409.1664 [hep-ex]].

[33] A. E. Dorokhov and M. A. Ivanov, Phys. Rev. D **75** (2007) 114007 [arXiv:0704.3498 [hep-ph]].

[34] A. E. Dorokhov, M. A. Ivanov and S. G. Kovalenko, Phys. Lett. B **677** (2009) 145 [arXiv:0903.4249 [hep-ph]].

[35] P. Vasko and J. Novotny, JHEP **1110** (2011) 122 [arXiv:1106.5956 [hep-ph]].

[36] P. Sanchez-Puertas and P. Masjuan, arXiv:1510.06564 [hep-ph].

[37] R. Escribano and J. M. Frere, JHEP **0506** (2005) 029 [hep-ph/0501072].

[38] T. Feldmann, P. Kroll and B. Stech, Phys. Rev. D **58** (1998) 114006 [hep-ph/9802409].

[39] G. Ecker, P. Masjuan and H. Neufeld, Phys. Lett. B **692** (2010) 184 [arXiv:1004.3422 [hep-ph]].

[40] S. S. Agaev, V. M. Braun, N. Offen, F. A. Porkert and A. Schäfer, Phys. Rev. D **90** (2014) 7, 074019 [arXiv:1409.4311 [hep-ph]].

[41] M. Benayoun, L. DelBuono and H. B. O'Connell, Eur. Phys. J. C **17** (2000) 593 [hep-ph/9905350].

Theory review of the muon $g - 2$

Kirill Melnikov[1,a]

[1] *Institute for Theoretical Particle Physics, Karlsruhe Institute of Technology, Karlsruhe, Germany*

Abstract. I discuss the current situation with the muon anomalous magnetic moment. I argue that a mistake in the theoretical predictions is a very unlikely explanation of the current discrepancy between the Standard Model value of the muon magnetic anomaly and its measured value.

1 Introduction

I was charged by the organizers of the conference with reviewing the theory of the muon anomalous magnetic moment. I will begin by summarizing the current situation with the muon $g - 2$ and then discuss the various contributions to the muon magnetic anomaly starting with the QED one and then moving on to the hadronic vacuum polarization and the hadronic light-by-light scattering contributions. I will not talk in any detail about the electroweak corrections to the muon magnetic anomaly since they are well understood by now.

The enormous interest in the muon magnetic anomaly in the particle physics community originates from the fact that the latest and most precise measurement of this quantity by an experiment E821 performed in Brookhaven National Laboratory left us with a persistent difference between the result of the measurement [1] and the theoretical prediction in the Standard Model [2]. The difference

$$\Delta a_\mu = a_\mu^{\rm exp} - a_\mu^{\rm th} = (261 \pm 80) \times 10^{-11}, \qquad (1)$$

amounts to 3.3 standard deviations. The uncertainty of the measured value of $g - 2$ is 63×10^{-11}; it is comparable to the uncertainty of the theoretical result, 50×10^{-11}.

The new experiment at the Fermi National Laboratory that will start operating in 2017 aims at reducing the experimental uncertainty by a factor four; if these plans succeed, the experimental error will be brought down to 16×10^{-11} [3]. If we assume no changes in central values of all the contributions and no changes in the theoretical uncertainty, the final discrepancy will be more than five standard deviations. By the standards of high-energy physics, this will qualify as a discovery of physics beyond the Standard Model. Hence, we see that the stakes are quite high. We have, on one hand, a clean electroweak observable that shows a surprising discrepancy between expected and measured values and, on the other hand, a forthcoming experiment that has a potential to increase the significance of the discrepancy further, to the point of a discovery. These prospects make the physics of muon $g-2$ quite an exciting and intriguing field for the next few years. At this point in time, the best thing that the $g-2$ theorists can do is to keep scrutinizing the theory predictions to ensure that theoretical results are sound and robust, at least at the level of the accepted uncertainties. On the other hand, it is important to realize that the current discrepancy shown in Eq.(1) is simply too large to be explained by a blunder on the theory side. This still leaves us with quite a few options to explain Δa_μ in Eq.(1); we will describe them below.

We begin with the discussion of the magnitude of the different contributions to $a_\mu^{\rm th}$ and their uncertainties. For definiteness, we will use results as compiled in Ref.[2]; they are shown in Table 2. The QED contribution is by far the largest one; it is known extremely precisely. For a proper perspective, it is useful to note that the uncertainty of the QED contribution is just $10^{-4} \times \Delta a_\mu$, i.e. completely irrelevant at the current level of discrepancy. The electroweak contribution is small [4], see Table 2. It is comparable to the current discrepancy between theory and experiment and its uncertainty is a factor of ten smaller than the expected precision of the FNAL experiment. The situation with hadronic contributions to the muon magnetic anomaly is very different. The so-called leading order hadronic vacuum polarization contribution is large and its uncertainty is sizable. The NLO hadronic vacuum polarization is rather small. Since its calculation uses the same non-perturbative input as the leading order hadronic vacuum polarization, we believe that this error estimate is justified and, for this reason, it plays no role in the current $g-2$ discussion. The last contribution is the so-called hadronic light-by-light scattering. It is relatively small but for a number of reasons that we discuss below it is considered to be quite uncertain.

If we put everything together, we arrive at the theoretical prediction for the muon $g - 2$ in the Standard Model

$$a_\mu^{\rm th} = 116591830(50) \times 10^{-11}. \qquad (2)$$

The largest contributors to the uncertainty are the leading order hadronic vacuum polarization contribution and

[a] e-mail: kirill.melnikov@kit.edu

© The Authors, published by EDP Sciences. This is an open access article distributed under the terms of the Creative Commons Attribution License 4.0 (http://creativecommons.org/licenses/by/4.0/).

the hadronic light-by-light. The uncertainties of QED and electroweak corrections and the hadronic vacuum polarization at next-to-leading order play a minor role.

Before discussing physics of the muon magnetic anomaly in detail, I emphasize that if we choose the current discrepancy between theory and experiment as a measure of how well the various contributions to the muon $g-2$ need to be known, we have to conclude that there is basically not a single one that requires refinements. Moreover, we can estimate the magnitude of all the different contributions to the muon magnetic anomaly using relatively simple physical considerations and in this way assure ourselves that we do not do anything particularly wrong. Of course, independent checks of the hadronic contributions are very welcome but, in my opinion, it is highly improbable that they will give us a different version of the $g-2$ story.

Therefore, given all the stress-tests that the theory of the muon $g-2$ has been subject to and the magnitude of the discrepancy Δa_μ in Eq.(1), I believe that there are three possible explanations of what is going on. They are as follows:

- the experimental result is wrong;
- several ingredients that appear in the theory predictions are "wrong" at the $O(1\sigma)$ level; for some reason that is not fully understood, the shifts are correlated;
- the discrepancy is real and is explained by physics beyond the Standard Model.

Thanks to the new experiment at Fermilab, we will be able to check the first item whereas a careful re-analysis of the many Standard Model contributions, the continued measurements of the $e^+e^- \to$ hadron cross sections and the anticipated progress in lattice computations will most likely allow us to either confirm or disprove the second one. Hence, it is possible to imagine that during the next decade we will be able to find a resolution of the muon $g-2$ puzzle and understand the cause of the current discrepancy.

In what follows we will discuss the different contributions starting from the QED one in Section 2 and continuing with the hadronic vacuum polarization in Section 3 and hadronic light-by-light in Section 4. We conclude in Section 5.

2 QED

A glance at the first entry in Table 2 shows that the QED contribution to the muon magnetic anomaly is known with the precision that exceeds the needs of the current theory-experiment comparison. For this reason, it is tempting to say that the QED corrections to $g-2$ are well-established and require no further discussion. However, the QED corrections to the muon magnetic anomaly are large and one can easily imagine that a tiny inconsistency in their evaluation will have important consequences for the interpretation of the $g-2$ puzzle. Since the state of the art computations of the muon magnetic anomaly [5] are incomprehensible for most people, except for a tiny number of

QED	116584718.95(8)
Electroweak	154 ± 2
Hadronic vacuum polarization, LO	$6949 \pm 37 \pm 21$
Hadronic vacuum polarization, NLO	-98.4
Hadronic light-by-light	105 ± 26

Table 1. Contributions to the muon anomalous magnetic moment, in units of 10^{-11}. The results are taken from Ref. [2].

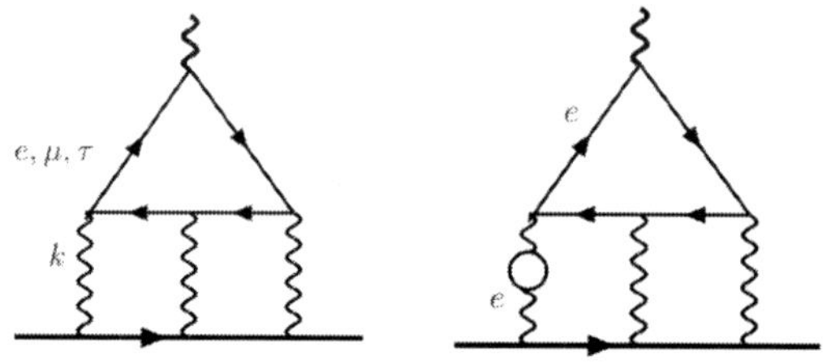

Figure 1. Enhanced three- and four-loop QED contirbutions to the muon $g-2$.

experts, it appears to be impossible to say how reliable these computations are without a completely independent confirmation of the final result. Unfortunately, such a confirmation has not yet fully happened although an impressive progress towards this goal occurred in recent years [6]. For this reason, we must take the results of the QED computations as they are currently reported and make our conclusions about the significance of the $g-2$ discrepancy based on this information. It is important to discuss why we believe that this is the sensible thing to do.

The answer to this question is actually quite simple: at the current level of discrepancy Δa_μ in Eq.(1) we do not need the results of the most advanced QED computations. In fact, all that is required are the three-loop QED corrections and the *leading* contributions at four-loops. Both the three-loop QED contributions to muon $g-2$ and the *leading* four-loop contributions are known since long ago and have been checked several times (see the discussion in Ref.[7] and references to the original papers therein). As we will see, these results *alone* allow us to predict the QED contribution to the muon magnetic anomaly with the precision that is better than 40×10^{-11} which is much smaller than Δa_μ.

The peculiarity of the QED contributions to the muon magnetic anomaly are the two enhancement parameters – the logarithm of the muon to electron mass ratio $\ln m_\mu/m_e \sim 5$ and $\pi^2 \sim 10$ that appear in certain cases. The logarithmic enhancement is a consequence of the renormalization-group running of either the QED coupling constant or effective operators that appear once diagrams that are sensitive to m_μ and m_e are written in such a way that contributions of disparate energy scales are separated. We note that it is somewhat unusual in perturbative computations to specifically address the enhancement by

π^2 factors[1] but in case of the muon magnetic anomaly there are cases when this enhancement can be sharply identified. This happens, for example, when π^2 and $\ln m_\mu/m_e$ appear in Feynman diagrams at the same time. Such situation occurs in the light-by-light scattering diagrams with the electron loop [8], see Fig. 1 The enhancement mechanism is very strong – at three loops, the light-by-light scattering diagrams with the electron loop provide 96% of the full result; all other diagrams give just about four percent.

At four loops, the situation is similar. We can not add another photon to connect the muon line and the electron loop since the resulting contribution vanishes due to Furry's theorem. Instead, we generate the logarithmic enhancement caused by the running of the coupling constant by inserting the electron vacuum polarization diagram into any of the photon propagators, see Fig. 1. This class of diagrams gives [9]

$$a_\mu^{(4),\text{leading}} = 117.4 \left(\frac{\alpha}{\pi}\right)^4. \tag{3}$$

It is instructive to compare the approximate and the exact results

$$a_\mu^{(4),\text{exact}} = (132.68 - 1.75) \left(\frac{\alpha}{\pi}\right)^4, \tag{4}$$

where the first term contains contributions from all diagrams enhanced by logarithms of m_μ/m_e ratio and the second represents contributions of all diagrams that are regular in the limit $m_e \to 0$. The difference between the full result and the leading contribution at four-loops translates into 40×10^{-11} shift at $g-2$; this number is much smaller than the current discrepancy in the muon magnetic anomaly. Moreover, other potentially enhanced contributions at the four-loop level were recently re-evaluated [6], confirming the results of the previous computations [10]. What remains unchecked at the four-loop level is the part of the computation that is regular in $m_e \to 0$ limit. We see from Eq.(4) that it only shifts the four-loop result by about 5×10^{-11} which is absolutely negligible at the current level of discrepancy. Finally, we note that the calculation of five-loop QED corrections to the magnetic anomaly was recently completed [5]. The result is consistent with a few earlier estimates [11] that utilized the possible enhancement mechanisms that we discussed earlier. The five-loop contribution shifts the muon anomalous magnetic moment by about 5×10^{-11} which is negligible compared to the current discrepancy.

Last year it was pointed out that the perturbative description of the QED corrections to the anomalous magnetic moment of an electron or a muon may be incomplete [12]. This discussion was motivated by an observation that contributions of QED bound states such as positronia, dimuonia etc. change the magnetic anomaly at the five-loop order, c.f. Fig. 2. However, since the bound states are non-perturbative in a sense that infinitely many Feynman diagrams need to be summed up to describe them, it seems hard to imagine a mechanism that makes these bound state contributions a part of conventional perturbative series. In fact such mechanism exists; it is known as "duality" in QCD. The point is that the magnetic anomaly is an Euclidean quantity that can be computed in QCD *after* the Wick rotation. If so, all the integrations over loop momenta can be performed over paths that are far from any of the physical singularities. For this reason, conventional perturbative expansion must be applicable. However, when one performs a computation *without* the Wick rotation and, in particular, uses dispersion representation for the photon vacuum polarization function, one has to integrate over sub-threshold positronia or dimuonia states. The non-perturbative $O(\alpha^5)$ correction to $g-2$ is then obtained. If one stops here, one concludes that there are contributions in QED that are not part of conventional perturbative series. However, such conclusion will be premature. Indeed, there is another non-perturbative contribution that appears in a form of a strong modification of the continuum scattering states *above* the e^+e^- or $\mu^+\mu^-$ production thresholds. These non-perturbative contributions in the continuum spectrum produce corrections to $g-2$ that *are equal and opposite in sign* to non-perturbative corrections produced by the bound states. When the sub- and above-threshold non-perturbative contributions are added, they cancel each other completely; the remaining part can be obtained within conventional perturbative expansion in QED [13, 14].[2] The reason for this is the Euclidean nature of the muon anomalous magnetic moment, as has been mentioned already.

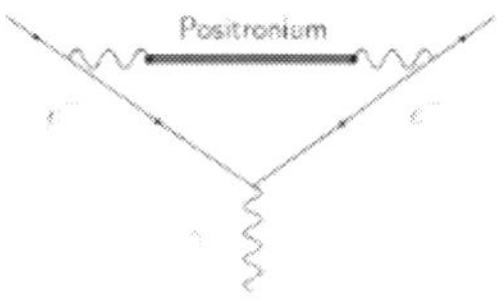

Figure 2. The contribution of a QED bound state to the anomalous magnetic moment of a lepton.

To summarize, I believe that QED corrections to the muon magnetic anomaly are understood so well that they can be discarded as a reason for the discrepancy in expected and measured values of the muon $g-2$. On the other hand, another measure of how well we need to control the QED corrections is provided by the expected precision of the new FNAL experiment, 15×10^{-11}. To match this precision, a re-calculation of the finite part of the four-loop muon $g-2$ probably becomes desirable. Given impressive recent developments in techniques for perturbative computations and their application to physics of the muon magnetic anomaly [6], there is no doubt that this task will be accomplished on the time scale of a few years.

[1] The usual argument against this is that π^2 may appear in many different ways in the final result and there are many known cases when large π^2 contributions are compensated by large rational contributions.

[2] Amusingly, this discussion was given in a more general context in full generality almost fifty years ago [15].

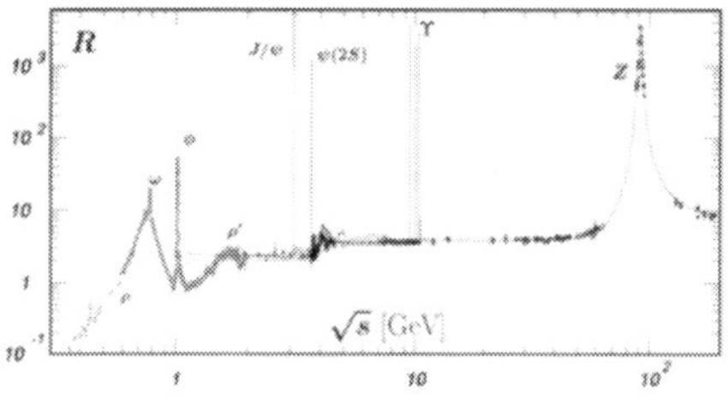

Figure 3. The spectral density $R(s)$ that appears in the calculation of the leading order hadronic vacuum polarization contribution to the muon magnetic anomaly.

3 Hadronic vacuum polarization contribution

The next largest correction to the muon anomalous magnetic moment is the hadronic vacuum polarization. The muon mass provides the energy scale for this contribution. This scale is smaller than the non-perturbative scale of strong interactions $\Lambda_{\rm QCD}$, so that perturbative QCD is not applicable and we have to resort to non-perturbative methods. The only viable option is to use the dispersion representation of the hadronic vacuum polarization which allows us to restore the full function from its imaginary part. Since the imaginary part of the hadronic contribution to photon vacuum polarization is proportional to $e^+e^- \to$ hadrons annihilation cross section, which is known thanks to many experimental studies, we can calculate the hadronic vacuum polarization non-perturbatively. Its contribution to the magnetic anomaly can be written as

$$a_\mu^{\rm hvp} = \frac{\alpha}{3\pi} \int\limits_{s_0}^{\infty} \frac{{\rm d}s}{s} R^{\rm hard}(s)\, a^{(1)}(s), \quad (5)$$

where $R(s) = \sigma_{e^+e^- \to {\rm hadr}}/\sigma_{e^+e^- \to \mu^+\mu^-}$ and $a_\mu^{(1)}$ is the one-loop anomalous magnetic moment that appears due to an exchange of a photon-like vector boson with the mass $\sqrt{s}$. Note that for the evaluation of the integral in Eq.(5) the low-energy region is of particular interest; this is so because the anomalous magnetic moment $a_\mu^{(1)}$ decreases for large vector boson masses $a_\mu^{(1)}(s) \sim m_\mu^2/s$ giving a stronger weight to the low-energy part of the integrand in Eq.(5).

Before I describe the results of the most recent computations, I would like to emphasize that it is possible to estimate $a_\mu^{\rm hvp}$ [7]. Such an estimate is interesting since, when contrasted with data-driven evaluations, it gives us an idea about how well such non-perturbative computations can be controlled. This is not very important for $a_\mu^{\rm hvp}$, where data is available at the first place, but it is crucial for the hadronic light-by-light scattering contribution $a_\mu^{\rm lbl}$, as we describe later.

The idea behind the estimate is quite simple. We try to model the essential features (see Fig.3) of the spectral density $R(s)$ at relatively low energies by accounting for three contributions:

- the chirally-enhanced two-pion threshold contribution $a_\mu^{\pi\pi}$ defined with an upper energy cut-off at $\sqrt{s} = m_\rho/2$;
- the contribution of ρ, ω, ϕ vector mesons $a_\mu^{\rho,\omega,\phi}$;
- the "continuum" contribution $a_\mu^{\rm cont}$ that starts above $\sqrt{s} \sim m_\phi$;

Numerically, we find $a_\mu^{\pi\pi} \sim 400 \times 10^{-11}$, $a_\mu^{\rho,\omega,\phi} = 5514 \times 10^{-11}$, $a_\mu^{\rm cont} = 1240 \times 10^{-11}$. Combining the three results we obtain the theoretical estimate of the leading order hadronic vacuum polarization contribution to the muon magnetic anomaly

$$a_\mu^{\rm hvp,th} = a_\mu^{\pi\pi} + a_\mu^{\rho,\omega,\phi} + a_\mu^{\rm cont} \approx 7160 \times 10^{-11}. \quad (6)$$

This theoretical estimate can be compared with one of the recent results of the data-driven evaluations [2]

$$a_\mu^{\rm hvp} = (6949 \pm 37.2 \pm 21.0) \times 10^{-11}. \quad (7)$$

The proximity of the two results is obvious and gives us confidence that we understand the physics of the hadronic vacuum polarization quite well; of course, the uncertainty of the theoretical estimate is hard to access a'priori.

Let us discuss now to what extent the existent data-driven evaluations are satisfactory, given the physics goals of the current and forthcoming muon $g-2$ studies.[3] The important point to emphasize in this respect is that the hadronic vacuum polarization contribution to the magnetic anomaly is large and it needs to be know quite precisely. In fact, currently, it is the largest contributor to the uncertainty of the theoretical prediction of the muon magnetic anomaly. The two uncertainties in Eq.(7) have the following origin: the first one is related to the uncertainty of the experimental data and how they are combined; the second one reflects the poor understanding of how QED radiative corrections are applied to analyses of available data in exclusive hadronic channels. Both of these uncertainties are relatively small compared to the current discrepancy between theory and experiment but, taken together, they are not negligible. Moreover, these uncertainties become quite substantial when compared to the expected precision of the FNAL experiment, which suggests that further effort is required to improve the measurements of the $e^+e^- \to$ hadrons annihilation cross sections at low energies. It is re-assuring that the corresponding program of measurements exists both at BEPC and at Novosibirsk, so that substantial improvements in our understanding of $a_\mu^{\rm hvp}$ can be expected on a few years time scale [16, 17].

There are several important issues that are debated currently in the context of the calculation of the hadronic vacuum polarization. One is the compatibility of data sets obtained in measurements of $e^+e^- \to$ hadrons by CMD, SND, KLOE, BABAR and BESS III experiments. All the experiments measure the contributions of the kinematic region around the ρ-meson to $a_\mu^{\rm hvp}$ so that the results of different experiments can be compared directly. Such a comparison is shown in Fig. 4. In principle, the results of the different measurements are consistent but there are unwelcome systematic trends: KLOE results are smaller than the BABAR results and somewhat smaller than SND results.

[3]For a detailed discussion of this question, see Ref. [16].

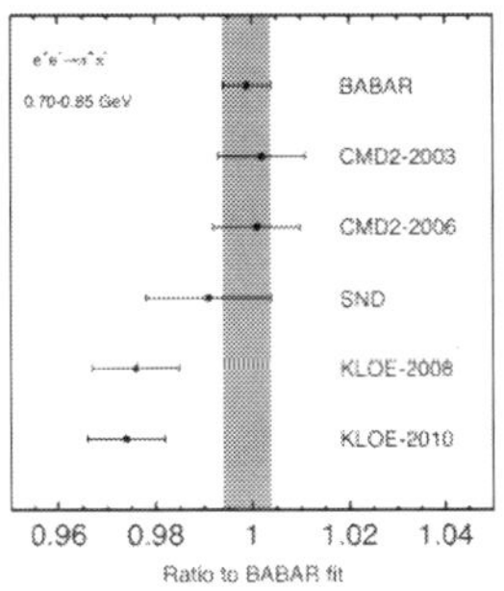

Figure 4. Comparison of recent evaluations of the ρ-meson contribution to $g-2$ that employ data from different experiments. The plot is taken from Ref. [18].

Figure 5. The so-called pion pole contribution to $a_\mu^{\rm hlbl}$ is the numerically dominant one.

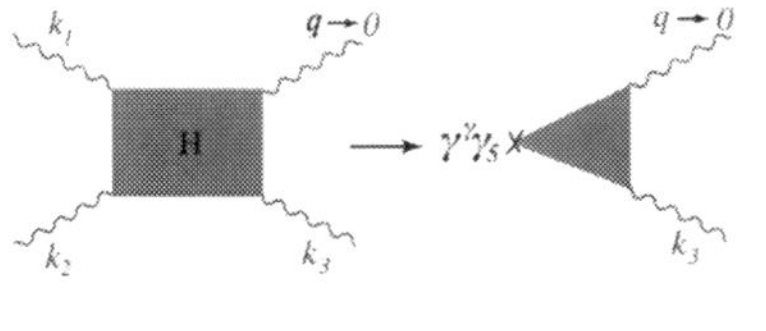

Figure 6. The OPE relation for the photon-photon scattering Green's function.

Also, the spectral densities $R(s)$ are somewhat tilted relative to each other in BABAR and KLOE measurements. The result of the very recent measurement by BESS III [19] is between the BABAR and KLOE results, but definitely closer to KLOE.

The other important topic that is being currently discussed is the question about the usefulness of τ data for the evaluation of the magnetic anomaly. The central issue is whether or not it is possible to control the isospin violating effects at the precision level that is required for the hadronic vacuum polarization to the muon magnetic anomaly. There were several talks about this issue during the Workshop; they seem to give an affirmative answer to this question [20, 21]. My personal opinion about this is that it is difficult to control the isospin violating effect at the level that is required to have meaningful description of the muon magnetic anomaly which implies that the use of τ data has the potential to create a lot of confusion. One thing, however, is clear. The recent work on the compatibility of the e^+e^- and the τ data seems to imply that, by and large, the data sets are quite compatible with each other [20, 21]; the situation therefore, is different compared to what was happening twelve years ago when the τ and e^+e^- data gave markedly different predictions for the muon $g-2$. Interestingly, one can check the compatibility of the τ data with the results of the $R(s)$ measurements by CMD, KLOE and BABAR; when this is done [21], the BABAR measurement appears to be disfavored.

4 Hadronic light-by-light scattering contribution

We will now discuss the current situation with the hadronic light-by-light scattering contribution to the muon anomalous magnetic moment. It is estimated to be $a_\mu^{\rm hlbl} = (105 \pm 26) \times 10^{-11}$ [22]. Clearly, $a_\mu^{\rm hlbl}$ is not large and its uncertainty, when compared to the current difference between theory and experiment, can be tolerated. However, given the persistent nature of the $g-2$ discrepancy, the volatile history of theoretical calculations of the hadronic light-by-light scattering contribution[4] and the fact that the calculation of $a_\mu^{\rm hlbl}$ only vaguely relies on the experimental data, there seems to be an uneasy feeling towards it. As the result, both the central value and the uncertainty estimate are being frequently questioned.

I begin by summarizing a few *solid facts* that we know about hadronic light-by-light scattering contribution. First of all, this contribution is non-perturbative since the energy scale set by the muon mass is small. Unfortunately, the use of robust non-perturbative methods such as e.g. the dispersion relations with experimentally-measured spectral densities, is very difficult in case of hadronic light-by-light.[5] As the consequence, most of the current computations of $a_\mu^{\rm hlbl}$ rely on models of low-energy hadron interactions. It is important to emphasize, however, that these models are constructed following *parametric* considerations. Indeed, to describe low-energy strong interactions physics, we can employ two parameters: 1) large number of colors N_c can be used to construct the $1/N_c$ expansion; 2) the apparent smallness of the pion mass m_π relative to the scale of strong interactions $\Lambda_{\rm QCD}$ allows us to treat a pion as a pseudo-Goldstone boson and uniquely predict its interactions with photons to leading order in the chiral expansion. The smallness of the pion mass suggests that loops of charged pions should provide the dominant contribution to the muon magnetic anomaly; the proximity of $N_c = 3$ to $N_c = \infty$ suggests that Green's functions relevant for the description of hadronic light-by-light scattering can be constructed as linear combination of contributions of non-interacting hadronic resonances. The existent studies of $a_\mu^{\rm hlbl}$ show that contributions that are enhanced by the large number of colors are (numerically) more important than the contributions that potentially exhibit the chiral enhancement.

Two statements can be made about hadronic light-by-light scattering contribution that are *exact* in the large-N_c

[4] The theory prediction for $a_\mu^{\rm hlbl}$ changed sign several times during its relatively short history.

[5] See Refs. [23–25] for the recent attempts to push forward with the data-driven evaluation of $a_\mu^{\rm hlbl}$.

approximation. First, we know that in the large-N_c approximation *and* in a situation where the mass gap between the pion and the ρ-meson is large, the hadronic light-by-light scattering contribution to the muon anomaly is given by the following formula [26]

$$a_\mu^{\rm lbl}[\pi^0] = \left(\frac{\alpha}{\pi}\right)^3 3\left(\frac{N_c}{3}\right)^2 \frac{m^2}{(4\pi F_\pi)^2} \ln^2\frac{m_\rho}{m_\pi} + ... \quad (8)$$

The ellipses stand for terms that are enhanced by a single power of $\ln m_\rho/m_\pi$ and for terms that are regular in $m_\rho \to \infty$ limit. This contribution originates from the diagrams in Fig.5 where π_0 is exchanged between the photon pairs. In addition, there is a constraint on the behavior of the photon-photon scattering amplitude at large virtualities of the photons that follows from perturbative QCD [27]. It is based on the OPE relation for the time-ordered product of two vector currents

$$i\int {\rm d}^4x{\rm d}^4y e^{-iq_1x-iq_2y} T j_{\mu_1}(x) j_{\mu_2}(y)$$
$$= \int {\rm d}^4z e^{-i(q_1+q_2)z} \frac{2i}{q^2}\epsilon_{\mu_1\mu_2\delta\rho}q^\delta f_5^\rho(z) + ... \quad (9)$$

Pictorially, this equation is illustrated in Fig. 6. The axial-vector current – that appears on the right-hand side of Eq.(10) – has a non-vanishing matrix element between the hadron vacuum and π_0; this allows to connect the short-distance computation of the photon-photon scattering amplitude where OPE is applicable and the long-distance description of the photon-photon scattering amplitude where π_0 exchange plays the primary role. The two constraints refer to opposite momenta scales, the very small and the very large; Eq.(10) shows that these constraints are connected and can be modeled by the π_0-exchange extrapolated to high-invariant masses in a consistent way [27].

Our best estimates of the large-N_c part of the hadronic light-by-light scattering contribution to the muon magnetic anomaly utilize models of the light-by-light scattering amplitudes based on pseudo-scalar (π_0, η, η') and pseudo-vector (a_1) exchanges , subject to two constraints mentioned above [27–31]. The current consensual result for the large-N_c part of the hadronic light-by-light scattering contribution to $g-2$ is [22]

$$a_\mu^{\rm hlbl,N_c} = (128 \pm 13) \times 10^{-11}. \quad (10)$$

We can check our understanding of $a_\mu^{\rm hlbl,N_c}$ in a number of different ways. Since the hadronic light-by-light scattering contribution to $g-2$ is Euclidean, we never need to integrate over the resonance regions when evaluating $a_\mu^{\rm hlbl,N_c}$. It is then possible to think that there should be a duality between hadronic light-by-light scattering contribution to $g-2$ calculated using hadronic models and the constituent quarks. The only parameter that a constituent quark can have – its mass – can be estimated by requiring that the same theory works for the leading order hadronic vacuum polarization. This approach, pioneered in Refs.[32, 33], gives $a_\mu^{\rm hlbl} \approx 130 \times 10^{-11}$.

One can check the stability of this result in a couple of ways. For example, one can include the "gluon" corrections to both light-by-light and hadronic vacuum polarization diagrams and check by how much the results for $a_\mu^{\rm hlbl}$ change if one changes the gluon-quark coupling [34]. Alternatively, one can combine the constituent quark loop with π_0 contribution to ensure that the chiral limit of the theory is correct [35]; in this case the quark loop contribution and the pion loop contribution must conspire to give the correct result for $a_\mu^{\rm hlbl}$. It turns out that whatever one does, the result comes out to be in in the range $a_\mu^{\rm hlbl,N_c} \approx (120-150) \times 10^{-11}$ which is perfectly consistent with the results of more detailed computations based on low-energy hadron models shown in Eq.(10).

Although further improvements in understanding the large-N_c part of the hadronic light-by-light scattering contributions are clearly warranted, the major current issue in the theory of $a_\mu^{\rm hlbl}$ is whether the sub-leading in N_c contributions are under control. The largest effect comes from loops of charged pions, which is the leading contribution in the chiral expansion. One finds $a_\mu^{\pi^+\pi^-} = -40 \times 10^{-11}$. However, it is interesting to point out that this particular contribution is not robust and that the pion loop changes strongly if pion-photon interactions are modified to account for contributions of vector mesons. Another illustration of this fact is that introducing such modifications in two different ways gives two results that differ by a significant amount. The reason for this behavior was analyzed in Ref. [27]. It was found in that reference that the pion loop is sensitive to larger values of the loop momenta than one naively expects. Moreover, these values of the loop momenta are sufficiently high to be sensitive to hadronic modifications of the interaction vertices and propagators. One can also learn from these results that first non-trivial power correction to the chiral limit is insufficient and one needs quite a number of such terms to obtain the reliable result.

A separate, but related issue is the question by how much pion polarizability modifies the pion loop contribution [38, 39]. The pion polarizability is a modification of the pion-photon interaction due to the following higher-dimensional operator in the low-energy Lagrangian $\mathcal{L}_{\rm pol} \sim F_\pi^{-2}\pi^+\pi^- F_{\mu\nu}F^{\mu\nu}$. This term is clearly not part of the leading chiral Lagrangian and it is also not generated when effects of heavy mesons are modeled by using vector meson dominance or similar models. It is argued in Refs. [38, 39] that effects of the pion polarizability are large and very uncertain; it is estimated in Refs. [38, 39] that the pion loop contribution, including the polarizability operator, is in between -10×10^{-11} and -70×10^{-11}.

It is important to stress that, as the consequence of these issues, the sub-leading in N_c contribution to muon $g-2$ is, currently, the leading source of the uncertainty in the final theoretical estimate of $a_\mu^{\rm hlbl}$. Given this situation, it is interesting to ask to what extent our understanding of the sub-leading in N_c contributions can be improved in the future and we have seen some positive signs of the possible improvements during the Workshop. Indeed, it was argued in [40] that out of the two standard references for the pion loop contributions – the Hidden Local Symmetry (HLS) result -4×10^{-11} of Ref. [36, 37] and the vector meson dominance (VMD) result -19×10^{-11} of Ref.[29], only the VMD calculation satisfies the short-distance con-

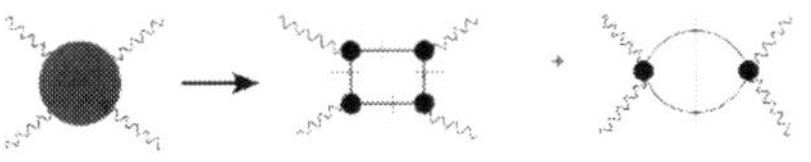

Figure 7. Decomposition of the chirally enhanced contributions to the photon-photon scattering amplitude in the dispersive approach of Refs.[23, 24].

straints on $\pi^+\pi^-\gamma\gamma$ vertex and leads, e.g. to the finite $\pi^+ - \pi^0$ mass difference. This fact would strongly favor the VMD result $a_\mu^{\pi^+\pi^-} = -19\times10^{-11}$ and would reduce the uncertainty since the previous uncertainty estimate was designed to cover both the VMD and the HLS model results. The analysis reported in Ref. [40] also seems to disfavor large effects of polarizability at least as long as one stays in the kinematic regime where the chiral expansion is well-behaved.

As we already mentioned, another interesting idea that was put forward recently is to use dispersion relations for the photon-photon scattering amplitude to make computations of hadronic light-by-light scattering less model-dependent [23–25]. For the subleading-N_c contributions, the decomposition of the photon-photon scattering amplitude shown in Fig. 7 is derived in Refs. [23, 24]. The first term on the right hand side, the four-pion cut, is simple and unique: it is chirally-enhanced pion box contribution modified by pion form factors on the external photon legs. The second term, the two-pion cut, is difficult; its analysis currently assumes the partial-wave expansion. The important point is that the four-pion cut is the same in VMD and HLS models and it is not affected by the pion polarizability issues. This means that all the "hot topics" in the current discussion of the hadronic light-by-light scattering contribution to $g-2$ reside in the two-pion cut contribution. This suggests that the dispersive approaches have a great chance to establish their credibility and usefulness by showing that the above issues (VMD vs. HLS and the polarizability) can be understood and clarified in a model-independent way. If anything, this is a simpler task than the complete model-independent analysis of hadronic light-by-light scattering contribution to the muon magnetic anomaly, but it is a very important one.

5 Conclusion

I would like to finish this discussion by re-iterating the following point. I believe that after more than ten years of searching for reasons for the discrepancy between the E821 result and the theoretical prediction for the muon magnetic anomaly, we can say with confidence that missed Standard Model effects as large as $\sim 260\times10^{-11}$ can be excluded as the reason for the discrepancy. The three logical explanation of the discrepancy are then 1) an experimental issue; 2) a coherent combination of small effects in theory and experiment that reduces the discrepancy to an "acceptable" level; 3) physics beyond the Standard Model.

The first item will be clarified by the new FNAL experiment. The second item will require some work on the theory side, new measurements of the $e^+e^- \to$ hadrons cross sections at Novosibirsk and Beijing and a better understanding of the hadronic light-by-light scattering contribution. The hope here is related to the new mode-independent approaches that are currently being developed, the improved measurements of photon transition form factors [41, 42] that appear to be possible at high-luminosity e^+e^- colliders and with continuous progress in applications of lattice QCD to hadronic light-by-light scattering that seems to be reaching the breakthrough moment [43].

The BSM contribution as an explanation of the discrepancy in muon $g-2$ is, arguably, the most exciting option. The likely candidate is still the supersymmetry with relatively small chargino and neutralino masses and relatively large value of $\tan\beta$. Here the interplay with direct measurements at the LHC is crucial but so far there is no contradiction between the LHC data and the masses of electroweakinos required to explain the muon magnetic anomaly [44].

Acknowledgements

I would like to thank the organizers for inviting me to this interesting Workshop and for creating nice and inspiring atmosphere.

References

[1] G. W. Bennett *et al.* [Muon g-2 Collaboration], Phys. Rev. D **73**, 072003 (2006).

[2] We take the results from K. Hagiwara, R. Liao, A.D. martin, D. Nomura and T. Teubner, J. Phys. **G38**, 085003 (2011).

[3] D. Hertzog, this conference, *Next Generation Muon g − 2 Experiments*

[4] A. Czarnecki, B. Krause and W. J. Marciano, Phys. Rev. Lett. **76**, 3267 (1996); S. Peris, M. Perrottet and E. de Rafael, Phys. Lett. B **355**, 523 (1995); A. Czarnecki, B. Krause and W. Marciano, Phys. Rev. **D**52, 2619 (1995); A. Czarnecki, W. J. Marciano and A. Vainshtein, Phys. Rev. **D**67, 073006 (2003); Erratum-ibid. **D**73 (2006) 119901; C. Gnendiger, D. Stöckinger and H. Stöckinger-Kim, Phys. Rev. **D**88, 053005 (2013).

[5] T. Aoyama, M. Hayakawa, T. Kinoshita and M. Nio, Phys. Rev. Lett. **109**, 111808 (2012).

[6] A. Kurz, T. Liu, P. Marquard, A. V. Smirnov, V. A. Smirnov and M. Steinhauser, arXiv:1508.00901 [hep-ph].

[7] K. Melnikov and A. Vainshtein, "Theory of the Muon Anomalous Magnetic Moment", Springer Tracts in Modern Physics, 2006.

[8] A. Yelkhovsky, On the higher order QED contributions to the muon $g-2$ value, preprint INP-88-50, (1988); A. Milstein and A. Yelkhovsky, Phys. Lett. B **233**, 11 (1989).

[9] C. Chlouber and M. Samuel, Phys. Rev. D**16**, 3596 (1977).

[10] T. Aoyama, M. Hayakawa, T. Kinoshita and M. Nio, Phys. Rev. Lett. **109**, 11808 (2012).

[11] S. Karshenboim, Phys. Atom. Nucl. **56**, 857 (1993).

[12] G. Mishima, Bound State Effect on the Electron g-2, arXiv:1311.7109 [hep-ph].

[13] K. Melnikov, A. Vainshtein and M. Voloshin, Phys. Rev. D **90**, 017301 (2014).

[14] M. Fael and M. Passera, Phys. Rev. D **90**, no. 5, 056004 (2014).

[15] M. A. Braun, Zh. Eksp. Teor. Fiz. 54, 1220 (1968) [Sov. Phys. JETP 27, 652 (1968)].

[16] S. Eidelman, this conference, *Muon $g-2$ and hadronic vacuum polarization: recent developments.*

[17] T. Blum, A. Denig, I. Logashenko, E. de Rafael, B. Lee Roberts, T. Teubner and G. Venanzoni, arXiv:1311.2198 [hep-ph].

[18] M. Davier and B. Malaescu, Eur. Phys. J. C **73**, no. 11, 2597 (2013).

[19] B. Kloss, talk at CIPANP 2015, arXiv:1510.02293[hep-ph].

[20] F. Jegerlehner, this conference, *Leading order hadronic contribution to the electron and muon $g-2$.*

[21] M. Benayoun, this conference, *VMD-HLS approach to the muon $g-2$: a solution to the $\tau - e^+e^-$ puzzle.*

[22] J. Prades, E. de Rafael and A. Vainshtein, "Hadronic Light-by-Light Scattering Contribution to the Muon Anomalous Magnetic Moment," (Advanced series on directions in high energy physics. 20) [arXiv:0901.0306 [hep-ph]].

[23] G. Colangelo, M. Hoferichter, M. Procura and P. Stoffer, JHEP **1409**, 091 (2014).

[24] G. Colangelo, M. Hoferichter, B. Kubis, M. Procura and P. Stoffer, Phys. Lett. B **738**, 6 (2014).

[25] V. Pauk and M. Vanderhaeghen, Phys. Rev. D **90**, no. 11, 113012 (2014).

[26] M. Knecht, A. Nyffeler, M. Perrottet and E. de Rafael, Phys. Rev. Lett. **88**, 071802 (2002).

[27] K. Melnikov and A. Vainshtein, Phys. Rev. D **70**, 113006 (2004).

[28] J. Bijnens, E. Pallante and J. Prades, Phys. Rev. Lett. **75**, 1447 (1995) [(E) Phys. Rev. Lett. **75**, 3781 (1995)].

[29] J. Bijnens, E. Pallante and J. Prades, Nucl. Phys. B **474**, 379 (1996).

[30] J. Bijnens, E. Pallante and J. Prades, Nucl. Phys. B **626**, 410 (2002).

[31] M. Knecht and A. Nyffeler, Phys. Rev. D **65**, 073034 (2002).

[32] J. Erler and G. Toledo Sanchez, Phys. Rev. Lett. **97**, 161801 (2006).

[33] A. A. Pivovarov, Phys. Atom. Nucl. **66**, 902 (2003) [Yad. Fiz. **66**, 934 (2003)].

[34] R. Boughezal and K. Melnikov, Phys. Lett. B **704**, 193 (2014).

[35] D. Greynat and E. de Rafael, JHEP **1207**, 020 (2012).

[36] M. Hayakawa, T. Kinoshita and A. I. Sanda, Phys. Rev. Lett. **75**, 790 (1995).

[37] M. Hayakawa, T. Kinoshita and A. I. Sanda, Phys. Rev. D **54**, 3137 (1996).

[38] K. T. Engel, H. H. Patel and M. J. Ramsey-Musolf, Phys. Rev. D **86**, 037502 (2012).

[39] K. T. Engel and M. J. Ramsey-Musolf, Phys. Lett. B **738**, 123 (2014).

[40] J. Bijnens, this conference, *Hadronic light-by-light scattering contribution to a_μ: extended Nambu-Jona-Lasino, chrial quark models and chiral Lagrangians.*

[41] A. Nyffeler, this conference, *On the precision of a data-driven estimate of the pseudoscalar-pole contribution to hadronic light-by-light scattering in the muon g-2.*

[42] A. Kupsc, this conference, *Transition form factors of light mesons.*

[43] C. Lehner, this conference, *Hadronic light-by-light contribution to $(g-2)_\mu$ from lattice QCD.*

[44] D. Stöckinger, this conference, *Magnetic moment $(g-2)_\mu$ and Supersymmetry.*

EPJ Web of Conferences **118**, 01021 (2016) DOI: 10.1051/epjconf/201611801021
FCCP2015

Design and status of the Mu2e experiment

Stefano Miscetti[1],[a] on behalf of the Mu2e Collaboration

[1] *Laboratori Nazionali di Frascati dell'INFN, Via Enrico Fermi 40, 00044, Frascati, Italy*

Abstract. The Mu2e experiment aims to measure the charged-lepton flavor violating neutrino-less conversion of a negative muon into an electron in the field of an aluminum nucleus. The objective is to improve the previous measurement by four orders of magnitude with the use of a similar technique. For the conversion process, the signal will appear as a mono-energetic electron very close to the muon rest mass. In the Standard Model these process have negligible rates. However, in many Beyond the Standard Model scenarios their rates are within the reach of next generation experiments. In this paper, we explain the sensitivity to new physics scale and the complementarity of approach and reach provided by MU2E with respect to Mu3e and MEG upgrade. Mu2e experimental technique, design and status will be the focus of this paper.

1 Introduction

In the Standard Model, SM, hadronic flavor transitions are allowed and the CKM matrix represents the probability for a quark of a specific family to turn into a different quark of another family. In the lepton sector, since SM neutrinos are massless, transitions between charged and neutral leptons preserve flavor. Nevertheless, flavor transitions between neutral leptons have been observed in neutrino oscillations that imply non conservation of lepton flavor number. Even assuming massive neutrinos, charged lepton flavor violation (CLFV) processes are extremely suppressed in the SM. The SM constraint on the branching ratio for $\mu \to e\gamma$ is $B(\mu \to e\gamma) \leq 10^{-52}$ [1]. Theories beyond the Standard Model, BSM, predict additional particles and interactions that can enhance such processes up to a measurable level (for a recent review see [2]).

Experimentally, the search for CLFV has a long history both in the muon and tau decays ($\mu \to e\gamma$, $\mu \to 3e$, $\tau \to \mu\gamma$, $\tau \to 3\mu$) as well as muon conversion processes, $\mu N \to eN$, in the presence of a nucleus. Starting with the pioneering measurements of cosmic rays in the late 1940's, the sensitivity has improved by many orders of magnitude and reached stringent limits for both $\mu \to e\gamma$ (with BR $< 5.7 \times 10^{-13}$ at 90 % C.L. [3]) and muon conversion on gold ($R_{\mu e} < 6 \times 10^{-13}$ [4]). An international search of CLFV continues today with the completion of MEG, its upgrade MEG-II and the proposed Mu3e search at PSI as well as the searches for muon to electron conversion with Mu2e at FNAL and COMET at J-PARC [5]. CLFV searches are of particular interest because of the possibility to carry out clean measurements free of theoretical background. In the case of observation they indicate a clear evidence of physics beyond the standard model while in the case of non observation they pose strong limits on the development of new theories.

A model independent description of the CLFV transitions for physics beyond the Standard Model is provided by an effective lagrangian [6] where the different processes are divided in dipole amplitudes and four-fermion, or contact term, operators. The $\mu \to e\gamma$ decay is mainly sensitive to the dipole amplitude, while $\mu \to e$ conversion and $\mu \to 3e$ receive contributions also from the four-fermion interactions. It is customary to parametrize the interplay between the two amplitudes by means of two parameters [6]: Λ, which sets the mass scale and κ, which governs the ratio of the four-fermion to the dipole amplitude. For $\kappa << 1 (>> 1)$ the dipole-type (four-fermion) operator dominates. Figure 1 summarizes the power of different searches to explore this parameter space. A scale $\Lambda < 700$ TeV is already excluded by present limits and poses serious constraints on Standard Model extensions, primarily, supersymmetry. Nevertheless, Λ is only an effective scale and is not immediately proportional to the mass of new particles accessible by direct TeV scale searches at LHC. The interpretation of an eventual direct observation of new physics at LHC will have to take into account precise measurements from MEG and Mu2e where the comparison between these determinations will provide the constraints necessary in pinning down the underlying theory.

An example of interplay between MEG and Mu2e is present in the scalar leptoquark searches. In Fig. 2*a*, the reach for the new coupling λ, as defined from eq. 14 of [7], is reported for a range of scalar leptoquark masses for both MEG upgrade and conversion experiments. The compilation is extracted from the Mu2e TDR. It is clear that while not excluding the Leptoquark existence at few TeV masses, the CLFV coverage extends up to masses of O(100 TeV). Another example is provided by the left-right

[a] e-mail: Stefano.Miscetti@LNF.INFN.IT

© The Authors, published by EDP Sciences. This is an open access article distributed under the terms of the Creative Commons Attribution License 4.0 (http://creativecommons.org/licenses/by/4.0/).

 DOI: 10.1051/epjconf/201611801021

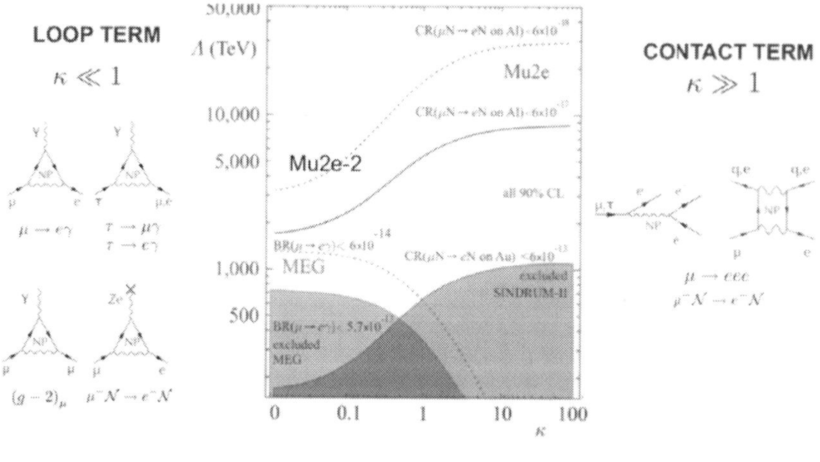

Figure 1. Sensitivity of $\mu \to e\gamma$, $\mu \to e$ transition and $\mu \to 3e$ to the scale of new physics Λ as a function of the parameter κ. The shaded areas are excluded by present limits. On the left (right) side, the dipole (four-fermion) diagrams are shown for the different CLFV processes.

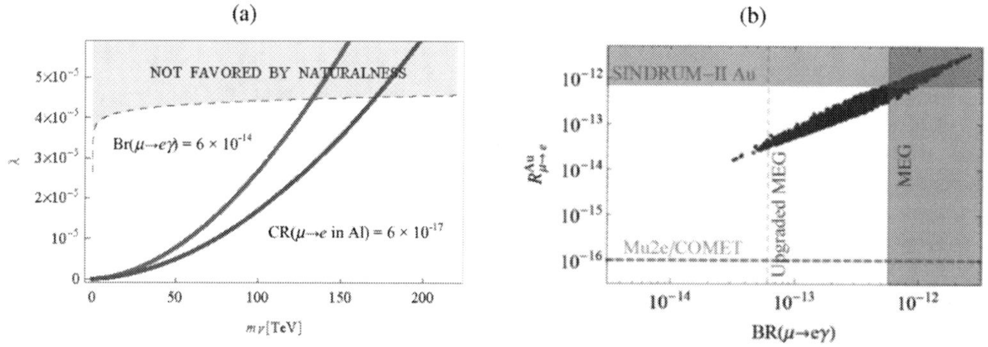

Figure 2. (*a*) Reach acheavable in scalar leptoquark search for Upgraded MEG and Mu2e as a function of the coupling λ and the scalar leptoquark mass. (*b*) Expected BR and $R_{\mu e}$ for Left-Right symmetric models for Upgraded MEG and muon conversion experiments.

symmetric models. A recent study [8] predicts the CLFV rates in this environment assuming a new mass breaking scale at around 5 TeV. The correlation between the BR for the MEG upgrade and $R_{\mu e}$ for the conversion experiments are shown in Fig. 2*b*. It is clear that the two experiments can cover the expectations for the full phase space of this theory and help constraining it.

2 Mu2e: search for μ to e conversion at FNAL

In presence of a nucleus, the measurement of the neutrinoless μ^- to e$^-$ conversion rate is defined relatively to the ordinary capture of the muon on the nucleus as follows:

$$R_{\mu e} = \frac{R(\mu^- + A(Z,N) \to e^- + A(Z,N))}{R(\mu^- + A(Z,N) \to \nu_\mu + A(Z-1,N)}). \quad (1)$$

The *conversion signal*, CE, consists of a monoenergetic electron (see Fig. 3(left)), close to the muon mass and corrected for binding energy and nuclear recoil, that has to be separated by means of high momentum resolution from the spectrum of the electrons produced by the *muon decay in orbit* process, DIO (see Fig. 3(right)). Stopped muons have a large chance of decaying when orbiting around the nucleus and their spectrum substantially differs from free decay due to the presence of a large recoil tail that falls rapidly as the energy approaches the kinematical endpoint. Unfortunately, in Sindrum-II additional background sources were dominating. In particular, after rejection of cosmics, a prompt background coming from *radiative pion capture*, RPC, $\pi + N \to \gamma^* + N'$, was found to be the most resistant background. In this process the electron positron pair, produced either by internal or external

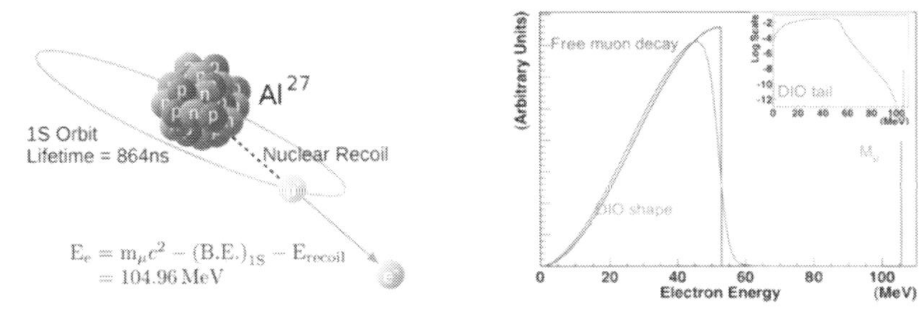

Figure 3. (Left) Sketch of the muon to electron conversion process. (Right) Energy spectrum for electrons produced from muon decays in orbit. The spectrum is a modified Michel spectrum with a long tail that falls quickly as $(E_e - E_{endpoint})^5$ and comes in proximity to the kinematic endpoint. Perfect resolution is assumed.

conversion, becomes a source of CE candidate when the e^- momentum gets in the right selection window. These observations have driven the design strategy of the new generation muon to electron conversion experiments like Mu2e. The Mu2e goal is to achieve a single CE event sensitivity of 2.5×10^{-17}, or constrain the R_μ value to below 6×10^{-17} at 90% C.L., four orders of magnitude better than the previous limit.

There are four ingredients to achieve: high intensity, pulsed beam structure, high proton extinction and a precise detector. *A high intensity muon beam.* The goal is to increase by a factor of four the nowadays target muon intensity to 10^{11}/sec. This can be achieved by means of a high muon production and a curved solenoidal system to allow for charge and momentum selection. The solenoidal system creates a selective transport channel that increases intensity while reducing the associated need for beam power (compare 1 MW at PSI for Sindrum-II and 8 kW for Mu2e). *A pulsed beam structure* allows for the rejection of the prompt background. The target has to be selected in a way that the muon lifetime, τ_μ, in the bound system is smaller than the bunch period. Mu2e utilizes an aluminum target (τ_μ = 864 ns) that well matches the bunch structure of the Fermilab accelerator complex. For the beam delivered to Mu2e the micro-bunch structure occur in a window of 1684 ns. The trick here is to wait for the *beam-flash* of particles travelling with the transported beam to disappear and waiting for the most relevant prompt backgrounds to decay. The start of the data acquisition is typically set at ~ 700 ns after the bunch arrival time. The effects of the unsynchronized particles traveling with the beam have to be negligible. This is referred to as the *proton extinction requirement.* Calculation by full simulation has shown that in order to suppress the prompt backgrounds generated by beam electrons, in-flight-decay muons, in-flight-decay pions and RPC requires a pulsed beam where the ratio of the beam between pulses to the beam contained in a pulse is below 10^{-10}. *A redundant high-precision detector* has to analyze the captured products in order to separate CE and DIO spectra and reduce to negligible contribution the additional background sources.

2.1 The Mu2e detector layout

The layout of the Mu2e experiment is shown in Figure 4. An array of superconducting solenoids forms a graded magnetic system composed of a Production Solenoid, PS, a Transport Solenoid, TS, and a Detector Solenoid, DS.

The PS contains a tungsten target that intercepts an 8 GeV pulsed proton beam brought from the FNAL delivery ring by means of a dedicated beam-line. The gradient field in the PS increases from 2.5 to 4.6 Tesla in the same direction of the incoming beam and opposite to the outgoing muon beam direction. This gradient field works as a magnetic lens used to focus low energy particles into a transport channel. The focused beam is constituted by muons, pions and a small number of protons and antiprotons. The S-shaped Transport Solenoid efficiently transfers low energy, negatively charged particles to the end of a beam-line while allowing for a large fraction of the pions to decay into muons. A midsection collimator attenuates nearly all positively charged particles while the antiproton residual background is reduced by a special absorber. The DS presents a graded field from 2 to 1 Tesla in the upstream region where the stopping target resides. This lens increases the acceptance for CE while aiding in the rejection of beam-related backgrounds. A uniform magnetic field of 1 Tesla occupies the region of the tracker and calorimeter systems. Approximately 50% of the muon beam is stopped by the target while the remaining beam is dumped at the end of the cryostat. The DS is covered externally by a Cosmic Ray Veto system.

When muons stop in the aluminum target, they are captured in an atomic excited state and thus prompty fall to the ground state with 39% decaying in orbit while the remaining 61% are captured on the nucleus. Low energy photons, neutrons and protons are emitted in the nuclear capture process and constitute an environmental background that produces an ionization dose and a neutron fluence on the detection systems as well as an accidental occupancy for the reconstruction program. This background is fully simulated in overlap with the signals coming from the apparatus.

The tracking detector is composed of low mass straw drift tubes oriented transverse to the solenoid axis and con-

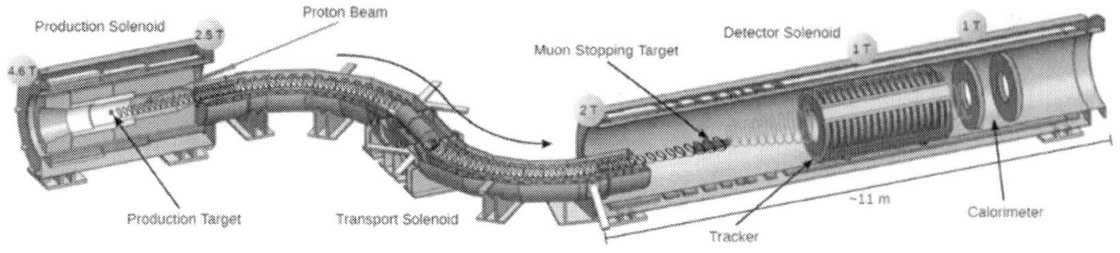

Figure 4. Layout of the Mu2e experiment.

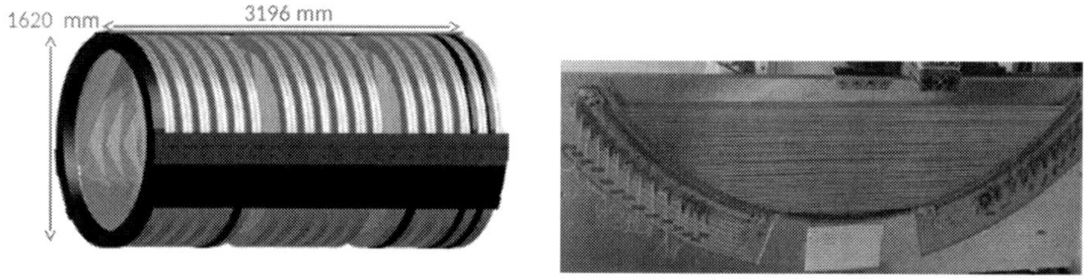

Figure 5. (Left) Sketch of the Mu2e straw tracker system. (Right) Picture of the first prototype built for straw tube panel.

sists of nearly 20000 straw tubes arranged in 18 station as shown in Fig. 5(left). Each tube is 5 mm in diameter and contains a 25 μm sense wire. The straw walls are made out of Mylar and have a thickness of 15 μm. The tracker is around 3 m long and placed inside a 1 T uniform magnetic field. It measures the momenta of the charged particles from the reconstructed trajectories using the hits detected in the straw. A circular hole inside the tracker allows it to be insensitive to charged particle momenta below 55 MeV/c. The tracking preparation proceeds well as shown by the picture of the first straw-panel prototype built (see Fig. 5(right)).

The electromagnetic calorimeter system [9] is composed by two disks of scintillating crystals and is placed downstream of the tracker. A single disk is composed of 830 BaF_2 crystals readout by two *solar blind* UV extended avalanche photo-diodes (APD). In Fig. 6(left) a drawing of the disks is shown. Similarly to the tracker, the inner circular hole allows to be insensitive to the DIO electrons up to 55 MeV/c momenta. The calorimeter tasks are that of providing a powerful particle identification between muons and electrons, an independent trigger system and a seed for tracking in a complicated reconstruction environment. In order to fulfill these requirements the calorimeter must provide an energy resolution of O(5%), a time resolution $<$ 500 ps and a position resolution better than 1 cm for 100 MeV electrons. The selected crystals should survive a dose of 100 krad and a fluence of up to 10^{12} n/cm^2. The photosensors are shielded by the crystals themselves and should only sustain a fluence of up to 3×10^{11} n_{1MeV}/cm^2. Prototypes for the solar blind APDs have been developed but the dark noise is too high and cooling them down to -10° C is needed. Backup options are being explored.

One major background source for Mu2e is related to cosmic ray muons producing false CE's when interacting with the detector materials. These events occur at a rate of approximately one/day. In order to reduce their contributions to below 0.1 event in the experiment lifetime the external area of the DS and a part of the TS are covered by a Cosmic Ray Veto, CRV, system. The requirement for the system is to obtain a vetoing efficiency of at least 99.99% for cosmic ray tracks while withstanding an intense radiation environment. Comprised of four staggered layers of scintillation counters (see 6(right)), the CRV utilizes two embedded Wavelength Shifting Fibers, each readout by means of 2×2 mm^2 Hamamatsu SIPM. Test beams on full size prototype have been carried out.

Further instrumentation exists, such as the monitor to measure the proton beam extinction and a High Precision Germanium Detector to monitor the stopping rate in the aluminum target. The latter one is needed to provide the normalization for the $R_{\mu e}$ conversion rate. Further information can be found in ref. [5] .

2.2 Reconstruction of CE candidates

At this energy the momentum resolution is dominated by fluctuations in the energy loss in the target caused by multiple scattering and bremsstrahlung in the tracker. By performing a full simulation of the tracker, a pattern recognition and a Kalman fitter allow us to obtain for the tracking a CE reconstruction efficiency of 9 % for good quality tracks and a minimum of 25 hits/track. The resolution is well parametrized by a Crystal Ball function with a negative bremsstrahlung tail, a gaussian core of 116 keV and a long exponential positive resolution tail. This re-

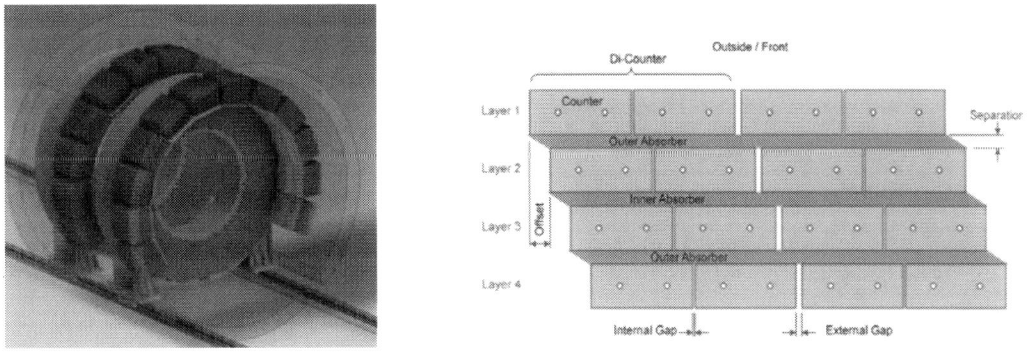

Figure 6. (Left) CAD drawings of the calorimeter disks. Calorimeter innermost (outermost) radius is of 351 mm (670 mm) with a crystal length of 200 mm. Layout of the FEE and digitization crates is also shown. (Right) Schematic drawing of the cross-sectional view of a CRV counter. Each counter is 20 mm thick with a width of 50 or 60 mm and a length up to 6.6 m.

sult is obtained in a hostile environment due to the large quantity of neutrons, photons and protons produced from the muon capture processes. Note that the muon stopping rate needed to arrive to the final goal of 6×10^{17} stopped muons is at around 10 GHz. In Fig. 7(left), an example of a single micro-bunch event in the tracker is shown. This has been obtained with a Geant-4 simulation displaying all hits generated by the environmental background. The irreducible background is provided by the DIO electrons that have to be distinguished by the mono-energetic CE. The finite tracking resolution and the positive reconstruction tail have a large effect on the falling spectrum of the DIO background that translates to a residual contamination in the signal region. Fig. 7(right) shows the signal and background distributions as seen by a full simulation of the experiment (pileup included) in the following conditions: (i) 3.6×10^{20} proton on target (ii) a corresponding number of 6×10^{17} muon stops and (iii) a $R_{\mu e}$ of 10^{-16}. After maximizing the signal over the background the best selection corresponds to events counted in a momentum window between 103.75 and 105 MeV/c. A DIO contribution of 0.199 events is estimated and 3.5 candidates are observed. Estimate of other backgrounds are presented in Tab.3 for a total background contribution of 0.37 events. This counting corresponds to set a limit on $R_{\mu e}$ below 6×10^{-17} at 90 % C.L. , in good agreement with the experiment design goal. A very important step for this measurement is the tracker calibration. To accurately predict the DIO rate, we should calibrate the tracker momentum i.e. determine the shifts, the scale and the resolution of the measured momentum compared to the true one. After a basic alignment, based on construction and installation survey data, an thorough calibration will be based both on the usage of the DIO sample and on independent processes of low energy two body decay, such as the one from stopped π^+ on target undergoing the $\pi \rightarrow e\nu$ decay. Selecting positive pions will help avoid nuclear capture but asks for reversing the momentum selection in the TS. Moreover, the data have to be collected at shorter rate and an extrapolation from lower energies to the conversion endpoint is still needed.

2.3 Experiment status and schedule

At the moment of writing the Mu2e experiment has succesfully completed the CD-2 review for all systems and the CD-3 reviews for the superconducting cables, civil construction and the Transport Solenoids.

The heart of the Mu2e apparatus is provided by a superconducting magnetic system whose design, fabbrication, assembly and commissioning drives the experiment schedule. The status of the magnet as well as the construction and testing of the superconducting cables is satisfactory. An international bid for the DS and PS has been concluded and the construction phase for the large magnets are started at General Atomics, San Diego, USA. For the TS, a prototype of one module has been built by ASG Superconducting in Genova in collaboration with the INFN group of Genova. Tests and quality control of the prototype have been carried out this summer with satisfactory results. The bid on the TS has also been completed and has been assigned to the ASG. Since the ground breaking cerimony for the experimental hall, held on April 2015, the civil construction is proceeding at a rapid pace. Upon completion of the CD-3 review expected for the summer of 2016, the construction of the detectors and accelerator will begin. The schedule foresees a completion of the installation and commissioning with beam at the end of 2020.

3 Conclusions and perspectives

The Mu2e experiment design and construction proceeds well and iis on schedule to be commissioned with a beam by the end of 2020. Its goal is to probe CLFV with a single event sensitivity of 2.5×10^{-16} or set an upper limit on the conversion rate $< 6 \times 10^{-17}$ at 90 % C.L., thus improving the previous sensitivity by four orders of magnitude. For the long term future (> 2025) the possibility of a Mu2e phase-2 is being explored with the goal of increasing the sensitivity by an additional factor of 10. This can be obtained with a higher beam intensity and a detector capable of handling the the accidental activity occuring from muon

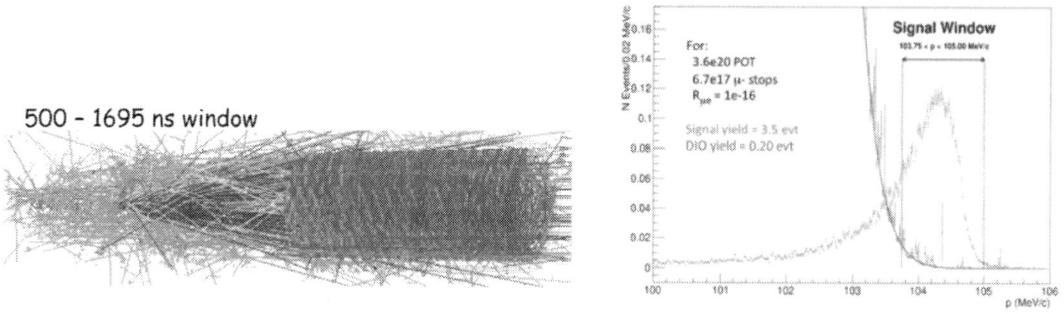

Figure 7. (Left) event display of a single micro-bunch event. (Right) full simulation of DIO and CE events for an assumed $R_{\mu e}$ of 10^{-16}.

Category	Background process	Estimated Yield
Intrinsic	Decay in orbir (DIO)	0.199 ± 0.092
	Muon Capture (RMC)	0.000 ± 0.004
Late Arriving	Pion capture (RPC)	0.023 ± 0.006
	Muon decay in flight	< 0.003
	Pion decay in flight	0.001 ± 0.001
	beam electrons	0.003 ± 0.001
Miscellaneous	pbar	0.047 ± 0.024
	Cosmic rays	0.092 ± 0.020
	Total	0.37 ± 0.10

Table 1. Expected background list as evaluated by full simulation. Extracted from the Mu2e TDR.

capture. In this context only an option based on the new Proton Improvement Plan, PIP-2, is being studied.

Acknowledgments

We are grateful for the vital contributions of the Fermilab staff and the technical staff of the participating institutions. This work was supported by the US Department onf Energy; the Italian Istituto Nazionale di Fisica Nucleare; the US National Science Foundation; the Ministry of Education and Science of the Russian Federation; the Thousand Talents Plan of China; the Helmholtz Association of Germany; and the EU Horizon 2020 Research and Innovation Progam under the Marie Sklodowska-Curie Grant Agreement N.690385. Fermilab is operated by Fermi Research Alliance, LLC under Contract No. De-AC02-07CH11349 with the US Department of Energy.

References

[1] S.M.Bilenky et al., *Phys. Lett.* **B67**, 309 (1977)

[2] F.Grancagnolo et al., *Nucl.Phys.Proc.Supp.* **248-250** 1-146 (2014).

[3] The MEG collaboration,*Phys.Rev.Lett.* **110** 201801 (2013).

[4] The SINDRUM II Collaboration, *Eur.Phys.J.* **C47** 337-346 (2006)

[5] The Mu2e Collaboration, *FERMILAB-TM-2594*, **ArXiv:1501.05241** (2015).

[6] A. De Gouvea et al., *Prog. Part. Nucl. Phys* **71** 75-92 (2013).

[7] J.M.Arnold et al., *Phys.Rev* **D88** 035009 (2013).

[8] C.H.Lee et al., *Phys.Rev* **D88** 092010 (2013).

[9] X.. et al., *Nucl. Instr. and Meth.* **A5857**, 10.1016/j.nima.2015.09.074.

Radiative corrections and Monte Carlo generators for physics at flavor factories

Guido Montagna[1,a]

[1] Dipartimento di Fisica, Università di Pavia and INFN, Sezione di Pavia, Via A. Bassi 6, 27100, Pavia, Italy

Abstract. I review the state of the art of precision calculations and related Monte Carlo generators used in physics at flavor factories. The review describes the tools relevant for the measurement of the hadron production cross section (via radiative return, energy scan and in $\gamma\gamma$ scattering), luminosity monitoring, searches for new physics and physics of the τ lepton.

1 Introduction

The physics program of e^+e^- colliders with center of mass (c.m.) energy between about 1 and 10 GeV (flavor factories) is broad and ranges from flavor physics to the measurement of the low-energy hadronic cross section, with its crucial implications for $g-2$ and the running of α, to searches for QCD exotic states and studies of QCD in the non-perturbative regime. It also includes precision tests of the SM at low energies and new physics (NP) searches (dark photon, light Higgs bosons), as well as measurements related to the physics of the τ lepton.

The reason why radiative corrections (RC) and accurate Monte Carlo (MC) generators are involved in most of these studies stems from the fact that the large statistics accumulated by these experiments at the intensity frontier necessarily requires precision calculations for a large spectrum of measurements and the description of physics phenomena. Examples are the simulation of the contribution due to initial-state radiation (ISR) and the measurement with 1% precision or better of the hadronic cross section via the mechanism of radiative return; the modeling of final-state radiation (FSR) and bremsstrahlung in decays (for example, in the leptonic decay of the τ); the measurement of the collider luminosity with per mille uncertainty and the control of the process of muon pair production in association with a photon as normalization of the hadronic cross section measurement via radiative return; and, finally, the simulation of QED processes, considered both as signals and backgrounds, depending on the measurement.

All these physics issues require that the precision calculations are made available in the form of MC generators, which are needed to allow for simulations and data-theory comparisons under complex selection criteria.

The theoretical methods used in this context mostly refer to the QED sector of the SM and include results at NLO in perturbation theory and, in some special cases, at NNLO. Also calculations addressing the resummation of large contributions due photon collinear emission are involved. QED resummation is performed according to different approaches, like collinear Structure Functions (SFs), Parton Shower (PS) or YFS exponentiation. In precision calculations for luminosity and QED processes, predictions based on the matching of NLO corrections with higher-order (h.o.) contributions are also available.

A working group (WG) on RC and MC for low energies was promoted in the middle of the 2000s, in order to stimulate and facilitate the collaboration between experimentalists and theorists working on physics at meson factories. It organizes two meetings per year (see `http://www.lnf.infn.it/wg/sighad/`) and led in 2010 to the publication of the review paper of Ref. [1], which stands as a reference in the field also today.

2 Hadron production cross section

At flavor factories, the hadronic cross section is measured via radiative return [2] (hadron production in association with an energetic ISR photon) and energy scan, as well as in $\gamma\gamma$ scattering.

In spite of the α reduction factor associated to photon radiation, the hadronic cross section via radiative return can be measured from threshold to the nominal c.m. energy with good precision (at the one per cent level or better), thanks to the large luminosity of flavor factories. In this context, the reference MC used by all the experiments as primary generator or to cross-check the accuracy of other tools is PHOKHARA [3–8]. It is the successor of EVA, the first MC developed for this kind of studies in Ref. [9] for the two pion + photon final state and in Ref. [10] for the 4 pions + photon signature, with a treat-

[a] e-mail: guido.montagna@pv.infn.it

© The Authors, published by EDP Sciences. This is an open access article distributed under the terms of the Creative Commons Attribution License 4.0 (http://creativecommons.org/licenses/by/4.0/).

ment of the ISR hard photon at LO supplemented by QED collinear SFs for the description of additional undetected radiation. [1] Now PHOKHARA can be used to simulate various hadronic final states, both for measurements with explicitly tagged or untagged photons. The treatment of ISR is at NLO accuracy for the all the final states, with an estimated precision of 0.5%, due to missing h.o. contributions beyond $O(\alpha)$ in the calculation of the radiator function. For two pions, two protons and K^+K^- final states also FSR is treated at NLO, using the point-like approximation. In the latest version of PHOKHARA, the complete set of NLO corrections is available for $\mu^+\mu^-\gamma$ production [14], a process often used as normalization in the measurement of the hadron + γ production cross section. The recent KLOE and BESIII measurements of the $\pi^+\pi^-\gamma$ cross section, of primary interest for the calculation of the leading order hadronic contribution to the $g-2$, were based on PHOKHARA, whereas in the analyses by BaBar the code AfkQED is adopted.

Concerning the measurements of the hadronic cross section with the method of energy scan, the situation of the MC generators is less established and more varied. What is available in the literature is MCGPJ [15], a generator developed by a Dubna-Novosibirsk collaboration and used at VEPP collider. It can presently simulates the $\pi\pi$ and KK production channels but the implementation of new multi-hadron final states is in progress. The code includes the complete set of NLO QED corrections plus h.o. contributions treated in terms of collinear SFs and makes use of Scalar QED for the calculation of FSR corrections. Therefore, this generator is quite accurate and has an estimated accuracy of 0.2%. A further available code is PHOKHARA 8.0 [16], that can be used to simulate the hadron cross section measured in energy scan for the same channels as in the PHOKHARA version for radiative return. Here, ISR is treated at NNLO accuracy, but without the contribution of multiple photon emission and FSR. The predictions of PHOKHARA 8.0 are in very good agreement, at a few per mille level, with those of MCGPJ for the two pion final state and with those of a benchmark MC, KKMC [17] [2], for muon pair production. Another interesting, ongoing project is `carlomat 3.0` [19], which is an automatic tool for the generation of the scattering amplitudes and cross sections related to complex topologies given by hadrons + photons + leptons. In `carlomat 3.0` the calculations are performed using the effective vertices of the Resonance Chiral Lagrangian ($R\chi L$) Theory or Hidden Symmetry model and, for the time being, RC are not included.

For $\gamma\gamma$ physics, of special interest for the measurements of the photon-meson transition form factors and tests of QCD, the various codes used till now by the experiments are all characterized by the common feature of using the Equivalent Photon Approximation (EPA) for the cross section calculation. A more precise generator, beyond the EPA, is EKHARA [20], which includes exact formulae and exact kinematics for both s- and t-channel amplitudes and their interference. It allows to obtain predictions for $e^+e^-\pi^+\pi^-$ production, as well as for $e^+e^-\pi^0$, $e^+e^-\eta/\eta'$ production, using different models for the photon-meson transition form factors. The implementation in EKHARA of the dominant QED corrections is under way and will available in future releases.

3 Luminosity measurement

The area where a particular progress occurred, both in the calculation of RC and realization of MC codes, is that related to the luminosity measurement. Indeed, at flavor factories, the luminosity is measured with very good precision, typically with an uncertainty between a few per mille and one per cent, by using the ratio between the observed number of events of a given reference process and the corresponding theoretical cross section through the relation $\mathcal{L} = N_{\rm obs}/\sigma_{\rm theory}$. To maintain as small as possible the total luminosity uncertainty, the best reference process is large-angle Bhabha (LABH) scattering, measured in the same detectors used for the measurement of the hadronic cross section, with similar acceptance conditions. Two photon production is also used a cross check or to obtain the luminosity as an average of the measurements obtained with the two processes. In the luminosity measurement, typically two independent generators are used to avoid loss of precision or introduce a bias in the measurement itself.

Table 1. The MC generators used in luminosity measurements.

Generator	Processes	Theory	Accuracy
BabaYaga 3.5	$e^+e^-, \gamma\gamma, \mu^+\mu^-$	QED PS	0.5%
BabaYaga@NLO	$e^+e^-, \gamma\gamma, \mu^+\mu^-$	$O(\alpha)$+QED PS	0.1%
BHWIDE	e^+e^-	$O(\alpha)$ YFS	0.1%
MCGPJ	$e^+e^-, \gamma\gamma, \mu^+\mu^-$	$O(\alpha)$+SFs	0.2%

The generators used to monitor the luminosity at meson factories are summarized in Table 1. The reference code adopted by most of the experimental collaborations is the generator BabaYaga [21–23]. In its most precise version, BabaYaga@NLO, the generator includes the exact NLO corrections matched to a QED PS for the simulation of exclusive multiple photon emission and has an estimated accuracy of 0.1%. Other programs used at flavor factories for simulations of the LABH process with a formulation and precision similar to that of BabaYaga@NLO are BHWIDE [24], developed during the LEP time, and MCGPJ, which, in addition to hadronic final states, provides predictions for lepton pair and two-photon [25] production.

Because of the precision requirements, a relevant question in the context of luminosity measurements is related

[1]The EVA approach is followed by BaBar collaboration in their analysis of hadron production via radiative return processes, using their own, not public event generator AfkQED. In the latter the muon final state is simulated independently according to the formulae of Ref. [11]. The usage of AfkQED in BaBar is supplemented by PHOTOS [12, 13] for the description of additional undetected FSR.

[2]KKMC is a precise MC tool for the simulation of several $2 \to 2$ processes [18]. It is based on a generalization of the YFS exponentiation approach (coherent exclusive exponentiation) and is used at flavor factories in various studies requiring simulations of lepton-pair production or QED radiative processes.

to the accuracy of the theoretical calculations and corresponding codes. As a consequence of the formulation underlying the most precise generators, featuring the matching of NLO corrections with QED exponentiation in factorized form, the bulk of the most important sub-leading $O(\alpha^2)$ corrections, i.e. $O(\alpha^2 L)$ photonic corrections (L being the large collinear logarithm), is effectively incorporated [26]. It follows that the main uncertainties affecting the theoretical precision of the luminosity generators come from two sources. The first one is related to the hadronic contribution to the vacuum polarization, which is treated using dispersion relations and is therefore parametric, driven the experimental uncertainty of the hadron cross section measurements. The second source of uncertainty is purely perturbative, due to the incomplete inclusion of QED corrections at NNLO. Fortunately, all the pieces of the Bhabha cross section at NNLO in QED have been computed over the last decade or so (see e.g. Ref. [1]) and therefore these calculations represent an important benchmark to assess the theoretical precision of the luminosity MC tools.

Within the WG on RC and MC for low energies, a particular effort was devoted to reach a reliable estimate of the theoretical uncertainty of the luminosity cross section calculation. Following similar work done during the LEP workshops in the 1990s, detailed comparisons between the predictions of the available NNLO calculations and the corresponding approximations present in BabaYaga@NLO led to the estimate of the total theoretical uncertainty summarized in Table 2. It quotes the official WG evaluation updated by the conclusions of the work on leptonic and hadronic pair corrections of Ref. [27] and by a less conservative estimate of the one-loop corrections to hard bremsstrahlung according to Ref. [28].

Source of unc. (%)	1–2 GeV	BESIII	BaBar/Belle
$\lvert\delta_{\rm VP}\rvert$ [Jegerlehner]	–	0.01	0.03
$\lvert\delta_{\rm VP}\rvert$ [HMNT]	0.02	0.01	0.02
$\lvert\delta^{\alpha^2}_{\rm photonic}\rvert$	0.02	0.02	0.02
$\lvert\delta^{\alpha^2}_{\rm pairs}\rvert$	0.03	0.02	0.03 ÷ 0.07
$\lvert\delta^{\alpha^2}_{\rm SV,H}\rvert$	0.05 / 0.03	0.05 / 0.03	0.05 / 0.03
$\lvert\delta^{\alpha^2}_{\rm HH}\rvert$	–	–	–
$\lvert\delta_{\rm total}\rvert$	0.07/0.05	0.06/0.04	~ 0.07 ÷ 0.09

Table 2. The total theoretical uncertainty of the LABH scattering cross section at flavor factories.

Note that the estimate of the total theoretical uncertainty in the luminosity measurement at flavor factories, between 0.05% and 0.1%, is comparable to that achieved at LEP in the 1990s. However, it slightly deteriorates in the proximity of the very narrow resonances because of the uncertainty induced by the hadronic contribution to the vacuum polarization.

4 Search for New Physics and τ-lepton physics

The generators involved in the search for NP and physics of the τ lepton concern the simulation of QED processes, the modeling of τ decays and the description of bremsstrahlung in τ decays.

Concerning NP searches, a particular attention was paid over the last few years in the search for a massive dark photon weakly coupled to the SM particles as a possible solution of the dark matter puzzle. One the best signals at meson factories is represented by the associated production of a dark photon and an energetic photon, with subsequent decay of the dark photon into electron or muon pairs. The signature consists of very narrow bumps in the lepton invariant mass spectrum, because of the very small coupling of the new particle to the leptons, over the continuum given by the QED background processes $e^+e^- \to e^+e^-\gamma, \mu^+\mu^-\gamma$. In this framework, an upgrade of BabaYaga [29] is used by KLOE for the simulation of both signal and backgrounds, whereas BaBar makes use of MadGraph/MadEvent [30] for signal simulations and of the tandem of generators BHWIDE/KKMC for the study of the QED radiative processes.

As far as physics of the τ lepton is concerned, the measurements giving access to the τ mass and related to τ pair production at threshold involve MC programs developed during the LEP era. The reference code used by BaBar, Belle and BESIII for the calculation of the $e^+e^- \to \tau^+\tau^-$ cross section is KKMC, but in some measurements also the older KORALB [31] generator is considered. For the measurements and physics studies related to τ decays, the reference tool is TAUOLA [32–34], which is a well-established program with a long tradition from LEP. Its usage is often accompanied by the process independent tool PHOTOS for the simulation of bremsstrahlung in τ leptonic decays. A recent interesting progress in TAUOLA has been the implementation of the hadronic currents using the $R\chi L$ theory, which supersedes previous theoretical parameterizations [35].

5 Summary

A wide spectrum of measurements and physics studies at flavor factories needs for precision calculations, including all the relevant RC, encoded into accurate and flexible MC generators. They are primarily required for simulations in the measurement of the hadron production cross section (via radiative return, energy scan and in $\gamma\gamma$ scattering), precise luminosity monitoring and control of the $\mu^+\mu^-\gamma$ normalization process. They are also involved in searches for NP and physics of the τ lepton.

Over the last decade or so, a significant progress occurred in theoretical predictions and related codes for the measurement of the hadronic cross section via radiative return and high-precision luminosity determination. The latter, in particular, took advantage of the computation of the full set of NNLO corrections to Bhabha process in QED. As a consequence, the accuracy of the presently available predictions for the luminosity cross section at flavor factories is at the sub-per mille level and robust.

More generally, theoretical results and generators used in simulations for luminosity, QED processes, NP searches and τ physics greatly benefited from LEP experience.

Further progress is expected in generators for the measurement of the hadron production cross section with energy scan and in $\gamma\gamma$ scattering processes. Presumably, new results will also come in the assessment of the theoretical uncertainty of the luminosity cross section as due to the hadronic contribution to the vacuum polarization close to ψ/Υ's resonance and one-loop corrections to hard bremsstrahlung in Bhabha scattering.

Acknowledgments

I am grateful to M. Passera and G. Venanzoni for their kind invitation to a very interesting workshop in a very beautiful place. I wish to thank C.M. Carloni Calame, O. Nicrosini and F. Piccinini for the collaboration which led to the results on the luminosity cross section calculations here summarized. Useful correspondence with H. Czyż on PHOKHARA is gratefully acknowledged. I also want to acknowledge the continuous effort by H. Czyż and G. Venanzoni in the organization of the activity of the WG on RC and MC for low energies which constitutes the forum for the results here presented.

References

[1] S. Actis *et al.* [Working Group on Radiative Corrections and Monte Carlo Generators for Low Energies Collaboration], Eur. Phys. J. C **66**, 585 (2010), arXiv:0912.0749 [hep-ph].
[2] A. B. Arbuzov, E. A. Kuraev, N. P. Merenkov and L. Trentadue, JHEP **9812**, 009 (1998), hep-ph/9804430.
[3] G. Rodrigo, H. Czyz, J. H. Kuhn and M. Szopa, Eur. Phys. J. C **24**, 71 (2002), hep-ph/0112184.
[4] H. Czyz, A. Grzelinska, J. H. Kuhn and G. Rodrigo, Eur. Phys. J. C **27**, 563 (2003), hep-ph/0212225.
[5] H. Czyz, A. Grzelinska, J. H. Kuhn and G. Rodrigo, Eur. Phys. J. C **33**, 333 (2004), hep-ph/0308312.
[6] H. Czyz, J. H. Kuhn, E. Nowak and G. Rodrigo, Eur. Phys. J. C **35**, 527 (2004), hep-ph/0403062.
[7] H. Czyz, A. Grzelinska, J. H. Kuhn and G. Rodrigo, Eur. Phys. J. C **47**, 617 (2006), hep-ph/0512180.
[8] H. Czyz, A. Grzelinska and J. H. Kuhn, Phys. Rev. D **81**, 094014 (2010), arXiv:1002.0279 [hep-ph].
[9] S. Binner, J. H. Kuhn and K. Melnikov, Phys. Lett. B **459**, 279 (1999), hep-ph/9902399.
[10] H. Czyz and J. H. Kuhn, Eur. Phys. J. C **18**, 497 (2001), hep-ph/0008262.
[11] A. B. Arbuzov, G. V. Fedotovich, E. A. Kuraev, N. P. Merenkov, V. D. Rushai and L. Trentadue, JHEP **9710**, 001 (1997), hep-ph/9702262.
[12] E. Barberio, B. van Eijk and Z. Was, Comput. Phys. Commun. **66**, 115 (1991).
[13] E. Barberio and Z. Was, Comput. Phys. Commun. **79**, 291 (1994).
[14] F. Campanario, H. Czyz, J. Gluza, M. Gunia, T. Riemann, G. Rodrigo and V. Yundin, JHEP **1402**, 114 (2014), arXiv:1312.3610 [hep-ph].
[15] A. B. Arbuzov, G. V. Fedotovich, F. V. Ignatov, E. A. Kuraev and A. L. Sibidanov, Eur. Phys. J. C **46**, 689 (2006), hep-ph/0504233.
[16] H. Czyz, M. Gunia and J. H. Kuhn, JHEP **1308** (2013) 110, arXiv:1306.1985 [hep-ph].
[17] S. Jadach, B. F. L. Ward and Z. Was, Phys. Rev. D **63**, 113009 (2001), hep-ph/0006359.
[18] S. Jadach, B. F. L. Ward and Z. Was, Phys. Rev. D **88** (2013) 11, 114022, arXiv:1307.4037 [hep-ph].
[19] K. Kolodziej, arXiv:1504.05915 [hep-ph].
[20] H. Czyz and S. Ivashyn, Comput. Phys. Commun. **182**, 1338 (2011), arXiv:1009.1881 [hep-ph].
[21] C. M. Carloni Calame, C. Lunardini, G. Montagna, O. Nicrosini and F. Piccinini, Nucl. Phys. B **584**, 459 (2000), hep-ph/0003268.
[22] G. Balossini, C. M. Carloni Calame, G. Montagna, O. Nicrosini and F. Piccinini, Nucl. Phys. B **758**, 227 (2006), hep-ph/0607181.
[23] G. Balossini, C. Bignamini, C. M. C. Calame, G. Montagna, O. Nicrosini and F. Piccinini, Phys. Lett. B **663**, 209 (2008), arXiv:0801.3360 [hep-ph].
[24] S. Jadach, W. Placzek and B. F. L. Ward, Phys. Lett. B **390**, 298 (1997), hep-ph/9608412.
[25] S. I. Eidelman, G. V. Fedotovich, E. A. Kuraev and A. L. Sibidanov, Eur. Phys. J. C **71** (2011) 1597, arXiv:1009.3390 [hep-ex].
[26] G. Montagna, O. Nicrosini and F. Piccinini, Phys. Lett. B **385** (1996) 348, hep-ph/9605252.
[27] C. Carloni Calame *et al.*, JHEP **1107**, 126 (2011), arXiv:1106.3178 [hep-ph].
[28] B. F. L. Ward, S. Jadach, M. Melles and S. A. Yost, Phys. Lett. B **450**, 262 (1999), hep-ph/9811245.
[29] L. Barze, G. Balossini, C. Bignamini, C. M. C. Calame, G. Montagna, O. Nicrosini and F. Piccinini, Eur. Phys. J. C **71**, 1680 (2011), arXiv:1007.4984 [hep-ph].
[30] J. Alwall *et al.*, JHEP **0709**, 028 (2007), arXiv:0706.2334 [hep-ph].
[31] S. Jadach and Z. Was, Comput. Phys. Commun. **36**, 191 (1985).
[32] S. Jadach, J. H. Kuhn and Z. Was, Comput. Phys. Commun. **64**, 275 (1990).
[33] M. Jezabek, Z. Was, S. Jadach and J. H. Kuhn, Comput. Phys. Commun. **70**, 69 (1992).
[34] S. Jadach, Z. Was, R. Decker and J. H. Kuhn, Comput. Phys. Commun. **76**, 361 (1993).
[35] I. M. Nugent, T. Przedzinski, P. Roig, O. Shekhovtsova and Z. Was, Phys. Rev. D **88** 9, 093012 (2013), arXiv:1310.1053 [hep-ph].

Review on $\gamma\gamma$ physics at KLOE-2

Dario Moricciani (for the KLOE-2 Collaboration)[1,a]

[1] *INFN - Roma "Tor Vergata", Via della Ricerca Scientifica 1, 00133 Rome (Italy)*

Abstract. The $\gamma\gamma$ physics at KLOE-2 experiment is made possible after the recent KLOE upgrade with new small angle tagging detectors along the DAFNE beam line. These taggers allow to detect the final leptons coming from process $e^+e^- \rightarrow e^+e^-\gamma^*\gamma^*$ giving the possibility to measure the transition form factor of the pseudoscalar mesons. These measurements provide constraints to the models used to calculate the hadronic light-by-light contribution to the muon anomaly. The KLOE-2 prospects for $\gamma\gamma$ physics of the new data-taking, started in November 2014 with the upgraded detector, are reviewed.

1 Introduction

The KLOE-2 data taking started in November 2014, with the goal to collect at least 5 fb^{-1} of integrated luminosity in 2 - 3 years. Until June 2015 DAΦNE provided 1 fb^{-1} of luminosity, that has been collected by KLOE-2 with an efficiency of about 80%. During this period the DAΦNE peak luminosity was about 2×10^{32} cm^{-2} s^{-1}, and the integrated luminosity collected in a day was about 10 pb^{-1}.

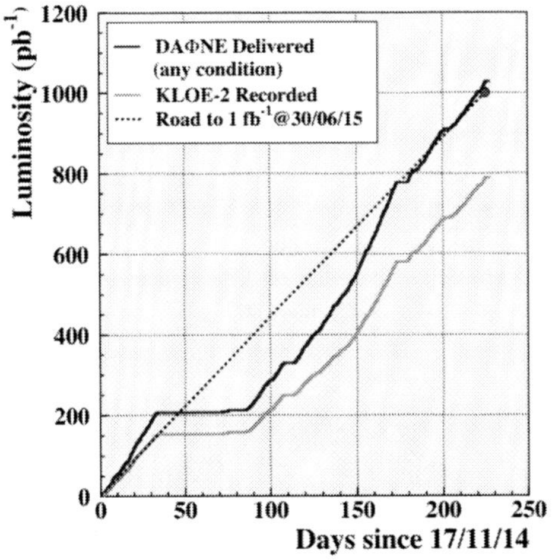

Figure 1. Integrated luminosity collected by KLOE-2 since the start of the new data-taking.

One of the main items of the KLOE-2 physics program [1] is the measurement of the Transition Form Factors (TFFs) of the pseudoscalar mesons both in the space-like and in time-like region of momentum transfer. The TFFs describe the coupling of mesons to photons and provide information about the nature of the mesons and their structure.

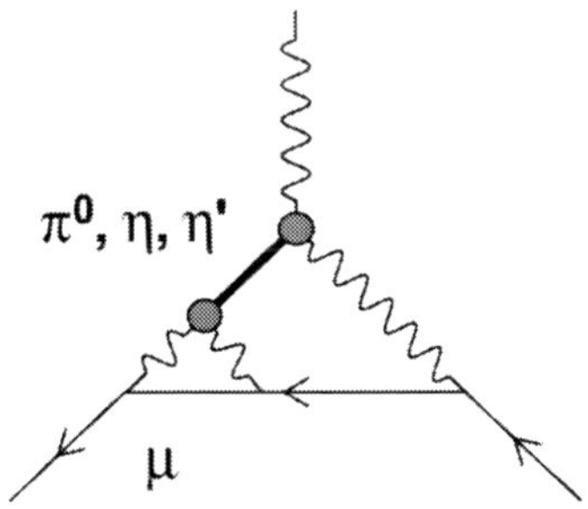

Figure 2. Dominant contribution to the hadronic LbL scattering for the (g-2)$_\mu$ theoretical calculation.

Recently the interest in the TFFs has been renewed since they are an essential ingredient in the calculation of the hadronic Light-by-Light (LbL) scattering contribution to the anomalous magnetic moment of the muon [2]. The leading contribution to the LbL scattering is the single pseudoscalar exchange where the TFFs enter at the vertices connecting the pseudoscalar to photons.

The calculation of this contribution is model dependent since the exchanged meson is off-shell, and the TFFs for off-shell meson are not measurable quantities.

Nevertheless any experimental information, both for space-like and time-like squared momentum transfer q^2, can help in constraining the models used in the calculations.

Another physics item that will be addressed by KLOE-2 is the $\gamma\gamma$ physics, *i.e.* processes like $e^+e^- \rightarrow e^+e^-\gamma^*\gamma^* \rightarrow e^+e^-X$, where X is a final state with even

[a] e-mail: Dario.Moricciani@roma2.infn.it

© The Authors, published by EDP Sciences. This is an open access article distributed under the terms of the Creative Commons Attribution License 4.0 (http://creativecommons.org/licenses/by/4.0/).

DOI: 10.1051/epjconf/201611801023

charge conjugation. The expected number of events as a function of the $\gamma\gamma$ energy $W_{\gamma\gamma}$ can be parameterized by the following expression:

$$\frac{dN}{dW_{\gamma\gamma}} = L_{int}\frac{dF}{dW_{\gamma\gamma}}\sigma_{\gamma\gamma\to X} \quad (1)$$

where L_{int} is the integrated luminosity, $\sigma_{\gamma\gamma\to X}$ the $\gamma\gamma$ cross-section, and $\frac{dF}{dW_{\gamma\gamma}}$ is the luminosity function. Since DAΦNE is operated at $\sqrt{s} \simeq M_\phi$, the accessible final state are either single pseudoscalar, $X = \eta$, π^0, or the double pion production, $X = \pi\pi$.
The cross-section for single pseudoscalar is:

$$\sigma_{\gamma\gamma\to X}(q_1^2, q_2^2) = \frac{8\pi^2}{M_X}\Gamma(X \to \gamma\gamma)\left|F(q_1^2, q_2^2)\right|^2 \quad (2)$$

with $(q_1^2 + q_2^2) = M_X^2$.

Then the radiative width $\Gamma(X \to \gamma\gamma)$ of the pseudoscalar meson, and the TFF $F(q_1^2, q_2^2)$ for space-like q^2 can be measured. Concerning the double pion final state, it is interesting to study the production of the lowest mass scalar meson $f_0(500)$, but it is also important for the new dispersive approach proposed for the hadronic LbL scattering [3].

2 Detector upgrade

As a first step of the detector upgrade, a tagger system for scattered electrons and positrons in $\gamma\gamma$ processes has been installed already in 2010. It consists of two different devices: the Low Energy Tagger (LET) and the High Energy Tagger (HET), referring to the energy of the detected electrons or positrons.

2.1 The Low Energy Tagger

The LET [4] has been designed to detect $e^\pm$ with energy between 150 and 350 MeV escaping from the beam-pipe, and it is placed at about 1 m from the IP. Since in this region there is no correlation between the energy and the scattering angle of the particles, a calorimetric device has been chosen. Then the LET consists of two calorimeters, each made of 4 × 5 LYSO crystals of 1.5 × 1.5 × 20 cm^3 dimensions. The crystals are readout by SiPM. The two calorimeters are placed symmetrically with respect to the IP, as shown in Fig. 3.

2.2 The High Energy Tagger

The HET [5] is designed to detect scattered $e^\pm$ of $E >$ 400 MeV. These particles escape the beam-pipe after the first bending dipole of DAΦNE, that can thus be used as spectrometer. The trajectories of the scattered electrons are strongly correlated with their energy. Then the HET is made of two scintillator hodoscopes readout by PMT, symmetrically placed 11 m far from the IP (Fig. 4).

The HET is acquired asynchronously with respect to the main KLOE detector, and for each KLOE trigger

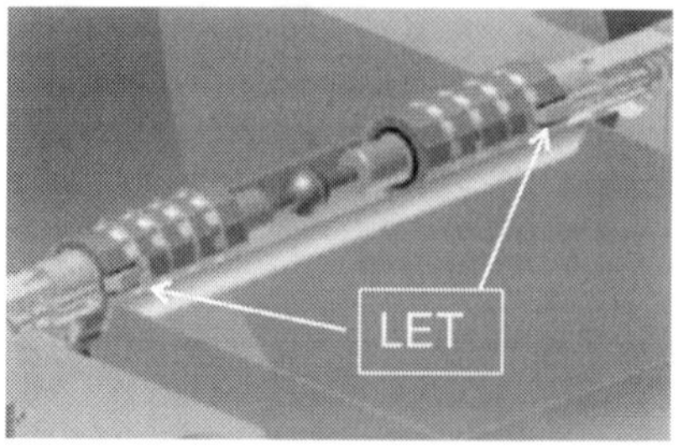

Figure 3. Sketch of the LET calorimeter positioning.

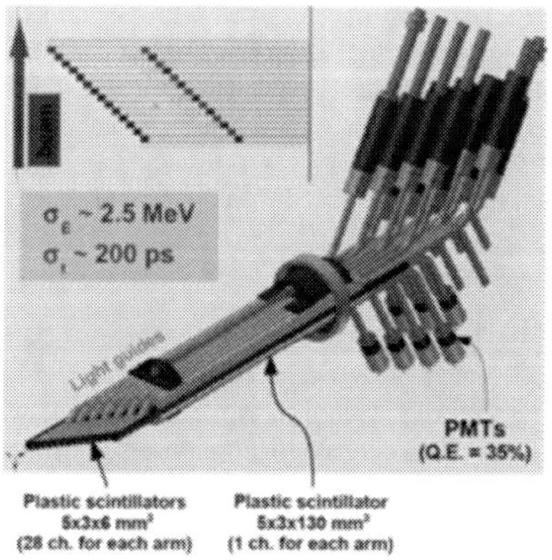

Figure 4. Sketch of the HET hodoscope.

the HET information concerning three DAΦNE beam revolutions is stored. The synchronisation is performed by using a machine signal. In Fig. 5 the time difference between the two HET stations is shown: the accelerator time structure of about 2.7 ns period is clearly visible. The superimposed histogram is the same distribution resulting from a run with separated beams in the IP, and shows that the level of background is less than 10%.

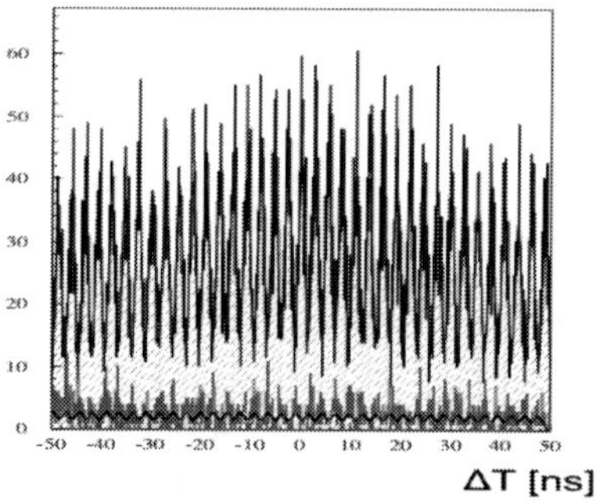

Figure 5. Time difference between the two HET stations. Black: colliding beams; red: no collisions.

3 Conclusion

A simulation of the KLOE-2 experiment with 5 fb^{-1} of data taking was performed. Numerical results indicate a feasibility of ~ 1% statistical error in the measurement of $\Gamma_{\pi^0\to\gamma\gamma}$ [6]. Such a precision is better than the current experimental world average and the theoretical accuracy. The π^0 electromagnetic transition form factor $F(Q^2)$ $(Q^2 = -q^2)$ in the region $0.01 < Q^2 < 0.1\,\mathrm{GeV}^2$ can be measured with a statistical error of < 6% in each bin. This low Q^2 measurement can test the consistency of the models which have been fitted so far to the data from CELLO, CLEO and BaBar at higher Q^2 and will serve as an important test of the strong interaction dynamics at low energies. The proposed measurements with the KLOE-2 experiment can also have an impact on the value and precision of the contribution of a neutral pion exchange to the hadronic light-by-light scattering in the muon $g - 2$.

References

[1] G. Amelino-Camelia *et al.* , Eur.Phys.J. C **68**, 619 (2010).
[2] F. Jegerlehner and A. Nyffeler, Phys.Rept. **477**, 1 (2009).
[3] G. Colangelo, M. Hoferichter, M. Procura and P. Stoffer JHEP 09074, (2015).
[4] D. Babusci *et al.* Nucl. Instrum. Meth. A **617**, 81 (2010).
[5] F. Archilli *et al.* Nucl. Instrum. Meth. A **617**, 266 (2010).
[6] D. Babusci *et al.*, Eur.Phys.J. C **72**, 1917 (2012).

EPJ Web of Conferences **118**, 01024 (2016) DOI: 10.1051/epjconf/201611801024
FCCP2015

On the precision of a data-driven estimate of the pseudoscalar-pole contribution to hadronic light-by-light scattering in the muon $g-2$

Andreas Nyffeler[1,a]

[1] *Institut für Kernphysik, Johann-Joachim-Becher-Weg 45, Johannes Gutenberg-Universität, D-55128 Mainz, Germany*

Abstract. The evaluation of the numerically dominant pseudoscalar-pole contribution to hadronic light-by-light scattering in the muon $g-2$ involves the pseudoscalar-photon transition form factor $\mathcal{F}_{P\gamma^*\gamma^*}(-Q_1^2,-Q_2^2)$ with $P=\pi^0,\eta,\eta'$ and, in general, two off-shell photons with spacelike momenta $Q_{1,2}^2$. We determine which regions of photon momenta give the main contribution for hadronic light-by-light scattering. Furthermore, we analyze how the precision of future measurements of the single- and double-virtual form factor impacts the precision of a data-driven estimate of this contribution to hadronic light-by-light scattering.

1 Introduction

The anomalous magnetic moment of the muon a_μ serves as an important test of the Standard Model (SM) [1]. Since several years, there is an intriguing discrepancy of $3-4\sigma$ between the experimental value [2] and the theoretical SM prediction [1, 3]. While this could be a sign of New Physics, the hadronic contributions from vacuum polarization (HVP) and light-by-light scattering (HLbL) have large uncertainties, which make it difficult to interpret the deviation as a clear sign of physics beyond the SM. The hadronic uncertainties need to be reduced and better controlled, also to fully profit from new planned $g-2$ experiments [4]. While the HVP contribution [5] can be improved systematically with measurements of $\sigma(e^+e^-\to$ hadrons), the often used estimates for HLbL

$$a_\mu^{\mathrm{HLbL}} = (105\pm 26)\times 10^{-11}, \quad [6] \qquad (1)$$

$$a_\mu^{\mathrm{HLbL}} = (116\pm 40)\times 10^{-11}, \quad [1,7] \qquad (2)$$

are both based on model calculations [8–12],[1] which suffer from uncontrollable uncertainties, see also [15].

In this situation, a dispersive approach to HLbL was proposed recently in Refs. [16, 17], which tries to relate the presumably numerically dominant contributions from the pseudoscalar-poles and the pion-loop to, in principle, measurable form factors and cross-sections, $\gamma^*\gamma^*\to\pi^0,\eta,\eta'$ and $\gamma^*\gamma^*\to\pi\pi$, with on-shell intermediate pseudoscalar states.[2] The hope is that this data-driven estimate for HLbL will allow a 10% precision for these contributions and that the remaining, hopefully smaller contributions, e.g. from axial-vectors (3π-intermediate state) and other heavier states, can be obtained within models with about 30% uncertainty to reach an overall, reliable precision goal of 20% ($\delta a_\mu^{\mathrm{HLbL}}\approx 20\times 10^{-11}$).

Most model evaluations of $a_\mu^{\mathrm{HLbL};\pi^0}$ (pion-pole defined in different ways [12, 16, 17] or pion-exchange with off-shell-pion form factors [1, 7]) and $a_\mu^{\mathrm{HLbL;P}}$, with $P=\pi^0,\eta,\eta'$, agree at the level of 15%, but the full range of estimates (central values) is much larger:

$$a_{\mu;\mathrm{models}}^{\mathrm{HLbL};\pi^0} = (65\pm 15)\times 10^{-11} \quad (\pm 23\%), \qquad (3)$$

$$a_{\mu;\mathrm{models}}^{\mathrm{HLbL;P}} = (87\pm 27)\times 10^{-11} \quad (\pm 31\%). \qquad (4)$$

In this paper we study, within the dispersive approach, which are the most important momentum regions for the pseudoscalar-pole contribution $a_\mu^{\mathrm{HLbL;P}}$. We also analyze what is the impact of the precision of current and future measurements of the single- and double-virtual pseudoscalar transition form factor $\mathcal{F}_{P\gamma^*\gamma^*}(-Q_1^2,-Q_2^2)$ (TFF) [19] on the uncertainty of a data-driven estimate of this contribution to HLbL. More details can be found in Ref. [20].

2 Pseudoscalar-pole contribution

The dominant contribution to HLbL arises, according to most model calculations, from the one-particle intermediate states of the light pseudoscalars π^0,η,η' shown in the Feynman diagrams in Fig. 1. We will evaluate only the pseudoscalar-pole contribution of these two-loop diagrams. In order to simplify the notation, we mainly discuss the neutral pion-pole contribution in this section. The generalization to η and η' is straightforward.

The Feynman diagrams for the pion-pole contribution involve the pion TFF $\mathcal{F}_{\pi^0\gamma^*\gamma^*}(q_1^2,q_2^2)$ which is defined by

[a] e-mail: nyffeler@kph.uni-mainz.de

[1] There are attempts ongoing to calculate the HLbL contribution from first principles in Lattice QCD. A first, still incomplete, result was obtained recently [13]. See also the approach proposed in [14].

[2] There have been objections raised at this meeting [18] about the implementation of the dispersive approach for the pion-pole contribution. There should be no form factor at the external vertex [12].

© The Authors, published by EDP Sciences. This is an open access article distributed under the terms of the Creative Commons Attribution License 4.0 (http://creativecommons.org/licenses/by/4.0/).

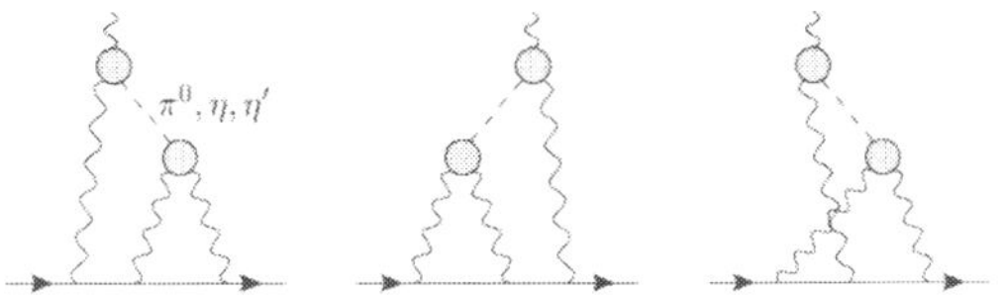

Figure 1. The pseudoscalar-pole contribution to HLbL scattering. The shaded blobs represent the transition form factor $\mathcal{F}_{\mathrm{P}\gamma^*\gamma^*}(q_1^2, q_2^2)$ where $\mathrm{P} = \pi^0, \eta, \eta'$.

the QCD vertex function:

$$i \int d^4x \, e^{iq_1 \cdot x} \langle 0 | T\{j_\mu(x) j_\nu(0)\} | \pi^0(q_1 + q_2) \rangle = \varepsilon_{\mu\nu\alpha\beta} \, q_1^\alpha \, q_2^\beta \, \mathcal{F}_{\pi^0\gamma^*\gamma^*}(q_1^2, q_2^2) \, . \quad (5)$$

Here $j_\mu(x) = (\overline{\psi} \hat{Q} \gamma_\mu \psi)(x)$ is the light quark part of the electromagnetic current ($\overline{\psi} \equiv (\overline{u}, \overline{d}, \overline{s})$ and $\hat{Q} = \mathrm{diag}(2, -1, -1)/3$ is the charge matrix). The form factor describes the interaction of an on-shell neutral pion with two off-shell photons with four-momenta q_1 and q_2. It is Bose symmetric $\mathcal{F}_{\pi^0\gamma^*\gamma^*}(q_1^2, q_2^2) = \mathcal{F}_{\pi^0\gamma^*\gamma^*}(q_2^2, q_1^2)$ and for real photons it is related to the decay width $\mathcal{F}^2_{\pi^0\gamma^*\gamma^*}(0,0) = 4\Gamma(\pi^0 \to \gamma\gamma)/(\pi\alpha^2 m_\pi^3)$. Often the normalization with the chiral anomaly $\mathcal{F}_{\pi^0\gamma^*\gamma^*}(0,0) = -N_c/(12\pi^2 F_\pi)$ is used.

The projection of the Feynman diagrams in Fig. 1 on the muon $g-2$ leads to a two-loop integral involving the propagators of the muon, the photons, the pion, the product of two transition form factors and some kinematical functions [10]. After a Wick rotation to Euclidean momenta and averaging over the direction of the muon momentum, one can perform for arbitrary form factors all angular integrations, except one over the angle θ between the four-momenta Q_1 and Q_2 which also appears through $Q_1 \cdot Q_2$ in the form factors [1]. Writing

$$a_\mu^{\mathrm{HLbL};\pi^0} = \left(\frac{\alpha}{\pi}\right)^3 \left[a_\mu^{\mathrm{HLbL};\pi^0(1)} + a_\mu^{\mathrm{HLbL};\pi^0(2)} \right], \quad (6)$$

where the first contribution arises from the first two Feynman diagrams in Fig. 1 and the second from the last graph, one obtains the following three-dimensional integral representation for the pion-pole contribution [1]

$$a_\mu^{\mathrm{HLbL};\pi^0(1)} = \int_0^\infty dQ_1 \int_0^\infty dQ_2 \int_{-1}^1 d\tau \; w_1(Q_1, Q_2, \tau) \times \mathcal{F}_{\pi^0\gamma^*\gamma^*}(-Q_1^2, -(Q_1+Q_2)^2) \mathcal{F}_{\pi^0\gamma^*\gamma^*}(-Q_2^2, 0), \quad (7)$$

$$a_\mu^{\mathrm{HLbL};\pi^0(2)} = \int_0^\infty dQ_1 \int_0^\infty dQ_2 \int_{-1}^1 d\tau \; w_2(Q_1, Q_2, \tau) \times \mathcal{F}_{\pi^0\gamma^*\gamma^*}(-Q_1^2, -Q_2^2) \mathcal{F}_{\pi^0\gamma^*\gamma^*}(-(Q_1+Q_2)^2, 0). \quad (8)$$

The integrations in Eqs. (7) and (8) run over the lengths of the two Euclidean four-momenta Q_1 and Q_2 and the angle θ between them $Q_1 \cdot Q_2 = Q_1 Q_2 \cos\theta$ and where we introduced $\tau = \cos\theta$.[3] From the definition of the form factor in Eq. (5) it follows that the weight functions $w_{1,2}(Q_1, Q_2, \tau)$ are dimensionless. Furthermore $w_2(Q_1, Q_2, \tau)$ is symmetric under $Q_1 \leftrightarrow Q_2$ [1]. Finally, $w_{1,2}(Q_1, Q_2, \tau) \to 0$ for $Q_{1,2} \to 0$ and for $\tau \to \pm 1$. Expressions for the precise behavior of $w_{1,2}(Q_1, Q_2, \tau)$ for $Q_{1,2} \to 0$ and $\tau \to \pm 1$, as well as for $Q_{1,2} \to \infty$, can be found in Ref. [20].

The three-dimensional integral representation in Eqs. (7) and (8) separates the generic kinematics in the pion-pole contribution to HLbL, described by the model-independent weight functions $w_{1,2}(Q_1, Q_2, \tau)$, from the dependence on the single- and double-virtual TFF $\mathcal{F}_{\pi^0\gamma^*\gamma^*}(-Q^2, 0)$ and $\mathcal{F}_{\pi^0\gamma^*\gamma^*}(-Q_1^2, -Q_2^2)$ in the spacelike region, which can in principle be measured, obtained from a dispersion relation (DR) [21, 22] or, as has been done so far for HLbL, modelled.

The only explicit dependence on the pseudoscalar appears in the weight functions $w_{1,2}(Q_1, Q_2, \tau)$ via the mass in the pseudoscalar propagators, i.e. a factor $1/(Q_2^2 + m_\mathrm{P}^2)$ in w_1 and a factor $1/((Q_1+Q_2)^2 + m_\mathrm{P}^2) = 1/(Q_1^2 + 2Q_1Q_2\tau + Q_2^2 + m_\mathrm{P}^2)$ in w_2. Of course, also the form factors will depend on the type of the pseudoscalar.

3 The weight functions $w_{1,2}(Q_1, Q_2, \tau)$

In Fig. 2 we have plotted the weight functions $w_1(Q_1, Q_2, \tau)$ and $w_2(Q_1, Q_2, \tau)$ for the light pseudoscalars π^0, η, η' as function of Q_1 and Q_2 for $\theta = 90°$ ($\tau = 0$). Three-dimensional plots for a selection of other values of θ and one-dimensional plots as function of $\tau = \cos\theta$ for some selected values of Q_1 and Q_2 can be found in Ref. [20]. Note that although the weight functions rise very quickly to the maxima in the plots in Fig. 2, the slopes along the two axis and along the diagonal $Q_1 = Q_2$ actually vanish for both functions [20]. We stress that these weight functions are completely independent of any models for the form factors.

We can immediately see from the weight functions for the pion that the low-momentum region, $Q_{1,2} \leq 0.5$ GeV, is the most important in the corresponding integrals (7) and (8) for $a_\mu^{\mathrm{HLbL};\pi^0}$. In $w_1(Q_1, Q_2, \tau)$ there is a peak around $Q_1 \sim 0.15 - 0.19$ GeV, $Q_2 \sim 0.09 - 0.16$ GeV. The values of the maxima of the weight functions for all light pseudoscalars and the locations of the maxima in the (Q_1, Q_2)-plane for a selection of θ-values have been collected in Ref. [20]. For $w_1(Q_1, Q_2, \tau)$ and $\theta \leq 150°$, a ridge develops along the Q_1 direction for $Q_2 \sim 0.18 - 0.26$ GeV (maximum along the line $Q_1 = 2$ GeV). For larger values of θ, the function looks more and more symmetric in Q_1 and Q_2. This ridge leads for a constant form factor to an ultraviolet divergence $(\alpha/\pi)^3 C \ln^2(\Lambda/m_\mu)$ for some momentum cutoff Λ with $C = 3(N_c/(12\pi))^2 (m_\mu/F_\pi)^2 = 0.0248$, see Refs. [10, 11]. Of course, realistic form factors fall off for large momenta and $a_\mu^{\mathrm{HLbL};\pi^0(1)}$ will be convergent.

The weight function $w_2(Q_1, Q_2, \tau)$ for the pion is about an order of magnitude smaller than $w_1(Q_1, Q_2, \tau)$. There is no ridge, since the function is symmetric under $Q_1 \leftrightarrow Q_2$. For τ near -1, the peak is around $Q_1 = Q_2 \sim 0.14$ GeV and broader than the one shown for $\theta = 90°$. The location

[3] We have written $Q_i \equiv |(Q_i)_\mu|, i = 1, 2$ for the length of the four vectors. Following [16] we changed the notation used in [1] and write $\tau = \cos\theta$ in order to avoid confusion with the Mandelstam variable t in the context of the dispersive approach.

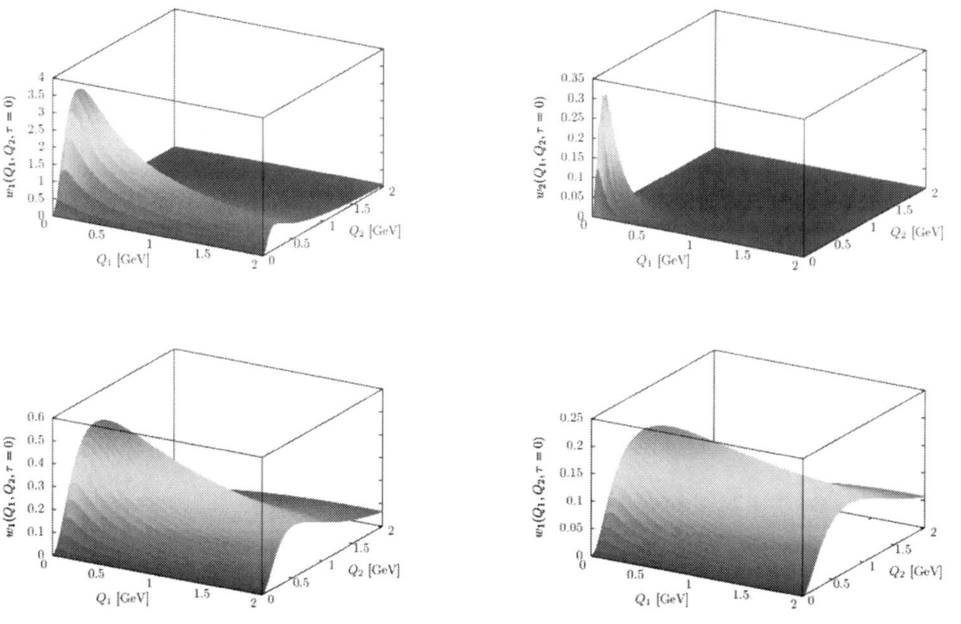

Figure 2. Model-independent weight functions for the π^0: $w_1(Q_1, Q_2, \tau)$ (top left) and $w_2(Q_1, Q_2, \tau)$ (top right) as function of the Euclidean momenta Q_1 and Q_2 for $\theta = 90°$ ($\tau = 0$). Weight functions $w_1(Q_1, Q_2, \tau)$ for η (bottom left) and η' (bottom right) for $\theta = 90°$. The weight functions $w_2(Q_1, Q_2, \tau)$ for η and η' have a similar shape as for the pion, but the peaks are broader.

of the peak moves to lower values $Q_1 = Q_2 \sim 0.04$ GeV when τ grows towards $+1$.

The dependence of the weight functions on the pseudoscalar mass through the propagators shifts the relevant momentum regions (peaks, ridges) in the weight functions, and thus also in HLbL, to higher momenta for η compared to π^0 and even higher for η', see Fig. 2. It also leads to a suppression in the absolute size of the weight functions due to the larger masses in the propagators. This pattern is also visible in the values for the contributions to a_μ. For the bulk of the weight functions (maxima, ridges) we observe the following approximate relations (not necessarily at the same values of the momenta and angles)

$$w_1|_\eta \approx \frac{1}{6}\, w_1|_{\pi^0}, \qquad w_1|_{\eta'} \approx \frac{1}{2.5}\, w_1|_\eta. \tag{9}$$

Of course, the ratio of the weight functions is given by the ratio of the propagators and is maximal at zero momenta and at that point equal to the ratio of the squares of the masses, but at zero momenta the weight functions themselves vanish. The combined effect is well described by the relations in Eq. (9). Furthermore, for both η and η', the weight function w_2 is about a factor 20 smaller than w_1.

The peaks for the weight function $w_1(Q_1, Q_2, \tau)$ for η and η' are less steep, compared to π^0, and the ridge is quite broad in the Q_2-direction, so that the weight function is still sizeable, compared to the maximum, for $Q_2 = 2$ GeV. Furthermore, the ridge falls off only slowly in the Q_1-direction. In particular for the η', the ridge for $\theta \leq 75°$ is almost as big as the maximum out to values of $Q_1 = 2$ GeV. For $w_2(Q_1, Q_2, \tau)$ the peaks are broader and larger momenta contribute, compared to the pion.

For the η-meson, the peak in the weight function $w_1(Q_1, Q_2, \tau)$ is around $Q_1 \sim 0.32 - 0.37$ GeV, $Q_2 \sim 0.22 - 0.33$ GeV. The peak for the weight function $w_2(Q_1, Q_2, \tau)$ is around $Q_1 = Q_2 \sim 0.14$ GeV for τ near -1 as for the pion. The location of the peak moves down to $Q_1 = Q_2 \sim 0.06$ GeV when τ is near $+1$. For the η', the peak in $w_1(Q_1, Q_2, \tau)$ now occurs for even higher momenta, $Q_1 \sim 0.41 - 0.51$ GeV, $Q_2 \sim 0.31 - 0.43$ GeV. The locations of the peaks of $w_2(Q_1, Q_2, \tau)$ in the (Q_1, Q_2)-plane follow a similar pattern as for the η meson.

4 Relevant momentum regions in $a_\mu^{\mathrm{HLbL}; P}$

In order to study the impact of different momentum regions on the pseudoscalar-pole contribution, we need, at least for the integral with the weight function $w_1(Q_1, Q_2, \tau)$ in Eq. (7), some knowledge on the form factor $\mathcal{F}_{P\gamma^*\gamma^*}(-Q_1^2, -Q_2^2)$, since the integral diverges for a constant form factor.[4] For illustration we take for the pion two simple models to perform the integrals: Lowest Meson Dominance with an additional vector multiplet, LMD+V model, based on the Minimal Hadronic Approximation to large-N_c QCD matched to certain QCD short-distance constraints from the operator product expansion

[4]The integral with the weight function $w_2(Q_1, Q_2, \tau)$ in Eq. (8) is finite and small for a constant form factor.

(OPE), see Refs. [10, 23] and references therein, and the well-known Vector Meson Dominance (VMD) model. Of course, in the end, the models have to be replaced as much as possible by experimental data on the double-virtual TFF or one can use a DR for the form factor itself [21].

The main difference of the models is a different behavior of the double-virtual form factor for large and equal momenta. The LMD+V model reproduces by construction the OPE, whereas the VMD TFF falls off too fast:

$$\mathcal{F}^{\mathrm{LMD+V}}_{\pi^0\gamma^*\gamma^*}(-Q^2,-Q^2) \sim \mathcal{F}^{\mathrm{OPE}}_{\pi^0\gamma^*\gamma^*}(-Q^2,-Q^2) \sim \frac{1}{Q^2}, \quad (10)$$

$$\mathcal{F}^{\mathrm{VMD}}_{\pi^0\gamma^*\gamma^*}(-Q^2,-Q^2) \sim \frac{1}{Q^4}, \quad \text{for large } Q^2. \quad (11)$$

Nevertheless for not too large momenta, $Q_1 = Q_2 = Q = 0.5$ [0.75] GeV, the form factors $\mathcal{F}_{\pi^0\gamma^*\gamma^*}(-Q^2,-Q^2)$ in the two models differ by only 3% [10%], see [20]. Furthermore, both models give an equally good description of the single-virtual TFF $\mathcal{F}_{\pi^0\gamma^*\gamma^*}(-Q^2,0)$ [23, 24].

The LMD+V model was developed in Ref. [23] in the chiral limit and assuming octet symmetry. This is certainly not a good approximation for the more massive η and η' mesons. For the η and η' meson we therefore simply take the usual VMD model as already done in Refs. [7, 10].

The two models yield the following results for the pole-contribution of the light pseudoscalars to HLbL (we only list here the central values)

$$a_{\mu;\mathrm{LMD+V}}^{\mathrm{HLbL};\pi^0} = 62.9 \times 10^{-11}, \quad (12)$$

$$a_{\mu;\mathrm{VMD}}^{\mathrm{HLbL};\pi^0} = 57.0 \times 10^{-11}, \quad (13)$$

$$a_{\mu;\mathrm{VMD}}^{\mathrm{HLbL};\eta} = 14.5 \times 10^{-11}, \quad (14)$$

$$a_{\mu;\mathrm{VMD}}^{\mathrm{HLbL};\eta'} = 12.5 \times 10^{-11}. \quad (15)$$

The results (12) and (13) for the pion-pole contribution in the two models are in the ballpark of many other estimates, but they also differ by 9.4%, relative to the LMD+V result, due to the different high-energy behavior for the double-virtual TFF in Eqs. (10) and (11) for $Q_{1,2} \geq 1$ GeV. In fact, the pattern of the contributions to $a_\mu^{\mathrm{HLbL};\pi^0}$ is to a large extent determined by the model-independent weight functions $w_{1,2}(Q_1,Q_2,\tau)$, which are concentrated below about 0.5 GeV, up to that ridge in w_1 along the Q_1 direction. As long as realistic form factor models for the double-virtual case fall off at large momenta and do not differ too much at low momenta, we expect similar results for the pion-pole contribution at the level of 15% which is in fact what is seen in the literature [1, 10, 15]. Nevertheless, due to the ridge-like structure in the weight function w_1, the high-energy behavior of the form factors is relevant at the precision of 10% one is aiming for.

For the η and η', the results in Eqs. (14) and (15) are as expected from the discussion of the relative size of the weight functions in Eq. (9). The result for η is about a factor 4 smaller than for the pion with VMD. The result for η' is only slighly smaller than for η. Note that the normalization of the TFF from $\Gamma(P \to \gamma\gamma)$ and the momentum dependence due to different vector meson masses for η and η' also play a role for the results in Eqs. (14) and (15).

For a more detailed analysis, we integrate in Eqs. (7) and (8) over individual momentum bins and all angles θ

$$\int_{Q_{1,\min}}^{Q_{1,\max}} dQ_1 \int_{Q_{2,\min}}^{Q_{2,\max}} dQ_2 \int_{-1}^{1} d\tau \quad (16)$$

and display the results, relative to the totals in Eqs. (12)-(15), in Fig. 3.

Since the absolute size of the weight function $w_1(Q_1,Q_2,\tau)$ is much larger than $w_2(Q_1,Q_2,\tau)$, the contribution from the integral $a_\mu^{\mathrm{HLbL};\pi^0(1)}$ in Eq. (7) dominates over $a_\mu^{\mathrm{HLbL};\pi^0(2)}$ in Eq. (8). Therefore the asymmetry seen in the (Q_1,Q_2)-plane in Fig. 3, with larger contributions below the diagonal, reflects the ridge-like structure of $w_1(Q_1,Q_2,\tau)$ in Fig. 2.

For the pion, the largest contribution comes from the lowest bin $Q_{1,2} \leq 0.25$ GeV since a large part of the peaks in the weight functions (for different angles θ) is contained in that bin. More than half of the contribution comes from the four bins with $Q_{1,2} \leq 0.5$ GeV. In contrast, for the η and η', it is not the bin $Q_{1,2} \leq 0.25$ GeV which yields the largest contribution, since the maxima of the weight functions are shifted to higher momenta, around $0.3-0.5$ GeV. Furthermore, more bins up to $Q_2 = 2$ GeV now contribute at least 1% to the total. This is different from the pattern seen for π^0. The plots of the weight functions for η and η' in Fig. 2 show that now the region $1.5-2.5$ GeV also is important for the evaluation of the η- and η'-pole contributions. The VMD model is, however, known to have a too fast fall-off at large momenta, compared to the OPE. Therefore the size of the contributions $a_\mu^{\mathrm{HLbL};\eta}$ and $a_\mu^{\mathrm{HLbL};\eta'}$ in Eqs. (14) and (15) might be underestimated by the VMD model, which could also affect the relative importance of the higher momentum region in Fig. 3.

Integrating both Q_1 and Q_2 from zero up to some upper momentum cutoff Λ and integrating over all angles θ, one obtains the results shown in Table 1. This amounts to summing up the individual bins shown in Fig. 3.

As one can see in Table 1, for the pion more than half of the final result stems from the region below $\Lambda = 0.5$ GeV (59% for LMD+V, 64% for VMD) and the region below $\Lambda = 1$ GeV gives the bulk of the total result (86% for LMD+V, 92% for VMD). The small difference between the form factor models for small momenta $Q_{1,2} \leq 0.5$ GeV is reflected in the small absolute difference for $a_\mu^{\mathrm{HLbL};\pi^0}$ in the two models for $\Lambda \leq 0.75$ GeV. For instance, for $\Lambda = 0.75$ GeV, the difference is only 0.8×10^{-11}, i.e. 1.6%. The faster fall-off of the VMD model at larger momenta beyond 1 GeV, compared to the LMD+V model, leads to a smaller contribution from that region to the total. Therefore we can see in Fig. 3 that the main contributions in the VMD model, relative to the total, are concentrated at lower momenta, compared to the LMD+V model, in particular below 0.75 GeV.

On the other hand, Table 1 shows that for η and η', the region below $\Lambda = 0.25$ GeV only gives a small contribution to the total (12% for η, 8% for η'). Up to $\Lambda = 0.5$ GeV, we get about half (one third) for η (η') and the bulk of the results comes from the region below $\Lambda = 1.5$ GeV, 96% for the η and 93% for the η'-meson.

 DOI: 10.1051/epjconf/201611801024

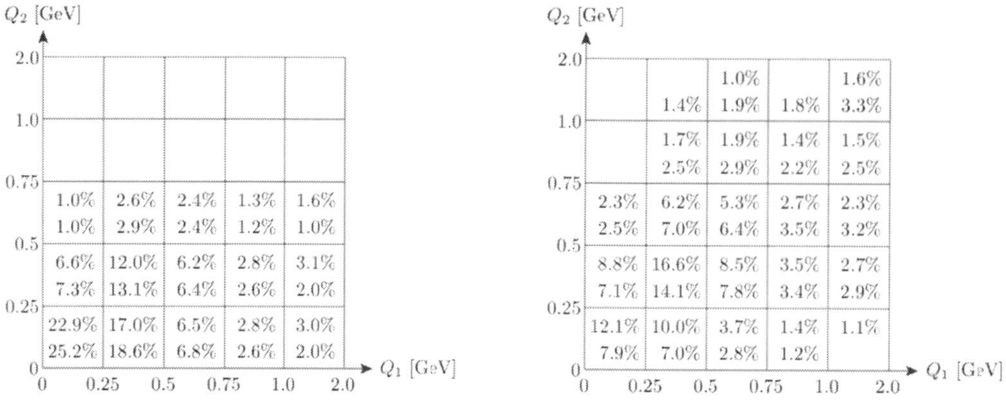

Figure 3. Left panel: Relative contributions to the total $a_\mu^{\mathrm{HLbL};\pi^0}$ from individual bins in the (Q_1, Q_2)-plane, integrated over all angles according to Eq. (16). Note the larger size of the bins with $Q_{1,2} \geq 1$ GeV. Top line in each bin: LMD+V model, bottom line: VMD model. Contributions smaller than 1% have not been displayed. For the LMD+V model there are further contributions bigger than 1% along the Q_1-axis. For 2 GeV $\leq Q_1 \leq$ 20 GeV: 1.1% in the bin $0 \leq Q_2 \leq 0.25$ GeV and 1.2% for 0.25 GeV $\leq Q_2 \leq$ 0.5 GeV. Right panel: Relative contributions to the total of $a_\mu^{\mathrm{HLbL};\eta}$ and $a_\mu^{\mathrm{HLbL};\eta'}$ with the VMD model from individual bins in the (Q_1, Q_2)-plane, integrated over all angles. Top line in each bin: η-meson, bottom line: η'-meson.

Table 1. Pseudoscalar-pole contribution $a_\mu^{\mathrm{HLbL;P}} \times 10^{11}$, P $= \pi^0, \eta, \eta'$ for different form factor models obtained with a momentum cutoff Λ. In brackets, relative contribution of the total obtained with $\Lambda = 20$ GeV.

Λ [GeV]	π^0 [LMD+V]	π^0 [VMD]	η [VMD]	η' [VMD]
0.25	14.4 (22.9%)	14.4 (25.2%)	1.8 (12.1%)	1.0 (7.9%)
0.5	36.8 (58.5%)	36.6 (64.2%)	6.9 (47.5%)	4.5 (36.1%)
0.75	48.5 (77.1%)	47.7 (83.8%)	10.7 (73.4%)	7.8 (62.5%)
1.0	54.1 (86.0%)	52.6 (92.3%)	12.6 (86.6%)	9.9 (79.1%)
1.5	58.8 (93.4%)	55.8 (97.8%)	14.0 (96.1%)	11.7 (93.1%)
2.0	60.5 (96.2%)	56.5 (99.2%)	14.3 (98.6%)	12.2 (97.4%)
5.0	62.5 (99.4%)	56.9 (99.9%)	14.5 (100%)	12.5 (99.9%)
20.0	62.9 (100%)	57.0 (100%)	14.5 (100%)	12.5 (100%)

5 Impact of transition form factor uncertainties on $a_\mu^{\mathrm{HLbL;P}}$

For the calculation of the pseudoscalar-pole contribution $a_\mu^{\mathrm{HLbL;P}}$ with P $= \pi^0, \eta, \eta'$ in Eqs. (7) and (8) the single-virtual form factor $\mathcal{F}_{\mathrm{P}\gamma^*\gamma^*}(-Q^2, 0)$ and the double-virtual form factor $\mathcal{F}_{\mathrm{P}\gamma^*\gamma^*}(-Q_1^2, -Q_2^2)$, both in the spacelike region, enter. We are interested here in the impact of uncertainties of experimental measurements of these form factors on the precision of $a_\mu^{\mathrm{HLbL;P}}$. See Ref. [19] for a brief overview of the various experimental processes where information on the transition form factors can be obtained. In the following, we will only quote the most relevant and most precise experimental references. Ref. [20] contains a detailed analysis of the current experimental situation.

For the single-virtual form factor $\mathcal{F}_{\mathrm{P}\gamma^*\gamma^*}(-Q^2, 0)$ the following experimental information is available. The normalization of the form factor can be obtained from the decay width $\Gamma(\mathrm{P} \to \gamma\gamma)$ [25, 26]. Another important experimental information is the slope of the form factor at the origin. In the spacelike region, the slope and the form factor itself have been measured in the process $e^+e^- \to e^+e^-\gamma^*\gamma^* \to e^+e^-\mathrm{P}$ [27]. The extraction of the slope requires, however, a model dependent extrapolation from rather large momenta $Q \sim 0.5$ GeV to $Q^2 = 0$. The slope and the TFF can also be obtained in the timelike region from the single Dalitz-decay $\mathrm{P} \to \gamma^*\gamma \to \ell^+\ell^-\gamma$ with $\ell = e, \mu$ [28]. Of course, for the pion only the decay with an electron pair is possible. For the pion the phase space is, however, rather small and the decay is not very sensitive to form factor effects. The corresponding determinations of the slope and the TFF are rather unprecise. The situation is much better for η and η'. However, one then still needs to perform an analytical continuation to obtain the form factor at spacelike momenta. Recently a DR has been proposed in Ref. [21] to determine the single- and double-virtual form factor for the pion. So far, only the single-virtual form factor has been evaluated in this dispersive framework with high precision at low momenta $Q \leq 1$ GeV. For the η and η', a dispersive approach for the single- and double-virtual TFF has been presented in Ref. [22].

For the single-virtual form factor we parametrize the measurement errors in Eqs. (7) and (8) as follows:

$$\mathcal{F}_{\mathrm{P}\gamma^*\gamma^*}(-Q^2, 0) \to \mathcal{F}_{\mathrm{P}\gamma^*\gamma^*}(-Q^2, 0)\,(1 \pm \delta_{1,\mathrm{P}}(Q)). \quad (17)$$

The momentum dependent errors $\delta_{1,\mathrm{P}}(Q)$ in different bins are displayed in Table 2, based on the analysis in Ref. [20]. There are currently no experimental data for the form factor available in the spacelike region in the lowest bin $0 \leq Q < 0.5$ GeV for π^0 and η, except from $\Gamma(\mathrm{P} \to \gamma\gamma)$, the slope and timelike data for the η-TFF. For the lowest bin we therefore assume an error, based on "extrapolating" the current data sets and the data that will be available soon from BES III [29] and maybe from KLOE-2 [24] for the pion. If one uses the DR from Ref. [21] below 1 GeV [30] and the current error on the normalization from the decay width, one obtains (conservatively) the uncertainties in the lowest two bins given in brackets in Table 2.

Table 2. Relative error $\delta_{1,\mathrm{P}}(Q)$ on the form factor $\mathcal{F}_{\mathrm{P}\gamma^*\gamma^*}(-Q^2, 0)$ for $\mathrm{P} = \pi^0, \eta, \eta'$ in different momentum regions. The errors for $\delta_{1,\pi^0}(Q)$ and $\delta_{1,\eta}(Q)$ below 0.5 GeV are based on assumptions. In brackets for π^0 the uncertainties with a DR for the TFF.

Region [GeV]	$\delta_{1,\pi^0}(Q)$	$\delta_{1,\eta}(Q)$	$\delta_{1,\eta'}(Q)$
$0 \leq Q < 0.5$	5% [2%]	10%	6%
$0.5 \leq Q < 1$	7% [4%]	15%	11%
$1 \leq Q < 2$	8%	8%	7%
$2 \leq Q$	4%	4%	4%

The second ingredient in Eqs. (7) and (8) is the double-virtual form factor $\mathcal{F}_{\mathrm{P}\gamma^*\gamma^*}(-Q_1^2, -Q_2^2)$. Currently, there are no direct experimental measurements available for this form factor at spacelike momenta. For the pion, from the double Dalitz decay $\pi^0 \to \gamma^*\gamma^* \to e^+e^-e^+e^-$, one can obtain the double-virtual form factor $|\mathcal{F}_{\pi^0\gamma^*\gamma^*}(q_1^2, q_2^2)|$ at small invariant momenta in the timelike region, but the results are inconclusive [31]. There is indirect information available on the double-virtual TFF from the loop-induced decay $\mathrm{P} \to \ell^+\ell^-$ ($\ell = e, \mu$) [25, 32]. Without a form factor at the $\mathrm{P} - \gamma^* - \gamma^*$-vertex, the loop integral is ultraviolet divergent. The relation between this decay and the pseudoscalar-pole contribution $a_\mu^{\mathrm{HLbL;P}}$ has been stressed in Ref. [11] and problems to explain both processes simultaneously with the same model have been pointed out [33].

In this situation, models have been used to describe the form factor $\mathcal{F}_{\mathrm{P}\gamma^*\gamma^*}(-Q_1^2, -Q_2^2)$ in the spacelike region and thus all current evaluations of $a_\mu^{\mathrm{HLbL;P}}$ are model dependent. Of course, it would be preferable to replace these model assumptions as much as possible by experimental data. In fact, it is planned to determine the double-virtual form factor, at least for the pion, at BES III for momenta $0.5 \leq Q_{1,2} \leq 1.5$ GeV and a first analysis is already in progress [29], based on existing data.

In analogy to the single-virtual form factor, we parametrize potential future measurement errors for the double-virtual form factor in Eqs. (7) and (8) in the following, simplifying way:

$$\mathcal{F}_{\mathrm{P}\gamma^*\gamma^*}(-Q_1^2, -Q_2^2) \to \mathcal{F}_{\mathrm{P}\gamma^*\gamma^*}(-Q_1^2, -Q_2^2)\,(1 \pm \delta_{2,\mathrm{P}}(Q_1, Q_2)), \quad (18)$$

where the assumed momentum dependent errors $\delta_{2,\mathrm{P}}(Q_1, Q_2)$ in different bins are shown in Fig. 4. The error estimate for the form factor in each bin has been obtained based on a Monte Carlo (MC) simulation [29] for the BES III detector using the LMD+V model in the EKHARA event generator [34] for the signal process $e^+e^- \to e^+e^-\gamma^*\gamma^* \to e^+e^-\pi^0$ and the VMD model for the production of η and η'. Since the number of events N_i in bin number i is proportional to the cross-section σ_i (in that bin) and since for the calculation of the cross-section the form factor enters squared, the statistical error on the form factor measurement is given according to Poisson statistics by $\sigma_i \sim \mathcal{F}^2_{\mathrm{P}\gamma^*\gamma^*} \Rightarrow \delta\mathcal{F}_{\mathrm{P}\gamma^*\gamma^*}/\mathcal{F}_{\mathrm{P}\gamma^*\gamma^*} = \sqrt{N_i}/(2N_i)$.

In the lowest momentum bin $Q_{1,2} \leq 0.5$ GeV, there are no events in the simulation, because of the acceptance of the detector. When both $Q_{1,2}^2$ are small, both photons are almost real and the scattered electrons and positrons escape detection along the beam pipe. As a further assumption, we have therefore taken the average of the uncertainties in the three neighboring bins as estimate for the error in that lowest bin. This "extrapolation" from the neighboring bins seems justified, since information along the two axis is (or will soon be) available and the value at the origin is known quite precisely from the decay width. Note that although the form factor for spacelike momenta is rather smooth, it is far from being a constant and some nontrivial extrapolation is needed. For instance, for the pion we get $\mathcal{F}_{\pi^0\gamma^*\gamma^*}(-(0.5\ \mathrm{GeV})^2, -(0.5\ \mathrm{GeV})^2)/\mathcal{F}_{\pi^0\gamma^*\gamma^*}(0,0) \approx 0.5$, for both the LMD+V and the VMD model [20].

The MC simulation [29] corresponds to a data sample of approximately half of the data collected at BES III so far. The simulation included only signal events. Based on a first preliminary analysis of the BES III data [29] with strong cuts to reduce the background from Bhabha events with additional photons, it seems possible, at least for the pion, that the number of events and the corresponding precision for $\mathcal{F}_{\pi^0\gamma^*\gamma^*}(-Q_1^2, -Q_2^2)$ shown in Fig. 4 could be achievable with the current data set plus a few more years of data taking. Of course, once experimental data will be available, e.g. event rates in the different momentum bins, there will still be the task to unfold the data to reconstruct the form factor $\mathcal{F}_{\mathrm{P}\gamma^*\gamma^*}(-Q_1^2, -Q_2^2)$ without introducing too much model dependence.

Taking the LMD+V and VMD models for illustration, the assumed momentum dependent errors from Table 2 and Fig. 4 impact the precision for the pseudoscalar-pole contributions to HLbL as follows

$$a_{\mu;\mathrm{LMD+V}}^{\mathrm{HLbL};\pi^0} = 62.9^{+8.9}_{-8.2} \times 10^{-11} \quad \left({}^{+14.1\%}_{-13.1\%}\right), \quad (19)$$

$$a_{\mu;\mathrm{VMD}}^{\mathrm{HLbL};\pi^0} = 57.0^{+7.8}_{-7.3} \times 10^{-11} \quad \left({}^{+13.7\%}_{-12.7\%}\right), \quad (20)$$

$$a_{\mu;\mathrm{VMD}}^{\mathrm{HLbL};\eta} = 14.5^{+3.4}_{-3.0} \times 10^{-11} \quad \left({}^{+23.4\%}_{-20.8\%}\right), \quad (21)$$

$$a_{\mu;\mathrm{VMD}}^{\mathrm{HLbL};\eta'} = 12.5^{+1.9}_{-1.7} \times 10^{-11} \quad \left({}^{+15.1\%}_{-13.9\%}\right). \quad (22)$$

While for the pion the absolute variations are different for the two models, as are the central values, the relative uncertainty for both models is around 14%. This will also be visible in the following more detailed analysis. We therefore expect that using other form factor models, and, eventually, using experimental data for the single- and double-virtual form factors, will not substantially change the following observations and conclusions.

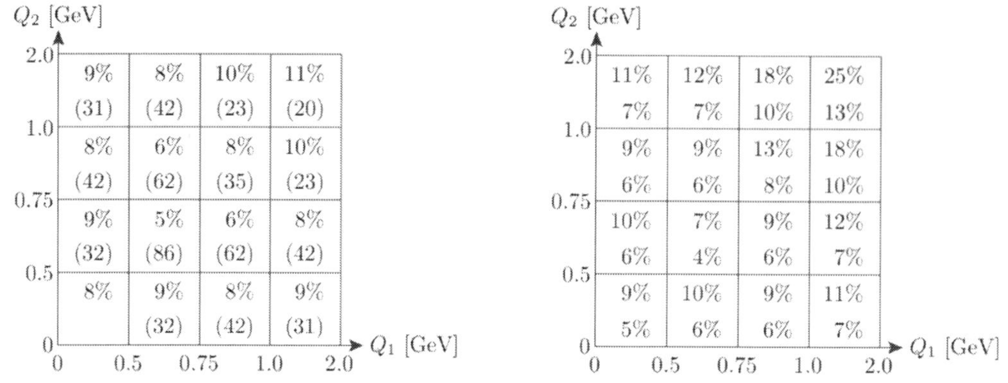

Figure 4. Left panel: Assumed relative error $\delta_{2,\pi^0}(Q_1, Q_2)$ on the pion TFF $\mathcal{F}_{\pi^0\gamma^*\gamma^*}(-Q_1^2, -Q_2^2)$ in different momentum bins. Note the unequal bin sizes. In brackets the number of MC events N_i in each bin according to the simulation with the LMD+V model for BES III. For the lowest bin, $Q_{1,2} \leq 0.5$ GeV, there are no events in the simulation due to the detector acceptance. In that bin, we assume as error the average of the three neighboring bins. For $Q_{1,2} \geq 2$ GeV, we take a constant error of 15%. Right panel: Assumed relative error $\delta_{2,\mathrm{P}}(Q_1, Q_2)$ on the form factor $\mathcal{F}_{\mathrm{P}\gamma^*\gamma^*}(-Q_1^2, -Q_2^2)$ for $P = \eta, \eta'$ in different momentum bins according to the MC simulations with the VMD form factor for η (top line) and η' (bottom line). For $Q_{1,2} \geq 2$ GeV, we assume a constant error of 25% for η and 15% for η'.

More details have been collected in Table 3, which contains the results for the relative uncertainties from Eqs. (19)-(22) in the first line. For the pion, the largest uncertainty of about 5% comes from the lowest bin $Q_{1,2} \leq 0.5$ GeV in the (Q_1, Q_2)-plane for δ_2 in Fig. 4 (fourth line in the table). Some improvement could be achieved, if the error in that lowest bin and the neighboring bins could be reduced, to a total error of about 12%, see the lines 8 and 9 in Table 3. The second largest uncertainty of 4.4% for the pion stems from the lowest bin $Q < 0.5$ GeV in δ_1 (second line in the table). Here the use of a dispersion relation for the single-virtual form factor $\mathcal{F}_{\pi^0\gamma^*\gamma^*}(-Q^2, 0)$ for $Q < 1$ GeV (see values in brackets in Table 2) could bring the total error of 14% down to 11%, see the sixth line.

For the η meson, the largest uncertainties of 7% originate from the region of δ_2 above 0.5 GeV (5th line) and from the lowest bin in δ_1 (2nd line). For the η', the largest uncertainty of 5% comes again from the region of δ_2 above 0.5 GeV (5th line). The second largest uncertainty of 4.5% comes from bins in δ_1 above 0.5 GeV (3rd line). For η and η' the errors go down to 20% and 13%, if the uncertainty in the two lowest bins in δ_1 could be reduced, see line 7 in Table 3. There is only a small reduction of the uncertainty by one percentage point, if the errors in the lowest few bins of δ_2 could be reduced further, see lines 8 and 9.

6 Conclusions

The three-dimensional integral representation for the pseudoscalar-pole contribution $a_\mu^{\mathrm{HLbL;P}}$ with $\mathrm{P} = \pi^0, \eta, \eta'$, from Ref. [1] allows one to separate the generic kinematics, described by model-independent weight functions $w_{1,2}(Q_1, Q_2, \tau)$, from the double-virtual transition form factors $\mathcal{F}_{\mathrm{P}\gamma^*\gamma^*}(-Q_1^2, -Q_2^2)$. From the weight functions one deduces that the relevant momentum regions are below 1 GeV for π^0 and below about 1.5 GeV for η and η'.

If the assumed measurement errors $\delta_1(Q)$ and $\delta_2(Q_1, Q_2)$ on the single- and double-virtual TFF can be achieved in the coming years (in particular by measurements of the double-virtual form factors at BES III), one could obtain the following, largely data driven, uncertainties for the pseudoscalar-pole contributions to HLbL:

$$\delta a_\mu^{\mathrm{HLbL};\pi^0} / a_\mu^{\mathrm{HLbL};\pi^0} = 14\% \quad [11\%], \tag{23}$$

$$\delta a_\mu^{\mathrm{HLbL};\eta} / a_\mu^{\mathrm{HLbL};\eta} = 23\%, \tag{24}$$

$$\delta a_\mu^{\mathrm{HLbL};\eta'} / a_\mu^{\mathrm{HLbL};\eta'} = 15\%. \tag{25}$$

The result in bracket for the pion uses the DR [21] for the single-virtual TFF $\mathcal{F}_{\pi^0\gamma^*\gamma^*}(-Q^2, 0)$. Compared to the range of estimates in the literature given in Eqs. (3) and (4) this would definitely be some progress. More work is needed, however, to reach a precision of 10% for all three contributions. Experimental data on the double-virtual form factors in the region $0 - 1.5$ GeV, e.g. from KLOE-2 or Belle 2, would be very helpful in this respect. A more detailed discussion can be found in [20].

Acknowledgements

I would like to thank the organizers for creating such a stimulating and nice atmosphere. I am grateful to Achim Denig, Christoph Redmer and Pascal Wasser for providing me with preliminary results of MC simulations for transition form factor measurements at BES III and to Martin Hoferichter and Bastian Kubis for sharing information about the precision of the DR approach to the pion TFF. I thank the Heinrich Greinacher Foundation, University of Bern, Switzerland, for financial support. This work was supported by Deutsche Forschungsgemeinschaft (DFG) through the Collaborative Research Center "The Low-Energy Frontier of the Standard Model" (SFB 1044).

Table 3. Impact of assumed measurement errors $\delta_{1,\mathrm{P}}(Q)$ and $\delta_{2,\mathrm{P}}(Q_1, Q_2)$ in the form factors $\mathcal{F}_{\mathrm{P}\gamma^*\gamma^*}(-Q^2, 0)$ and $\mathcal{F}_{\mathrm{P}\gamma^*\gamma^*}(-Q_1^2, -Q_2^2)$ on the relative precision of the pseudoscalar-pole contributions (first line). Lines 2 – 5 show the effects of uncertainties in different momentum regions below and above 0.5 GeV. Lines 6 – 9 show the impact of potential improvements of some of the assumed errors.

$\frac{\delta a_{\mu;\mathrm{LMD+V}}^{\mathrm{HLbL};\pi^0}}{a_{\mu;\mathrm{LMD+V}}^{\mathrm{HLbL};\pi^0}}$	$\frac{\delta a_{\mu;\mathrm{VMD}}^{\mathrm{HLbL};\pi^0}}{a_{\mu;\mathrm{VMD}}^{\mathrm{HLbL};\pi^0}}$	$\frac{\delta a_{\mu;\mathrm{VMD}}^{\mathrm{HLbL};\eta}}{a_{\mu;\mathrm{VMD}}^{\mathrm{HLbL};\eta}}$	$\frac{\delta a_{\mu;\mathrm{VMD}}^{\mathrm{HLbL};\eta'}}{a_{\mu;\mathrm{VMD}}^{\mathrm{HLbL};\eta'}}$	Comment
+14.1% −13.1%	+13.7% −12.7%	+23.4% −20.8%	+15.1% −13.9%	Given δ_1, δ_2
+4.3% −4.2%	+4.4% −4.3%	+6.9% −6.8%	+3.4% −3.3%	Bin $Q < 0.5$ GeV in δ_1 as given, rest: $\delta_{1,2} = 0$
+1.1% −1.0%	+1.0% −0.9%	+4.4% −4.3%	+4.5% −4.4%	Bins $Q \geq 0.5$ GeV in δ_1 as given, rest: $\delta_{1,2} = 0$
+4.5% −4.4%	+4.9% −4.8%	+4.0% −4.0%	+1.7% −1.7%	Bin $Q_{1,2} < 0.5$ GeV in δ_2 as given, rest: $\delta_{1,2} = 0$
+3.9% −3.8%	+3.2% −3.1%	+7.0% −6.8%	+5.1% −5.0%	Bins $Q_{1,2} \geq 0.5$ GeV in δ_2 as given, rest: $\delta_{1,2} = 0$
+10.9% −10.5%	+10.6% −10.1%	–	–	π^0: given δ_1, δ_2, but two lowest bins in δ_{1,π^0} from DR: 2%, 4%
–	–	+20.4% −18.5%	+13.4% −12.5%	η, η': given δ_1, δ_2, but two lowest bins in $\delta_{1,\eta}$: 8%, 10% and $\delta_{1,\eta'}$: 5%, 8%
+12.4% −11.6%	+11.8% −11.0%	+22.4% −20.0%	+14.8% −13.6%	For π^0, η, η': given δ_1, δ_2, but lowest bin in δ_2: 8%, 9%, 5% $\rightarrow$ 5%, 7%, 4%
+12.0% −11.2%	+11.4% −10.6%	+21.9% −19.6%	+14.4% −13.4%	In addition, bins in δ_2 close to lowest bin: 9%, 10%, 6% $\rightarrow$ 5%, 7%, 4%

References

[1] F. Jegerlehner and A. Nyffeler, Phys. Rept. **477**, 1 (2009).

[2] G. W. Bennett *et al.*, Phys. Rev. D **73**, 072003 (2006).

[3] T. Blum *et al.*, arXiv:1311.2198; talks at this meeting: M. Knecht; K. Melnikov.

[4] D. Hertzog, talk at this meeting.

[5] Talks on HVP at this meeting: M. Benayoun, arXiv:1511.01329; S. Eidelman; F. Jegerlehner, arXiv:1511.04473; M. Petschlies; M. Steinhauser; L. Trentadue; Z. Zhang, arXiv:1511.05405.

[6] J. Prades, E. de Rafael and A. Vainshtein, Adv. Ser. Direct. High Energy Phys. **20**, 303 (2009) [arXiv:0901.0306].

[7] A. Nyffeler, Phys. Rev. D **79**, 073012 (2009).

[8] M. Hayakawa, T. Kinoshita and A. I. Sanda, Phys. Rev. Lett. **75**, 790 (1995); M. Hayakawa and T. Kinoshita, Phys. Rev. D **57**, 465 (1998) [**66**, 019902(E) (2002)].

[9] J. Bijnens, E. Pallante and J. Prades, Phys. Rev. Lett. **75**, 1447 (1995) [**75**, 3781(E) (1995)]; Nucl. Phys. B **474**, 379 (1996); Nucl. Phys. B **626**, 410 (2002).

[10] M. Knecht and A. Nyffeler, Phys. Rev. D **65**, 073034 (2002).

[11] M. Knecht *et al.*, Phys. Rev. Lett. **88**, 071802 (2002).

[12] K. Melnikov and A. Vainshtein, Phys. Rev. D **70**, 113006 (2004).

[13] T. Blum *et al.*, Phys. Rev. Lett. **114**, 012001 (2015); T. Blum *et al.*, arXiv:1510.07100.

[14] J. Green *et al.*, arXiv:1507.01577; J. Green *et al.*, arXiv:1510.08384.

[15] Talks on HLbL at this meeting: J. Bijnens, arXiv:1510.05796; L. Cappiello; D. Greynat; C. Lehner; P. Masjuan; M. Procura.

[16] G. Colangelo *et al.*, JHEP **1409**, 091 (2014); G. Colangelo *et al.*, Phys. Lett. B **738**, 6 (2014); G. Colangelo *et al.*, JHEP **1509**, 074 (2015).

[17] V. Pauk and M. Vanderhaeghen, arXiv:1403.7503; Phys. Rev. D **90**, 113012 (2014).

[18] A. Vainshtein, talk at this meeting.

[19] E. Czerwinski *et al.*, arXiv:1207.6556; talks at this meeting: A. Kupsc; P. Masjuan.

[20] A. Nyffeler, work in preparation.

[21] M. Hoferichter *et al.*, Eur. Phys. J. C **74**, 3180 (2014).

[22] C. Hanhart *et al.*, Eur. Phys. J. C **73**, 2668 (2013) [**75**, 242(E) (2015)]; C. W. Xiao *et al.*, arXiv:1509.02194.

[23] M. Knecht and A. Nyffeler, Eur. Phys. J. C **21**, 659 (2001).

[24] D. Babusci *et al.*, Eur. Phys. J. C **72**, 1917 (2012); D. Moricciani, talk at this meeting.

[25] K. A. Olive *et al.* [Particle Data Group], Chin. Phys. C **38**, 090001 (2014).

[26] I. Larin *et al.*, Phys. Rev. Lett. **106**, 162303 (2011); D. Babusci *et al.*, JHEP **1301**, 119 (2013).

[27] H. Aihara *et al.*, Phys. Rev. Lett. **64**, 172 (1990); H. J. Behrend *et al.*, Z. Phys. C **49**, 401 (1991); J. Gronberg *et al.*, Phys. Rev. D **57**, 33 (1998); M. Acciarri *et al.*, Phys. Lett. B **418**, 399 (1998); B. Aubert *et al.*, Phys. Rev. D **80**, 052002 (2009); P. del Amo Sanchez *et al.*, Phys. Rev. D **84**, 052001 (2011); S. Uehara *et al.*, Phys. Rev. D **86**, 092007 (2012).

[28] R. Arnaldi *et al.*, Phys. Lett. B **677**, 260 (2009); P. Aguar-Bartolome *et al.*, Phys. Rev. C **89**, 044608 (2014); M. Ablikim *et al.*, Phys. Rev. D **92**, 012001 (2015).

[29] A. Denig, C. Redmer and P. Wasser, private comm.

[30] M. Hoferichter and B. Kubis, private comm.

[31] E. Abouzaid *et al.*, Phys. Rev. Lett. **100**, 182001 (2008).

[32] E. Abouzaid *et al.*, Phys. Rev. D **75**, 012004 (2007).

[33] M. J. Ramsey-Musolf and M. B. Wise, Phys. Rev. Lett. **89**, 041601 (2002); A. E. Dorokhov and M. A. Ivanov, JETP Lett. **87**, 531 (2008); P. Masjuan and P. Sanchez-Puertas, arXiv:1504.07001.

[34] H. Czyż and S. Ivashyn, Comput. Phys. Commun. **182**, 1338 (2011); H. Czyż *et al.*, Phys. Rev. D **85**, 094010 (2012).

EPJ Web of Conferences **118**, 01025 (2016) DOI: 10.1051/epjconf/201611801025
FCCP2015

Status of the NA62 Experiment

Vito Palladino[1,a] on behalf of NA62 Collaboration.

[1] *European Organization for Nuclear Research*

Abstract. The rare decays $K^+ \to \pi^+\nu\bar{\nu}$ are excellent processes to make tests of new physics at the highest scale complementary to LHC thanks to their theoretically cleaness. The NA62 experiment at CERN SPS aims to collect of the order of 100 events in two years of data taking, keeping the background at the level of 10%. Part of the experimental apparatus has been commissioned during a technical run in 2012. The physics prospects and the status of the experiment will be reviewed after the commissioning run of 2014 and the data taking in 2015.

1 Introduction

NA62 is the last generation kaon experiment at CERN SPS aiming to study the decay $K^+ \to \pi^+\nu\bar{\nu}$. The goal of the experiment is to measure the decay branching ratio ($O(10^{-10})$) with 10% accuracy, collecting about 100 events in two years of data taking and assuming a 10% signal acceptance. The available intensity of the proton beam extracted from the SPS in the north area at CERN fit perfectly such demanding. The NA62 detector must be able to reject background events from decay channels with branching ratios up to 10 orders of magnitude higher than the signal and with similar experimental signature, such as $K^+ \to \mu^+\nu$. To this purpose, good kinematic rejection, PID (Particle IDentification) and veto systems are required. The required resolution on the measurement of kinematic parameters (momenta, missing mass) is obtained tracking both the decaying kaon with a silicon pixel detector (GTK) and the parent charged particles with a straw spectrometer. The key elements of the PID in NA62 are: a differential Cherenkov detector (KTAG) to identify kaons in the charged beam, a Ring Imaging Cherenkov detector (RICH) to identify π^+ and μ^+ in the momentum range between 15 and 35 GeV/c and three scintillators detector acting as hadron calorimeters and muon counters at the end of the detector region. Veto systems working at different angles with respect to the beam direction, especially to suppress background coming from kaon decays with photons in the final state, complete the experimental setup: Large Photon Vetos (LAV) made by lead glass at large angle, a Liquid Krypton calorimeter (LKr) at intermediate angles and two shashlik detectors at small angles. The detectors have to stand a rate of about 10 MHz of events, high performances read-out and trigger systems have been developed for this purpose. High resolution on time measurement is also required in order to reduce background produced by event overlaps. In the following sections, after a short theoretical introduction, a review of the detectors and of the results obtained during the 2014 pilot run is presented.

[a] e-mail: Vito.Palladino@cern.ch

2 Theoretical framework and experimental status of the art

The decay $K \to \nu\bar{\nu}$ are flavor-changing neutral current (FCNC) processes that probe the $s \to d\nu\bar{\nu}$ transition via the Z-penguin and the box diagram shown in Figure 1. They are highly GIM suppressed and their Standard Model (SM) rates are very small. For the reasons listed below the SM calculation for their branching ratios (BRs) is particularly clean [2]:

- The loop amplitudes are dominated by the top-quark contributions. The neutral decay violates CP; its amplitude involves the top-quark contribution only. Small corrections to the amplitudes from the lighter quarks come into play for the charged channel.
- The hadronic matrix element for these decays can be factorized and obtained from the precise experimental measurement of the K_{e3} rate.
- There are no long-distance contributions from processes with intermediate photons.

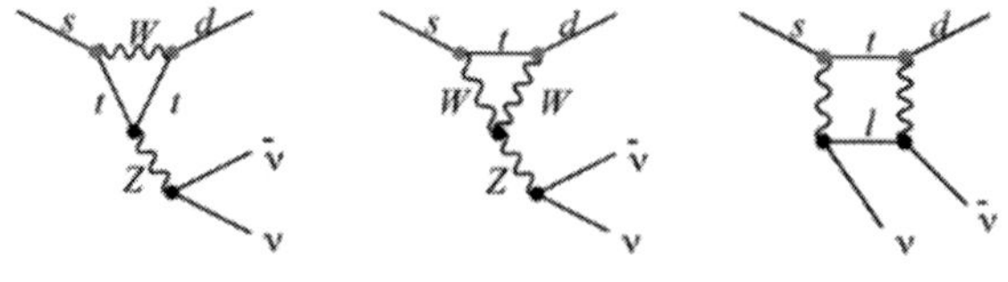

Figure 1. Z-penguin diagrams and box diagram contributing to the process $K \to \pi\nu\bar{\nu}$.

In the SM, BR($K_L \to \pi^0\nu\bar{\nu}$) = $(3.00 \pm 0.30) \times 10^{-11}$ and BR($K^+ \to \pi^+\nu\bar{\nu}$) = $(9.11 \pm 0.72) \times 10^{-11}$ [3]. Giv-

© The Authors, published by EDP Sciences. This is an open access article distributed under the terms of the Creative Commons Attribution License 4.0 (http://creativecommons.org/licenses/by/4.0/).

ing the corrections from lighter-quark contributions, these are slightly larger for the charged channel. Because the SM rates are small and predicted very precisely, the BRs for these decays are sensitive probes for new physics. In evaluating the rates for the FCNC kaon decays, the different terms of the operator product expansion are differently sensitive to modifications from a given new-physics scenario. If BR($K_L \to \pi^0 \nu\bar{\nu}$) and BR($K^+ \to \pi^+ \nu\bar{\nu}$) are ultimately both measured, and one or both BRs is found to differ from its SM value, it may be possible to characterize the physical mechanism involved [3], e.g., a mechanism with minimal flavor violation [4], manifestations of supersymmetry [5], a fourth generation of fermions [6], Higgs compositeness as in the littlest Higgs model [7], or an extra- dimensional mechanism such as in the Randall-Sundrum model [8]. The most precise experimental result for the BR($K^+ \to \pi^+ \nu\bar{\nu}$) has been obtained by the dedicated experiments E787 and E949 at the Brookhaven National Laboratory [9] and is based on a total of 7 events using a decay-at-rest technique; the combined measurement is BR($K^+ \to \pi^+ \nu\bar{\nu}$) = $(17.3 \pm 11.5) \times 10^{-11}$. For the neutral channel only an upper limit for the BR has been measured [10], BR($K^0 \to \pi^0 \nu\bar{\nu}$) $< 2.6 \times 10^{-8}$ at 90% confidence level.

3 The NA62 detector

NA62 uses a decay-in-flight technique, differently from the kaon decay at rest approach which was at the basis of the previous experiments.

The presence of two undetectable neutrinos in the final state reduces the signal signature to one high momentum charged track with nothing else, which has to be discriminated against background coming from all other kaon decays. The high momentum of the incoming beam (75 GeV/c) improves the background rejection and sets the longitudinal scale of the experiment. To achieve the required background suppression different principles have to be combined and the resulting requirements are outlined here:

- High intensity and good timing: a high intensity kaon beam is essential in order to reach sensitivity to a branching ratio of $\sim 10^{-10}$. The incoming secondary beam from the SPS provides a particle rate of 750 MHz, containing about 6% of kaons delivering roughly $45 \cdot 10^6$ decays in the fiducial region per spill. Precise timing (in the range 100 - 150ps) of the K^+ and the π^+ allows precise matching of the particles in the decay. The time resolution is essential to keep wrong associations below 1%.
- Low-mass tracking: track position and momenta have to be measured with high accuracy in low mass detectors. They are essential because inelastic scattering of beam particles in the detector material can mimic an isolated π^+ appearing like a signal event and hence contribute to the background.
- Hermetic vetoing for photons and muons: the kinematic rejection must be accompanied by direct vetoing for photons (in particular for the $K^+ \to \pi^+\pi^0$ background) requiring a typical inefficiency of 10^{-4} for high energetic photons. About two third of K^+ decays contain muons in the final state, therefore a muon veto system is mandatory both in the trigger and off-line.
- Particle ID: several detectors must complement the event information with direct evidence on the particle species.

The experimental setup, shown in Figure 2, consists of a 100 m long beam line to select the appropriate secondary beam component produced by protons from the SPS CERN accelerator, followed by a 80 m long evacuated volume which defines the decay region. Along the beam line, different detectors are distributed as listed below.

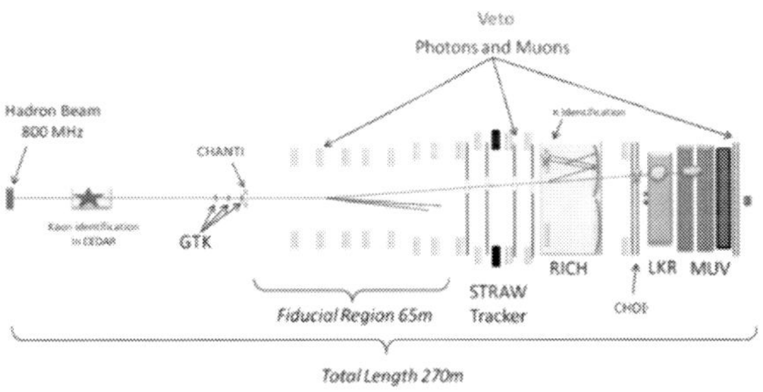

Figure 2. Schematic view of the NA62 experiment showing the main sub-detectors (not to scale). The main elements for the detection of the K^+ decay products are located along a 170 m long region starting about $\sim$ 100 m downstream of the beryllium target. Useful K^+ decays will be detected within a 65 m long fiducial decay region.

3.1 Differential Cherenkov counter (KTAG)

The KTAG identifies the K^+ component in the beam with respect to the other beam particles by employing an upgraded differential Cherenkov counter (CEDAR). The CEDAR filled with nitrogen gas is placed in the incoming beam to positively identify the kaon component in the high rate environment. It is designed to identify particles of a specific mass by making the detector blind to the Cherenkov light produced by particles of different mass. The KTAG upgrade of NA62 acts in the Cherenkov light detection stage: to cope with the expected 45 MHz kaon rate, 384 photomultipliers (PMTs) grouped in 8 light boxes are placed behind the 8 annular slits. A preliminary analysis of 2014 data shows a kaon-identification efficiency better than 95% and a time resolution below 100 ps.

3.2 Beam tracker (GTK)

The GigaTracker (GTK) comprises three stations measuring time, direction and momentum of the beam particles before entering the decay region. The three stations are placed along the beam line just before the fiducial region. Each GTK station is a hybrid silicon micro-pixel detector with a total number of 18000 300×300 μm^2 pixels grouped in 10 read-out chips. The expected performances on the track measurement are a 0.4% resolution for the momentum, 16 mrad for the direction and 200 ps for the timing. One chip per station has been commissioned during the 2014 run.

3.3 Guard-Ring detector (CHANTI)

This detector is a guard-ring counter following the last GTK station to veto the particles produced by the beam particles inelastic interaction in the GTK third station. It consists of six stations of x-y plastic scintillator bars surrounding the beam, covering an angle between 0.026 and 1.38 rad. The expected efficiency in vetoing signal-like events is 99%. Preliminary 2014 data analysis shows a time resolution of the order of 1 ns.

3.4 Pion spectrometer (STRAW)

The STRAW Tracker measures the coordinates and momentum of secondary charged particles originating from the decay region. To minimise multiple scattering the chambers are built of ultra-light material and are installed inside the vacuum tank. Two straw chambers are located on each side of a large aperture dipole magnet, providing a vertical B-field of 0.36T. Each chamber is equipped with 1792 straw tubes which are arranged in four "views" providing measurements of four coordinates x, y, u, v. The requirements for the detector performances are a relative momentum resolution of 1%, a spatial resolution of 130 μm per coordinate and a very low track reconstruction inefficiency. The detector has been fully commissioned during the 2014 run.

3.5 Pion Timing (CHOD)

The CHOD is a Charged-particle HODoscope (inherited from NA48) to detect tracks with precise measurements of the arrival time and impact point, able to provide a fast signal used to drive the trigger. It is composed of two planes made of 64 plastic scintillators, one with vertical and one with horizontal slabs. The CHOD can provide the timing of charged decay products with a resolution of about 200 ps.

3.6 Ring Image CHerenkov detector (RICH)

The RICH detector is placed downstream of the last straw chamber. It consists of a 17.5 m long cylindrical vessel with a diameter varying from 3.4 to 4 m filled with neon at atmospheric pressure and room temperature. An aluminum beam pipe spanning the length of the RICH keeps non-decayed beam particles in vacuum. The Cherenkov light is reflected by a mosaic of 18 hexagonal and 2 semi-hexagonal (close to the beam pipe) mirrors with 17 m focal length, made of 2.5 cm thick glass. They are placed at the downstream end of the vessel and they cover a total surface of about 3 m diameter. The reflected light is collected by 2 × 960 photomultiplier tubes (PMTs) with 18 mm diameter size and suitable quantum efficiency and timing performances. The PMTs are assembled in a compact hexagonal packing into two aluminum disks placed at the entrance window. Winston cones carved into the disks and covered with aluminized mylar are used to convey the light onto the active PMT area, thus increasing light collection. Quartz windows are used to separate the PMTs from the neon in order to avoid electrical discharges at the working voltage (900 V). The mirror segments are individually hanged on a light aluminum honeycomb structure and moved by means of two piezo-motors each, in order to align the light toward the two PMTs disks. The RICH allows an additional rejection of the muon background. The requested μ suppression factor is 10^{-2} between 15 and 35 GeV/c (the momentum range of interest), and a time resolution of 70 ps has been already measured.

3.7 Large-Angle Photon Vetoes (LAV)

The LAV system provides full coverage for decay photons with polar angles from 8.5 to 50 mrad, with an inefficiency of $10^{-3} \div 10^{-4}$ on photons down to 150 MeV. It consists of 12 stations, whose diameter increases with distance from the target, containing 4 or 5 layers of azimuthally staggered lead glass blocks, for a total of 2500 channels. The time resolution observed in 2014 run is about 1 ns.

3.8 Small-Angle Photon Vetoes (IRC and SAC)

A Small Angle Calorimeter (SAC) and an Intermediate Ring Calorimeter (IRC) give photon coverage below 1 mrad. Both detectors are made of consecutive lead and plastic scintillator plates. The inefficiency for 1 GeV photons is below 10^{-4}. A time resolution of 3 ns was measured in the 2014 run.

3.9 Quasi-homogeneous liquid kripton calorimeter (LKr)

The LKr Calorimeter (inherited from NA48) measures precisely the electro-magnetic energy with the possibility to discriminate between positrons and pions using their shower properties. It covers intermediate angle between 1 and 8.5 mrad. It is divided in 13248 cells and the electromagnetic showers are fully detected through ionisation of low energy charged particles. The reconstructed shower time resolution is 500 ps, and the space resolution of the order of 1 mm; the total inefficiency for 10 GeV photons is below 10^{-5}. The new readout has been validated during the 2014 run.

3.10 Hadron and muon detectors (MUV)

NA62 uses a fast scintillator (MUV3) for direct muon vetoing and 2 sampling hadron calorimeters (MUV1+2) which measure the deposited hadron energy in the event, distinguishing hadron showers from muons. MUV1 and MUV2 modules are iron-scintillator sandwich calorimeters with scintillator strips alternately oriented in the horizontal and vertical directions. This system supplements and provides redundancy with respect to the RICH in the detection and rejection of muons. Only one of these was installed in 2014. The MUV3 is a vertical array of fast scintillator tiles placed after an iron wall and is used to detect non-showering muons. Each tile is read by 2 PMTs to avoid the effect of Cherenkov photons on time resolution. Data from the 2014 run show a time resolution below 500 ps.

3.11 Trigger and data acquisition

The experiment makes use of an integrated trigger and data acquisition system with three trigger levels. The lowest level, level 0, is implemented directly in the digital readout card for each detector subsystem. The detector hits are resolved into quantities that are used in trigger logic to decide which events will be read out for level 1. Level 0 will process about 10 MHz of detector hits; about 1 MHz of events will be read out for level 1. The level 1 trigger is implemented in software. It is the first asynchronous trigger level and will reduce the rate of events seen by level 2 by an order of magnitude. The level 2 trigger is implemented in the event builder running on the acquisition PC farm; it is the first trigger level which makes use of the full event information. The O(100 kHz) of events input to level 2 are reduced to a few kHz of events ultimately written to disk.

4 Analysis strategy

NA62 detectors are designed to measure the K^+ and the secondary particles from kaon decays occurring in the decay volume. The signature of the signal is one track in the final state matched with one K^+ track in the beam. The measurable quantities are $\vec{p}_{K^+}$ (the kaon momentum-vector), $\vec{p}_{\pi^+}$ (the pion momentum-vector) and the angle $\theta_{\pi K}$ (angle between the pion and the kaon).

The main kinematic variable to discriminate the signal from the overwhelming background is the squared missing mass m^2_{miss}. It is defined for the events, when only one charged track from Kaon decay is detected assuming the track belongs to a charged pion,

$$m^2_{miss} = (p_{K^+} - p_{\pi^+}). \qquad (1)$$

Figure 3 (bottom) shows the m^2_{miss} distribution in log scale for the main decay channels and the $K^+ \to \pi^+\nu\bar{\nu}$ decay (red curve). The signal decay differential width is magnified by 10^{10}. Two signal regions (I and II), on each side of the $K^+ \to \pi^+\pi^0$ peak, are chosen, where more than 90% of the main K^+ decays do not contribute. The kinematics background suppression is of the order of 10^5. The further suppression of background at the order of 10^7 is done by means of the detector related cuts (timing, particle identification, muon and photon veto).

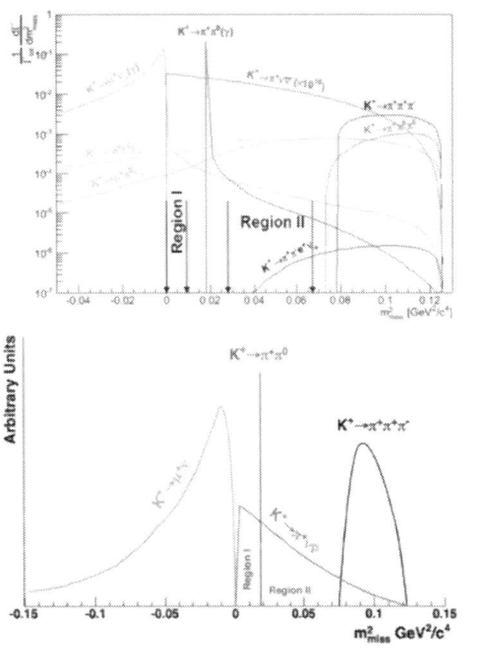

Figure 3. Top: Distribution of m^2_{miss} for the signal and the main background sources after kinematic cuts. Bottom: Distribution of m^2_{miss} for the main kaon decays and the signal, with the hypothesis that the charged decay product is a pion. The two signal regions, $[0, 0.01]GeV^2/c^4$ and $[0.026, 0.068]GeV^2/c^4$, are indicated.

Table 1 shows the expected number of signal and background events per year of data taking estimated from MC simulation. Given this expectation, NA62 will be able to reach better than 10% precision in a measurement of $BR(K^+ \to \pi^+\nu\bar{\nu})$. The main background remains the $K^+ \to \pi^+\pi^0(\gamma)$ decay, where two photons from π^0 decay (and FSR photon) are not detected and the signal event topology is mimicked. To suppress this background very good photon veto ability is required. The aimed π^0 suppression factor is 10^8. The second important background comes from the $K^+ \to \mu^+\nu_\mu(\gamma)$ decay. The designed overall suppression factor for muons is of the order of 10^5.

Decay	event/year
$K^+ \to \pi^+\nu\bar{\nu}$	45
$K^+ \to \pi^+\pi^0$	5
$K^+ \to \pi^+\pi^+\pi^-$	1
$K^+ \to \pi^+\pi^-e^+\nu_e$	< 1
$K^+ \to \mu^+\nu_\mu\gamma$	1.5
other rare decays	0.5
Total background	< 10

Table 1. Expected signal and background from K^+ decays estimated from NA62 sensitivity studies.

In the analysis, the π^+ momentum will be required to be less than 35 GeV/c. In this way, the momentum of the

π^0 amounts to at least 40 GeV/c. Such a large energy deposit can hardly be missed in the calorimeters. In addition only events with the K^+ vertex reconstructed in the fiducial volume are selected. The m^2_{miss} distribution for signal and the three main background sources after the kinematic cuts is shown in Figure 3 (bottom).

5 2014 pilot run

Data recorded during the pilot run in 2014 are very successfully used to understand the performance of the NA62 sub-detectors. However, only a small part of it was recorded with stable run conditions and is useful for preliminary physics studies. The beam intensity was 5% of the nominal one and two triggers have been used to collect a minimum bias sample and a $\pi\nu\bar{\nu}$-like without photon rejection sample. Only about 1% of the available data have been studied. The experiment was at a preliminary stage: data were not yet readout from GTK and the analysis used the mean nominal direction and momentum of the kaon beam; LAV information was not yet used for photon veto; only preliminary detector time alignment and energy calibration were used.

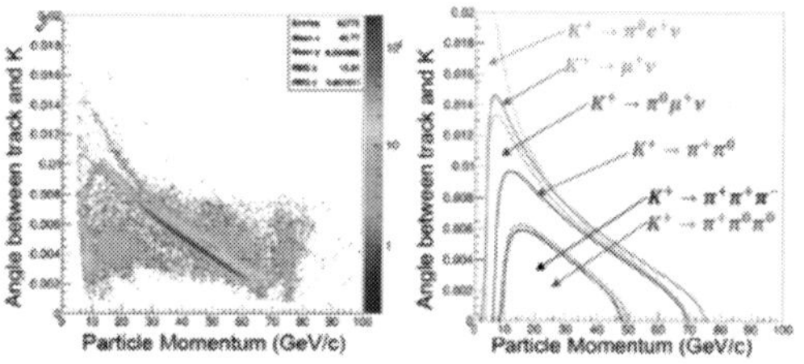

Figure 4. Angle between the reconstructed track and the nominal kaon direction as a function of particle momentum [GeV /c] for data (left) and as expected from pure kinematics (right).

The distribution of the angle between the kaon and the secondary track as a function of the track momentum is shown in Figure 4 (left), for events with only 1 track in the spectrometer and after requiring the presence of a kaon in the KTAG. A reasonable agreement can be observed if compared with the expectations for the main kaon decays (right): the suppression of the $K^+ \rightarrow \mu^+\nu$ component is due to the muon rejection at trigger level. The intersection between the track and the nominal beam direction is used to reconstruct the vertex and suppresses the background from kaon interactions. After requiring the vertex position to be reconstructed in the fiducial decay region, and selecting the track momentum between 15 and 35 GeV/c, the squared missing mass distribution shown in Figure 5 is obtained. The $K^+ \rightarrow \mu^+\nu$ background is well suppressed with the help of RICH and muon veto system. The three-track background, mainly coming from the $K^+ \rightarrow \pi^+\pi^+\pi^-$ decay is also well suppressed and does not contribute to the signal selected region. The main background contribution comes from the $K^+ \rightarrow \pi^+\pi^0$ decay. This particular background dominates in the signal region after the final selection, especially in region I. LAV information is not yet exploited both at trigger level and analysis level, preventing an additional suppression of $K^+ \rightarrow \pi^+\pi^0$ while the use of mean beam momentum and direction, instead of event by event GTK information, worsens the m^2_{miss} resolution.

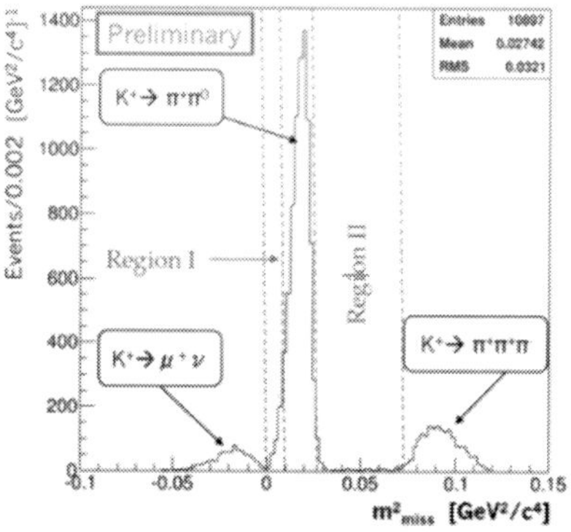

Figure 5. the m^2_{miss} distribution after the signal event selection. Two regions for the final selection of signal events are indicated by the vertical red-dashed lines. The kaon decays contributing to the distribution are indicated.

References

[1] na62.web.cern.ch/na62/Documents/TechnicalDesign.html, 2010.

[2] J. Brod, M. Gorbahn and E. Stamou, Phys. Rev. D 83, 034030 (2011).

[3] A.J. Buras, D. Bu-azzo, J. Girrbach-Noe and R.Knegjens, arXiv:1503.02693.

[4] T. Hurth, G. Isidori, J. F. Kamenik and F. Mescia, Nucl. Phys. B 808, 326 (2009).

[5] G. Isidori, F. Mescia, P. Paradisi, C. Smith and S. Trine, JHEP 0608, 064 (2006).

[6] A. J. Buras, B. Duling, T. Feldmann, T. Heidsieck and C. Promberger, JHEP 1009, 104 (2010).

[7] M. Blanke, A. J. Buras, B. Duling, S. Recksiegel and C. Tarantino, Acta Phys. Polon. B 41, 657 (2010) [arXiv:0906.5454 [hep-ph]].

[8] M. Blanke, A. J. Buras, B. Duling, K. Gemmler and S. Gori, JHEP 0903, 108 (2009) [arXiv:0812.3803 [hep-ph]].

[9] E949 COLLABORATION; ARTAMONOV, A.V. ET AL. Phys. Rev. Lett. 101 191802, 2008.

[10] E391a Collaboration (J.K. Ahn et al.) Phys.Rev. D81 (2010) 072004.

EPJ Web of Conferences **118**, 01026 (2016) DOI: 10.1051/epjconf/201611801026
FCCP2015

On the interrelationship among leptonic $g-2$, EDMs and lepton flavor violation

Paride Paradisi[1,a]

[1] *Dipartimento di Fisica ed Astronomia "G.Galilei", Univarsita' degli Studi di Padova, and Istituto Nazionale di Fisica Nucleare, Sezione di Padova, Via Marzolo 8, 35131 Padova, Italy*

Abstract. We present a concise review of the status of charged lepton flavor violation (cLFV) in scenarios beyond the SM. We emphasize that the current experimental resolutions on cLFV processes are already testing territories of new physics (NP) models well beyond the LHC reach. On the other hand, with the expected sensitivities of next-generation experiments, cLFV will become the most powerful probe of NP signals at our disposal. Finally, the interrelationship among leptonic $g-2$, EDMs and cLFV will turn out to be of outmost importance to disentangle among different NP scenarios.

1 Introduction

The origin of flavor remains, to a large extent, an open problem. However, significant progress has been achieved in the phenomenological investigation of the sources of flavour symmetry breaking which are accessible at low energies, ruling out models with significant misalignments from the SM Yukawa couplings at the TeV scale.

The search for LFV in charged leptons is probably the most interesting goal of flavour physics in the next years. The observation of neutrino oscillations has clearly demonstrated that lepton flavour is not conserved. The question is whether LFV effects can be visible also in other sectors of the theory, or if we can observe LFV in processes that conserve total lepton number. The most promising LFV low-energy channels are probably $\mu \to e\gamma$, $\mu \to eee$, $\mu \to e$ conversion in Nuclei as well as τ LFV processes which will be further investigated at the Super-Belle machine. The future sensitivities of next-generation experiments are collected in table 1.

Moreover, the flavour-conserving component of the same diagrams generating $\mu \to e\gamma$ induces non-vanishing contributions to the anomalous magnetic moment of leptons as well as to the leptonic EDMs. In this context, the current anomaly for the muon $(g-2)$, reinforces the expectation of detecting $\mu \to e\gamma$ within the reach of the MEG experiment. Once some clear deviation from the SM is established, the next most important step is to identify correlations among different non-standard effects that can reveal the flavour-breaking pattern of the new degrees of freedom providing, at the same time, a powerful tool to disentangle among different New Physics scenarios. The above program represents one of the most exciting proofs of the synergy and interplay existing between the LHC, i.e. the *high-energy frontier*, and high-precision low-energy experiments, i.e. the *high-intensity frontier*.

LFV Process	Present Bound	Future Sensitivity
$\mu \to e\gamma$	5.7×10^{-13} [1]	$\approx 6 \times 10^{-14}$ [2]
$\mu \to 3e$	1.0×10^{-12}[3]	$\approx 10^{-16}$ [4]
$\mu^-\mathrm{Au} \to e^-\mathrm{Au}$	7.0×10^{-13} [5]	?
$\mu^-\mathrm{Ti} \to e^-\mathrm{Ti}$	4.3×10^{-12} [6]	?
$\mu^-\mathrm{Al} \to e^-\mathrm{Al}$	–	$\approx 10^{-16}$ [7, 8]
$\tau \to e\gamma$	3.3×10^{-8} [9]	$\sim 10^{-8} - 10^{-9}$ [10]
$\tau \to \mu\gamma$	4.4×10^{-8} [9]	$\sim 10^{-8} - 10^{-9}$ [10]
$\tau \to 3e$	2.7×10^{-8}[11]	$\sim 10^{-9} - 10^{-10}$ [10]
$\tau \to 3\mu$	2.1×10^{-8}[11]	$\sim 10^{-9} - 10^{-10}$ [10]
Lepton EDM	Present Bound	Future Sensitivity
d_e(e cm)	8.7×10^{-29} [12]	?
d_μ(e cm)	1.9×10^{-19} [13]	?

Table 1. Present and future experimental sensitivities for relevant low-energy observables.

2 Leptonic $g-2$, EDMs and LFV: a model-independent analysis

The physics responsible for neutrino masses and mixing might or might not be related to the physics related to cLFV. On general grounds, we can say that:

- neutrino masses might be naturally explained within see-saw scenarios which introduce heavy right-handed Majorana neutrinos typically at the grand-unification (GUT) scale. These scenarios can also explain the baryon-antibaryon asymmetry in the universe through the leptogenesis mechanism. The new interactions of the model generally violate lepton-number $L = L_e + L_\mu + L_\tau$ (LNV).

[a] e-mail: paride.paradisi@pd.infn.it

© The Authors, published by EDP Sciences. This is an open access article distributed under the terms of the Creative Commons Attribution License 4.0 (http://creativecommons.org/licenses/by/4.0/).

- In the Standard Model (SM) with massive neutrinos, where the only source of LFV is coming from the operators responsible for the neutrino masses, the LFV effects are loop suppressed and proportional to the GIM factor $(m_\nu/M_W)^4$, therefore, completely negligible. For instance, it turns out that $\mathrm{Br}(\mu \to e\gamma) \sim 10^{-54}$.
- On the other hand, generic models for new physics (NP) at the TeV scale contain new sources for LFV (but not necessarily for LNV), leading to decay rates accessible with future experiments.

From the low-energy point of view, these observations can be accounted for by considering the SM as an effective theory and extending its Lagrangian,

$$\mathcal{L}_{\rm eff} = \mathcal{L}_{\rm SM} + \frac{1}{\Lambda_{\rm LNV}} O^{\rm dim-5} + \frac{1}{\Lambda^2_{\rm LFV}} O^{\rm dim-6} + \dots \quad (1)$$

Here, the dimension-5 operator responsible for the neutrino masses is uniquely given in terms of the lepton doublets L^i and the Higgs doublet H in the SM,

$$O^{\rm dim-5} = (g_\nu)^{ij} (\bar{L}^i \widetilde{H})(\widetilde{H}^\dagger L^j)^c + \text{h.c.} \quad (2)$$

and the misalignment between the flavour matrix g_ν and the Yukawa coupling matrix Y_E in the charged-lepton sector leads to a non-trivial mixing matrix $U_{\rm PMNS}$ for neutrino oscillations.

For instance, within scenarios with right-handed Majorana neutrinos (type-I see saw), one can identify $g_\nu/\Lambda_{\rm LNV} = Y_\nu M^{-1} Y_\nu^T$, where Y_ν is the Yukawa matrix in the neutrino sector, and M the Majorana mass matrix.

Examples for a dimension-6 operator are

$$\bar{\mu}_R \sigma^{\mu\nu} H e_L F_{\mu\nu}\,, \quad (\bar{\mu}_L \gamma^\mu e_L)\left(\bar{f}_L \gamma^\mu f_L\right)\,, \quad (\bar{\mu}_R e_L)\left(\bar{f}_R f_L\right)\,, \quad (3)$$

where $f = e, u, d$ and the first dipole-operator leads to LFV decays like $\mu \to e\gamma$ while the second and third ones generate, at the leading order, only processes like $\mu \to eee$ and $\mu \leftrightarrow e$ conversion in Nuclei. Obviously, the underlying dipole-transition $\mu \to e\gamma^\star$ with a virtual γ also contributes to $\mu \to eee$ and $\mu \leftrightarrow e$ conversion in Nuclei.

In particular, within NP theories where the dominant LFV effects are captured by the dipole-operator, the following model-independent relations hold

$$\mathrm{BR}(\ell_i \to \ell_j \ell_k \bar{\ell}_k) \simeq \frac{\alpha_{el}}{3\pi}\left(\log\frac{m^2_{\ell_i}}{m^2_{\ell_k}} - 3\right)\mathrm{BR}(\ell_i \to \ell_j\gamma)\,,$$

$$\mathrm{CR}(\mu \to e \text{ in N}) \simeq \alpha_{\rm em} \times \mathrm{BR}(\mu \to e\gamma)\,, \quad (4)$$

and therefore, the current MEG bound $\mathrm{BR}(\mu \to e\gamma) \sim 5\times 10^{-13}$ already implies that $\mathrm{BR}(\mu \to eee) \lesssim 3\times10^{-15}$ and $\mathrm{CR}(\mu \to e \text{ in N}) \lesssim 3\times 10^{-15}$.

However, it is worth stressing that in many NP scenarios non-dipole operators, such as those shown in eq. 3, provide the dominant sources of LFV effects. Therefore, in such cases, $\mu \to eee$ and $\mu \leftrightarrow e$ conversion in Nuclei represent the best probes of LFV.

Dipole transitions $\ell \to \ell'\gamma$ in the leptonic sector are accounted for by means of the effective Lagrangian

$$\mathcal{L} = e\,\frac{m_\ell}{2}\left(\bar{\ell}_R \sigma_{\mu\nu} A_{\ell\ell'} \ell'_L + \bar{\ell}'_L \sigma_{\mu\nu} A^\star_{\ell\ell'} \ell_R\right) F^{\mu\nu}\,, \quad (5)$$

where $\ell, \ell' = e, \mu, \tau$. Starting from eq. (5), we can evaluate LFV processes, such as $\mu \to e\gamma$,

$$\frac{\mathrm{BR}(\ell \to \ell'\gamma)}{\mathrm{BR}(\ell \to \ell'\nu_\ell\bar{\nu}_{\ell'})} = \frac{48\pi^3\alpha}{G_F^2}\left(|A_{\ell\ell'}|^2 + |A_{\ell'\ell}|^2\right). \quad (6)$$

The underlying $\ell \to \ell'\gamma$ transition can generate, in addition to LFV processes, also lepton flavor conserving processes like the anomalous magnetic moments Δa_ℓ as well as leptonic electric dipole moments (EDMs, d_ℓ). In terms of the effective Lagrangian of eq. (5) we can write Δa_ℓ and d_ℓ as

$$\Delta a_\ell = 2m_\ell^2\,\mathrm{Re}(A_{\ell\ell})\,, \qquad \frac{d_\ell}{e} = m_\ell\,\mathrm{Im}(A_{\ell\ell})\,. \quad (7)$$

On general grounds, one would expect that, in concrete NP scenarios, Δa_ℓ, d_ℓ and $\mathrm{BR}(\ell \to \ell'\gamma)$, are correlated. In practice, their correlations depend on the unknown flavor and CP structure of the NP couplings and thus we cannot draw any firm conclusion.

Parametrizing the amplitude $A_{\ell\ell'}$ as $A_{\ell\ell'} = c_{\ell\ell'}/\Lambda^2$, where Λ refers to the NP scale, we can evaluate which are the values of Λ probed by $\mu \to e\gamma$. We find that

$$\mathrm{BR}(\mu \to e\gamma) \approx 10^{-12}\left(\frac{500\ \mathrm{TeV}}{\Lambda}\right)^4\left(|c_{\mu e}|^2 + |c_{e\mu}|^2\right), \quad (8)$$

and therefore, for $c_{\mu e} \sim 1$ and/or $c_{e\mu} \sim 1$, we are left with $\Lambda > 500$ TeV.

Since the anomalous magnetic moment of the muon $a_\mu = (g-2)_\mu/2$ exhibits a $\sim 3.5\sigma$ discrepancy between the SM prediction and the experimental value [14] $\Delta a_\mu = a_\mu^{\rm EXP} - a_\mu^{\rm SM} = 2.90\,(90)\times 10^{-9}$, it is interesting to monitor the implications for $\mathrm{BR}(\ell \to \ell'\gamma)$ assuming that such a discrepancy is due to NP. In particular, we find that

$$\mathrm{BR}(\mu \to e\gamma) \approx 10^{-12}\left(\frac{\Delta a_\mu}{3\times 10^{-9}}\right)^2\left(\frac{\theta_{e\mu}}{2\times 10^{-5}}\right)^2,$$

$$\mathrm{BR}(\tau \to \ell\gamma) \approx 10^{-8}\left(\frac{\Delta a_\mu}{3\times 10^{-9}}\right)^2\left(\frac{\theta_{\ell\tau}}{5\times 10^{-3}}\right)^2. \quad (9)$$

where $\theta_{\ell\ell'} = \sqrt{|c_{\ell\ell'}|^2 + |c_{\ell'\ell}|^2}/c_{\mu\mu}$. Therefore, we learn that the a_μ anomaly can be accommodated while satisfying the $\mathrm{BR}(\mu \to e\gamma)$ bound only for extremely small flavor mixing angles $\theta_{e\mu}$.

Similarly, from eq. 7, we find that $d_e \simeq 10^{-24} \times [\mathrm{Im}(c_{ee})/\mathrm{Re}(c_{\mu\mu})]\ e$ cm whenever $\Delta a_\mu \approx 3\times 10^{-9}$. Therefore, also the electron EDM exceeds the current experimental bound by many orders of magnitudes unless there exists a dynamical mechanism suppressing the relevant CP violating phases.

3 Specific NP Models

The phenomenology of cLFV observables has been worked out in a number of well motivated NP scenarios. Among the most important questions are (i) which are the best probes among cLFV processes for any given NP model, (ii) how the predictions compare with the present/foreseen experimental bounds, (iii) what the constraints are on new sources of LFV and new-particle masses, (iv) what are the correlations among different LFV observables.

Concerning the latter point, it should be stressed that 1) ratios for branching ratios of processes such as $\mu \to e\gamma$ and $\tau \to \mu\gamma$ would provide a direct access to the flavor structure of the NP model while 2) a comparative analysis of processes with the same underlying flavor transition (such as $\mu \to e\gamma$ and $\mu \to eee$) would provide information about the operators which are generating potential LFV signals.

In the following, we briefly discuss two classes of NP models: supersymmetric (SUSY) extensions of the SM and strongly interacting models based on the partial-compositeness paradigm.

3.1 SUSY models

In SUSY models, new sources for LFV stem from the soft SUSY-breaking sector since the lepton and slepton mass matrices are generally misaligned. The leading effects for cLFV processes arise from sneutrino-chargino and slepton-neutralino loops. In the generic MSSM, it is useful to stick to the mass-insertion approximation, assuming small off-diagonal entries in the slepton mass matrices $(\delta^{ij}_{AB})_f = (m^2_{\tilde{A}\tilde{B}})_{ij}/m^2_{\tilde{\ell}}$, where $A, B = L, R$ and $m_{\tilde{\ell}}$ is an average slepton mass.

A scenario which has received particular attention after the discovery of the Higgs-like boson at the LHC is the so-called "disoriented A-terms" scenario [15]. The assumption of disoriented A-terms is that flavor violation is restricted to the trilinear terms

$$(\delta^{ij}_{LR})_f \sim \frac{A_f \theta^f_{ij} m_{f_j}}{m_{\tilde{f}}} \qquad f = u, d, \ell\,, \tag{10}$$

where θ^f_{ij} are generic mixing angles. This pattern can be obtained when the trilinear terms have the same hierarchical pattern as the corresponding Yukawa matrices but they do not respect exact proportionality. A natural realization of this ansatz arises in scenarios with partial compositeness [16], where also the SM flavor puzzle can be accounted for. Interestingly, the structure of eq. (10) allows us to naturally satisfy the very stringent flavor bounds of the down-sector thanks to the smallness of down-type quark masses. On the other hand, sizable A-terms help to account for a Higgs boson with mass around 125 GeV while keeping the SUSY scale not too far from the TeV.

The bounds from the lepton sector can be satisfied under the (natural) assumption that the unknown leptonic flavor mixing angles are of the form $\theta^\ell_{ij} \sim \sqrt{m_i/m_j}$ [16]. In particular, we get the following predictions [17]

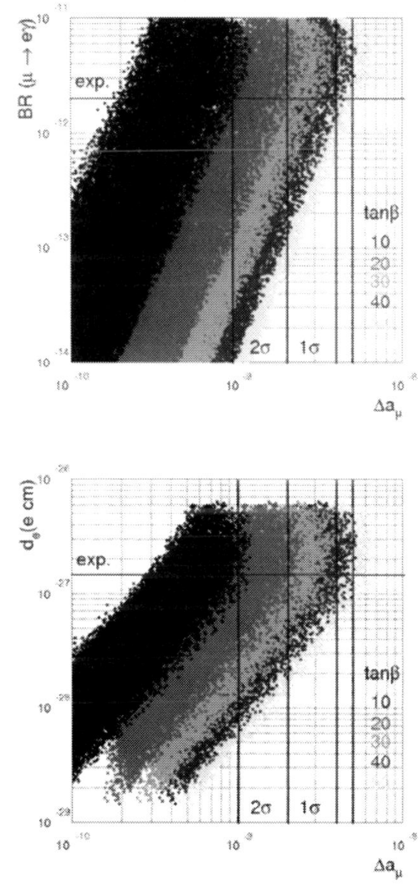

Figure 1. Predictions of the disoriented A-term scenario [17]. Left: $\mu \to e\gamma$ vs. Δa_μ. Right: d_e vs. Δa_μ.

$$\mathrm{BR}(\mu \to e\gamma) \approx 6 \times 10^{-13} \left| \frac{A_\ell}{\mathrm{TeV}} \frac{\theta^\ell_{12}}{\sqrt{m_e/m_\mu}} \right|^2 \left(\frac{\mathrm{TeV}}{m_{\tilde{\ell}}} \right)^4,$$
$$d_e \approx 4 \times 10^{-28}\, \mathrm{Im}\left(\frac{A_\ell\, \theta^\ell_{11}}{\mathrm{TeV}} \right) \left(\frac{\mathrm{TeV}}{m_{\tilde{\ell}}} \right)^2 e\,\mathrm{cm},$$
$$\Delta a_\mu \approx 1 \times 10^{-9} \left(\frac{\mathrm{TeV}}{m_{\tilde{\ell}}} \right)^2 \left(\frac{\tan\beta}{30} \right). \tag{11}$$

where we have assumed that the only possible sources of CP violation arise from A terms, as well. These estimates are fully confirmed by the numerical analysis shown in fig. 1 which has been obtained by means of the following scan: $0.5 \leq |A_e|/\tilde{m} \leq 2$ with $\sin\phi_{A_e}=1$, $\tilde{m} \leq 2$ TeV, $(M_2, \mu, M_1) \leq 1$ TeV and $10 \leq \tan\beta \leq 50$ [17].

It is interesting that disoriented A-terms can account for $(g-2)_\mu$, satisfy the bounds on $\mu \to e\gamma$ and d_e, while giving predictions within experimental reach [17].

3.2 Composite Higgs Models

Besides low-energy supersymmetry, a class of attractive SM extensions addressing the gauge hierarchy problem is provided by composite Higgs models [18, 19], where fermion masses and mixing angles are described by partial compositeness [20]. Light fermions obtain hierarchical masses from the mixing between an elementary sector and a composite one.

As a toy model, let us consider, for each SM fermion, a pair of heavy fermions allowing a Dirac mass term of the order of the compositeness scale and a mixing term with the SM fields [21]

$$\mathcal{L}_Y = -\sum_{i,j=1}^{3}\left(\bar{\ell}_{Li}\Delta_{ij}L_{Rj} - \bar{\tilde{e}}_{Ri}\tilde{\Delta}_{ij}\tilde{E}_{Lj}\right) + h.c.$$
$$-\sum_{i=1}^{3}\left(\bar{L}_i m_i L_i + \bar{\tilde{E}}_i \tilde{m}_i \tilde{E}_i\right)$$
$$-\sum_{i,j=1}^{3}\left(\bar{L}_{Ri}\varphi Y^*_{L_{ij}}\tilde{E}_{Lj} + \bar{L}_{Li}\varphi Y^*_{R_{ij}}\tilde{E}_{Rj}\right) + h.c. \qquad (12)$$

The first line represents the mixing between the elementary composite sectors, the second line contains Dirac mass terms for the fermions of the composite sector and the third line shows the Yukawa interactions which are restricted to the composite sector only with $1 \leq |Y^*_R|, |Y^*_L| \leq 4\pi$. By integrating out the composite sector under the assumptions $m_i = m$, $\tilde{m}_i = \tilde{m}$ and $\tilde{m}, m \gg v$, we get the SM-like Yukawa interaction

$$\mathcal{L}^{eff}_Y = -\bar{\ell}_L\varphi\, \hat{y}^{SM}_\ell\, \tilde{e}_R + \ldots \qquad y^{SM}_\ell = XY^*_R\tilde{X}^\dagger \ , \quad (13)$$

where $X \equiv \Delta m^{-1}$, $\tilde{X} \equiv \tilde{\Delta}\tilde{m}^{-1\dagger}$ and dots stand for contributions of higher order in v/m. The remarkable feature of this pattern is that hierarchical fermion masses and mixing angles can be explained by the mixing matrices X and $\tilde{X}$, even in the presence of anarchical matrices Y^*_R and Y^*_L. At the one-loop level, summing over the h, Z and W amplitudes, we get the main contribution to the electromagnetic dipole operator:

$$A_{\ell\ell'} \sim \frac{1}{16\pi^2}\frac{1}{m\tilde{m}}\left(XY^*_R Y^{*\dagger}_L Y^*_R \tilde{X}^\dagger\right)_{\ell\ell'} . \qquad (14)$$

If Y^*_R and Y^*_L are anarchical matrices we see that, in general, y^{SM}_ℓ and $A_{\ell\ell'}$ are not diagonal in the same basis. From the present bounds on $BR(\mu \to e\gamma)$ and d_e we find that, with reasonable assumptions on the mixing matrices X and $\tilde{X}$, we need $m/\langle Y\rangle$ and $\tilde{m}/\langle Y\rangle$ well above 10 TeV, $\langle Y\rangle$ denoting an average absolute value of the Yukawa matrices. If we postulate that $Y^*_L = 0$, charged lepton masses do not vanish while the leading order contribution (14) does and we find [22]

$$\text{BR}(\mu \to e\gamma) \approx 3 \times 10^{-13}\left(\frac{1.5\,\text{TeV}}{m}\right)^8 |Y^*_R|^8 , \qquad (15)$$

where Y^*_R now stands for an average element of the anarchic matrix $Y^*_{R_{ij}}$, implying that $\mu \to e\gamma$ saturates its current experimental bound for $m \approx 1.5$ TeV and $Y^*_R \approx 1$. A quite similar behavior is expected for the electron EDM. Indeed, setting $Y^*_L = 0$ and assuming $O(1)$ CP-violating phases, it turns out that [22]

$$\frac{|d_e|}{e} \approx 10^{-28} cm \left(\frac{3\,\text{TeV}}{m}\right)^4 Y^{*4}_R . \qquad (16)$$

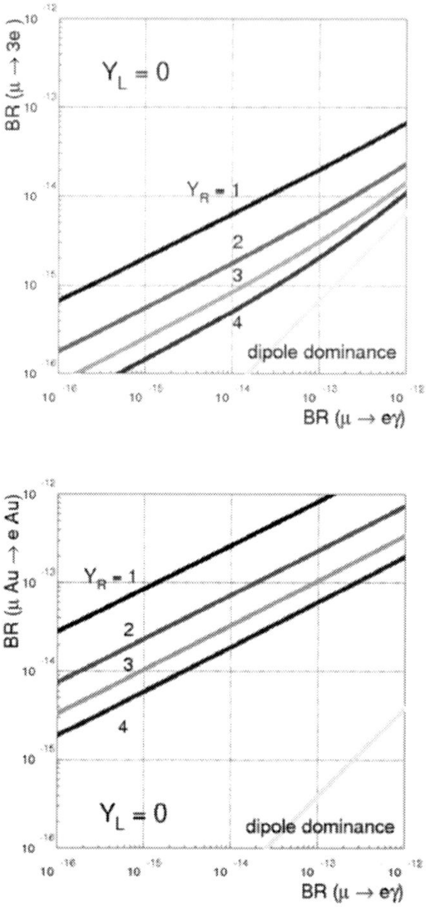

Figure 2. Branching ratio of $\mu \to 3e$ (left) and $\mu^- Au \to e^- Au$ (right) versus the branching ratio of $\mu \to e\gamma$ for $Y^*_L = 0$. The case of dominance of the dipole operator is shown in yellow.

Moreover, it turns $\mu \to 3e$ and $\text{BR}(\mu^- Au \to e^- Au)$ are dominated by non-dipole operators and the correlation of eq. (4) is significantly violated as it is explicitly shown in fig. 2. This is a relevant result, as within composite Higgs models with $Y^*_L \neq 0$, as well as in supersymmetric scenarios, eq. (4) holds to an excellent approximation. In particular, we find [22]

$$\text{BR}(\mu \to 3e) \approx 5 \times 10^{-13}\left(\frac{1\,\text{TeV}}{m}\right)^4 |Y^*_R|^2 , \qquad (17)$$

$$\text{BR}(\mu^- Au \to e^- Au) \approx 4 \times 10^{-13}\left(\frac{3\,\text{TeV}}{m}\right)^4 |Y^*_R|^2 , \qquad (18)$$

and therefore $\mu^- Au \to e^- Au$ is a better probe than $\mu \to 3e$ of the scenario in question.

4 Conclusions

Despite of the fact that the origin of flavor remains a major open problem, significant progress has been achieved in the phenomenological investigation of the sources of flavour symmetry breaking which are accessible at low energies, ruling out models with significant misalignments from the SM Yukawa couplings at the TeV scale.

The search for LFV in charged leptons is probably the most interesting goal of flavour physics in the next years (see table 1). The observation of neutrino oscillations has clearly demonstrated that lepton flavour is not conserved. The question is whether LFV effects can be visible also in other sectors of the theory. The most promising LFV low-energy channels are probably $\mu \to e\gamma$, $\mu \to eee$, $\mu \to e$ conversion in Nuclei as well as τ LFV processes. The current experimental resolutions on cLFV processes are already testing territories of new physics (NP) models well beyond the LHC reach. On the other hand, with the expected sensitivities of next-generation experiments, cLFV will become the most powerful probe of NP signals at our disposal and the interrelationship among leptonic $g-2$, EDMs and cLFV will be of outmost importance to disentangle among different NP scenarios.

Acknowledgments

This work was supported by the ERC Advanced Grant No.267985 *Electroweak Symmetry Breaking, Flavour and Dark Matter: One Solution for Three Mysteries (DaMeSyFla)*.

References

[1] J. Adam *et al.* [MEG Collaboration], Phys. Rev. Lett. **110** (2013) 20, 201801 [arXiv:1303.0754 [hep-ex]].

[2] A. M. Baldini, F. Cei, C. Cerri, S. Dussoni, L. Galli, M. Grassi, D. Nicolo and F. Raffaelli *et al.*, arXiv:1301.7225 [physics.ins-det].

[3] U. Bellgardt *et al.* [SINDRUM Collaboration], Nucl. Phys. B **299** (1988) 1.

[4] A. Blondel, A. Bravar, M. Pohl, S. Bachmann, N. Berger, M. Kiehn, A. Schoning and D. Wiedner *et al.*, arXiv:1301.6113 [physics.ins-det].

[5] W. H. Bertl *et al.* [SINDRUM II Collaboration], Eur. Phys. J. C **47** (2006) 337.

[6] C. Dohmen *et al.* [SINDRUM II. Collaboration], Phys. Lett. B **317** (1993) 631.

[7] COMET Collaboration, http://comet.kek.jp/Documents.html

[8] R. J. Abrams *et al.* [The mu2e Project Collaboration], arXiv:1211.7019 [physics.ins-det].

[9] BaBar Collaboration Collaboration, B. Aubert et al., *Phys.Rev.Lett.* **104** (2010) 021802.

[10] Belle and Belle II Collaborations, K. Hayasaka, *J.Phys.Conf.Ser.* **408** (2013) 012069.

[11] K. Hayasaka, K. Inami, Y. Miyazaki, K. Arinstein, V. Aulchenko, et al., *Phys.Lett.* **B687** (2010) 139.

[12] J. Baron *et al.* [ACME Collaboration], Science **343** (2014) 6168, 269 [arXiv:1310.7534 [physics.atom-ph]].

[13] G. W. Bennett *et al.* [Muon (g-2) Collaboration], Phys. Rev. D **80** (2009) 052008 [arXiv:0811.1207 [hep-ex]].

[14] F. Jegerlehner and A. Nyffeler, Phys. Rept. **477** (2009) 1.

[15] G. F. Giudice, G. Isidori and P. Paradisi, JHEP **1204** (2012) 060.

[16] B. Keren-Zur, P. Lodone, M. Nardecchia, D. Pappadopulo, R. Rattazzi and L. Vecchi, arXiv:1205.5803 [hep-ph].

[17] G. F. Giudice, P. Paradisi and M. Passera, JHEP **1211** (2012) 113 [arXiv:1208.6583 [hep-ph]].

[18] For a review see: B. Bellazzini, C. Csaki and J. Serra, Eur. Phys. J. C **74** (2014) 5, 2766 [arXiv:1401.2457 [hep-ph]].

[19] For a review see: G. Panico and A. Wulzer, arXiv:1506.01961 [hep-ph].

[20] D. B. Kaplan, Nucl. Phys. B **365** (1991) 259.

[21] R. Contino, T. Kramer, M. Son and R. Sundrum, JHEP **0705** (2007) 074 [hep-ph/0612180].

[22] F. Feruglio, P. Paradisi and A. Pattori, Eur. Phys. J. C **75** (2015) 12, 579 doi:10.1140/epjc/s10052-015-3807-9 [arXiv:1509.03241 [hep-ph]].

EPJ Web of Conferences **118**, 01027 (2016) DOI: 10.1051/epjconf/201611801027
FCCP2015

The muon $g-2$ and degenerate supersymmetry

Debtosh Chowdhury[1,a], Ketan M. Patel[2,b], Xerxes Tata[3,c], and Sudhir K. Vempati[4,d]

[1] *Istituto Nazionale di Fisica Nucleare, Sezione di Roma, Piazzale Aldo Moro 2, I-00185, Rome, Italy.*
[2] *Istituto Nazionale di Fisica Nucleare, Sezione di Padova, I-35131 Padova, Italy.*
[3] *Department of Physics and Astronomy, University of Hawaii, Honolulu, HI 96822, USA.*
[4] *Centre for High Energy Physics, Indian Institute of Science, Bangalore 560 012, India.*

Abstract. A degenerate supersymmetric particle spectrum can escape constraints from flavor physics and at the same time evade limits from the direct searches. If such a spectrum is light enough, it can also account for the observed value of the anomalous magnetic moment of the muon. Inspired by this, we consider a scenario where all the soft terms have approximately a common mass scale while allowing for small splittings. We study this scenario considering the constraints from Higgs mass, various B meson decays and the dark matter relic density. We find that, with superpartners $\sim 800-1000$ GeV, it is still possible to escape the present limits from the first run of LHC and flavor physics and can account for muon $g-2$ within 2σ.

1 Introduction

Low energy Supersymmetry (SUSY) has been considered as one of the most compelling solutions of the problem of naturalness of the electroweak scale. It provides dynamical origin of electroweak symmetry breaking and stabilizes the Higgs mass parameter against the large quantum corrections. In its simplest realization, this feature strongly prefers the existence of SUSY breaking scale below TeV. While the Large Hadron Collider (LHC) has conclusively established an existence of the Standard Model (SM) like Higgs boson with mass 126 GeV [1, 2], it hasn't indicated any statistically significant signal of new physics yet.

The mass spectrum of sparticles is not clearly known. Because of the presence of several ideas of SUSY breaking and its mediation to the SM superpartners, no clear prediction for sparticle masses can be made. The SUSY mass spectrum is mostly discussed in the context of rather simplified models such as minimal supergravity or the constrained minimal supersymmetric standard model (CMSSM) or some extended versions of it. These models typically generate a widely spread SUSY mass spectrum that leads to experimentally detectable large visible or missing energies. Both the ATLAS and CMS detectors at the LHC have put significant constraints on these kind of spectrum [3, 4]. The gluino and squarks as heavy as 1.8 TeV have been ruled out by inclusive searches from the data collected with integrated luminosity 20.3/fb with 8 TeV center of mass energy by ATLAS [3]. Further in generic models, squarks up to 850 GeV and gluino up to 1.3 TeV have been ruled out if the lightest supersymmetric particle (LSP) is assumed to be massless [5]. There exist relatively weaker constraints on sleptons and electroweak gauginos due to their small production cross-sections at the LHC. In conclusion, the low energy SUSY with wide mass spectrum is seemingly pushed to a corner by the LHC and, according to some authors, losing its appeal as a solution to the hierarchy problem: see however Ref. [6–8], and references therein.

A very interesting possibility to evade the strong LHC bounds on low energy SUSY is to consider all the sparticles nearly degenerate in masses. The degeneracy among the sparticles and conservation of R-parity makes them hidden from the experimental searches because of small missing and/or visible energy release in sparticle decays. If the sparticle spectrum is compressed, in particular if the mass difference between the colored superpartners such as the squarks and gluino and LSP is very small, there will be less energetic jets/leptons/missing transverse energy leading to low detection efficiency. Such a weak signal would be hidden below the SM backgrounds. Indeed, there exist relatively weak constraints on these kind of spectrum from LHC run-I. For example, the lower limits on stop reduces to 300 GeV if the mass difference between stop and LSP is lower than the top quark mass [3]. A similar reduction on gluino mass constraint holds in case of compressed spectrum.

The phenomenological implications of degenerate SUSY spectra has been discussed in details in [9, 10]. It has been mostly studied from the context of direct searches and collider physics experiments. In [11], we study various indirect constraints and implications of (approximately) degenerate SUSY spectrum. Specifically, we

[a] e-mail: debtosh.chowdhury@roma1.infn.it
[b] e-mail: ketan.patel@pd.infn.it
[c] e-mail: tata@phys.hawaii.edu
[d] e-mail: vempati@cts.iisc.ernet.in

© The Authors, published by EDP Sciences. This is an open access article distributed under the terms of the Creative Commons Attribution License 4.0 (http://creativecommons.org/licenses/by/4.0/).

consider the bounds from B meson decays, Higgs mass and muon anomalous magnetic moment, *i.e.* $(g-2)_\mu$, and their implications on compressed SUSY spectrum. Here, we mostly focus on the SUSY resolution of the $(g-2)_\mu$ anomaly. At present, the discrepancy between the SM prediction $a_\mu^{\rm SM}$ and experimental measurements $a_\mu^{\rm exp}$ of muon anomalous magnetic moment is given by [12]

$$\Delta a_\mu = a_\mu^{\rm exp} - a_\mu^{\rm SM} = (2.87 \pm 0.8) \times 10^{-9}\,, \qquad (1)$$

where $a_\mu \equiv \frac{1}{2}(g-2)_\mu$. Clearly, there is more than 3σ deviation between the SM prediction and experimental observation in the value of a_μ. In this talk, we assume SUSY as the solution of this discrepancy and examine phenomenologically viable SUSY spectrum which can account for the muon magnetic moment anomaly.

2 Degenerate Supersymmetry

Before we discuss the various constraints on the degenerate SUSY, we briefly discuss the possible model origin of SUSY breaking for compressed SUSY spectra. Note that one requires degeneracy between only colored sparticles and the LSP in order to evade the most stringent LHC constraints. However models leading to such degeneracy frequently imply degeneracy among the complete spectrum including sleptons and electroweak gauginos. A well motivated class of models leading to such compressed spectrum arise from the SUSY breaking by extra spatial dimension based on the Scherk-Schwarz mechanism [13]. An example of such model is described in [14]. An extra dimension is compactified on an orbifold S^1/Z_2. The 5D N=1 (or equivalently, N=2 in 4D) SUSY is completely broken by Z_2 parity and a non-trivial twist, parametrized by twist parameter α. One can also use twisting to break $SU(2)_H$ global symmetry of two Higgs doublets. Denoting this twisting by parameter γ, one obtains complete MSSM spectrum in terms of only three parameters: α, γ and the compactification scale $1/R$. At the leading order, the soft SUSY breaking terms are given as:

$$M_1 = M_2 = M_3 = \frac{\alpha}{R}\,, \qquad (2)$$

$$m_{H_u}^2 = m_{H_d}^2 = m_{\tilde Q}^2 = m_{\tilde U}^2 = m_{\tilde D}^2 = m_{\tilde L}^2 = m_{\tilde E}^2 = \left(\frac{\alpha}{R}\right)^2\,, \qquad (3)$$

$$A = -3\frac{\alpha}{R},\;\; \mu = \frac{\gamma}{R},\;\; \mu B = -2\frac{\alpha\gamma}{R^2}\,. \qquad (4)$$

Here $M_{1,2,3}$ represents the gaugino masses and m_s are the soft masses of various scalars in the MSSM. Note that the twisting which breaks SUSY is purely a geometrical effect and does not distinguish among flavours, colours and even between the gauginos and scalars. Further, α and γ are real parameters and do not introduce SUSY CP and flavour effects. The renormalization group effects between the mediation scale $1/R$ and SUSY breaking scale α/R remains small if they are not very far from each other or $\alpha/R \approx 1/R \approx$ TeV. There exists also other variants of the above model with additional freedom among the Higgs sector parameters where the conditions in Eqs. (2,3,4) get modified depending on the location of the matter, gauge and Higgs fields in the extra dimension.

We now review the implications of degenerate soft masses on the physical mass spectrum of Degenerate Minimal Supersymmetric Standard Model (DMSSM). Adopting a phenomenological approach, we set a common mass scale M_D for the soft masses for gauginos, squarks and sleptons.

$$M_1 \approx M_2 \approx M_3 \equiv M_D, \qquad (5)$$

$$m_{\tilde Q}^2 \approx m_{\tilde U}^2 \approx m_{\tilde D}^2 \approx m_{\tilde L}^2 \approx m_{\tilde E}^2 \equiv M_D^2\,. \qquad (6)$$

We also parametrize

$$|\mu|^2 = k_\mu\, M_D^2, \;\; \text{and} \;\; m_A^2 = k_A\, M_D^2\,, \qquad (7)$$

where $k_{\mu,A}$ are real and positive and are $O(1)$ parameter while m_A is the mass of pseudoscalar Higgs. Note that the above relations can arise from the conditions in Eqs. (2,3,4) using specific choices of α, γ and taking the radiative corrections into account.

The physical masses of first and second generations of sfermions turn out to be degenerate with M_D while the third generations of squarks and sleptons receive large corrections form large trilinear terms. The chargino and neutralino mass spectrum also remains degenerate with M_D if $k_\mu \approx O(1)$. The Higgs spectrum implies one light CP even neutral Higgs which should be identified with the observed signal of Higgs boson while the other CP even state, a CP odd and the charged Higgs masses turn out to be $\sqrt{k_A}M_D$. For more details of physical mass spectrum of DMSSM, we refer reader to [11].

3 Muon $(g-2)$ in DMSSM

The SUSY contributes to $(g-2)_\mu$ at one loop through dominant neutralino-charge slepton and chargino-sneutrino loops. It is estimated as [15]:

$$\begin{aligned}\Delta a_\mu &= \frac{\alpha\, m_\mu^2\, \mu M_1 \tan\beta}{4\pi \cos^2\theta_W (m_R^2 - m_L^2)}\left(\frac{f_N[M_1^2/m_R^2]}{m_R^2} - \frac{f_N[M_1^2/m_L^2]}{m_L^2}\right)\\ &+ \frac{\alpha\, m_\mu^2\, \mu\, M_2 \tan\beta}{4\pi \sin^2\theta_W m_L^2}\left(\frac{f_\chi[M_2^2/m_L^2] - f_\chi[\mu^2/m_L^2]}{(M_2^2-\mu^2)}\right). \end{aligned} \qquad (8)$$

The loop integration function are given by

$$\begin{aligned} f_\chi[x] &= \frac{x^2 - 4x + 3 + 2\ln(x)}{(1-x)^3} \Rightarrow \lim_{x\to 1} f_\chi[x] = -\frac{2}{3}\\ f_N[x] &= \frac{x^2 - 1 - 2x\ln(x)}{(1-x)^3} \Rightarrow \lim_{x\to 1} f_N[x] = -\frac{1}{3} \end{aligned} \qquad (9)$$

Assuming the degeneracy $M_1 \approx M_2 \approx m_L \approx m_R \equiv M_D$, Eq. (8) gets simplified to

$$\Delta a_\mu = \frac{\alpha\, m_\mu^2\, \mu\, \tan\beta}{4\pi \sin^2\theta_W M_D^3}\left(\frac{1}{(1-x_r^2)}\left(-\frac{2}{3} - f_\chi[x_r^2]\right) + \frac{1}{3\cot^2\theta_W}\right) \qquad (10)$$

with $x_r = \mu/M_D$. We choose the sign of μ to be positive to account for the correct Δa_μ. Note that $\mu \approx M_D$ is required by the complete degeneracy of neutralino and

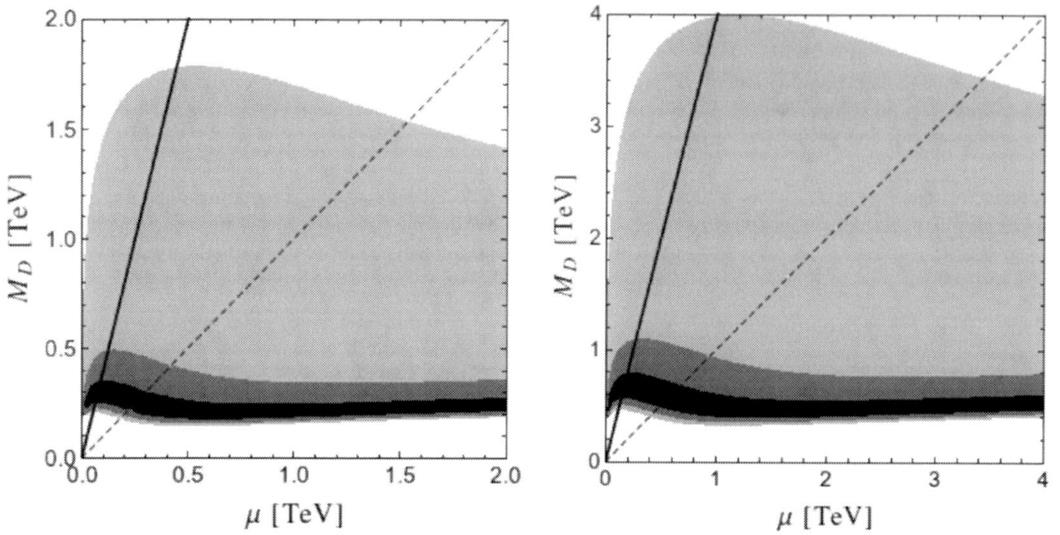

Figure 1. The μ-M_D plane allowed by the 1σ (black), 2σ (green) and 3σ (orange) ranges of the Δa_μ for $\tan\beta = 10$ and $\tan\beta = 50$ in the left and right panel respectively. The solid (dashed) line corresponds to $\mu = \frac{1}{4}M_D$ ($\mu = M_D$). We have set $k_A = 1$.

chargino spectrum. In this limit, the first term dominates over the second and gives finite contribution to $(g-2)_\mu$.

In Fig. 1, we show the parameter space in the $M_D - \mu$ plane (with $k_A = 1$) preferred by Δa_μ as required by Eq. (1) at different levels of statistical significance. Our purpose here is only to delineate the ranges of parameters preferred by the magnetic moment constraint. As expected, low values of M_D (and hence, of sparticle masses) 300-500 GeV (400-1000 GeV) for low (high) values of $\tan\beta$ are preferred at the 2σ level. Note also that the constraint weakens considerably if we allow for deviations up to 3σ. We emphasize that although the direct constraints on squark and gluino masses are satisfied, much of the favoured parameter ranges in Fig. 1 are not viable because they do not yield the correct value of Higgs mass or violate other low energy constraints, as discussed in the next section.

4 Other indirect constraints on DMSSM

Next, we consider the constraints arising from the Higgs mass and most relavant B meson decays, such as $B \to X_s\gamma$ and $B_s \to \mu^+\mu^-$ on the DMSSM. As it is well known, the tree level value of Higgs mass in MSSM is restricted to be $\leq M_Z$ and large radiative corrections are needed to account for the observed Higgs mass. The dominant stop, sbottom and stau 1-loop correction to Higgs mass in the large $\tan\beta$ limit is given by [16]

$$\begin{aligned} m_h^2 &= m_Z^2\cos^2 2\beta + \delta m_h^2\,, \\ \delta m_h^2 &= \frac{3}{4\pi^2}\frac{m_t^4}{v^2}\left(\log\left(\frac{M_{\rm SUSY}^2}{m_t^2}\right) + \frac{X_t^2}{M_{\rm SUSY}^2} - \frac{X_t^4}{12M_{\rm SUSY}^4}\right) \\ &- \frac{3}{48\pi^2}\frac{m_b^4}{v^2}\frac{\tan\beta^4}{(1+\epsilon_b\tan\beta)^4}\frac{\mu^4}{m_{\tilde b}^4} \\ &- \frac{1}{48\pi^2}\frac{m_\tau^4}{v^2}\frac{\tan\beta^4}{(1+\epsilon_l\tan\beta)^4}\frac{\mu^4}{m_{\tilde\tau}^4}\,, \end{aligned} \tag{11}$$

where $X_t = A_t - \mu\cot\beta$. The various ϵ parameters and $M_{\rm SUSY}$ are given in [11] in the limit of degenerate SUSY mass spectrum. As can be seen, large X_t arising from large A_t is required to reproduce the correct Higgs mass. We estimate the Higgs mass using the above simplified formula and in the degenerate SUSY limit parametrized by Eq. (5,7) and using $\mu \approx M_D$. The window 124.4-125.8 GeV in Higgs mass is allowed by experimental measurements at 3σ [17]. In addition to this, we allow $\pm$ 2 GeV as a theoretical uncertainty in the estimation of the Higgs mass due to the simplified 1-loop expression we use. Therefore the conservative Higgs mass range we consider is 122.4 - 127.8 GeV.

We evaluate 1-loop contribution to $B \to X_s\gamma$ in the DMSSM. It is parametrized as:

$$R_{bs\gamma} \equiv \frac{{\rm BR}(B\to X_s\gamma)}{{\rm BR}(B\to X_s\gamma)_{\rm SM}} = 1 - 2.45C_7^{NP} - 0.59C_8^{NP}, \tag{12}$$

Simplified expressions of the Wilson coefficients $C_{7,8}^{NP}$ in the limit of degenerate SUSY spectrum are also given in [11]. In the presence of nonzero A_t (as it is required by Higgs mass) the largest contribution arises from the Higgsino-stop loop and it leads to large flavour violation.

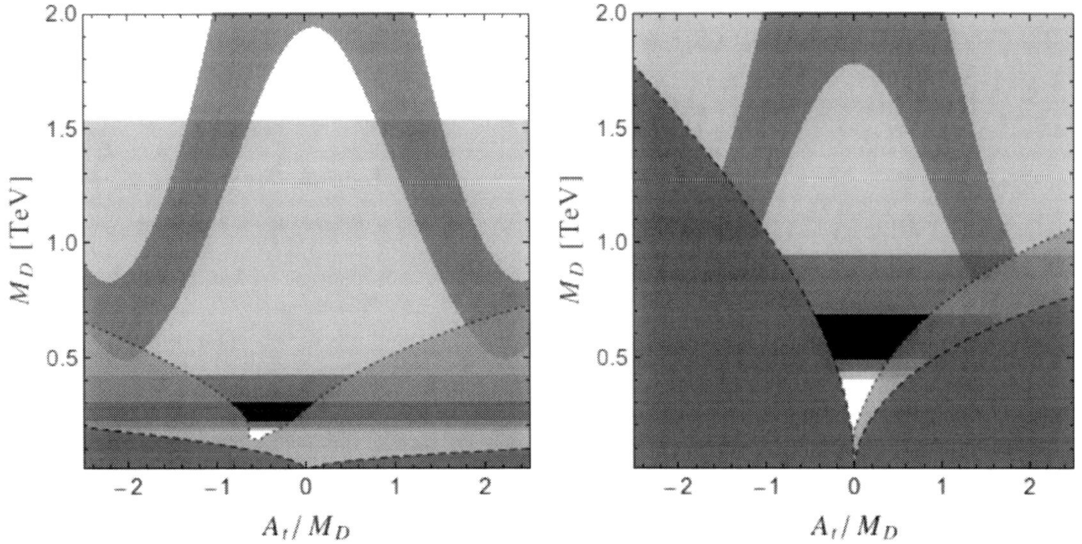

Figure 2. The region in A_t-M_D plane by different constraints for $\tan\beta = 10$ (left panel) and $\tan\beta = 50$ (right panel). The horizontal black, green and orange bands show the favored values of M_D by δa_μ at 1σ, 2σ and 3σ respectively. The red band corresponds to a valid Higgs mass (122.4-127.8 GeV) region. The lighter and darker gray regions are excluded by BR($B \to X_s\gamma$) and BR($B_s \to \mu^+\mu^-$) respectively at 2σ.

The allowed room for new physics in this channel is [16]

$$R_{bs\gamma} = 1.02 \pm 0.10\,. \quad (13)$$

We also impose the $B_s \to \mu^+\mu^-$ constraints whose recent measurement at LHCb is in good agreement with the SM predictions and, therefore, leaves very little room for new physics. The DMSSM contribution to this channel is very sensitive to pseusdoscalar Higgs mass and it is largest when $k_A \lesssim 1$, *i.e.* when the pseudoscalar Higgs is degenerate with SUSY particles.

After imposing all these constraints together with $(g-2)_\mu$, the results are shown in Fig. 2. For this analysis we consider $k_\mu \approx k_A \approx 1$ and various constraints are displayed in $A_t - M_D$ plane for two values of $\tan\beta$. As can be seen, large values of A_t is essentially required for correct Higgs mass. On the contrary, such large A_t introduces large flavour violations particularly in $B \to X_s\gamma$ channel through stop-Higgsino loop. There exist a tiny room for $M_D \approx 800 - 1000$ GeV in case of large $\tan\beta$ and $A_t/M_D \approx 1.5 - 1.8$ which can solve the current discrepancy in $(g-2)_\mu$ at 2σ being consistent with other constraints considered in this paper. Large regions are possible if we relax the $(g-2)_\mu$ requirement to be satisfied at 3σ. Note though, that the size of the grey regions excluded by flavour constraints will be sensitive to our choice of k_A and k_μ.

5 Summary

The LHC constraints on SUSY can be evaded if the sparticle spectrum is approximately degenerate in masses. We show that degenerate sparticles as light as 800 - 1000 GeV can remain hidden from collider searches and can provide a viable solution to the discrepancy between the SM prediction and experimentally observed value of muon anomalous magnetic moment, also respecting various indirect constraints such as the Higgs mass and most relevant B meson decays. As we have shown using simplified analytical estimation, large A_t would be needed to reproduce the correct Higgs mass and large $\tan\beta$ is favoured for large enough contribution to $(g-2)_\mu$. The $(g-2)_\mu$ at 2σ sets an upper limit on degenerate scale, $M_D \lesssim 1$ TeV and requires larger values of $\tan\beta$.

We would like to emphasize that the results presented in this talk are based on simplified analytical formulas and assuming complete degeneracy in the soft masses. There can be various effects which can introduce small departure from the exact degeneracy in soft masses at weak scale. We therefore also perform a full numerical analysis introducing small deviation from the degeneracy and estimate various observables numerically using full 1-loop calculations. Also, we allow m_A to be a free parameter. Our preliminary results of such analysis are in qualitative agreement with the simplified analysis presented in this talk. However, there are significant quantitative changes mainly because of small deviations introduced from the exact degeneracy limit. We also discuss in detail the sparticle spectrum and dark matter constraints. The complete results of numerical analysis will be reported in [11].

Acknowledgments

We acknowledge useful discussions with Biplob Bhattacherjee and Mikołaj Misiak. KMP thanks the organizers of workshop on Flavour Changing and Conserving Processes (FCCP-2015) for an invitation and for their excellent arrangements and kind hospitality during the workshop. He also acknowledges partial support from the European Union network FP7 ITN INVISIBLES (Marie Curie Actions, PITN-GA-2011-289442). XT was supported in part by the US Dept. of Energy, Grant No. DE-SC-0010504. SKV thanks IUSTFF Grant and Department of Physics and Astronomy, University of Hawaii for hospitality where this work was started. DC is supported by the European Research Council under the European Union's Seventh Framework Programme (FP/2007-2013)/ERC Grant Agreement N° 279972. SKV is supported by DST Ramanujan fellowship SR/S2/RJN-25/2008.

References

[1] G. Aad *et al.* [ATLAS Collaboration], Phys. Lett. B **716** (2013) 1 [arXiv:1207.7214 [hep-ex]].

[2] S. Chatrchyan *et al.* [CMS Collaboration], Phys. Lett. B **716** (2012) 30 [arXiv:1207.7235 [hep-ex]].

[3] G. Aad *et al.* [ATLAS Collaboration], JHEP **1510** (2015) 054 [arXiv:1507.05525 [hep-ex]].

[4] V. Khachatryan *et al.* [CMS Collaboration], Phys. Lett. B **745** (2015) 5 [arXiv:1412.4109 [hep-ex]].

[5] G. Aad *et al.* [ATLAS Collaboration], JHEP **1409** (2014) 176 [arXiv:1405.7875 [hep-ex]].

[6] A. Mustafayev and X. Tata, Indian J. Phys. **88** (2014) 991 [arXiv:1404.1386 [hep-ph]].

[7] H. Baer, V. Barger and M. Savoy, Phys. Scripta **90** (2015) 068003 [arXiv:1502.04127 [hep-ph]].

[8] X. Tata, Phys. Scripta **90** (2015) 10, 108001 [arXiv:1506.07151 [hep-ph]].

[9] T. J. LeCompte and S. P. Martin, Phys. Rev. D **84** (2011) 015004 [arXiv:1105.4304 [hep-ph]].

[10] T. J. LeCompte and S. P. Martin, Phys. Rev. D **85** (2012) 035023 [arXiv:1111.6897 [hep-ph]].

[11] D. Chowdhury, K. M. Patel, X. Tata and S. K. Vempati, (in preparation).

[12] T. Blum, A. Denig, I. Logashenko, E. de Rafael, B. Lee Roberts, T. Teubner and G. Venanzoni, arXiv:1311.2198 [hep-ph].

[13] J. Scherk and J. H. Schwarz, Nucl. Phys. B **153** (1979) 61.

[14] R. Barbieri, L. J. Hall and Y. Nomura, Phys. Rev. D **66** (2002) 045025 [hep-ph/0106190].

[15] T. Moroi, Phys. Rev. D **53** (1996) 6565 [Phys. Rev. D **56** (1997) 4424] [hep-ph/9512396].

[16] W. Altmannshofer, M. Carena, N. R. Shah and F. Yu, JHEP **1301** (2013) 160 [arXiv:1211.1976 [hep-ph]].

[17] G. Aad *et al.* [ATLAS and CMS Collaborations], Phys. Rev. Lett. **114** (2015) 191803 [arXiv:1503.07589 [hep-ex]].

EPJ Web of Conferences **118**, 01028 (2016) DOI: 10.1051/epjconf/201611801028
FCCP2015

Status of the Mu3e Experiment at PSI

Ann-Kathrin Perrevoort[1,a] for the Mu3e Collaboration

[1] *Physikalisches Institut, Heidelberg University, Im Neuenheimer Feld 226, 69120 Heidelberg, Germany*

Abstract. The Mu3e experiment aims to search for the lepton-flavour violating decay $\mu^+ \rightarrow e^+e^+e^-$ with a sensitivity to one signal decay in 10^{16} muon decays at a muon stopping rate of 2×10^9 muons/s. With currently available rates of 10^8 muons/s, a sensitivity on the branching ratio of 10^{-15} is the aim of the first phase. This will allow for tests of new physics models with enhanced branching ratios for lepton-flavour violating processes with an unprecedented precision.
The experiment must be operated at very high muon rates all the while suppressing the background of the decay $\mu^+ \rightarrow e^+e^-e^+\overline{\nu}_\mu\nu_e$ and accidental electron-positron combinations. Therefore, a tracking detector made of thin pixel sensors with additional scintillating fibres and tiles for precise time measurement will be built. The development of the subdetectors is ongoing while detector construction is still in preparation.

1 Introduction

The Mu3e experiment is dedicated to search for physics beyond the Standard Model (BSM) via the lepton flavour-violating decay $\mu^+ \rightarrow e^+e^+e^-$. In principle, this process can be mediated via neutrino mixing (Fig. 1(a)), but as it is suppressed to a branching ratio below 10^{-54}, any observation would be a clear sign of new physics. In fact, many BSM models naturally predict enhanced branching ratios for charged lepton flavour violating processes like $\mu \rightarrow eee$ [1].
In BSM models, the $\mu \rightarrow eee$ decay can be mediated at lowest order by dipole-like interactions (Fig. 1(b)) or four-fermion contact interactions (Fig. 1(c)) depending on the particular model.
In [2], the $\mu \rightarrow eee$ and $\mu \rightarrow e\gamma$ decay have been studied by the means of effective field theory. Depending on the underlying interaction, a search for $\mu \rightarrow eee$ with a sensitivity of 10^{-16} on the branching ratio is able to test the effective mass scale of lepton-flavour violating operators in the range of several thousand TeV (Fig. 2).
The current limit on the process $\mu \rightarrow eee$ has been set by the SINDRUM experiment to $BR_{\mu\rightarrow eee} < 1.0 \times 10^{-12}$ at 90 % confidence level (CL) [3]. The Mu3e experiment aims to search for $\mu \rightarrow eee$ with a sensitivity on the branching ratio of 10^{-15} at 90 % CL in the first stage of the experiment and 10^{-16} at 90 % CL in the second stage [4].

2 Signal and Background

Experimentally the $\mu^+ \rightarrow e^+e^+e^-$ decay is identified by measuring two positrons and an electron from a common vertex in space and time. In the Mu3e experiment the antimuons are stopped and decay at rest. Therefore, the energies of the electron and positrons sum up to the muon mass, whereas the sum of their momenta vanishes. Any other process that mimics this signature is a potential source of background and needs to be suppressed in the experiment to below the sensitivity level.
One source of background can arise from the accidental combination of any two positrons and an electron that within the detector resolution shows the characteristics of the decay signal. As an example, this could be a positron from the dominant Michel decay $\mu^+ \rightarrow e^+\overline{\nu}_\mu\nu_e$ in combination with a positron and electron from a Bhabha scattering event or photon conversion. The rate of the accidental combination background is dependent on the rate of incoming muons; higher antimuon rates result in more accidental combinations. This source of background can be controlled by optimized time, vertex and momentum resolutions.
Furthermore, the radiative decay of an antimuon with internal conversion $\mu^+ \rightarrow e^+e^-e^+\overline{\nu}_\mu\nu_e$ has a similar signature as the decay signal. The undetected neutrinos create missing energy, thus an excellent momentum resolution is crucial in order to control this type of background. This is illustrated in Fig. 3 which shows the integrated branching ratio for the internal conversion decay for different cut values on the sum of the energies of the three decay particles E_{tot}. Taking this into account, a momentum resolution of ~0.5 MeV is required to sufficiently suppress internal conversion decays. This is particularly challenging to achieve as the electrons[1] have a low momentum and thus are strongly affected by multiple Coulomb scattering which in turn affects the momentum measurement.

[a] e-mail: perrevoort@physi.uni-heidelberg.de

[1] Here and in the following, electrons and positrons will simply be denoted as electrons. In the same way, muons and antimuons will be referred to as muons.

© The Authors, published by EDP Sciences. This is an open access article distributed under the terms of the Creative Commons Attribution License 4.0 (http://creativecommons.org/licenses/by/4.0/).

DOI: 10.1051/epjconf/201611801028

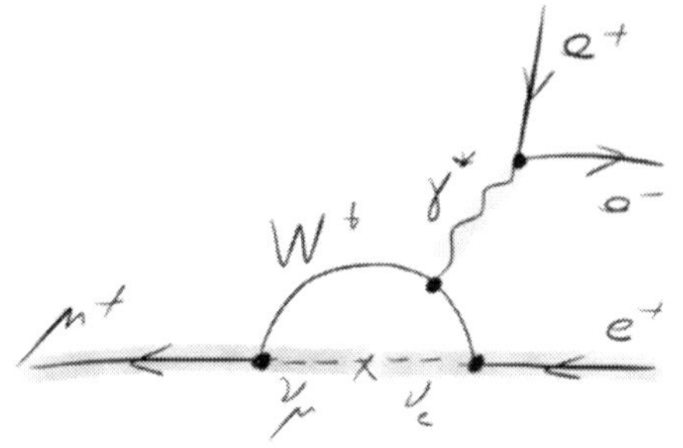

(a) The decay μ → eee mediated by neutrino mixing in the Standard Model.

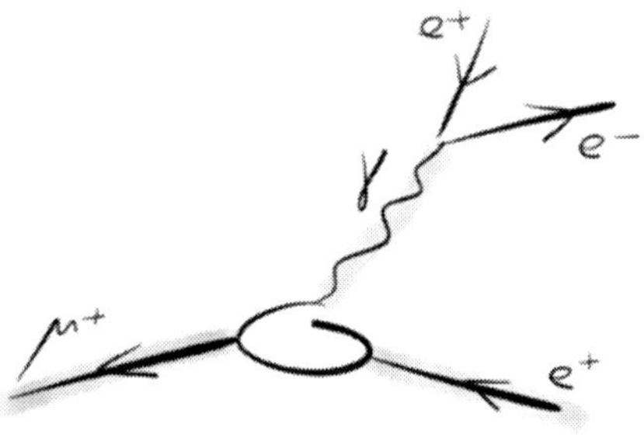

(b) The μ → eee decay at loop level in BSM models.

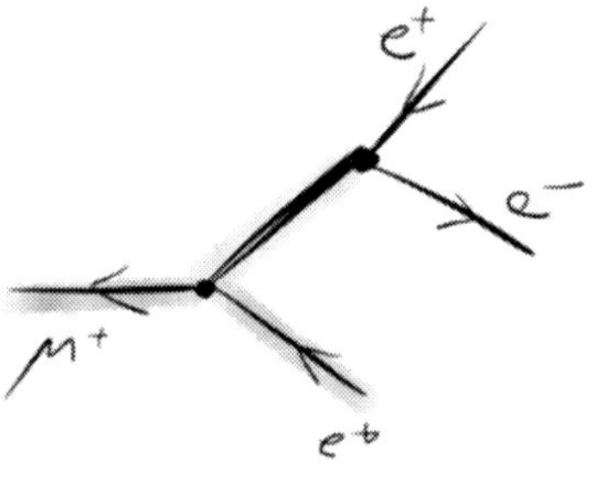

(c) The μ → eee decay at tree level in BSM models.

Figure 1. Diagrams of the decay μ → eee (a) mediated in the Standard Model via neutrino oscillations, and at (b) loop and (c) tree level in BSM models.

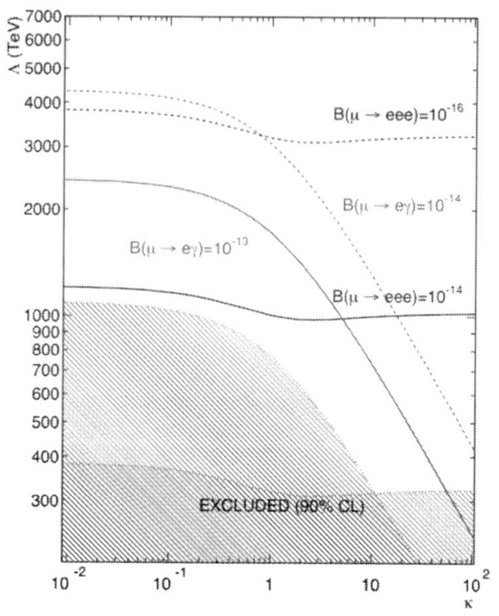

Figure 2. Comparison of the reach in effective mass scale Λ of new physics theories in searches of the decay μ → eγ and μ → eee. Low values of κ represent a dominating dipole contribution whereas high κ values represent dominating four-fermion contact interactions [2].

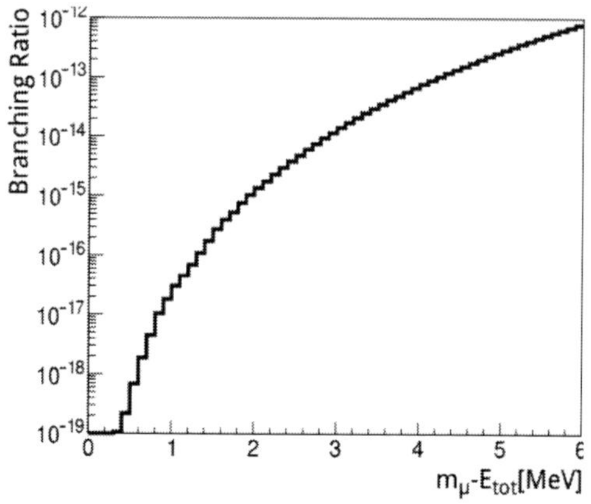

Figure 3. Integrated branching ratio for the radiative muon decay with internal conversion μ → eeν$\overline{\nu}$ for different cut values on the total energy E_{tot} of the three decay electrons. Adapted from [5], matrix element by [6].

For the Mu3e experiment a vertex resolution of about 200 μm and a time resolution of ~100 ps is envisaged.

3 Experimental Concept

The detector in the Mu3e experiment has to be able to cope with high muon stopping rates while the amount of material in the active detector volume has to be kept to a minimum. With thin pixel sensors, a high momentum resolution will be achieved, while scintillating fibres and tiles will allow for precise time measurements.

The Mu3e detector will be built in stages corresponding to an increase in the muon stopping rate.

3.1 Muon Beam

At the Paul-Scherrer Institute (PSI) a proton beam with 590 MeV is provided with a very high intensity of 2.2 mA. Pions are produced on carbon targets and decay subsequently into muons. The πE5 beam line, currently in use

by the MEG experiment [7], is one of the highest intensity low energy muon beam lines at PSI with attainable rates of up to 10^8 muons/s of 28 MeV/c muons. It is planned that the beam time at πE5 will be shared between the MEG and Mu3e experiments.
The envisaged sensitivity of 10^{-16} on the branching ratio, however, can only be reached by muon rates of about 2×10^9 muons/s. This would require a new beam line – a possibility that is currently under investigation at PSI.

3.2 Detector Design

In the Mu3e experiment the incoming 28 MeV/c muon beam will be stopped in a hollow double-cone shaped target made of Mylar. The shape of the target is chosen such that the decay vertices will spread over a large area which aids in the reduction of background from accidental combinations. As the target and detectors will be placed in a solenoidal magnetic field of 1 T, the momentum of the decay electrons will be measured by determining the bending radius of their trajectories. The pixel detector will be arranged in a barrel-type geometry around the beam axis (Fig. 4).
In the first phase of the experiment (phase IA), there will be a double-layer of pixel sensors surrounding the target, and one double-layer at larger radii (see the central detector part in Fig. 4). The detector will be 36 cm long with a diameter of 18 cm, and will initially be operated at 2×10^7 muons/s. At this stage of the experiment a sensitivity to $\mu \rightarrow eee$ of 10^{-13} to 10^{-14} can be achieved depending on the run time.
In order to improve the momentum resolution the electrons will be re-measured as they move on a helical trajectory and pass again through the detector. In phase IB of the experiment, the acceptance for the re-curling tracks will be increased by installing so-called recurl-stations upstream and downstream of the central detector. Similarly to the central detector the recurl-stations will consist of a double-layer of pixel sensors and will have the same dimensions.
An operation at 1×10^8 muons/s will become feasible in phase IB by installing additional timing detectors: scintillating fibres in the central detector, and scintillating tiles in the additional recurl-stations. Thus, accidental combinations will be efficiently suppressed. A target sensitivity of up to 10^{-15} in branching ratio is attainable in this particular detector configuration.
Given a new beam line, the Mu3e experiment will be operated in phase II studying 2×10^9 muons/s. In order to increase the acceptance for recurling tracks, an additional pair of recurl-stations will be installed, one upstream and another downstream of the existing detector. In this configuration the Mu3e experiment can reach its target sensitivity of 10^{-16}.

3.3 Pixel Sensors

The tracking detector in the Mu3e experiment will be built from silicon pixel sensors as they provide the high granularity necessary at high muon stopping rates [8].

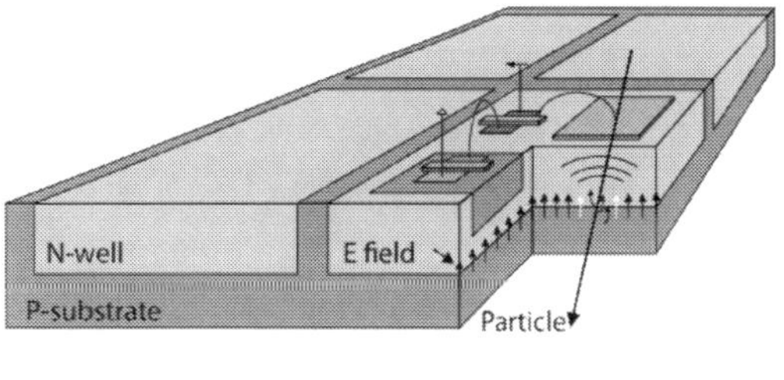

Figure 5. Schematic of a pixel sensor in HV-MAPS technology [9].

The High-Voltage Monolithic Active Pixel Sensors (HV-MAPS) [9–13] technology has been chosen for the Mu3e experiment. The pixels are implemented as deep N-wells in a p-doped substrate and are reversely biased by about 60 V to 90 V (Fig. 5). Thus, charges generated by traversing particles are fastly collected via drift within $O(1\text{ ns})$.
Moreover, the charge collection is limited to a thin depleted zone close to the surface. Therefore, the sensors can be thinned to 50 μm which allows for a very lightweight pixel detector.
Another advantage of the HV-MAPS is the possibility to implement analogue and digital logic directly on the sensor allowing for amplification and digitization of the signals already on the chip. No additional readout chip is needed as in standard hybrid pixel sensor designs.
The Mu3e experiment plans to use HV-MAPS chips of a size of $2 \times 2\,\text{cm}^2$ with a pixel size of $80 \times 80\,\mu\text{m}^2$ and a thickness of 50 μm.
The latest HV-MAPS prototype for the Mu3e experiment is the MuPix7. It is a small sensor with a size of $2.9 \times 3.2\,\text{mm}^2$ and a pixel size of $103 \times 80\,\mu\text{m}^2$, but features already the full functionality of the final chip including an on-chip readout state machine. The output of the MuPix7 is zero-suppressed hit addresses and corresponding time stamps transmitted via a fast serial link at 1.25 Gbit/s.
The MuPix prototypes have been extensively tested in the laboratory and in testbeam campaigns [14, 15]. For MuPix7, efficiencies above 99 % for the entire test system at a noise rate of less than 100 Hz per pixel have been measured as well as a time resolution of better than 12 ns. Thus, already the MuPix7 prototype meets all the requirements on the pixel sensors.

3.4 Lightweight Mechanics and Cooling

For the 50 μm thin pixel sensors 100 μm thick Kapton based flexprints with aluminum or copper traces will be utilized to connect the signals to the front-end electronics and to power the sensors. In addition, the sensors with flexprints will be glued to a mechanical support structure built of 25 μm Kapton foil. Thus, one layer of the pixel detector will contribute to the overall material budget with about 1 ‰ of a radiation length.
The sensors and electronics dissipate heat and need to be cooled. In the case of the Mu3e experiment, gaseous helium is chosen as a coolant in order to minimize multiple

 DOI: 10.1051/epjconf/201611801028

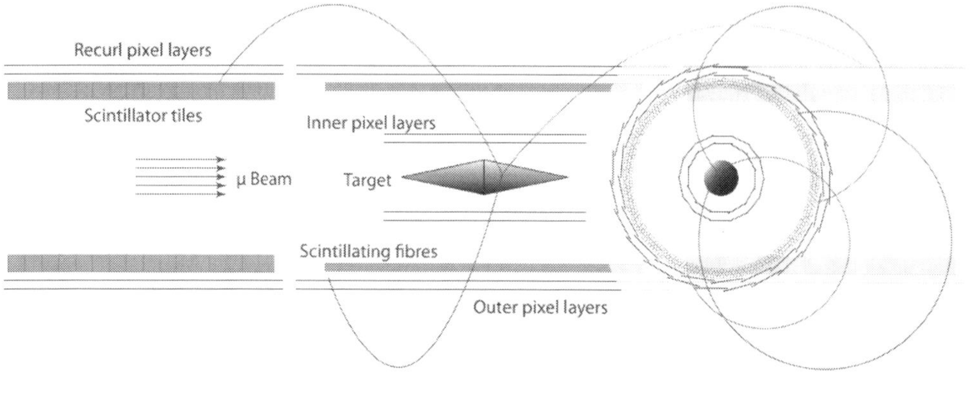

Figure 4. Schematic layout of the Mu3e detector in phase IB to be operated at 10^8 muons/s.

Coulomb scattering in the detector volume.
The cooling concept has been studied in simulation and tested with mechanical prototypes built of Kapton and thin glass plates [16].

3.5 Scintillating Fibres and Tiles

In order to improve the time resolution, the detector will be equipped with scintillating detectors from phase IB onward. In the most central part of the detector, three layers of 250 μm thick scintillating fibres will be used with silicon photo-multiplier (SiPM) readout at both ends. Fibres with round and squared cross sections are currently under study. Current prototypes of scintillating fibre modules yield a time resolution of below 1 ns [17].
In the recurl-stations the amount of material is no longer an issue and therefore scintillating tiles of about 0.2 cm^3 will be used. The tiles will also be read out by SiPMs. Current prototypes yield a time resolution of better than 100 ps [18].
The readout of the SiPMs of both fibres and tiles will be performed with a custom-designed ASIC, the STiC chip [19, 20].

3.6 Data Acquisition

In the final phase of Mu3e the detector will consist of nearly 300 million pixels, 4000 scintillating fibres and 7000 scintillating tiles.
The data acquisition will be performed without a hardware trigger [21]. All sub-detectors will continuously send zero-suppressed data to the readout chain. At 2×10^9 muons/s, this will account for a data rate of about 1 Tbit/s. The data will then be forwarded to a filter farm consisting of PCs equipped with graphics processing units. On the filter farm, fast online track reconstruction will be performed and signal-like topologies will be searched for.

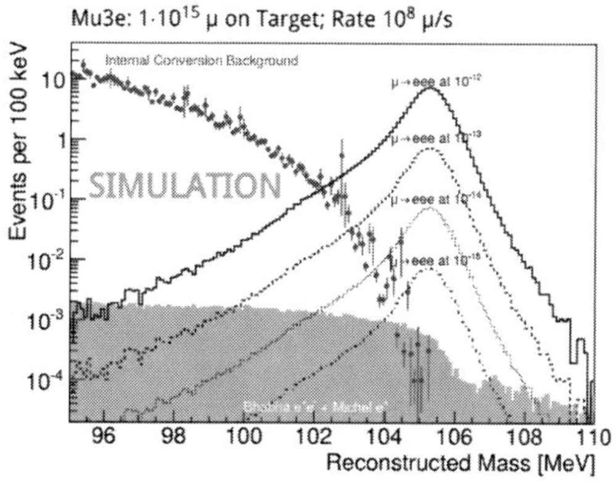

Figure 6. Reconstructed mass of three electrons from simulated internal conversion events, accidental combinations of Bhabha scattering events and Michel decays. In comparison, the $\mu \rightarrow eee$ decay at various branching ratios is shown.

By this means, the data rate written to mass storage will be reduced to less than 100 MB/s.

4 Sensitivity Studies

In order to estimate the detector performance, a full Geant4 [22, 23] based simulation of the Mu3e detector has been established.
The result for the phase IB detector configuration with 10^{15} muons on target at a rate of 10^8 muons/s can be seen in Fig. 6. It shows the reconstructed invariant mass of three electrons in the signal region from different sources of background, namely the radiative decay with internal conversion and accidental combinations of electrons from Bhabha scattering and the Michel decay, as well as signal events implemented as three body decays at different

branching ratios. The detector performs well enough to separate signal decays from background even at a branching ratio of 10^{-15}.

5 Status and Outlook

The research proposal [4] for the Mu3e experiment was approved at PSI in 2013. At the moment, the technical design report is being prepared and will be presented at PSI in 2016.
Research and development on the subdetectors as well as the preparation of the detector construction is ongoing. Commissioning of the detector and first data collection is expected for 2017.

Acknowledgements

The author acknowledges support by the DFG Research Training Group on Particle Physics beyond the Standard Model.

References

[1] Y. Kuno, Y. Okada, Rev.Mod.Phys. **73**, 151-202 (2001)
[2] A. Gouvêa, P. Vogel, Prog.Part.Nucl.Phys. **71**, 75-92 (2013)
[3] U. Bellgardt et al., Nucl.Phys. **B299**, 1-6 (1988)
[4] A. Blondel et al., *Mu3e Research Proposal* (2013)
[5] R. Djilkibaev, R. Konoplich, Phys.Rev. **D79**, 073004 (2009)
[6] A. Signer et al., *private communication* (2015)
[7] J. Adam et al., Eur.Phys.J. **C73**, 2365 (2013)
[8] N. Berger et al., Nucl.Instrum.Meth. **A732**, 61-65 (2013)
[9] I. Perić, Nucl.Instrum.Meth. **A582**, 876-885 (2007)
[10] I. Perić, C. Takacs, Nucl.Instrum.Meth. **A624**, 504-508 (2010)
[11] I. Perić, C. Kreidl, P. Fischer, Nucl.Instrum.Meth. **A650**, 158-162 (2010)
[12] I. Perić et al., Nucl.Instrum.Meth. **A731**, 131-136 (2013)
[13] I. Perić et al., JINST **10**, C05021 (2015)
[14] S. Shrestha, PoS **TTIP2014**, 047 (2014)
[15] D. Wiedner et al., JINST **10**, C03044 (2015)
[16] D. Wiedner et al., JINST **9**, C08023 (2014)
[17] A. Bravar, Nucl.Part.Phys.Proc. **260**, 155-159 (2015)
[18] P. Eckert, *PhD Thesis* (Heidelberg University, 2015)
[19] W. Shen et al., NSS/MIC **C10-10-30**, 406-408 (2010)
[20] T. Harion et al., JINST **9**, C02003 (2014)
[21] D. Wiedner et al., JINST **9**, C01011 (2014)
[22] S. Agostinelli et al., Nucl.Instrum.Meth. **A506**, 250-303 (2003)
[23] J. Allison et al., IEEE Trans.Nucl.Sci. **53**, 270 (2006)

Leading-order hadronic contributions to the lepton anomalous magnetic moments from the lattice

Florian Burger[1], Xu Feng[2], Karl Jansen[4], Marcus Petschlies[5,a], Grit Pientka[3], and Dru B. Renner[6]

[1] OakLabs GmbH, Neuendorfstr. 20B, D-16761 Hennigsdorf, Germany
[2] Columbia University, Department of Physics, 538 West 120th Street, 804 Pupin Hall, New York, NY 10027, USA
[3] Humboldt-Universität zu Berlin, Institut für Physik, Newtonstr. 15, D-12489 Berlin, Germany
[4] NIC, DESY, Platanenallee 6, D-15738 Zeuthen, Germany
[5] Rheinische Friedrich-Wilhelms-Universität Bonn, Institut für Strahlen- und Kernphysik, Nußallee 14-16, D-53115 Bonn, Germany
[6] Los Alamos National Laboratory (LANL), P.O. Box 1663, Los Alamos, NM 87545, USA

Abstract. The hadronic leading-order (hlo) contribution to the lepton anomalous magnetic moments a_l^{hlo} of the Standard Model leptons still accounts for the dominant source of the uncertainty of the Standard Model estimates. We present the results of an investigation of the hadronic leading order anomalous magnetic moments of the electron, muon and tau lepton from first principles in twisted mass lattice QCD. With lattice data for multiple pion masses in the range 230 MeV $\lesssim m_{PS} \lesssim$ 490 MeV, multiple lattice volumes and three lattice spacings we perform the extrapolation to the continuum and to the physical pion mass and check for all systematic uncertainties in the lattice calculation. As a result we calculate a_l^{hlo} for the three Standard Model leptons with controlled statistical and systematic error in agreement with phenomenological determinations using dispersion relations and experimental data. In addition, we also give a first estimate of the hadronic leading order anomalous magnetic moments from simulations directly at the physical value of the pion mass.

1 Introduction

Lattice QCD offers the opportunity to study hadronic contributions to electroweak observables from first principles. This conference contribution focuses on the hadronic leading order (in the electroweak couplings) contribution to the anomalous magnetic moments a_l^{hlo} of the Standard Model leptons electron, muon and tau. Beyond these important quantities, the calculation shown here can be adapted to a broad range of applications for a variety of quantities such as the next-to-leading order vacuum-polarization type contributions to the lepton anomalous magnetic moments, the hadronic running of the electroweak couplings α_{QED} and $\sin^2(\theta)$, the Adler function or Λ_{QCD} amongst others [25].

As a crucial prerequisite, the nonperturbative hadron dynamics entering all these quantities can be calculated entirely in Euclidean spacetime. This makes them accessible to ab-initio lattice calculations which simulate the strong interaction of quarks and gluons on a Euclidean spacetime lattice. For the muon anomalous magnetic moment in particular, this has been worked out in the pioneering work by Blum [1]. On the other hand, the regularization of QCD by the finite lattice volume L^3, the lattice spacing a and the simulation at unphysically large pion masses introduce a systematic dependence on the infrared and ultraviolet cutoff, $1/L$ and $1/a$, respectively, and the light pseudoscalar mass, which needs to be extrapolated with care to the corresponding limits, $L \to \infty$, $a \to 0$ and $m_{PS} \to m_\pi$.

The simulation effort by the European Twisted Mass Collaboration (ETMC) has matured to a state, at which we can calculate the above mentioned quantities for multiple lattice volumes, lattice spacings and pion masses close to or even at the physical point with contributions from up, down, strange and charm quarks. Within this framework we are now able to reach for high-precision results from lattice QCD with control over systematic uncertainties. We report our progress for the hadronic leading order's contribution to the anomalous magnetic moments of the Standard Model leptons with a goal of reaching the current precision obtained in phenomenological analyses using experimental data.

The essential ingredient for this calculation is the hadronic vacuum polarization tensor and function

$$\Pi_{\mu\nu}^{AB}(Q) = \int d^4x \, \langle J_\mu^A(x)\, J_\nu^B(y)\rangle \, e^{iQ(x-y)} = \left(Q_\mu Q_\nu - \delta_{\mu\nu} Q^2\right) \Pi^{AB}(Q^2)\,. \quad (1)$$

determined from lattice QCD. Here J_μ^A denotes a flavor diagonal isospin component of the vector current or a linear combination of such components. In particular a_l^{hlo} requires the vacuum polarization function from the 2-point correlator of electromagnetic currents, $A = \gamma$, as part of the integrand for the QED-1-loop integral (cf. figure [1])

[a] e-mail: marcus.petschlies@hiskp.uni-bonn.de

© The Authors, published by EDP Sciences. This is an open access article distributed under the terms of the Creative Commons Attribution License 4.0 (http://creativecommons.org/licenses/by/4.0/).

[1]

$$a_l^{\mathrm{hlo}} = 4\alpha^2 \int \frac{dQ^2}{Q^2}\,\Pi_R^{\gamma\gamma}(Q^2)\,w\big(Q^2/m_l^2\big)\,. \tag{2}$$

with the subtracted vacuum polarization function

$$\Pi_R^{\gamma\gamma}(Q^2) = \Pi^{\gamma\gamma}(Q^2) - \Pi^{\gamma\gamma}(0)\,.$$

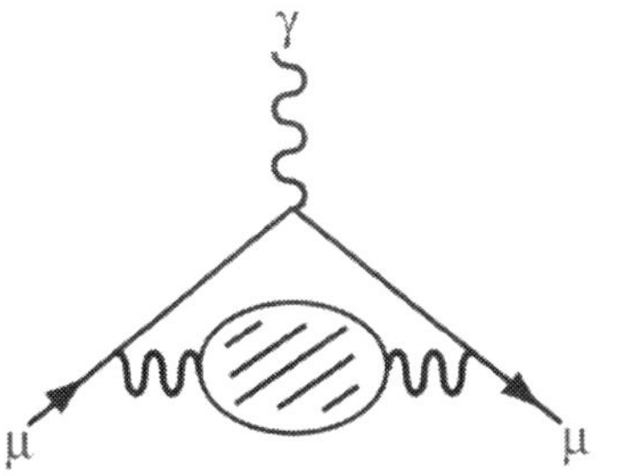

Figure 1. Hadronic leading order correction to the muon-photon-vertex

The muon anomalous magnetic moment has been determined both experimentally and theoretically to an uncertainty of approximately 0.5 ppm with a discrepancy of 2 to 4 standard deviations between the experimental and the Standard Model value [3]

$$a_\mu(\mathrm{EXP}) = 11659208.9\,(6.3)\cdot 10^{-10}\,,$$
$$a_\mu(\mathrm{SM}) = 11659180.4\,(4.2)\,(2.6)\cdot 10^{-10}\,.$$

$a_\mu(\mathrm{SM})$ is not a pure Standard Model number, though, since its extraction partly relies on experimental data and some amount of modeling. Thus, it has become a focal point of numerous studies in lattice QCD with an enhanced effort to investigate the systematic uncertainties which arise naturally in the framework of a lattice QCD calculation, such as the finite lattice volume.

Though not immediately apparent from the definition in equation (2), the extension of the lattice QCD calculation to the electron and tau hadronic leading order anomalous magnetic moments is a non-trivial step. From the practical point of view, the calculation is universal for all three leptons since the specific lepton property only enters via the lepton mass in the analytically known weight function $w\big(Q^2/m_l^2\big)$ in the integrand. However, the ratios of the lepton masses $m_e^2 : m_\mu^2 : m_\tau^2 \approx 1 : 4\cdot 10^4 : 10^7$ extending over 7 orders of magnitude imply a significantly different weighting of the momentum dependence of the polarization function in the momentum integration for the individual leptons. As a consequence the individual lattice artifacts will have significantly different bearings on a_l^{hlo}, in each case, which we discuss in more detail below.

2 Twisted mass lattice QCD calculation

We use the twisted-mass lattice regularization of QCD [9] for a mass degenerate light quark doublet $\chi_l = (u,d)^T$ and a non-degenerate heavy quark doublet $\chi_h = (c,s)^T$ together with the Iwasaki gauge action [22]. The sea-quark action is given by

$$S_{\mathrm{sea}} = \sum_x \bar\chi_l(x)\left(D_W + i\mu_l\gamma_5\tau^3\right)\chi_l(x)$$
$$+ \sum_x \bar\chi_h(x)\left(D_W + i\mu_\sigma\gamma_5\tau^1 + \mu_\delta\tau^3\right)\chi_h(x)\,. \tag{3}$$

D_W is the Wilson-Dirac operator

$$D_W = \frac{1}{2}\gamma_\mu\left(\nabla_\mu^f + \nabla_\mu^b\right) - \left(\frac{ar}{2}\nabla_\mu^b\nabla_\mu^f - m_{\mathrm{cr}}\right) + (m_0 + m_{\mathrm{cr}}) \tag{4}$$

where ∇^f, ∇^b denote the gauge-covariant, discrete forward and backward derivative, m_0 the bare quark mass and $\tau^{1,2,3}$ the Pauli matrices. The twisted mass term for the heavy quark doublet $i\gamma_5\tau^1\mu_\sigma$ breaks the vector part of isospin symmetry such that the flavor diagonal isovector currents, and in particular also the electromagnetic current, are no longer conserved. We thus use a mixed action ansatz with a valence quark action that has improved isospin symmetry [8, 10, 19],

$$S_{\mathrm{val}} = \sum_{f=l,s,c}\sum_x \bar\psi_f(x)\left(D_W + i\mu_f\gamma_5\tau^3\right)\psi_f(x) \tag{5}$$

$$\psi_f = \left(\psi_f^+, \psi_f^-\right)^T\,. \tag{6}$$

Using this mixed action on the lattice has been shown to guarantee the automatic absence of lattice artifacts of $O(a)$ in the continuum limit of physical, on-shell observables [10] by solely tuning the bare quark mass m_0 to its critical value $m_0 \to m_{\mathrm{cr}}$ without any further improvement of lattice action or operators. This property remains true for the hadronic vacuum polarization function $\Pi^{\gamma\gamma}$ in the presence of off-shell, short distance contributions in equation (1) [20]. The renormalized strange and charm valence quark masses in the mixed action are fixed by requiring $2m_K^2 - m_{PS}^2$ and the D-meson mass calculated in the valence sector to attain their physical value.

The gauge field ensembles with $N_f = 2+1+1$ dynamical quarks produced by the ETMC and used for this calculation are listed in table {1}. Details about their production and properties are given in references [6, 7]. Additionally, we consider one ETMC ensemble at the physical pion mass [16, 17, 21], whose parameters are listed in the last row of the table.

For the valence quark action eq. (5) the electromagnetic current on the lattice follows as the Noether current from the vector flavor variation

$$\delta\psi(x) = iQ^{\mathrm{em}}\psi(x)\,,\quad \delta\bar\psi(x) = -i\bar\psi(x)\,Q^{\mathrm{em}} \tag{7}$$

generated by the electromagnetic charge matrix $Q^{\mathrm{em}} = \mathrm{diag}\,(+2/3, -1/3, \ldots)$ acting in flavor space. Eq. (7) defines the electromagnetic current as the usual sum of single-flavor quark currents

$$J_\mu^\gamma = \frac{2}{3}J_\mu^u - \frac{1}{3}J_\mu^d + \frac{2}{3}J_\mu^c - \frac{1}{3}J_\mu^s$$

Table 1. Gauge field ensembles generated by the European Twisted Mass Collaboration with $N_f = 2 + 1 + 1$ dynamical quarks [6, 7] and with $N_f = 2$ dynamical quark flavors at physical pion mass [16, 17] (last row).

Ensemble	a[fm]	m_{PS}[MeV]	L[fm]	$m_{PS} \cdot L$
D15.48	0.061	227	2.9	3.3
D30.48	0.061	318	2.9	4.7
D45.32sc	0.061	387	1.9	3.7
B25.32t	0.078	274	2.5	3.5
B35.32	0.078	319	2.5	4.0
B35.48	0.078	314	3.7	5.9
B55.32	0.078	393	2.5	5.0
B75.32	0.078	456	2.5	5.8
B85.24	0.078	491	1.9	4.7
A30.32	0.086	283	2.8	4.0
A40.32	0.086	323	2.8	4.6
A50.32	0.086	361	2.8	5.1
cA2.09.48	0.091	$\approx m_\pi$	4.4	3.0

and the lattice vacuum polarization tensor

$$\Pi_{\mu\nu}^{\gamma\gamma}(x,y) = \langle J_\mu^\gamma(x)\, J_\nu^\gamma(y)\rangle - a^{-3}\,\delta_{x,y}\,\delta_{\mu\nu}\,\langle S_\nu(y)\rangle \quad (8)$$

as the 2-point current correlator with a lattice contact term. The tensor fulfills the lattice version of exact transversality at non-zero lattice spacing

$$\partial_\mu^b\, \Pi_{\mu\nu}^{\gamma\gamma}(x,y) = 0 \quad (9)$$

with the discrete backward partial derivative ∂_μ^b.

The discrete 4-dimensional Fourier transform translates this transversality to momentum space

$$\Pi_{\mu\nu}(Q) = \sum_x \Pi_{\mu\nu}(x,y)\, e^{iQ(x+a\hat{\mu}/2-y-a\hat{\nu}/2)}$$

$$\hat{Q}_\mu\, \Pi_{\mu\nu}(Q) = 0\,, \quad \hat{Q}_\mu = \frac{2}{a}\,\sin\left(aQ_\mu/2\right)$$

with $Q_i \in 2\pi/L \cdot \{0,\ldots,L/a-1\}$ and $Q_0 \in 2\pi/T \cdot \{0,\ldots,T/a-1\}$ for the spatial and temporal momentum components. We extract the polarization function according to

$$\Pi^{\gamma\gamma}\left(\hat{Q}^2\right) = \frac{1}{\#\left[\hat{Q}^2\right]} \sum_{Q\in[\hat{Q}^2]} \mathrm{Re}\left(\sum_{\mu,\nu} P_{\mu\nu}(Q)\, \Pi_{\mu\nu}^{\gamma\gamma}(Q)\right) \times \left(\sum_{\mu,\nu} P_{\mu\nu}(Q)\, P_{\mu\nu}(Q)\right)^{-1}, \quad (10)$$

$$P_{\mu\nu}(Q) = \hat{Q}_\mu\,\hat{Q}_\nu - \delta_{\mu\nu}\,\hat{Q}^2\,.$$

The definition in eq. (10) includes the projection with $P_{\mu\nu}$, which is non-trivial for the index ν, and discarding the imaginary part, which is a lattice artifact due to flavor symmetry breaking. With a $\left[\hat{Q}^2\right]$ we denote the set of all lattice momenta Q with the same $\hat{Q}^2$ and with $\#\left[\hat{Q}^2\right]$ the number of such momenta.

The vacuum polarization tensor and function decompose into a dominant quark-connected and a small quark-disconnected contribution, which are illustrated by the left-hand and right-hand diagram in figure [2], respectively.

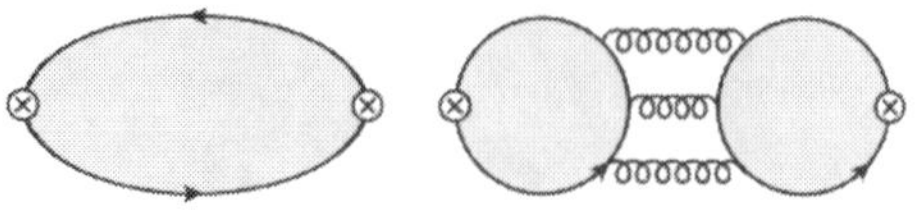

Figure 2. Diagrammatic representation of the quark-connected (left) and (exemplary) quark-disconnected contribution to the vacuum polarization; black lines denote valence quark loops and gray shading represents full, non-perturbative QCD interaction.

For illustration we show exemplary data for the connected and disconnected contribution to the vacuum polarization function in the left and right panel of figure [3], respectively. The left-hand panel shows the typical smooth momentum dependence of the connected contribution from up and down quarks for ensemble B25.32t (cf. table {1}) between the infrared and ultraviolet lattice cut-off $(2\sin(a\pi/T)/a)^2 \leq \hat{Q}^2 \leq 16/a^2$. The gray band in the plot shows our standard fit to the data, which will be discussed below. In the detail plot we magnify the crucial region at small momenta, where lattice data points become sparse. The right-hand panel shows an order-of-magnitude comparison of the signal for the light-quark connected polarization function (blue band) and the disconnected contribution from the 2-point correlation of the isovector components of the electromagnetic current (dark-red band labeled isospin 1) and of the isoscalar components (yellow band labeled isospin 0). The data shown was obtained with particularly high statistics for ensemble B55.32, cf. [29]. While the contribution from the isovector currents is a lattice artifact in the twisted mass regularization, the physically meaningful contribution from the isoscalar vector current correlator appears about two orders of magnitude smaller compared to the connected contribution. In fact, with the presently available data we cannot resolve a statistically significant contribution from the disconnected vacuum polarization function to a_l^{hlo}. In the following we will thus be concerned only with the connected vacuum polarization function.

Momentum dependence of the polarization function

The lattice calculation generically provides us with data points for the vacuum polarization function for a finite set of discrete, nonzero momenta. According to equation (2) the subtracted polarization function enters the integrand and the integral extends beneath the lowest non-zero lattice momentum to zero. We thus need to extrapolate the existing lattice data to all $Q^2 < Q^2_{\mathrm{min}} = 4/a^2\sin^2(\pi/L)$. Secondly, the infinite upper integration limit is relaxed to the lattice cut-off $Q^2_{\mathrm{max}} = 16/a^2$. To appreciate the significance of these two issues we display the integral saturation in the left-hand panel of figure [4]: shown is the ratio

$$R_l\left(Q^2_{\mathrm{max}}\right) = \frac{a_l^{\mathrm{hlo}}\left(Q^2_{\mathrm{max}}\right)}{a_l^{\mathrm{hlo}}\left(100\,\mathrm{GeV}^2\right)}\,.$$

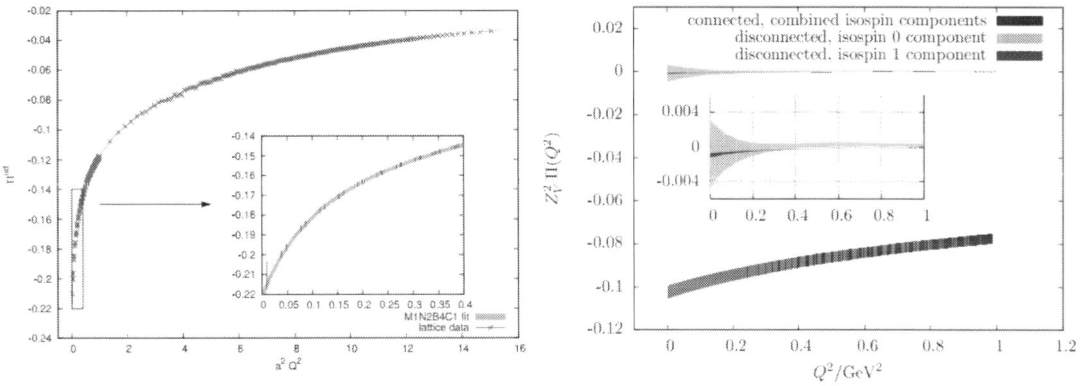

Figure 3. Left: quark-connected contribution from up and down quark to the hadronic vacuum polarization function; the data shown belongs to ensemble B25.32t (m_{PS} = 274 MeV, L = 2.5 fm and a = 0.078 fm). Right: quark-disconnected contribution from up and down quark for ensemble B55.32 (m_{PS} = 393 MeV, L = 2.5 fm and a = 0.078 fm).The detail plots show a magnification of the small-momentum region.

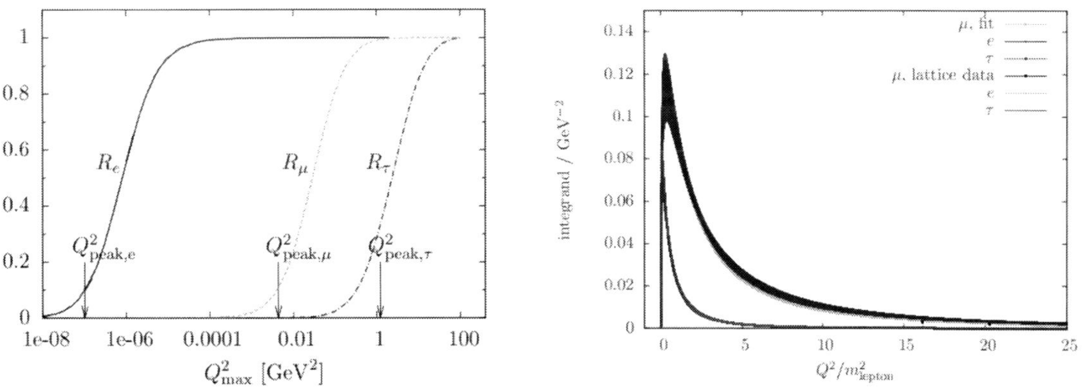

Figure 4. Left: saturation of the integral defining a_l^{hlo} for $l = e, \mu, \tau$; the data shown corresponds to ensemble B55.32. Right: the full integrand for ensemble D15.48 in a universal plot for all three leptons.

of a_l^{hlo} integrated from zero up to Q^1_{max} over the a_l^{hlo} integrated from zero up to 100 GeV^2. The plot shows that the upper integration limit Q^2_{max} on the lattice is not problematic since for electron and muon the integral saturates long before reaching the lattice momentum cut-off and also for the tau the change of the integral value with $Q^2_{\mathrm{max}} \lesssim 100\,\mathrm{GeV}^2$ is insignificant compared to the statistical and other sources of systematic uncertainties.

In the same plot $Q^2_{\mathrm{peak},l}$ marks the maximum of the weight function $w\left(Q^2/m_l^2\right)$ for each lepton, which roughly locates the main support area for the integral. As is well known, for the electron and the muon this main support region is far below the lowest attainable lattice momentum for currently available lattice volumes. In particular $Q^2_{\mathrm{min}} \gtrsim 0.1\,\mathrm{GeV}^2$ for our ensembles in table {1}. In the extreme case of the electron, a_l^{hlo} is essentially determined by the $d\Pi^{\gamma\gamma}/dQ^2(0)$ alone. The right panel in figure [4] shows the location of lattice data in comparison to the fitted integrand exemplarily for ensemble D15.48. While for the electron all lattice data is outside the displayed momentum range, for the muon the lattice data points (black) start at $Q^2/m_\mu^2 \approx 16$. For the tau lattice data (red) stops short of the peak on the right-hand slope of the integrand (brown band).

To fit the momentum dependence of the polarization function numerous fit ansätze can be constructed based on the analyticity and spectral properties of the polarization function, resonances and other features. In particular, fit functions of the Padé type have been shown to be a suitable set [11, 23, 24] of functions with correct convergence properties to reproduce the analytic structure of $\Pi^{\gamma\gamma}$ in the limit of infinite volume and data precision. Here we use the *MNBC* fit ansatz which is conceptually close to the Padé type fits. Since at this point we consider only the connected part of $\Pi^{\gamma\gamma}$ we can decompose it immediately into single-quark-flavor components

$$\Pi^{\gamma\gamma} = \frac{5}{9}\,\Pi^{u+d} + \frac{1}{9}\,\Pi^{s} + \frac{4}{9}\,\Pi^{c}\,.$$

Further on, we decompose the integration interval in two regions: low momentum $0 \leq Q^2 \leq Q^2_{\text{match}}$ and high momentum $Q^2_{\text{match}} < Q^2 \leq Q^2_{\text{max}}$. Then for each individual quark flavor $f = u + d, s, c$ we use the family of fit functions

$$\Pi^f_{\text{low}}\left(Q^2\right) = \sum_{j=1}^{M} g_j^{f2} \frac{m_j^{f2}}{m_j^{f2} + Q^2} + \sum_{j=0}^{N-1} a_j^f \left(Q^2\right)^j , \quad (11)$$

$$\Pi^f_{\text{high}}\left(Q^2\right) = \log\left(Q^2\right) \sum_{j=0}^{B-1} b_j^f \left(Q^2\right)^j + \sum_{j=0}^{C-1} c_j^f \left(Q^2\right)^j \quad (12)$$

and the full fit function is given by the sum of low- and high-momentum part weighted by Heaviside step functions

$$\Pi^f\left(Q^2\right) = \left(1 - \Theta\left(Q^2 - Q^2_{\text{match}}\right)\right) \Pi^f_{\text{low}}\left(Q^2\right) + \\ + \Theta\left(Q^2 - Q^2_{\text{match}}\right) \Pi^f_{\text{high}}\left(Q^2\right) . \quad (13)$$

The set of pairs $\left\{\left(g_i^f, m_i^f\right) \mid i = 1, \ldots, M\right\}$ of couplings and masses are determined from the corresponding connected, time-dependent single-quark-flavor vector current 2-point function at zero momentum. The remaining polynomial in Π_{low} can be understood as the contributions from poles with position above Q^2_{match} and for these terms we can safely use a Taylor expansion around the origin. It is important to note that Π_{high} is only relevant for interpolation of the lattice data.

Finite volume effects related to the infrared momentum cut-off and the dependence on the specific fit function are then probed by varying the number of parameters M, N, B, C. In practice, we can resolve $M = 1$ or 2 poles with $N = 1, 2$ polynomial terms. Beyond this the fit parameters both from the multi-exponential fit of the vector current 2-point function as well as the fit of the vacuum polarization function become highly correlated. Moreover, the variation of B and C leads to insignificant variations of the integral value.

External scales problem and modified observables

As a peculiarity of the lattice QCD regularization, which excludes the dynamics of leptons and photons, the lepton mass in equation (2) enters as an external scale which is not related to the lattice parameters. In particular, there is no inherent value for the lepton mass in lattice units, such that the lattice spacing is explicitly required to convert the lepton mass to lattice units. By this procedure, however, the originally dimensionless a_l^{hlo} becomes effectively dependent on the lattice QCD scale setting, which can be monitored by considering its effective dimension at a fixed bare coupling [25, 26],

$$\mathrm{d}_{\text{eff}}\left[a_l^{\text{hlo}}\right] = -\left(\frac{a}{a_l^{\text{hlo}}}\right)\left(\frac{\partial a_l^{\text{hlo}}}{\partial a}\right)\Bigg|_{g_0=\text{fixed}} = O(-2 \sim -1)$$

The latter ranges from approximately -2 for the electron to -1 for the tau, such that a_l^{hlo} behaves rather like an inverse mass, $M^{\mathrm{d}_{\text{eff}}}$, than a dimensionless quantity. This external scale problem has been solved by introducing a new family of modified observables, $a_{\bar{l}}^{\text{hlo}}$ defined as follows. We denote with $\tilde{Q}$ the momentum in lattice units and can rewrite the integral in equation (2) in terms of $\tilde{Q}$ and define for any QCD scale H, which can be calculated on the lattice

$$a_{\bar{l}}^{\text{hlo}} = 4\alpha^2 \int \frac{d\tilde{Q}^2}{\tilde{Q}^2} \Pi_R^{\gamma\gamma}\left(\tilde{Q}^2\right) w\left(\frac{\tilde{Q}^2}{a^2 H^2} \cdot \frac{H^2_{\text{phys}}}{m_l^2}\right) \quad (14)$$

with the consistency condition

$$\lim_{m_{PS} \to m_\pi} H(m_{PS}) = H_{\text{phys}}(m_\pi)$$

such that at the physical point the modified observable agrees with the standard definition for each choice of a consistent H

$$\lim_{m_{PS} \to m_\pi} a_{\bar{l}}^{\text{hlo}}(m_{PS}) = a_l^{\text{hlo}}(m_\pi) . \quad (15)$$

The introduction of a QCD-scale H redefines the lepton mass in the lattice calculation

$$m_{\bar{l}} = m_l / H_{\text{phys}} \cdot H ,$$

such that the modified lepton mass becomes dependent on the lattice scale setting and

$$\mathrm{d}_{\text{eff}}\left[m_{\bar{l}}^2\right] = 2 \quad \text{and} \quad \mathrm{d}_{\text{eff}}\left[a_{\bar{l}}^{\text{hlo}}\right] = 0 .$$

In practice we use the empirically motivated choice $H = m_V$ which denotes the mass of the ground state in the light vector channel. This exploits the strong statistical correlation of the light vector meson mass and $a_{\bar{l}}^{\text{hlo}}$.

Extrapolation and results

To extrapolate in the light pseudoscalar mass to the physical point $m_{PS} \to m_\pi$ and to the continuum we use the ansatz

$$a_{\bar{l}}^{\text{hlo}}(m_{PS}, a) = A + B_1 m_{PS}^2 \left(+B_2 m_{PS}^4 + \ldots\right) + C a^2 . \quad (16)$$

These extrapolations of the lattice data for $a_{\bar{l}}^{\text{hlo}}$ are shown in figure [5] for electron (top), muon (center) and tau lepton (bottom). A quick inspection of the data points strongly suggests a linear dependence of the modified observable $a_{\bar{l}}^{\text{hlo}}$ on the pion mass and as it turns out, optimizing only B_1 already provides a good fit. With (at most) B_2 we probe for potential non-linearity in the pion mass dependence of our lattice data, which turns out to be insignificant. The three dashed lines in each plot show the evaluation of the fit function for the corresponding lattice spacing. The result for zero lattice spacing and physical pion mass is given by the red downward triangle shifted slightly off to the left from the physical pion mass marked by the vertical dashed line.

Additionally, in figure [6] we show the light-quark contribution (using only Π^{u+d}) for $a_{\bar{l}}^{\text{hlo}}$ for all three leptons, that is, we compare the results from the $N_f = 2+1+1$ simulations, which are extrapolated to the physical pion mass ([18, 19]), with recent calculations on an ETMC $N_f = 2$

DOI: 10.1051/epjconf/201611801029

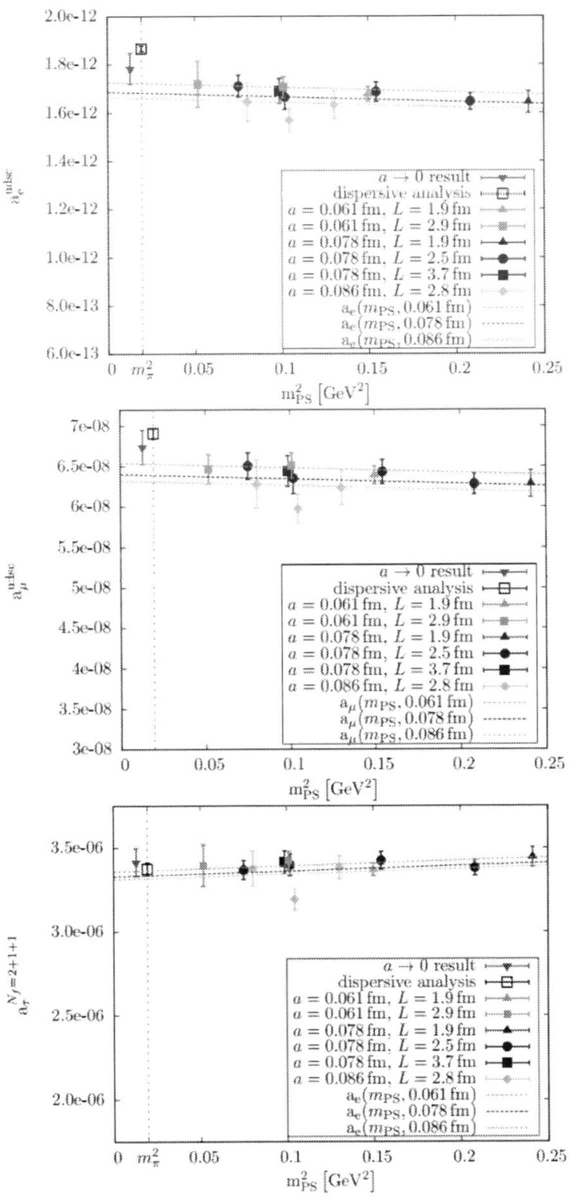

Figure 5. Combined extrapolation in pion mass and lattice spacing for a_l^{hlo} for electron (top), muon (center) and tau (bottom) using free fit parameters A, B_1, C in equation (16).

ensemble directly at physical pion mass [21]. The results for the simulation directly at the physical pion mass are obtained using the standard definition of the observables (cf. eq. (2) or (14) with $H = 1$). Within presently reachable statistical uncertainties, there is full agreement of extrapolated results and results obtained from the simulation at physical pion mass.

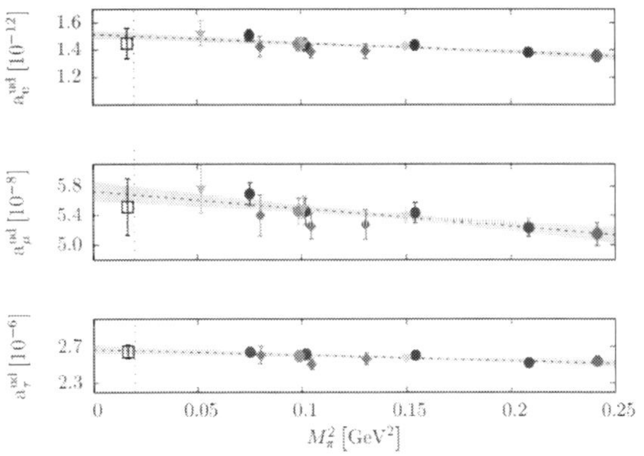

Figure 6. Light-quark contribution to a_l^{hlo} at physical pion mass for electron (top), muon (center) and tau (bottom) from the standard definition with $H = 1$ (cf. eq. (14)); shown is the comparison of the m_{PS}^2- extrapolated data from [18, 19] with the result calculated directly at the physical pion mass [21].

Systematic uncertainties and error budget

We investigate comprehensively the impact of all systematic uncertainties X in our lattice calculation and call the corresponding estimate Δ_X. In detail we probe

(i) finite size effects, Δ_{FSE}, by restricting the range in $m_{PS} \cdot L$ of lattice data entering the combined continuum and chiral extrapolation to $m_{PS} \cdot L > 3.8$ and by comparing a_l^{hlo} for ensembles with different lattice volumes at fixed lattice spacing and pion mass;

(ii) the impact of data from larger pion masses, Δ_χ, by restricting to lattice data in the range $m_{PS} < 400$ MeV;

(iii) the impact Δ_V of varying the fit range for the vector meson 2-point function;

(iv) the dependence on the fit function Δ_{MNBC} by varying the number of parameters M, N, B, C (cf. eq. (11), (12));

(v) the impact of disconnected contributions, Δ_{disc};

(vi - viii) the uncertainty from matching the Osterwalder-Seiler quark masses, Δ_{OS}, from varying the transition point from low- to high-momentum region, $\Delta_{Q^2_{\text{match}}}$, and the uncertainty from mistuned strange and charm sea quark masses, $\Delta_{(\mu_\sigma,\mu_\delta)}$.

Though some of these effects are likely correlated, in particular (i) and (iv), we take a conservative approach and check each uncertainty individually, thus potentially overestimating the overall uncertainty. As a general observation, the contributions from (vi) to (viii) are negligible and we can set $\Delta_{\text{OS}} = \Delta_{Q^2_{\text{match}}} = \Delta_{(\mu_\sigma,\mu_\delta)} = 0$. For (v) we showed above (cf. figure [3]), that with presently available statistics, the impact of disconnected contributions is also negligible in our calculation as they do not lead to a statistically significant shift of the result. Moreover, the lattice data

Table 2. Estimates for systematic uncertainties Δ_V and Δ_{MNBC} for our lattice calculation.

a_l^{hlo}	Δ_V	Δ_{MNBC}
e	$0.035 \cdot 10^{-12}$	$0.078 \cdot 10^{-12}$
μ	$0.13 \cdot 10^{-8}$	$0.09 \cdot 10^{-8}$
τ	$0.046 \cdot 10^{-6}$	$0.032 \cdot 10^{-6}$

Table 3. Comparison of results for a_l^{hlo} from $N_f = 2 + 1 + 1$ lattice QCD (center) and dispersive analyses (right).

a_l^{hlo}	tmLQCD	disp. analyses
e	$1.782\,(64)\,(85) \cdot 10^{-12}$ [18]	$1.866\,(10)\,(05) \cdot 10^{-12}$ [13]
μ	$6.78\,(24)\,(16) \cdot 10^{-8}$ [19]	$6.91\,(01)\,(05) \cdot 10^{-8}$ [14]
τ	$3.41\,(8)\,(6) \cdot 10^{-6}$ [18]	$3.38\,(4) \cdot 10^{-6}$ [15]

for a_l^{hlo} in figure [5] covering $3 \lesssim m_{PS} \cdot L \lesssim 6$ and the extrapolations indicate small finite size effects, which is confirmed again by the statistical analysis. Already above we commented on the apparently highly dominant linear dependence of the modified a_l^{hlo} on the m_{PS}^2. This too leads to statistically non-discernible effects when probing for terms of higher than first order in m_{PS}^2 with the present data. We list the estimates for the remaining two sources of systematic errors in table {2}. Finally, in table {3} we list the complete results for a_l^{hlo} for electron, muon and tau and compare to the results from phenomenological analyses based on dispersion relations and using experimental data. Where available, we give the decomposition of the total error $(\Delta_{\mathrm{stat}})\,(\Delta_{\mathrm{sys}})$ into statistical uncertainty in the first bracket and the second bracket quotes the systematic uncertainty. The systematic uncertainty for the lattice results is obtained as $\Delta_{\mathrm{sys}} = \sqrt{\sum_X \Delta_X^2}$.

3 Discussion and outlook

With this contribution we report on the determination of the hadronic leading order anomalous magnetic moments of all three Standard Model leptons by the ETM Collaboration with $N_f = 2 + 1 + 1$ dynamical quarks. The center column of table {3} gives our estimates for a_l^{hlo} in the continuum and at the physical pion mass with statistical and a conservatively determined overall systematic uncertainty. We find agreement with the estimates from phenomenological analyses within presently reached uncertainties. With the lattice data available for this analysis the uncertainties from the lattice calculation are still significantly larger than those obtained from the dispersive analyses. As a result of this analysis we reach a precision of roughly 6% and 4% for the case of the electron and muon, respectively, and of 3% for the tau lepton with combined statistical and systematic error. The smaller uncertainty for the case of the tau may be expected, since the main support for the integral in equation (2) is shifted towards a region of momenta, where lattice data for the vacuum polarization tensor is actually available. Yet we emphasize, that also in this case there is a critical dependence on the extrapolation of the vacuum polarization function in the small-momentum region: it appears through the subtraction of $\Pi^{\gamma\gamma}(Q^2 = 0)$, and this latter value is predicted from the extrapolation of the lattice data. The determination of a_e^{hlo} on the lattice is a strong test of the lattice determination of the vacuum polarization function at smallest momenta, since a_e^{hlo} is essentially determined by the derivative of the polarization function at the origin, $d\Pi^{\gamma\gamma}(Q^2)/dQ^2\big|_{Q^2=0}$. Thus the agreement obtained for a_e^{hlo} is both remarkable and encouraging. With the results from our recent determination of the hadronic leading order contribution to the running of the electroweak couplings [27], this test has been successfully taken another step further: in this reference we demonstrate the agreement of the renormalized vacuum polarization function determined from $N_f = 2 + 1 + 1$ twisted mass lattice QCD with the dispersive relation result in a continuous interval $\left[0,\ 10\,\mathrm{GeV}^2\right]$, albeit again with somewhat larger uncertainty on the lattice result.

The use of the modified observables a_l^{hlo} (cf. equation (14)) at unphysically large pion masses and non-zero lattice spacing has proven superior to the standard definition of a_l^{hlo} in equation (2) in our lattice calculation. The reason is twofold. Firstly, it is due to the introduction of the scale H, that the strong pion mass dependence of a_l^{hlo} is tamed to the level shown in figure [5]. As a result we are able to perform a controlled chiral extrapolation with insignificant variations of the result when probing the pion mass dependence beyond the leading, linear term. With the standard definition a controlled extrapolation from the range of pion masses we have available (cf. table {1}) will require some modeling or otherwise lead to a prohibitively large uncertainty at the physical point (cf. [25, 26] for a more detailed discussion). Secondly, by an optimized choice of the scale H, which for us is $H = m_V$, the light vector meson mass, the per-ensemble estimate of a_l^{hlo} benefits from a partial cancellation of statistical errors due to the correlation of the vector meson mass and coupling and the polarization function. Both advantageous features of this mechanism have been put to work in [27] as well with the above described positive outcome. Moreover, the direct comparison of the light-quark contribution from extrapolation in $m_{PS}^2 \to m_\pi^2$ and direct calculation at the physical pion mass, which is done without any modification of the observable, provides further confirmation of the present results' consistency, but this test has to be made more stringent by achieving higher statistical accuracy for the result from direct simulation at the physical pion mass.

The more accurate ab-initio determination of a_l^{hlo} directly at the physical pion mass is an on-going effort within the ETMC. The evaluation of the crucial light quark contribution for $N_f = 2$ dynamical fermions at physical pion mass along the lines of [21] for a larger volume with $m_{PS} \cdot L = 4$ is on-going. Moreover, with the continuing simulation activity of the ETMC aiming for gauge field ensembles with fully dynamical up, down, strange and charm quark with physical values for the pion, kaon and D-meson mass we have for the near future the prospect of calculating a_l^{hlo} in a setup including QCD-dynamics from the first two families of quark flavors with quark masses at their physical values.

We are then basically left with the systematic effect of the finite volume in the form of the infrared cut-off $\sim 1/L$ of the lattice data. Lattice simulations in larger physical volumes are an essential yet at present computationally cost-intensive way to constrain further the polarization function at smallest momenta. Furthermore, alternative methods have been proposed and are being developed (e.g. [30–32]) to exploit further the analytic structure of the polarization function in finite volume to constrain its shape in the small momentum region.

Acknowledgments

We thank all members of the organizing committee for organizing this highly informative and stimulating conference. M.P. is grateful for the opportunity to attend the conference and present this work. We are most grateful to Fred Jegerlehner for very enlightening discussions while carrying out these computations. We thank the European Twisted Mass Collaboration for generating the gauge field ensembles used for the calculations. This work has been supported in part by the DFG Corroborative Research Center SFB/TR9. G.P. gratefully acknowledges the support of the German Academic National Foundation (Studienstiftung des deutschen Volkes e.V.) and of the DFG-funded Graduate School GK 1504.

References

[1] T. Blum, Phys. Rev. Lett. **91** (2003) 052001 [hep-lat/0212018].

[2] G. F. Giudice, P. Paradisi and M. Passera, JHEP **1211** (2012) 113 [arXiv:1208.6583 [hep-ph]].

[3] N. Brambilla *et al.*, Eur. Phys. J. C **74** (2014) 10, 2981 [arXiv:1404.3723 [hep-ph]].

[4] M. Fael, L. Mercolli and M. Passera, Nucl. Phys. Proc. Suppl. **253-255** (2014) 103 [arXiv:1301.5302 [hep-ph]].

[5] B. Oberhof [BaBar Collaboration], Nucl. Part. Phys. Proc. **260** (2015) 12 [arXiv:1502.01810 [hep-ex]].

[6] R. Baron *et al.*, JHEP **1006** (2010) 111 [arXiv:1004.5284 [hep-lat]].

[7] R. Baron *et al.* [European Twisted Mass Collaboration], Comput. Phys. Commun. **182** (2011) 299 [arXiv:1005.2042 [hep-lat]].

[8] K. Osterwalder and E. Seiler, Annals Phys. **110** (1978) 440.

[9] R. Frezzotti and G. C. Rossi, JHEP **0408** (2004) 007 [hep-lat/0306014].

[10] R. Frezzotti and G. C. Rossi, JHEP **0410** (2004) 070 [hep-lat/0407002].

[11] C. Aubin, T. Blum, M. Golterman and S. Peris, Phys. Rev. D **86** (2012) 054509 [arXiv:1205.3695 [hep-lat]].

[12] F. Jegerlehner, Nucl. Phys. Proc. Suppl. **181-182** (2008) 26.

[13] D. Nomura and T. Teubner, Nucl. Phys. B **867** (2013) 236 [arXiv:1208.4194 [hep-ph]].

[14] F. Jegerlehner and R. Szafron, Eur. Phys. J. C **71** (2011) 1632 [arXiv:1101.2872 [hep-ph]].

[15] S. Eidelman and M. Passera, Mod. Phys. Lett. A **22** (2007) 159 [hep-ph/0701260].

[16] A. Abdel-Rehim *et al.*, PoS LATTICE **2013** (2014) 264 [arXiv:1311.4522 [hep-lat]].

[17] A. Abdel-Rehim *et al.*, PoS LATTICE **2014** (2015) 119 [arXiv:1411.6842 [hep-lat]].

[18] F. Burger, G. Hotzel, K. Jansen and M. Petschlies, arXiv:1501.05110 [hep-lat].

[19] F. Burger *et al.* [ETM Collaboration], JHEP **1402** (2014) 099 [arXiv:1308.4327 [hep-lat]].

[20] F. Burger, G. Hotzel, K. Jansen and M. Petschlies, JHEP **1503** (2015) 073 [arXiv:1412.0546 [hep-lat]].

[21] A. Abdel-Rehim *et al.* [ETM Collaboration], arXiv:1507.05068 [hep-lat].

[22] Y. Iwasaki, Nucl. Phys. B **258** (1985) 141.

[23] M. Golterman, K. Maltman and S. Peris, Phys. Rev. D **90** (2014) 7, 074508 [arXiv:1405.2389 [hep-lat]].

[24] M. Golterman, K. Maltman and S. Peris, Phys. Rev. D **88** (2013) 11, 114508 [arXiv:1309.2153 [hep-lat]].

[25] D. B. Renner, X. Feng, K. Jansen and M. Petschlies, PoS LATTICE **2011** (2012) 022 [arXiv:1206.3113 [hep-lat]].

[26] X. Feng, K. Jansen, M. Petschlies and D. B. Renner, Phys. Rev. Lett. **107** (2011) 081802 [arXiv:1103.4818 [hep-lat]].

[27] F. Burger, K. Jansen, M. Petschlies and G. Pientka, arXiv:1505.03283 [hep-lat].

[28] F. Jegerlehner, Nuovo Cim. C **034S1** (2011) 31 [arXiv:1107.4683 [hep-ph]].

[29] C. Michael *et al.* [ETM Collaboration], Phys. Rev. Lett. **111** (2013) 18, 181602 [arXiv:1310.1207 [hep-lat]].

[30] D. Bernecker and H. B. Meyer, Eur. Phys. J. A **47** (2011) 148 [arXiv:1107.4388 [hep-lat]].

[31] X. Feng, S. Hashimoto, G. Hotzel, K. Jansen, M. Petschlies and D. B. Renner, Phys. Rev. D **88** (2013) 034505 [arXiv:1305.5878 [hep-lat]].

[32] B. Chakraborty, C. Davies, P. Gonçalves de Oliveira, J. Koponen, G. P. Lepage, PoS LATTICE **2015** (2015) Talk given at the 33rd International Symposium on Lattice Field Theory, Kobe, Japan, 2015.

DOI: 10.1051/epjconf/201611801030

Dispersion relation for hadronic light-by-light scattering

Massimiliano Procura[1,a], Gilberto Colangelo[2,b], Martin Hoferichter[3,c], and Peter Stoffer[4,d]

[1] CERN Theory Division, 1211 Geneva 23, Switzerland
[2] Albert Einstein Center for Fundamental Physics, Institute for Theoretical Physics, University of Bern, Sidlerstrasse 5, 3012 Bern, Switzerland
[3] Institute for Nuclear Theory, University of Washington, Seattle, WA 98195-1550, USA
[4] Helmholtz-Institut für Strahlen- und Kernphysik (Theory) and Bethe Center for Theoretical Physics, University of Bonn, 53115 Bonn, Germany

Abstract. The largest uncertainties in the Standard Model calculation of the anomalous magnetic moment of the muon $(g-2)_\mu$ come from hadronic contributions. In particular, in a few years the subleading hadronic light-by-light (HLbL) contribution might dominate the theory uncertainty. We present a dispersive description of the HLbL tensor, which is based on unitarity, analyticity, crossing symmetry, and gauge invariance. This opens up the possibility of a data-driven determination of the HLbL contribution to $(g-2)_\mu$ with the aim of reducing model dependence and achieving a reliable error estimate.

Our dispersive approach defines unambiguously the pion-pole and the pion-box contribution to the HLbL tensor. Using Mandelstam's double-spectral representation, we have proven that the pion-box contribution coincides exactly with the one-loop scalar QED amplitude, multiplied by the appropriate pion vector form factors.

1 Introduction

The anomalous magnetic moment of the muon $(g-2)_\mu$ has been measured [1] and computed to very high precision of about 0.5 ppm (see e.g. [2]). For more than a decade, a discrepancy has persisted between the experiment and the Standard Model prediction, now of about 3σ. Forthcoming measurements at FNAL [3] and J-PARC [4] will update the experimental value. The aim is to increase the precision by a factor of 4 and check for systematic errors.

The main uncertainty of the theory prediction is due to strong interaction effects. At present, the largest error arises from hadronic vacuum polarisation, which, however, forthcoming data from e^+e^- experiments [2] may help reduce. Thus in a few years, the subleading[1] hadronic light-by-light contribution might dominate the theory error. In present calculations of the HLbL contribution, systematic errors are difficult, if not impossible, to quantify, due to model dependence. A new strategy is required to provide a solid estimate of the theory uncertainties and reduce them. In the recent past, lattice QCD has made remarkable progress in this direction, and may play a leading role in this field in the near future [8–12]. In [13, 14], we have presented the first dispersive description of the HLbL tensor.[2] By making use of the fundamental principles of unitarity, analyticity, crossing symmetry, and gauge invariance, we provide an approach that reduces model dependence and allows for a more data-driven determination of the HLbL contribution to $(g-2)_\mu$.

Here, we report on an improvement of our dispersive framework [16, 17]. We have constructed a generating set of Lorentz structures for the HLbL tensor that is free of kinematic singularities and zeros. This simplifies significantly the calculation of the HLbL contribution to $(g-2)_\mu$. Within our dispersive formalism, the definitions of both the pion-pole and pion-box topologies are unambiguous. By constructing a Mandelstam representation for the scalar functions, we prove that the box topologies are equal to the scalar QED (sQED) contribution multiplied by pion vector form factors. First numerical results for the pion-box topologies are shown and future steps are discussed.

2 Lorentz structure of the HLbL tensor

In order to study the HLbL contribution to $(g-2)_\mu$, we need a description of the HLbL tensor, namely the hadronic Green's function of four quark electromagnetic currents,

[a] e-mail: massimiliano.procura@cern.ch
[b] e-mail: gilberto@itp.unibe.ch
[c] e-mail: mhofer@uw.edu
[d] e-mail: stoffer@hiskp.uni-bonn.de

[1] Even higher-order hadronic contributions have been considered in [5, 6].

[2] A different approach, which aims at a dispersive description of the muon vertex function instead of the HLbL tensor, has been presented in [15].

© The Authors, published by EDP Sciences. This is an open access article distributed under the terms of the Creative Commons Attribution License 4.0 (http://creativecommons.org/licenses/by/4.0/).

evaluated in pure QCD:

$$\Pi^{\mu\nu\lambda\sigma}(q_1, q_2, q_3) = -i \int d^4x\, d^4y\, d^4z\, e^{-i(q_1 \cdot x + q_2 \cdot y + q_3 \cdot z)} \times \langle 0|T\{j^\mu_{\rm em}(x) j^\nu_{\rm em}(y) j^\lambda_{\rm em}(z) j^\sigma_{\rm em}(0)\}|0\rangle. \tag{1}$$

Gauge invariance requires the HLbL tensor to satisfy the Ward-Takahashi (WT) identities

$$\{q_1^\mu, q_2^\nu, q_3^\lambda, q_4^\sigma\}\Pi_{\mu\nu\lambda\sigma}(q_1, q_2, q_3) = 0, \tag{2}$$

where $q_4 = q_1 + q_2 + q_3$. The HLbL tensor can be written a priori in terms of 138 basic Lorentz structures built out of the metric tensor and the four-momenta [18]. Our first task is to write the HLbL tensor in terms of Lorentz structures that satisfy the WT identities, while at the same time the scalar functions that multiply these structures must be free of kinematic singularities and zeros. A recipe for the construction of these structures has been given by Bardeen, Tung [19], and Tarrach [20] for generic photon amplitudes. Gauge invariance imposes 95 linear relations between the 138 initial scalar functions. A generating set[3] consisting of 43 elements can be constructed following Bardeen and Tung [19]. However, as it was shown by Tarrach [20], such a set is not free of kinematic singularities and has to be supplemented by additional structures. We find a redundant generating set of dimension 54:

$$\Pi^{\mu\nu\lambda\sigma}(q_1, q_2, q_3) = \sum_{i=1}^{54} T_i^{\mu\nu\lambda\sigma}\Pi_i(s, t, u), \tag{3}$$

where the scalar functions Π_i are free of kinematic singularities and zeros, and therefore suited for a dispersive description. The Mandelstam variables are defined by $s = (q_1 + q_2)^2$, $t = (q_1 + q_3)^2$, $u = (q_2 + q_3)^2$. There are only seven distinct Lorentz structures:

$$\begin{aligned}
T_1^{\mu\nu\lambda\sigma} &= \epsilon^{\mu\nu\alpha\beta}\epsilon^{\lambda\sigma\gamma\delta} q_{1\alpha} q_{2\beta} q_{3\gamma} q_{4\delta}, \\
T_4^{\mu\nu\lambda\sigma} &= \left(q_2^\mu q_1^\nu - q_1 \cdot q_2 g^{\mu\nu}\right)\left(q_4^\lambda q_3^\sigma - q_3 \cdot q_4 g^{\lambda\sigma}\right), \\
T_7^{\mu\nu\lambda\sigma} &= \left(q_2^\mu q_1^\nu - q_1 \cdot q_2 g^{\mu\nu}\right)\Big(q_1 \cdot q_4 (q_1^\lambda q_3^\sigma - q_1 \cdot q_3 g^{\lambda\sigma}) \\
&\quad + q_4^\lambda q_1^\sigma q_1 \cdot q_3 - q_1^\lambda q_1^\sigma q_3 \cdot q_4\Big), \\
T_{19}^{\mu\nu\lambda\sigma} &= \left(q_2^\mu q_1^\nu - q_1 \cdot q_2 g^{\mu\nu}\right)\Big(q_2 \cdot q_4 (q_1^\lambda q_3^\sigma - q_1 \cdot q_3 g^{\lambda\sigma}) \\
&\quad + q_4^\lambda q_2^\sigma q_1 \cdot q_3 - q_1^\lambda q_2^\sigma q_3 \cdot q_4\Big), \\
T_{31}^{\mu\nu\lambda\sigma} &= \left(q_2^\mu q_1^\nu - q_1 \cdot q_2 g^{\mu\nu}\right)\left(q_2^\lambda q_1 \cdot q_3 - q_1^\lambda q_2 \cdot q_3\right) \\
&\quad \times \left(q_2^\sigma q_1 \cdot q_4 - q_1^\sigma q_2 \cdot q_4\right), \\
T_{37}^{\mu\nu\lambda\sigma} &= \left(q_3^\mu q_1 \cdot q_4 - q_4^\mu q_1 \cdot q_3\right) \\
&\quad \times \Big(q_3^\nu q_4^\lambda q_2^\sigma - q_4^\nu q_2^\lambda q_3^\sigma + g^{\lambda\sigma}(q_4^\nu q_2 \cdot q_3 - q_3^\nu q_2 \cdot q_4) \\
&\quad + g^{\nu\sigma}(q_2^\lambda q_3 \cdot q_4 - q_4^\lambda q_2 \cdot q_3) \\
&\quad + g^{\lambda\nu}(q_3^\sigma q_2 \cdot q_4 - q_2^\sigma q_3 \cdot q_4)\Big), \\
T_{49}^{\mu\nu\lambda\sigma} &= q_3^\sigma\Big(q_1 \cdot q_3 q_2 \cdot q_4 q_4^\mu g^{\lambda\nu} - q_2 \cdot q_3 q_1 \cdot q_4 q_4^\nu g^{\lambda\mu} \\
&\quad + q_4^\mu q_4^\nu (q_1^\lambda q_2 \cdot q_3 - q_2^\lambda q_1 \cdot q_3) \\
&\quad + q_1 \cdot q_4 q_3^\mu q_4^\nu q_2^\lambda - q_2 \cdot q_4 q_4^\mu q_3^\nu q_1^\lambda \\
&\quad + q_1 \cdot q_4 q_2 \cdot q_4 (q_3^\nu g^{\lambda\mu} - q_3^\mu g^{\lambda\nu})\Big) \\
&\quad - q_4^\lambda\Big(q_1 \cdot q_4 q_2 \cdot q_3 q_3^\mu g^{\nu\sigma} - q_2 \cdot q_4 q_1 \cdot q_3 q_3^\nu g^{\mu\sigma} \\
&\quad + q_3^\mu q_3^\nu (q_1^\sigma q_2 \cdot q_4 - q_2^\sigma q_1 \cdot q_4) \\
&\quad + q_1 \cdot q_3 q_4^\mu q_3^\nu q_2^\sigma - q_2 \cdot q_3 q_3^\mu q_4^\nu q_1^\sigma \\
&\quad + q_1 \cdot q_3 q_2 \cdot q_3 (q_4^\nu g^{\mu\sigma} - q_4^\mu g^{\nu\sigma})\Big) \\
&\quad + q_3 \cdot q_4\Big((q_1^\lambda q_4^\mu - q_1 \cdot q_4 g^{\lambda\mu})(q_3^\nu q_2^\sigma - q_2 \cdot q_3 g^{\nu\sigma}) \\
&\quad - (q_2^\lambda q_4^\nu - q_2 \cdot q_4 g^{\lambda\nu})(q_3^\mu q_1^\sigma - q_1 \cdot q_3 g^{\mu\sigma})\Big).
\end{aligned} \tag{4}$$

All the remaining ones are just crossed versions of the seven structures above. As each structure fullfils the WT identities, both crossing symmetry and gauge invariance are implemented in a manifest way in the set $\{T_i^{\mu\nu\lambda\sigma}\}$.

3 HLbL contribution to $(g-2)_\mu$

The HLbL contribution to $a_\mu = (g-2)_\mu/2$ can be extracted with the help of well-known Dirac projector techniques [23]. With our decomposition of the HLbL tensor in 54 structures, this amounts to the calculation of the following two-loop integral:

$$\begin{aligned}
a_\mu^{\rm HLbL} &= -\frac{e^6}{48 m_\mu}\int \frac{d^4 q_1}{(2\pi)^4}\frac{d^4 q_2}{(2\pi)^4}\frac{1}{q_1^2 q_2^2 (q_1+q_2)^2} \\
&\quad \times \frac{1}{(p+q_1)^2 - m_\mu^2}\frac{1}{(p-q_2)^2 - m_\mu^2} \\
&\quad \times \mathrm{Tr}\Big((\not{p} + m_\mu)[\gamma_\rho, \gamma_\sigma](\not{p} + m_\mu)\gamma_\mu \\
&\quad (\not{p} + \not{q}_1 + m_\mu)\gamma_\lambda(\not{p} - \not{q}_2 + m_\mu)\gamma_\nu\Big) \\
&\quad \times \sum_{i=1}^{54}\left(\frac{\partial}{\partial q_{4\rho}} T_i^{\mu\nu\lambda\sigma}(q_1, q_2, q_4 - q_1 - q_2)\right)\Bigg|_{q_4=0} \\
&\quad \times \Pi_i(q_1, q_2, -q_1 - q_2).
\end{aligned} \tag{5}$$

After a Wick rotation of the momenta, five of the eight loop integrals can be carried out with the technique of

[3] In 4 space-time dimensions, there are two additional linear relations, hence a basis consists of 41 elements [21, 22].

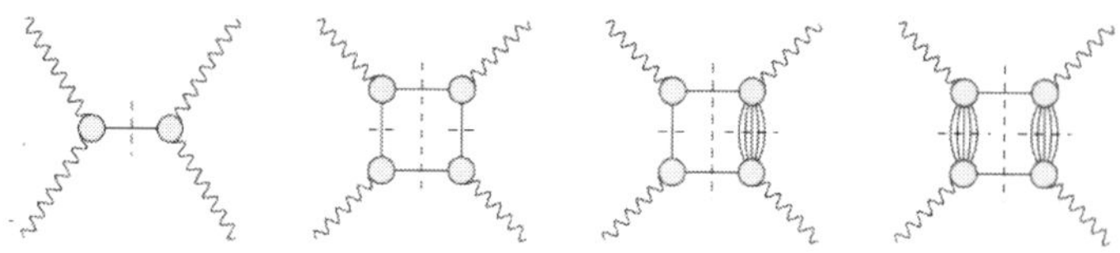

Figure 1. Unitarity diagrams according to the Mandelstam representation. Crossed diagrams are omitted.

Gegenbauer polynomials [24]. We have checked that this Wick rotation is justified even in the presence of anomalous thresholds in the scalar functions Π_i. In analogy to the pion-pole contribution [25], a master formula for the full HLbL contribution to $(g-2)_\mu$ can be worked out:

$$a_\mu^{\text{HLbL}} = \frac{2\alpha^3}{3\pi^2}\int_0^\infty dQ_1 \int_0^\infty dQ_2 \int_{-1}^1 d\tau \sqrt{1-\tau^2} Q_1^3 Q_2^3 \times \sum_{i=1}^{12} T_i(Q_1, Q_2, \tau)\bar{\Pi}_i(Q_1, Q_2, \tau), \tag{6}$$

where $\alpha = e^2/(4\pi)$ and the T_i are integration kernels. With Q_i we denote Euclidean momenta. Only twelve independent linear combinations of the hadronic scalar functions Π_i contribute, denoted by $\bar{\Pi}_i$ [17]. They have to be evaluated for the reduced kinematics

$$s = -Q_3^2, \quad t = -Q_2^2, \quad u = -Q_1^2,$$
$$q_1^2 = -Q_1^2, \quad q_2^2 = -Q_2^2,$$
$$q_3^2 = -Q_3^2 = -Q_1^2 - 2Q_1Q_2\tau - Q_2^2, \quad q_4^2 = 0. \tag{7}$$

4 Mandelstam representation

Although the scalar functions in the master formula (6) are needed only for the reduced kinematics (7), where the limit $q_4 \to 0$ is taken, we define the dispersion relation in the Mandelstam variables of the four-point function with general kinematics and evaluate it only afterwards for the special case $q_4 \to 0$. This procedure has the following advantage: the HLbL contribution to $(g-2)_\mu$ splits into contributions from different topologies, each of them linked to a specific sub-process, which is either data input or again a dispersively reconstructed quantity. These different contributions are discussed in the following.

Gauge invariance, encoded in the decomposition (3), leads to Lorentz structures $T_i^{\mu\nu\lambda\sigma}$ of mass dimension 4, 6, and 8. Hence, we expect the scalar functions Π_i to be rather strongly suppressed at high energies. Thus we write down unsubtracted double-spectral (Mandelstam) representations for the Π_i [26], *i.e.* parameter-free dispersion relations. The input to the dispersion relation are the residues at poles (due to single-particle intermediate states) and the discontinuities along branch cuts (due to two-particle intermediate states). Both are defined by the unitarity relation, in which the intermediate states are always on-shell. We neglect contributions from intermediate states consisting of more than two particles in the primary cut. Heavier intermediate states are expected to be suppressed by higher thresholds and smaller phase space, in agreement with the outcome of model calculations.

In the Mandelstam representation, the sum over intermediate states in the unitarity relations (for the primary and secondary cuts) translates into a splitting of the HLbL tensor into several topologies, shown in fig. 1. The first topology consists of the pion pole, *i.e.* the terms arising from a single pion intermediate state. This contribution is well-known [25] and given by

$$\bar{\Pi}_1^{\pi^0\text{-pole}} = -\frac{\mathcal{F}_{\pi^0\gamma^*\gamma^*}(-Q_1^2, -Q_2^2)\mathcal{F}_{\pi^0\gamma^*\gamma^*}(-Q_3^2, 0)}{Q_3^2 + M_\pi^2},$$
$$\bar{\Pi}_2^{\pi^0\text{-pole}} = -\frac{\mathcal{F}_{\pi^0\gamma^*\gamma^*}(-Q_1^2, -Q_3^2)\mathcal{F}_{\pi^0\gamma^*\gamma^*}(-Q_2^2, 0)}{Q_2^2 + M_\pi^2}, \tag{8}$$

where $\mathcal{F}_{\pi^0\gamma^*\gamma^*}$ denotes the pion transition form factor (for off-shell photons but an on-shell pion).

The other topologies are obtained by selecting two-pion intermediate states in the primary cut. The sub-process $\gamma^*\gamma^* \to \pi\pi$ is again cut in the crossed channel. If we single out the pion-pole contribution in both of the sub-processes, we obtain the box topologies for HLbL. For higher intermediate states in the crossed channel of $\gamma^*\gamma^* \to \pi\pi$, we obtain boxes with multi-particle cuts instead of poles in the sub-processes.

By explicitly constructing the Mandelstam representation, we have shown that the box topologies in the sense of unitarity have the same analytic structure as the one-loop sQED contribution, multiplied with pion electromagnetic form factors $F_\pi^V(q_i^2)$ for each of the off-shell photons (FsQED). The dispersion relation defines unambiguously this particular q_i^2 dependence. With the construction of the Mandelstam representation, we prove that FsQED and box topologies are the same. Note that the sQED loop contribution in terms of Feynman diagrams consists of boxes, triangles, and bulbs, but that the corresponding unitarity diagrams are just box topologies. This can be understood as follows: in sQED, the appearance of triangle and bulb diagrams is due to the seagull vertex, needed to ensure gauge invariance. In our formalism, gauge invariance is already encoded in the tensor decomposition (3). If the sQED contribution is projected on this tensor decomposition, which separates kinematics from dynamics, one can check that the dynamical singularities of the scalar functions Π_i in sQED are the ones of a box topology.

The equivalence of the pion-box topologies with FsQED allows us to derive compact expressions for the contribution to the scalar functions Π_i in terms of two-dimensional Feynman parameter integrals. In the limit

$q_4 \to 0$, they are given by

$$\Pi_i^{\pi\text{-box}}(q_1^2, q_2^2, q_3^2) = F_\pi^V(q_1^2)F_\pi^V(q_2^2)F_\pi^V(q_3^2) \times \frac{1}{16\pi^2}\int_0^1 dx \int_0^{1-x} dy\, I_i(x,y), \quad (9)$$

where

$$\begin{aligned}
I_1(x,y) &= -\frac{2}{3}\frac{(1-2y)(1-2x-2y)(1-6x(1-x))}{\Delta_{123}^2},\\
I_4(x,y) &= -\frac{2}{3}\frac{(1-2x)(1+2x(1-3x(1-2y)-6y(1-y)))}{\Delta_{123}^2},\\
I_7(x,y) &= -\frac{4}{3}\frac{(1-2x)^2(1-2y)^2 y(1-y)}{\Delta_{123}^3},\\
I_{16}(x,y) &= \frac{4}{3}\frac{x(1-2x)y(1-2y)}{\Delta_{213}\Delta_{13}}\left(\frac{1}{\Delta_{213}}+\frac{1}{\Delta_{13}}\right),\\
I_{19}(x,y) &= -\frac{4}{3}\frac{(1-2x)^2(1-2y)y(1-y)(1-2x-2y)}{\Delta_{213}^3},\\
I_{31}(x,y) &= -\frac{8}{3}\frac{x^2(1-x)(1-2x)^3 y(1-2y)}{\Delta_{213}\Delta_{13}} \times\left(\frac{1}{\Delta_{213}^2}+\frac{1}{\Delta_{213}\Delta_{13}}+\frac{1}{\Delta_{13}^2}\right),\\
I_{39}(x,y) &= \frac{4}{3}\frac{(1-2x)(1-2y)^2 y(1-y)(1-2x-2y)}{\Delta_{123}^3},\\
I_{42}(x,y) &= -\frac{4}{3}\frac{x(1-2x)y(1-2y)(1-6y(1-y))}{\Delta_{321}\Delta_{21}} \times\left(\frac{1}{\Delta_{321}}+\frac{1}{\Delta_{21}}\right),\\
I_{50}(x,y) &= 0, \qquad (10)
\end{aligned}$$

and

$$\begin{aligned}
\Delta_{ijk} &= M_\pi^2 - xyq_i^2 - x(1-x-y)q_j^2 - y(1-x-y)q_k^2,\\
\Delta_{ij} &= M_\pi^2 - x(1-x)q_i^2 - y(1-y)q_j^2. \qquad (11)
\end{aligned}$$

The remaining functions Π_i that enter the master formula can be obtained with crossing relations permuting only q_1, q_2, and q_3, which are still valid in the limit $q_4 \to 0$.

For a numerical analysis of the pion box contribution, the only input needed is the pion vector form factor in the space-like region. In fig. 2 we show two fits: one is based on a modified Omnès representation which accounts for the prominent singularities in the low-energy region [27, 28], the other on vector meson (here ρ-) dominance. Correspondently we obtain

$$\begin{aligned}
a_\mu^{\pi\,\text{box}} &\simeq -15.9 \cdot 10^{-11}\\
a_\mu^{\pi\,\text{box,VMD}} &\simeq -16.4 \cdot 10^{-11} \qquad (12)
\end{aligned}$$

Error estimate is in progress but associated uncertainties are expected to be very small.

We treat the contribution from topologies with higher intermediate states in a partial-wave picture. This means that the a multi-particle cut is approximated by a polynomial, as illustrated in fig. 3. The dispersive formulation

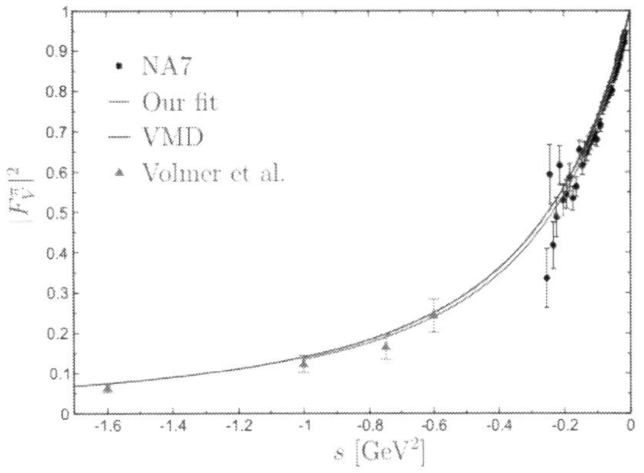

Figure 2. Different parameterization of the pion vector form factor for space-like kinematics used in our numerical analysis of the pion box contribution to a_μ.

allows us to describe here the effect of two-pion rescattering in the primary channel. In [13], we already discussed the S-wave contribution. The new Lorentz decomposition allows us to include also higher partial waves. The contribution of these topologies is given by dispersion integrals over products of $\gamma^*\gamma^{(*)} \to \pi\pi$ helicity partial waves. The Born terms of the sub-process have to be properly subtracted to avoid double-counting with the box topologies.

5 Conclusion and outlook

Using the Mandelstam representation for the hadronic scalar functions Π_i, we have split a_μ^{HLbL} into three contributions: pion-pole contributions, box topologies, and $\pi\pi$-rescattering contributions:

$$a_\mu^{\text{HLbL}} = a_\mu^{\pi^0\text{-pole}} + a_\mu^{\text{box}} + a_\mu^{\pi\pi} + \ldots, \qquad (13)$$

where the dots denote neglected higher intermediate states in the primary cut. We stress that in our approach we need to solve the dispersion relation for the HLbL tensor at *fixed photon virtualities*. The input quantities in our dispersive description are the pion transition form factor $\mathcal{F}_{\pi^0\gamma^*\gamma^*}(q_i^2, q_j^2)$, the pion electromagnetic form factor $F_\pi^V(q_i^2)$, and the $\gamma^*\gamma^* \to \pi\pi$ helicity partial waves. In the absence of experimental data on the doubly-virtual processes, these quantities will be reconstructed again dispersively [13, 29–36].

Our dispersive formalism defines unambiguously both the pion-pole and pion-box contribution. They are treated without any approximation. For the two-pion rescattering contribution a partial-wave expansion is employed.

We have limited the discussion to pions although the formalism can be extended to higher pseudoscalar poles (η, η') or KK intermediate states [37–40]. Future work will include model estimates of the contribution from intermediate states with more than two pseudoscalar particles, and the incorporation of high-energy constraints from perturbative QCD.

 DOI: 10.1051/epjconf/201611801030

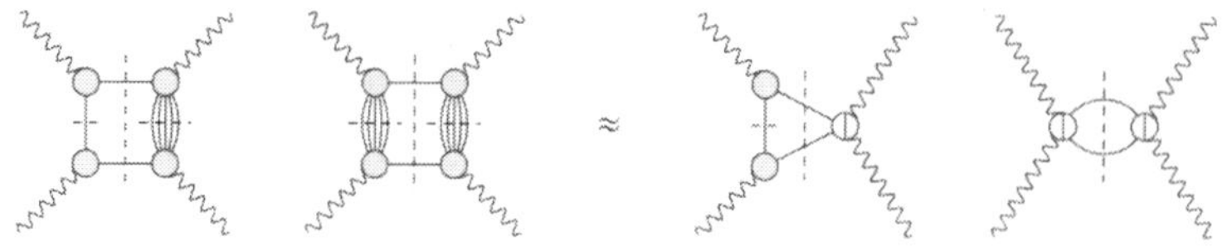

Figure 3. Partial-wave approximation of multi-particle intermediate states in the secondary cut.

Our dispersive approach is expected to help substantially reduce model dependence in HLbL scattering through a more data-driven evaluation of the HLbL contribution to $(g-2)_\mu$. A careful numerical study is currently under way, with the final goal to identify the experimental input with the largest impact on the theory uncertainty.

The speaker (MP) thanks the organizing committee of the Workshop FCCP2015 for the kind invitation. We thank J. Bijnens, J. Gasser, M. Knecht, B. Kubis, H. Leutwyler, A. Nyffeler, and S. Scherer for useful discussions. We gratefully acknowledge financial support by the Swiss National Science Foundation, the DFG (CRC 16, "Subnuclear Structure of Matter"), the European Community's 7th Framework Programme (Marie Curie Intra-European Fellowship) and the DOE (Grant No. DE-FG02-00ER41132).

References

[1] G. W. Bennett *et al.* [Muon g-2 Collaboration], Phys. Rev. D **73** (2006) 072003 [hep-ex/0602035].

[2] T. Blum, A. Denig, I. Logashenko, E. de Rafael, B. Lee Roberts, T. Teubner and G. Venanzoni, arXiv:1311.2198 [hep-ph].

[3] J. Grange *et al.* [Muon g-2 Collaboration], arXiv:1501.06858 [physics.ins-det].

[4] N. Saito [J-PARC g-'2/EDM Collaboration], AIP Conf. Proc. **1467** (2012) 45.

[5] A. Kurz, T. Liu, P. Marquard and M. Steinhauser, Phys. Lett. B **734** (2014) 144 [arXiv:1403.6400 [hep-ph]].

[6] G. Colangelo, M. Hoferichter, A. Nyffeler, M. Passera and P. Stoffer, Phys. Lett. B **735** (2014) 90 [arXiv:1403.7512 [hep-ph]].

[7] C. Lehner, these proceedings.

[8] M. Hayakawa, T. Blum, T. Izubuchi and N. Yamada, PoS LAT **2005** (2006) 353 [hep-lat/0509016].

[9] T. Blum, M. Hayakawa and T. Izubuchi, PoS LATTICE **2012** (2012) 022 [arXiv:1301.2607 [hep-lat]].

[10] T. Blum, S. Chowdhury, M. Hayakawa and T. Izubuchi, Phys. Rev. Lett. **114** (2015) 1, 012001 [arXiv:1407.2923 [hep-lat]].

[11] J. Green, O. Gryniuk, G. von Hippel, H. B. Meyer and V. Pascalutsa, arXiv:1507.01577 [hep-lat].

[12] T. Blum, N. Christ, M. Hayakawa, T. Izubuchi, L. Jin and C. Lehner, arXiv:1510.07100 [hep-lat].

[13] G. Colangelo, M. Hoferichter, M. Procura and P. Stoffer, JHEP **1409** (2014) 091 [arXiv:1402.7081 [hep-ph]].

[14] G. Colangelo, M. Hoferichter, B. Kubis, M. Procura and P. Stoffer, Phys. Lett. B **738** (2014) 6 [arXiv:1408.2517 [hep-ph]].

[15] V. Pauk and M. Vanderhaeghen, Phys. Rev. D **90** (2014) 11, 113012 [arXiv:1409.0819 [hep-ph]].

[16] P. Stoffer, arXiv:1412.5171 [hep-ph].

[17] G. Colangelo, M. Hoferichter, M. Procura and P. Stoffer, JHEP **1509** (2015) 074 [arXiv:1506.01386 [hep-ph]].

[18] R. A. Leo, A. Minguzzi and G. Soliani, Nuovo Cim. A **30** (1975) 270.

[19] W. A. Bardeen and W. K. Tung, Phys. Rev. **173** (1968) 1423 [Phys. Rev. D **4** (1971) 3229].

[20] R. Tarrach, Nuovo Cim. A **28** (1975) 409.

[21] G. Eichmann, C. S. Fischer, W. Heupel and R. Williams, arXiv:1411.7876 [hep-ph].

[22] G. Eichmann, C. S. Fischer and W. Heupel, Phys. Rev. D **92** (2015) 5, 056006 [arXiv:1505.06336 [hep-ph]].

[23] J. Aldins, T. Kinoshita, S. J. Brodsky and A. J. Dufner, Phys. Rev. D **1** (1970) 2378.

[24] J. L. Rosner, Annals Phys. **44** (1967) 11.

[25] M. Knecht and A. Nyffeler, Phys. Rev. D **65** (2002) 073034 [hep-ph/0111058].

[26] S. Mandelstam, Phys. Rev. **112** (1958) 1344.

[27] H. Leutwyler, hep-ph/0212324.

[28] G. Colangelo, Nucl. Phys. Proc. Suppl. **131** (2004) 185 [hep-ph/0312017].

[29] F. Niecknig, B. Kubis and S. P. Schneider, Eur. Phys. J. C **72** (2012) 2014 [arXiv:1203.2501 [hep-ph]].

[30] S. P. Schneider, B. Kubis and F. Niecknig, Phys. Rev. D **86** (2012) 054013 [arXiv:1206.3098 [hep-ph]].

[31] M. Hoferichter, G. Colangelo, M. Procura and P. Stoffer, Int. J. Mod. Phys. Conf. Ser. **35** (2014) 1460400 [arXiv:1309.6877 [hep-ph]].

[32] M. Hoferichter, B. Kubis, S. Leupold, F. Niecknig and S. P. Schneider, Eur. Phys. J. C **74** (2014) 3180 [arXiv:1410.4691 [hep-ph]].

[33] R. Garcia-Martin and B. Moussallam, Eur. Phys. J. C **70** (2010) 155 [arXiv:1006.5373 [hep-ph]].

[34] M. Hoferichter, D. R. Phillips and C. Schat, Eur. Phys. J. C **71** (2011) 1743 [arXiv:1106.4147 [hep-ph]].

[35] B. Moussallam, Eur. Phys. J. C **73** (2013) 2539 [arXiv:1305.3143 [hep-ph]].

[36] M. Hoferichter, B. Kubis and D. Sakkas, Phys. Rev. D **86** (2012) 116009 [arXiv:1210.6793 [hep-ph]].
[37] F. Stollenwerk, C. Hanhart, A. Kupsc, U. G. Meissner and A. Wirzba, Phys. Lett. B **707** (2012) 184 [arXiv:1108.2419 [nucl-th]].
[38] C. Hanhart, A. Kup?c, U.-G. Meißner, F. Stollenwerk and A. Wirzba, Eur. Phys. J. C **73** (2013) 12, 2668 [Eur. Phys. J. C **75** (2015) 6, 242] [arXiv:1307.5654 [hep-ph]].
[39] B. Kubis and J. Plenter, Eur. Phys. J. C **75** (2015) 6, 283 [arXiv:1504.02588 [hep-ph]].
[40] C. W. Xiao, T. Dato, C. Hanhart, B. Kubis, U.-G. Meißner and A. Wirzba, arXiv:1509.02194 [hep-ph].

EPJ Web of Conferences **118**, 01031 (2016) DOI: 10.1051/epjconf/201611801031
FCCP2015

Lepton-flavour violating decays in theories with dimension 6 operators

Giovanni Marco Pruna[1,a] and Adrian Signer[1,2,b]

[1] *Paul Scherrer Institut, CH-5232 Villigen PSI, Switzerland*
[2] *Physik-Institut, Universität Zürich, CH-8057 Zürich, Switzerland*

Abstract. Despite a large experimental effort, so far no evidence for flavour-violating decays of charged leptons such as $l_i \to l_j\gamma$ and $l_i \to l_j l_k l_k$ has been found. The absence of a signal puts very severe constraints on many extensions of the Standard Model. Here we apply a model independent approach by studying such decays in the Standard Model effective field theory. Going beyond leading order in the Standard Model couplings and considering all dimension 6 operators that might lead to lepton-flavour violation, we are able to extract limits on a large number of Wilson coefficients of such operators. We are also able to compare the impact of particular searches and find, for example, that flavour-violating decays of the Z-boson $Z \to \mu e$ are much more constrained from low-energy experiments $\mu \to e\gamma$ than from the limits of current and future direct searches at high energy.

1 Introduction

It is well established that the Standard Model (SM) of particle physics provides a structure of gauge symmetries that accidentally conserves the leptonic flavour. This opens up a possibility for very stringent tests of the SM and powerful searches for physics beyond the SM. For this reason, a longstanding experimental effort has been devoted to the search of lepton-flavour violating (LFV) signals both in the neutral and charged sectors.

In the neutrino sector, the evidence for flavour violation is by now established beyond any doubt [1–3]. Thus, the SM with only left-handed neutrinos needs to be extended. Concerning the charged lepton sector, searches for LFV in μ and τ decays at low energy experiments as well as in Z boson decays at high energy experiments have been analysed for decades, resulting in ever more stringent limits on various LFV branching ratios (BRs) of the aforementioned particles. One possibility to study and interpret the absence (so far) of any charged LFV signals is to consider specific models beyond the SM and a large effort has been made in this direction (see e.g., [4–10]).

In this proceeding, the adopted strategy follows from a bottom-up approach established in [11] and refined in the past thirty years [12, 13]. The SM is considered to be an effective theory valid up to an (unknown) large energy scale Λ and the Lagrangian consists of a systematic dimensional expansion in $1/\Lambda$ that includes all operators constructed from SM fields (without right-handed neutrinos) that respect the SM gauge symmetries, i.e.

$$\mathcal{L} = \mathcal{L}_{\rm SM} + \frac{1}{\Lambda}\sum_k C_k^{(5)} Q_k^{(5)} + \frac{1}{\Lambda^2}\sum_k C_k^{(6)} Q_k^{(6)} + O\left(\frac{1}{\Lambda^3}\right). \quad (1)$$

The coefficients $C_k^{(d)}$ are couplings and, very much like the usual couplings of the SM, they have to be determined by experiment. Beyond tree level, a renormalisation scheme has to be chosen to give the couplings a well-defined meaning.

Once an ultraviolet (UV) complete theory is known that describes the effects of physics beyond the scale Λ, the coefficients $C_k^{(d)}$ of the Lagrangian can be expressed in terms of the parameters of the UV complete theory. Such a matching has to be done at the scale Λ. To obtain the coefficients at the relevant low energy scale, renormalisation-group (RG) techniques have to be used to determine the running. We want to stress that due to operator mixing the RG evolution leads to qualitatively new effects and, at least for the time being, is not primarily a question of precision.

Performing an expansion in $1/\Lambda$ of the most general gauge-invariant Lagrangian one is allowed to write only one Dim-5 operator [14–16]. This operator provides both neutrino mass terms and LFV in the neutrino sector. Going beyond tree level, it also causes LFV in the charged sector. However, the smallness of the neutrino masses results in very severe constraints on its coefficients. These constraints in turn imply that the amount of charged LFV provided by the Dim-5 operator is far below any foreseeable experimental capability. Hence, the next step consists both in the scrutiny of the 19 Dim-6 operators that yield LFV and in the interpretation of the current BR limits with respect to such extended parameter space. Given the complete lack of information on the structure of the

[a] e-mail: Giovanni-Marco.Pruna@psi.ch
[b] e-mail: Adrian.Signer@psi.ch

© The Authors, published by EDP Sciences. This is an open access article distributed under the terms of the Creative Commons Attribution License 4.0 (http://creativecommons.org/licenses/by/4.0/).

UV complete theory, we make no assumption whatsoever on the relative importance of the various Dim-6 operators.

In the literature, muon decays are often studied using an effective theory that stops one step before, by integrating out any dynamics at the electroweak scale. The resulting effective theory is simpler and contains fewer parameters. However, we refrain from doing this and work directly with the Lagrangian of Eq. (1). This provides us with a framework to combine constraints from low-energy experiments such as $\mu^+ \to e^+\gamma$ and $\mu^+ \to e^+e^-e^+$ with LFV searches at higher energies, such as LFV decays of the Z boson or $H \to \tau\mu$.

On the low energy side, very accurate limits on the LFV muon decays $\mu^+ \to e^+\gamma$ and $\mu^+ \to e^+e^-e^+$ have been made available by the MEG [17] and SINDRUM [18] collaborations, respectively; in the near future, an upgrade of the MEG experiment [19] will take place almost in parallel with Mu3e, a new experiment devoted to the $\mu \to 3e$ searches [20]. In the tauonic sector, a big effort is ongoing to continuously improve the bounds on several decay channels in various experiments [21, 22].

On the high energy side, interesting searches have been performed at LEP in the context of LFV decays of the Z boson [23–26], and further improvements have either been made or are expected from the ATLAS [27] and CMS [28] experiments at the LHC collider. Moreover, after the discovery of the Higgs boson [29, 30], new limits on the $H \to \tau\mu$ BR are currently under investigation [31, 32].

In this proceeding, the constraints on various Dim-6 coefficients are reviewed in light of the latest bounds on LFV μ and τ decays and in the assumption that the relevant operators are not correlated. Due to the very good accuracy of the bounds on $l_i \to l_j\gamma$ transitions, these processes will be considered at the one-loop level, while the rest of our study will be performed at the tree level. The tables supplied in this note are neither meant to be complete nor to exhaust the general discussion on LFV decays in theories with Dim-6 operators. For example LFV transitions in nuclei, where a substantial improvement [33, 34] in the experimental sensitivity with respect to the current best limit [35] is expected, is not considered. At this stage, we also do not include decays $l_i \to l_j 2l_k$ with $k \neq j$. Furthermore, we also neglect correlations among operators and potentially important two loop Barr-Zee effects. The main two messages that will be delivered in the following sections are: (i) the richness of the information on the coefficients $C_k^{(d)}$ that can be gained from low-energy experiments if Eq. (1) is taken seriously as a quantum field theory and (ii) the fact that the limits on LFV decays of the Z boson from low-energy experiments are far more stringent than those from the high-energy experiments.

Lastly, a remark is required: these proceedings are largely based on the elements collected in Ref. [36], to which the reader is redirected for further details.

2 LFV dimension 6 operators

Among the 19 Dim-6 operators that trigger LFV transitions, the relevant ones for the study of $l_i \to l_j\gamma$ (at the one loop level), $l_i \to 3l_j$, $Z \to l_il_j$ and $H \to l_il_j$ are collected in Table 1.

$\psi^2 X\varphi$		$\psi^2\varphi^2 D$	
Q_{eW}	$(\bar{l}_p\sigma^{\mu\nu}e_r)\tau^I\varphi W^I_{\mu\nu}$	$Q^{(1)}_{\varphi l}$	$(\varphi^\dagger i\overleftrightarrow{D}_\mu\varphi)(\bar{l}_p\gamma^\mu l_r)$
Q_{eB}	$(\bar{l}_p\sigma^{\mu\nu}e_r)\varphi B_{\mu\nu}$	$Q^{(3)}_{\varphi l}$	$(\varphi^\dagger i\overleftrightarrow{D}^I_\mu\varphi)(\bar{l}_p\tau^I\gamma^\mu l_r)$
		$Q_{\varphi e}$	$(\varphi^\dagger i\overleftrightarrow{D}_\mu\varphi)(\bar{e}_p\gamma^\mu e_r)$
$\psi^2\varphi^3$			
$Q_{e\varphi}$		$(\varphi^\dagger\varphi)(\bar{l}_p e_r\varphi)$	
$(\bar{l}l)(\bar{l}l)$		$(\bar{l}l)(\bar{q}q)$	
Q_{ll}	$(\bar{l}_p\gamma_\mu l_r)(\bar{l}_s\gamma^\mu l_t)$	$Q^{(1)}_{lequ}$	$(\bar{l}^j_p e_r)\varepsilon_{jk}(\bar{q}^k_s u_t)$
Q_{ee}	$(\bar{e}_p\gamma_\mu e_r)(\bar{e}_s\gamma^\mu e_t)$	$Q^{(3)}_{lequ}$	$(\bar{l}^j_p\sigma_{\mu\nu}e_r)\varepsilon_{jk}(\bar{q}^k_s\sigma^{\mu\nu}u_t)$
Q_{le}	$(\bar{l}_p\gamma_\mu l_r)(\bar{e}_s\gamma^\mu e_t)$		

Table 1. LFV Dim-6 operators.

Working in the physical basis rather than in the gauge basis, the dipole operators (set $\psi^2X\varphi$) are rewritten using

$$Q_{eB} \to Q_{e\gamma}c_W - Q_{eZ}s_W, \tag{2}$$

$$Q_{eW} \to -Q_{e\gamma}s_W - Q_{eZ}c_W, \tag{3}$$

where $s_W = \sin(\theta_W)$ and $c_W = \cos(\theta_W)$ are the sine and cosine of the weak mixing angle. The term

$$\mathcal{L}_{e\gamma} \equiv \frac{C_{e\gamma}}{\Lambda^2}Q_{e\gamma} + \text{h.c.} = \frac{C^{pr}_{e\gamma}}{\Lambda^2}(\bar{l}_p\sigma^{\mu\nu}e_r)\varphi F_{\mu\nu} + \text{h.c.}, \tag{4}$$

where $F_{\mu\nu}$ is the electromagnetic field-strength tensor, is then the only term in the Dim-6 Lagrangian that induces a $l_i \to l_j\gamma$ transition at tree level.

In the Feynman gauge, the combination of $Q_{e\varphi}$ with the Dim-4 SM Yukawa terms gives

$$\begin{aligned}
&\mathcal{L}_{\text{Yukawa}} + \mathcal{L}_{e\varphi} = \\
&= \frac{v}{\sqrt{2}}\left(-y_{pr} + \frac{v^2}{2\Lambda^2}C^{pr}_{e\varphi}\right)\bar{e}_p e_r \\
&+ \frac{1}{\sqrt{2}}\left(-y_{pr} + \frac{v^2}{2\Lambda^2}C^{pr}_{e\varphi}\right)\bar{e}_p e_r h + \frac{v^2}{\sqrt{2}\Lambda^2}C^{pr}_{e\varphi}\bar{e}_p e_r h \\
&+ \frac{i}{\sqrt{2}}\left(-y_{pr} + \frac{v^2}{2\Lambda^2}C^{pr}_{e\varphi}\right)\bar{e}_p e_r \widehat{Z} \\
&+ i\left(-y_{pr} + \frac{v^2}{2\Lambda^2}C^{pr}_{e\varphi}\right)\bar{e}_p \nu_r \widehat{W}^+ + [\ldots].
\end{aligned} \tag{5}$$

From Eq. (5), it is understood that any 3-point off-diagonal interaction involving Goldstone bosons is not physical, i.e. it can be removed by an orthogonal transformation[1]. However, this procedure leaves a residual term with a physical Higgs supporting LFV currents, which is the only tree-level contribution to the $H \to l_il_j$ transition.

[1] In this proceeding, any impact of such diagonalisation on the charged lepton eigenstates and their masses will be neglected.

Concerning the decay $Z \to l_i l_j$, such transition is triggered at the tree level by the following operators: Q_{eZ} (the dipole contribution), $Q_{\varphi l}^{(1)}$, $Q_{\varphi l}^{(3)}$ and $Q_{\varphi e}$. In addition, μ and τ three body decays are also produced by the point-like four-fermion operators Q_{ll}, Q_{le} and Q_{ee}.

The impact of $Q_{lequ}^{(1)}$ and $Q_{lequ}^{(3)}$ in the $l_i \to l_j\gamma$ transition is of a different kind: they contribute in the running of the dipole operators, hence they must be carefully included when different energy scales are considered, as was shown in [36–39].

3 Branching Ratios

In the limit where $m_1 \gg m_2$ and *no correlation is considered among different operators*, the partial widths of the aforementioned processes are the following:

- for the two-body decay $l_1^\mp \to l_2^\mp\gamma$, one has

$$\Gamma(l_1^\mp \to l_2^\mp\gamma) = \frac{m_1^3}{4\pi\Lambda^4}\left(|C_{TL}|^2 + |C_{TR}|^2\right), \qquad (6)$$

where the contributions to C_{TL} and C_{TR} (at the energy scale $\lambda = m_Z$, in the assumption that the coefficients are real) are given in Table 2, according to [36] (see also [40]). The coefficients $C_k^{(d)}$ in Table 2 are to be interpreted as renormalised in the $\overline{\text{MS}}$-scheme;

- for the three-body decay $l_1^\mp \to l_2^\mp l_2^\mp l_2^\pm$, one has

$$\begin{aligned}\Gamma(l_1^+ \to l_2^+ l_2^- l_2^+) = &\Bigg(40e^2v^2\left(\left|C_{e\gamma}^{12}\right|^2 + \left|C_{e\gamma}^{21}\right|^2\right)\left(8\ln\left[\frac{m_1}{m_2}\right] - 11\right)\\ &+ \frac{2m_1^4}{m_Z^2}\Big(\left(5 - 20s_W^2 + 36s_W^4\right)\left|C_{eZ}^{12}\right|^2\\ &\qquad + 4\left(1 - 4s_W^2 + 9s_W^4\right)\left|C_{eZ}^{21}\right|^2\Big)\\ &+ \frac{15m_2^2m_1^2v^2\left(\left|C_{e\varphi}^{12}\right|^2 + \left|C_{e\varphi}^{21}\right|^2\right)}{8m_H^4}\\ &+ 10m_1^2\left(1 - 4s_W^2 + 12s_W^4\right)\left|C_{\varphi e}^{12}\right|^2\\ &+ 20m_1^2\left(1 - 4s_W^2 + 6s_W^4\right)\left(\left|C_{\varphi l(1)}^{12}\right|^2 + \left|C_{\varphi l(3)}^{12}\right|^2\right) +\\ &+ 10m_1^2\left(\left|C_{le}^{1112}\right|^2 + \left|C_{le}^{1211}\right|^2\right)\\ &+ 80m_1^2\left(\left|C_{ee}^{1112}\right|^2 + \left|C_{ll}^{1112}\right|^2\right)\Bigg)\frac{m_1^3}{30(8\pi)^3\Lambda^4}, \end{aligned} \qquad (7)$$

where the integration over the phase space of the photonic dipole contribution gives rise to a logarithmic term, in agreement with [41]. The result in Eq. (7) is also valid for $\Gamma(l_1^- \to l_2^- l_2^+ l_2^-)$;

- flavour-violating Z decays can be parametrised at the tree level by means of the following four operators:

$$\begin{aligned}\Gamma(Z \to l_1^\mp l_2^\mp) = &\frac{m_Z^3 v^2}{12\pi\Lambda^4}\Big(\left|C_{eZ}^{12}\right|^2 + \left|C_{eZ}^{21}\right|^2\\ &+ \left|C_{\varphi e}^{12}\right|^2 + \left|C_{\varphi l(1)}^{12}\right|^2 + \left|C_{\varphi l(3)}^{12}\right|^2\Big), \end{aligned} \qquad (8)$$

and as one can see from the last equation, all of their contributions occur at the same order. In Eq. (8) we have summed over the two possible final states, $l_1^+ l_2^-$ and $l_1^- l_2^+$;

- for the Higgs boson decay $H \to l_1^\mp l_2^\mp$, one has

$$\Gamma(H \to l_1^\mp l_2^\mp) = \frac{m_H v^4}{16\pi\Lambda^4}\left(\left|C_{e\varphi}^{12}\right|^2 + \left|C_{e\varphi}^{21}\right|^2\right), \qquad (9)$$

where only one operator contributes at tree level. As for the Z decays, in Eq. (9) we have summed over the two possible decays $l_1^+ l_2^-$ and $l_1^- l_2^+$.

Op.	C_{TL} or $C_{TR}(1 \longleftrightarrow 2)$
$Q_{e\gamma}$	$-C_{e\gamma}^{12}\frac{v}{\sqrt{2}}$
Q_{eZ}	$-C_{eZ}^{12}\frac{em_Z}{16\sqrt{2}\pi^2}\left(3 - 6c_W^2 + 4c_W^2\log\left[c_W^2\right]\right)$
$Q_{\varphi l}^{(1)}$	$-C_{\varphi l(1)}^{12}\frac{em_1\left(1 + s_W^2\right)}{24\pi^2}$
$Q_{\varphi l}^{(3)}$	$C_{\varphi l(3)}^{12}\frac{em_1\left(3 - 2s_W^2\right)}{48\pi^2}$
$Q_{\varphi e}$	$C_{\varphi e}^{12}\frac{em_2\left(3 - 2s_W^2\right)}{48\pi^2}$
$Q_{e\varphi}$	$C_{e\varphi}^{12}\frac{ev\,\text{Max}(m_1^2, m_2^2)}{96\sqrt{2}m_H^2\pi^2}\left(4 + 3\log\left[\frac{\text{Max}(m_1^2, m_2^2)}{m_H^2}\right]\right)$
Q_{le}	$\frac{e}{16\pi^2}\sum_{i=e,\mu,\tau} m_i C_{le}^{1ii2}$

Table 2. Leading order results (up to one loop) for the contributions of the various (real) Dim-6 operators to the $l_1 \to l_2\gamma$ decay at the energy scale $\lambda = m_Z$.

In order to numerically evaluate the BRs, the Particle Data Group (PDG) values [42] for the physical constants and for the μ and τ total decay width were used.

4 Results

Here, the elements collected in Section 3 are combined with the experimental limits on LFV transitions at high and low energies. Since the category of transitions $l_i \to l_j\gamma$ was studied at the one-loop level, the limits on $\mu \to e\gamma$ and $\tau \to \mu\gamma$ from MEG and BaBar were interpreted at the energy scale $\lambda = m_Z$ by means of the QED running of the Wilson coefficients in Eq. (6). Being absolutely rigorous, also the limits on three-body decays should be studied at the one-loop level and reinterpreted at the Z-boson mass scale. However, it was proved by direct computation that such running, in absence of correlations among operators, only affects the limits at the 10% level. Hence it is not relevant in the following discussion where differences of orders of magnitude are involved.

4.1 Limits at a fixed energy $\lambda = m_Z$

Under the assumption that only one operator at a time is non-vanishing and the corresponding coefficients are real, the numerical limits of Tables 3-4 are obtained. Note that $C_{\varphi e}^{21} = (C_{\varphi e}^{12})^* \doteq C_{\varphi e}^{12}$ (and similar for $C_{\varphi l(1)}$ and $C_{\varphi l(3)}$) since we treat the coefficients as real. We also have $C_{ee}^{prst} = C_{ee}^{stpr} = (C_{ee}^{tsrp})^* \doteq C_{ee}^{tsrp}$ (and similar for C_{ll}), where again the last step assumes the coefficients to be real.

Coeff. $\lambda = m_Z$	$\mu^+ \to e^+\gamma$ BR≤ $5.7\cdot10^{-13}$	$Z \to e^\pm\mu^\mp$ BR≤ $7.5\cdot10^{-7}$	$\mu^+ \to e^+e^-e^+$ BR≤ $1.0\cdot10^{-12}$
$C_{e\gamma}^{21/12}$	$2.5\cdot10^{-16}$		$3.8\cdot10^{-15}$
$C_{eZ}^{21/12}$	$1.4\cdot10^{-13}$	$3.9\cdot10^{-8}$	$4.0\cdot10^{-8}$
$C_{\varphi l(1)}^{12}$	$2.6\cdot10^{-10}$	$3.9\cdot10^{-8}$	$3.5\cdot10^{-11}$
$C_{\varphi l(3)}^{12}$	$2.5\cdot10^{-10}$	$3.9\cdot10^{-8}$	$3.5\cdot10^{-11}$
$C_{\varphi e}^{12}$	$2.5\cdot10^{-10}$	$3.9\cdot10^{-8}$	$3.7\cdot10^{-11}$
$C_{e\varphi}^{21/12}$	$2.8\cdot10^{-8}$		$8.7\cdot10^{-6}$
$C_{le}^{2111/1112}$	$4.4\cdot10^{-8}$		$3.1\cdot10^{-11}$
$C_{le}^{2221/1222}$	$2.1\cdot10^{-10}$		
$C_{le}^{2331/1332}$	$1.2\cdot10^{-11}$		
C_{ee}^{2111}			$1.1\cdot10^{-11}$
C_{ll}^{2111}			$1.1\cdot10^{-11}$

Table 3. Limits on the Wilson coefficients in $[\mathrm{GeV}]^{-2}$ contributing to the LFV muonic transitions. $\mu \to 3e$ will lead to better limits on C_{eZ} once NLO corrections are taken into account.

Focusing on LFV muonic transitions, the first thing to notice in Table 3 is that the best limits are currently always obtained from the constraints established by low energy experiments. The constraints from LFV decays $Z \to e^\pm\mu^\mp$ on the coefficients of the operators Q_{eZ}, $Q_{\varphi l(1,3)}$ and $Q_{\varphi e}$ are considerably less stringent than from $\mu^+ \to e^+\gamma$ and $\mu^+ \to e^+e^-e^+$.

Even if at a future high energy lepton colliders (e.g. FCC-ee [43]) the LEP limits for BR($Z \to e^\pm\mu^\mp$) could be improved by several orders of magnitude it is still the case that the information one can get from such future machine on $Z \to e^\pm\mu^\mp$ is only complementary, and surely never competitive with the planned future low energy experiments. Indeed, if no correlation is assumed, from Eq. (8) one deduces that in order to reach a probing power equivalent to the current low energy experiments, the required branching ratios for $Z \to e^\pm\mu^\mp$ are BR≤ $9.5\cdot10^{-18}$ for C_{eZ}^{21}, BR≤ $5.9\cdot10^{-13}$ for $C_{\varphi l(1,3)}^{21}$ and BR≤ $6.6\cdot10^{-13}$ for $C_{\varphi e}^{21}$, respectively. Hence, the operator Q_{eZ}^{21} will be out of the reach of FCC-ee, while the coefficients of $Q_{\varphi l(1)}^{12}$, $Q_{\varphi l(3)}^{12}$ and $Q_{\varphi e}^{12}$ could be constrained with (more or less) the same precision of the dated SINDRUM experiment. The interplay between a high-luminosity Z-factory and low-energy experiments has also been studied with concrete BSM models (see e.g. [44]).

Comparing the limits obtained from $\mu^+ \to e^+\gamma$ with $\mu^+ \to e^+e^-e^+$ it is clear that both experiments provide very valuable information. The standard statement is that $\mu \to 3e$ is more powerful in constraining four-fermion operators (contact interactions) whereas $\mu \to e\gamma$ is more sensitive to the dipole operators. However, a closer look reveals that this is only partially true. Some four-fermion operators have stringent limits due to $\mu \to e\gamma$ (at one loop) but are not constrained by $\mu \to 3e$ (at tree level). Furthermore, in the long term $\mu \to 3e$ might well be competitive in putting limits on the photon dipole operator. The BR for $\mu \to 3e$ is expected to be improved by several orders of magnitude. A limit BR($\mu \to 3e$) $\le 4\cdot10^{-15}$ is as constraining to $C_{e\gamma}^{21}$ as the current BR($\mu \to e\gamma$) $\le 5.7\cdot10^{-13}$. Of course, the BR for $\mu \to e\gamma$ is also expected to improve and in the longer term, there will also be very strong constraints from future muon conversion experiments. Regarding the operators with a Higgs, $\mu \to 3e$ provides the stronger limits except for $C_{e\varphi}$. However, even stronger limits on this coefficient can be obtained by looking at muon conversion [45, 46]. Let us mention again that our limit on $C_{e\varphi}$ is indicative at best, since we have not taken into account numerically important two-loop contributions from Barr-Zee diagrams [47, 48]. We will come back to this when discussing tau decays.

Note that the limits for C_{ll}^{2111} and C_{ee}^{2111} differ by a factor 2 from the corresponding limits given in [49], since we write the terms containing Q_{ll} (and Q_{ee}) in the Lagrangian as

$$\mathcal{L}_{ll} = \sum_{prst} \frac{C_{ll}^{prst}}{\Lambda^2}(\bar{l}_p\gamma_\nu l_r)(\bar{l}_s\gamma^\nu l_t)$$
$$= 2\frac{C_{ll}^{2111}}{\Lambda^2}(\bar{l}_2\gamma_\nu l_1)(\bar{l}_1\gamma^\nu l_1) + [\ldots], \qquad (10)$$

where we exploited the symmetry of the operator.

Coeff. $\lambda = m_Z$	$\tau^+ \to \mu^+\gamma$ BR≤ $4.4\cdot10^{-8}$	$Z \to \mu^\pm\tau^\mp$ BR≤ $1.2\cdot10^{-5}$	$\tau^+ \to \mu^+\mu^-\mu^+$ BR≤ $2.1\cdot10^{-8}$
$C_{e\gamma}^{32/23}$	$2.7\cdot10^{-12}$		$3.8\cdot10^{-11}$
$C_{eZ}^{32/23}$	$1.5\cdot10^{-9}$	$1.5\cdot10^{-7}$	$8.7\cdot10^{-7}$
$C_{\varphi l(1)}^{23}$	$1.7\cdot10^{-7}$	$1.5\cdot10^{-7}$	$1.3\cdot10^{-8}$
$C_{\varphi l(3)}^{23}$	$1.6\cdot10^{-7}$	$1.5\cdot10^{-7}$	$1.3\cdot10^{-8}$
$C_{\varphi e}^{23}$	$1.6\cdot10^{-7}$	$1.5\cdot10^{-7}$	$1.3\cdot10^{-8}$
Coeff. $\lambda = m_Z$		$H \to \mu^\pm\tau^\mp$ BR≤ $1.8\cdot10^{-2}$	
$C_{e\varphi}^{32/23}$	$1.9\cdot10^{-6}$	$9.0\cdot10^{-8}$	$1.6\cdot10^{-5}$
$C_{le}^{3112/2113}$	$4.8\cdot10^{-4}$		
$C_{le}^{3222/2223}$	$2.3\cdot10^{-6}$		$1.1\cdot10^{-8}$
$C_{le}^{3222/2223}$	$1.4\cdot10^{-7}$		
C_{ee}^{3222}			$4.0\cdot10^{-9}$
C_{ll}^{3222}			$4.0\cdot10^{-9}$

Table 4. Limits on the Wilson coefficients in $[\mathrm{GeV}]^{-2}$ contributing to the LFV tauonic transitions.

The analysis can easily be adapted to the tau sector and the corresponding limits on the coefficients are given in Table 4. We should note that similar results can be found

DOI: 10.1051/epjconf/201611801031

in the literature [40, 50–52] and a review of studies of the tauonic limits can be found e.g. in [49]. The pattern is very similar to muonic decays. However, the limits from LFV Z decays play a more important role and if they were to be improved by several orders of magnitude at a future high energy lepton colliders, they could provide the most stringent limits on $C^{23}_{\varphi l(1)}$, $C^{23}_{\varphi l(3)}$ and $C^{23}_{\varphi e}$.

A new feature of the tau sector is the appearance of the $H \to \tau\mu$ decay, which has caused a huge theoretical activity [53–59]. At tree level, the only operator that induces such a LFV decay in the Lagrangian Eq. (1) is $Q^{32/23}_{e\varphi}$. The non-observation of this decay at the LHC is placing a limit on the $C^{32/23}_{e\varphi}$ coefficient that is more stringent then any limit extracted from the low energy experiment. This decay has been considered extensively in the literature, often with phenomenologically motivated Lagrangians [53], where the LFV interactions are given in terms of generalised Yukawa matrices. In terms of Dim-6 operators, the deviations from the SM Yukawa interactions are induced by the operator $Q_{e\varphi}$, as can be seen in Eq. (5). Our limits on $C^{32/23}_{e\varphi}$ in Table 4 indicate that $H \to \tau\mu$ provides the strongest constraint, followed by $\tau \to \mu\gamma$ and $\tau \to 3\mu$. This is in qualitative agreement with [54]. However, we should mention once more that the limits extracted here from low-energy observables do not consider the impact of Barr-Zee type two-loop contributions, which is particularly important in this case [54].

In Table 4, the ATLAS & CMS data on $H \to \tau\mu$ are considered only in their function of placing new limits on such a transition. However, if one wants to consider the possibility that the picture corresponds to a signal compatible with BR($H \to \tau\mu$) = 0.84% (with a significance of 2.4 σ), then a trivial computation leads to $C^{32/23}_{e\varphi}/\Lambda^2 \sim 8.7 \cdot 10^{-8}$ GeV^{-2} or, in the same way, to $\Lambda \sim 3.4\sqrt{C^{32/23}_{e\varphi}}$ TeV. From this, it follows that an effective coefficient that preserves the perturbative behaviour $C^{32/23}_{e\varphi} \leq 1$ would imply a limit on the UV completion scale $\Lambda \geq 3.4$ TeV, that in turn preserves the perturbative behaviour of the expansion in Eq. (1), being $C_{e\varphi}(v/\Lambda)^2 \leq O(10^{-2})$. Here, no further speculation will be carried out about the possible nature of the UV completion and we refer to e.g. [60, 61] for a deeper discussion on the topic.

4.2 Limits from RGE

In the previous subsection we have given limits that were obtained by comparing NLO (or LO) calculations using Eq. (1) to experimental limits on branching ratios. However, the most direct link of the Wilson coefficients $C^{(d)}_k$ of Eq. (1) to the (unknown) UV complete theory is at the large scale Λ. In order to obtain the Wilson coefficients at the low scale (m_Z or m_l) that are relevant for the branching ratios, RG evolution has to be applied. This has two effects. First, the limits on $C^{(d)}_k(m_Z)$ will be translated into limits $C^{(d)}_k(\Lambda)$. Second, and more important, qualitatively new effects happen. It might well be possible, that a certain UV complete theory does result in a vanishing coefficient $C_{e\gamma}(\Lambda) = 0$ at the large scale. Nevertheless, the corresponding theory can lead to a LFV decay $l^i \to l^j\gamma$. In fact, other operators that are generated by the UV-complete theory at Λ can mix under RG evolution with $Q_{e\gamma}$ and result in a non-vanishing Wilson coefficient $C_{e\gamma}(m_z) \neq 0$.

In principle, the complete set of Dim-6 operators has to be taken into account to study $C_{e\gamma}(m_z)$. Restricting ourselves to operators that either directly or indirectly mix into $Q_{e\gamma}$, the RG evolution of the corresponding coefficients is given in [36]:

$$\begin{aligned}
16\pi^2 \frac{\partial C^{ij}_{e\gamma}}{\partial \log\lambda} &\simeq \left(\frac{47e^2}{3} + \frac{e^2}{4c_W^2} - \frac{9e^2}{4s_W^2} + 3Y_t^2\right) C^{ij}_{e\gamma} \\
&\quad + 6e^2\left(\frac{c_W}{s_W} - \frac{s_W}{c_W}\right) C^{ij}_{eZ} + 16eY_t C^{(3)}_{ij33}, \\
16\pi^2 \frac{\partial C^{ij}_{eZ}}{\partial \log\lambda} &\simeq -\frac{2e^2}{3}\left(\frac{2c_W}{s_W} + \frac{31s_W}{c_W}\right) C^{ij}_{e\gamma} \\
&\quad + 2e\left(\frac{3c_W}{s_W} - \frac{5s_W}{c_W}\right) Y_t C^{(3)}_{ij33} \\
&\quad + \left(-\frac{47e^2}{3} + \frac{151e^2}{12c_W^2} - \frac{11e^2}{12s_W^2} + 3Y_t^2\right) C^{ij}_{eZ}, \\
16\pi^2 \frac{\partial C^{(3)}_{ij33}}{\partial \log\lambda} &\simeq \frac{7eY_t}{3} C^{ij}_{e\gamma} + \frac{eY_t}{2}\left(\frac{3c_W}{s_W} - \frac{5s_W}{3c_W}\right) C^{ij}_{eZ} + \\
&\quad + \left(\frac{2e^2}{9c_W^2} - \frac{3e^2}{s_W^2} + \frac{3Y_t^2}{2} + \frac{8g_S^2}{3}\right) C^{(3)}_{ij33} \\
&\quad + \frac{e^2}{8}\left(\frac{5}{c_W^2} + \frac{3}{s_W^2}\right) C^{(1)}_{ij33}, \\
16\pi^2 \frac{\partial C^{(1)}_{ij33}}{\partial \log\lambda} &\simeq \left(\frac{30e^2}{c_W^2} + \frac{18e^2}{s_W^2}\right) C^{(3)}_{ij33} \\
&\quad + \left(-\frac{11e^2}{3c_W^2} + \frac{15Y_t^2}{2} - 8g_S^2\right) C^{(1)}_{ij33}.
\end{aligned} \tag{11}$$

We would like to point out the role of $C^{(3)}_{ij33}$ and $C^{(1)}_{ij33}$, the Wilson coefficients of the operators $Q^{(3)}_{lequ}$ and $Q^{(1)}_{lequ}$ respectively. These operators enter in the running of $C^{ij}_{e\gamma}$ with a coefficient proportional to the Yukawa coupling. Hence we restrict ourselves to the contribution of the top quarks in Eq. (11). Note that $Q^{(3)}_{lequ}$ mixes directly into $Q_{e\gamma}$, resulting in a rather stringent limit on $C^{(3)}_{ij33}(\Lambda)$ from BR($l_i \to l_j\gamma$). The constraint on $C^{(1)}_{ij33}(\Lambda)$ is weaker, since $Q^{(1)}_{lequ}$ enters only indirectly, i.e. through $Q^{(3)}_{lequ}$.

In Tables 5 and 6 we list the limits on the various coefficients of Eq. (11) obtained from $\mu \to e\gamma$ and $\tau \to \mu\gamma$, respectively. Obviously, very similar limits can be derived for $\tau \to e\gamma$. The limits are given for several choices of the large scale Λ. As can be seen from the entries of Tables 5 and 6, the effect of the RG evolution is very modest for $C^{ij}_{e\gamma}$ and more pronounced for C^{ij}_{eZ}. We also point out that the limits obtained for C^{ij}_{eZ} through the effect of the RG evolution is considerably more stringent than the corresponding limit from a standard NLO calculation, as can be seen by comparing the entries in Tables 3 and 4 with those in Tables 5 and 6, respectively. More important, however, is the fact that without taking the RG into ac-

$\mu \to e\gamma$			
Coeff.	$\Lambda = 10^3$ GeV	$\Lambda = 10^5$ GeV	$\Lambda = 10^7$ GeV
$C_{e\gamma}^{21}$	$2.7 \cdot 10^{-10}$	$2.9 \cdot 10^{-6}$	$3.1 \cdot 10^{-2}$
C_{eZ}^{21}	$2.5 \cdot 10^{-8}$	$1.0 \cdot 10^{-4}$	$7.1 \cdot 10^{-1}$
$C_{lequ(3)}^{2133}$	$3.6 \cdot 10^{-9}$	$1.4 \cdot 10^{-5}$	$9.8 \cdot 10^{-2}$
$C_{lequ(1)}^{2133}$	$1.9 \cdot 10^{-6}$	$2.5 \cdot 10^{-3}$	n/a
$C_{lequ(3)}^{2122}$	$4.8 \cdot 10^{-7}$	$1.9 \cdot 10^{-3}$	n/a
$C_{lequ(1)}^{2122}$	$2.6 \cdot 10^{-4}$	$3.3 \cdot 10^{-1}$	n/a

Table 5. Limits on the Wilson coefficients defined at the scale $\lambda = \Lambda$ for three choices of $\Lambda = 10^3, 10^5, 10^7$ GeV.

count, it would be impossible to get meaningful limits on the coefficients of $C_{ij33}^{(3)}(\Lambda)$ and $C_{ij33}^{(1)}(\Lambda)$. For illustration, we have also included in the Tables limits on these coefficients in the charm sector, $C_{ij22}^{(3)}(\Lambda)$ and $C_{ij22}^{(1)}(\Lambda)$. Since the corresponding Yukawa coupling is smaller, the limits are less constraining.

$\tau \to \mu\gamma$			
Coeff.	$\Lambda = 10^3$ GeV	$\Lambda = 10^4$ GeV	$\Lambda = 10^5$ GeV
$C_{e\gamma}^{32}$	$3.0 \cdot 10^{-6}$	$3.1 \cdot 10^{-4}$	$3.2 \cdot 10^{-2}$
C_{eZ}^{32}	$2.8 \cdot 10^{-4}$	$1.5 \cdot 10^{-2}$	~ 1.1
$C_{lequ(3)}^{3233}$	$4.0 \cdot 10^{-5}$	$2.2 \cdot 10^{-3}$	$1.6 \cdot 10^{-1}$
$C_{lequ(1)}^{3233}$	$2.1 \cdot 10^{-2}$	$5.9 \cdot 10^{-1}$	n/a
$C_{lequ(3)}^{3222}$	$5.4 \cdot 10^{-3}$	$3.0 \cdot 10^{-1}$	n/a
$C_{lequ(1)}^{3222}$	~ 2.8	n/a	n/a

Table 6. Limits on the Wilson coefficients defined at the scale $\lambda = \Lambda$ for three choices of $\Lambda = 10^3, 10^4, 10^5$ GeV.

5 Conclusion

In these proceedings we have described the first steps towards taking the Standard Model effective field theory, that is the Lagrangian given in Eq. (1), seriously as a quantum field theory. This allows us to perform well defined calculations beyond leading order in the Standard Model couplings and apply renormalisation group techniques. At the current stage, the motivation to go beyond leading order is not to obtain limits on the coefficients that are more precise by ~10%. Rather, performing NLO calculations and taking RG evolution effects into account, we obtain qualitatively new results. This is the case in particular when considering the very stringent limits of low-energy LFV experiments.

We take the attitude that Λ, the scale of new physics responsible for LFV effects is much larger than the electroweak scale. Hence we restrict ourselves to terms suppressed by at most two powers of Λ, i.e. to single insertions of Dim-6 operators. Ideally, the coefficients of these operators could all be measured experimentally. This would provide a huge amount of information about the nature of the UV complete theory valid beyond the scale Λ. In practise, a relatively small number of experiments related to LFV in the charged sector will be carried out at energy scales well below Λ and typically result in exclusion limits. In order to extract the maximal possible amount of information from these experiments it is essential to perform computations beyond tree level and take RG evolution into account.

A complete experimental determination of the various coefficients is certainly not a realistic prospect, but even partial information on the coefficients will provide extremely useful guidance for the search of the UV complete theory. However, for this it is essential to know $C_k^{(6)}(\Lambda)$, the coefficients at the high scale Λ. It is at this scale where there is the most direct link between the effective theory and the underlying theory. Even if the underlying theory does not produce an operator $Q_{e\gamma}$ when integrating out the heavy modes it can still result in the LFV decay $\mu \to e\gamma$. To describe this effect, the inclusion of RG evolution is at least as important as the inclusion of one-loop corrections. In particular, if Λ is considered to be far above the electroweak scale, RG evolution leads to qualitatively new effects and can lead to substantial improvements in the limits that are extracted from low-energy measurements.

In the results presented here we have considered the decay $l_i \to l_j\gamma$ at NLO and have taken RG effects into account. However, other processes such as muon conversion have been completely ignored and for others, e.g. the decay $l_i \to 3\, l_j$ only a tree-level analysis has been made. Furthermore, the rather unrealistic assumption has been made that only a single coefficient at a time is non-vanishing. Clearly, there is considerable room for improvement. We are convinced that with a more complete analysis, LFV processes can play an even more important role in the search for physics beyond the Standard Model.

Acknowledgements

The work of GMP is supported by the Swiss National Science Foundation (SNF) under contract 200021_160156.

References

[1] Y. Fukuda et al. (Super-Kamiokande), Phys. Rev. Lett. **81**, 1562 (1998), hep-ex/9807003
[2] Q.R. Ahmad et al. (SNO), Phys. Rev. Lett. **87**, 071301 (2001), nucl-ex/0106015
[3] Q.R. Ahmad et al. (SNO), Phys. Rev. Lett. **89**, 011301 (2002), nucl-ex/0204008
[4] A. Crivellin, A. Kokulu, C. Greub, Phys. Rev. **D87**, 094031 (2013), 1303.5877
[5] A. Abada, M.E. Krauss, W. Porod, F. Staub, A. Vicente, C. Weiland, JHEP **11**, 048 (2014), 1408.0138
[6] L. Calibbi, P. Paradisi, R. Ziegler, Eur. Phys. J. **C74**, 3211 (2014), 1408.0754
[7] B. Gripaios, M. Nardecchia, S.A. Renner, JHEP **05**, 006 (2015), 1412.1791

[8] J. Kersten, J.h. Park, D. Stöckinger, L. Velasco-Sevilla, JHEP **08**, 118 (2014), 1405.2972
[9] M. Beneke, P. Moch, J. Rohrwild (2015), 1508.01705
[10] F. Feruglio, P. Paradisi, A. Pattori (2015), 1509.03241
[11] W. Buchmuller, D. Wyler, Nucl. Phys. **B268**, 621 (1986)
[12] B. Grzadkowski, M. Iskrzynski, M. Misiak, J. Rosiek, JHEP **10**, 085 (2010), 1008.4884
[13] L. Lehman, Phys. Rev. **D90**, 125023 (2014), 1410.4193
[14] S.T. Petcov, Sov. J. Nucl. Phys. **25**, 340 (1977), [Erratum: Yad. Fiz.25,1336(1977)]
[15] P. Minkowski, Phys. Lett. **B67**, 421 (1977)
[16] S. Weinberg, Phys. Rev. Lett. **43**, 1566 (1979)
[17] J. Adam et al. (MEG), Phys. Rev. Lett. **110**, 201801 (2013), 1303.0754
[18] U. Bellgardt et al. (SINDRUM), Nucl. Phys. **B299**, 1 (1988)
[19] A.M. Baldini et al. (2013), 1301.7225
[20] N. Berger (Mu3e), Nucl. Phys. Proc. Suppl. **248-250**, 35 (2014)
[21] B. Aubert et al. (BaBar), Phys. Rev. Lett. **104**, 021802 (2010), 0908.2381
[22] K. Hayasaka et al., Phys. Lett. **B687**, 139 (2010), 1001.3221
[23] D. Decamp et al. (ALEPH), Phys. Rept. **216**, 253 (1992)
[24] O. Adriani et al. (L3), Phys. Lett. **B316**, 427 (1993)
[25] R. Akers et al. (OPAL), Z. Phys. **C67**, 555 (1995)
[26] P. Abreu et al. (DELPHI), Z. Phys. **C73**, 243 (1997)
[27] G. Aad et al. (ATLAS), Phys. Rev. **D90**, 072010 (2014), 1408.5774
[28] Tech. Rep. CMS-PAS-EXO-13-005, CERN, Geneva (2015), https://cds.cern.ch/record/2019863
[29] G. Aad et al. (ATLAS), Phys. Lett. **B716**, 1 (2012), 1207.7214
[30] S. Chatrchyan et al. (CMS), Phys. Lett. **B716**, 30 (2012), 1207.7235
[31] G. Aad et al. (ATLAS) (2015), 1508.03372
[32] V. Khachatryan et al. (CMS), Phys. Lett. **B749**, 337 (2015), 1502.07400
[33] R.K. Kutschke (2011), 1112.0242
[34] Y. Kuno (COMET), PTEP **2013**, 022C01 (2013)
[35] W.H. Bertl et al. (SINDRUM II), Eur. Phys. J. **C47**, 337 (2006)
[36] G.M. Pruna, A. Signer, JHEP **10**, 14 (2014), 1408.3565
[37] E.E. Jenkins, A.V. Manohar, M. Trott, JHEP **10**, 087 (2013), 1308.2627
[38] E.E. Jenkins, A.V. Manohar, M. Trott, JHEP **01**, 035 (2014), 1310.4838
[39] R. Alonso, E.E. Jenkins, A.V. Manohar, M. Trott, JHEP **04**, 159 (2014), 1312.2014
[40] A. Crivellin, S. Najjari, J. Rosiek, JHEP **04**, 167 (2014), 1312.0634
[41] Y. Kuno, Y. Okada, Rev. Mod. Phys. **73**, 151 (2001), hep-ph/9909265
[42] K.A. Olive et al. (Particle Data Group), Chin. Phys. **C38**, 090001 (2014)
[43] M. Bicer et al. (TLEP Design Study Working Group), JHEP **01**, 164 (2014), 1308.6176
[44] A. Abada, V. De Romeri, S. Monteil, J. Orloff, A.M. Teixeira, JHEP **04**, 051 (2015), 1412.6322
[45] A.A. Petrov, D.V. Zhuridov, Phys. Rev. **D89**, 033005 (2014), 1308.6561
[46] A. Crivellin, M. Hoferichter, M. Procura, Phys. Rev. **D89**, 093024 (2014), 1404.7134
[47] S.M. Barr, A. Zee, Phys. Rev. Lett. **65**, 21 (1990), [Erratum: Phys. Rev. Lett.65,2920(1990)]
[48] D. Chang, W.S. Hou, W.Y. Keung, Phys. Rev. **D48**, 217 (1993), hep-ph/9302267
[49] F. Feruglio (2015), 1509.08428
[50] B.M. Dassinger, T. Feldmann, T. Mannel, S. Turczyk, JHEP **10**, 039 (2007), 0707.0988
[51] A. Celis, V. Cirigliano, E. Passemar (2014), 1409.4439
[52] A. Celis, V. Cirigliano, E. Passemar, Phys. Rev. **D89**, 095014 (2014), 1403.5781
[53] G. Blankenburg, J. Ellis, G. Isidori, Phys. Lett. **B712**, 386 (2012), 1202.5704
[54] R. Harnik, J. Kopp, J. Zupan, JHEP **03**, 026 (2013), 1209.1397
[55] J. Kopp, M. Nardecchia, JHEP **10**, 156 (2014), 1406.5303
[56] D. Aristizabal Sierra, A. Vicente, Phys. Rev. **D90**, 115004 (2014), 1409.7690
[57] L. de Lima, C.S. Machado, R.D. Matheus, L.A.F. do Prado (2015), 1501.06923
[58] I. Doršner, S. Fajfer, A. Greljo, J.F. Kamenik, N. Košnik, I. Nišandžic, JHEP **06**, 108 (2015), 1502.07784
[59] Y. Omura, E. Senaha, K. Tobe, JHEP **05**, 028 (2015), 1502.07824
[60] W. Altmannshofer, S. Gori, A.L. Kagan, L. Silvestrini, J. Zupan (2015), 1507.07927
[61] A. Crivellin, G. D'Ambrosio, J. Heeck, Phys. Rev. **D91**, 075006 (2015), 1503.03477

 DOI: 10.1051/epjconf/201611801032

Storage Ring EDM Experiments

Yannis K. Semertzidis[1,2,a]

[1] Center for Axion and Precision Physics Research, IBS, Daejeon 305-701, Republic of Korea
[2] Dept. of Physics, KAIST, Daejeon 305-701, Republic of Korea

Abstract. Dedicated storage ring electric dipole moment (EDM) methods show great promise advancing the sensitivity level by a couple orders of magnitude over currently planned hadronic EDM experiments. We describe the present status and recent updates of the field.

1 Introduction

Electric dipole moments of fundamental particles violate separately the discrete symmetries of parity (P) and time (T) reversal. Assuming exact CPT symmetry, where (C) denotes the symmetry with charge, T-symmetry violation also means CP-violation. CP-violation is one of Sakharov's necessary conditions to develop a matter-dominated universe from an originally matter-antimatter symmetric one. The CP-violation present in electro-weak interactions turns out to be too small to account for the present observed asymmetry by roughly eight orders of magnitude. However, physics beyond the standard model (SM), e.g., super-symmetry (SUSY), or other extensions of the SM can very well fill this void.

The physics reach of the storage ring EDM method for the proton at 10^{-29} e · cm is at the 10^3 TeV level, much higher than the reach of the Large Hadron Collider (LHC) at CERN located in Geneva, Switzerland. The LHC looks for physics beyond the SM in a direct way with a mass reach of up to 1 TeV or double that value when high luminosity becomes a reality. Other experimental efforts using ultra-cold neutron sources (UCN) such as the neutron EDM at the spallation neutron source (SNS) at Oak Ridge, USA, and the experiment at the Paul Scherer Institute (PSI) in Zurich, Switzerland are all suffering from statistical limitations in obtaining enough UCN intensity. Nonetheless, they are very promising experiments since they are currently setting the limits on SUSY and when operational they might discover a non-zero EDM value.

1.1 Spin and EDM of fundamental particles

Particles with spin $\vec{s}$ have a magnetic dipole moment given by

$$\vec{\mu} = g\frac{q}{2m}\vec{s} \tag{1}$$

with the gyromagnetic ratio $g = 2$ for point-like particles in a world without quantum fluctuations. For leptons, e.g., electrons and muons, the g-factor is very close to 2 and the difference, or the anomalous magnetic moment as it is called $a = (g-2)/2 \approx 10^{-3}$, being due to quantum field fluctuations. For particles with internal structure, e.g., hadrons, the g-factor differs wildly from the ideal value of 2, with the anomalous magnetic moment represented by $G = (g-2)/2$ and of order ± 1, with the proton value $G_p = 1.79$ and the deuteron $G_d = -0.143$. There are no particles with an anomalous magnetic moment equal to exactly zero, which is unfortunate as this would be a chance to design a very sensitive EDM experiment. Nonetheless, we can recreate the conditions where a particle will effectively behave as if it has a zero anomalous magnetic moment and we can design a very sensitive experiment. In this paper we will present such a case for the proton in an all-electric storage ring.

The particle electric dipole moment is given by

$$\vec{d} = \eta\frac{q}{2mc}\vec{s} \tag{2}$$

with η playing the same role as the g-factor in the magnetic dipole moment case.

The interaction Hamiltonian of magnetic and electric dipole moments at rest in magnetic $\vec{B}$ and electric $\vec{E}$ fields is given by

$$H = -\vec{\mu}\cdot\vec{B} - \vec{d}\cdot\vec{E} \tag{3}$$

which after applying the parity (P) operator becomes

$$H = -\vec{\mu}\cdot\vec{B} - \vec{d}\cdot(-\vec{E}) = -\vec{\mu}\cdot\vec{B} + \vec{d}\cdot\vec{E} \tag{4}$$

or the time (T) operator

$$H = -(-\vec{\mu})\cdot(-\vec{B}) - (-\vec{d})\cdot\vec{E} = -\vec{\mu}\cdot\vec{B} + \vec{d}\cdot\vec{E} \tag{5}$$

or the charge (C) operator

$$H = -(-\vec{\mu})\cdot(-\vec{B}) - (-\vec{d})\cdot(-\vec{E}) = -\vec{\mu}\cdot\vec{B} - \vec{d}\cdot\vec{E} \tag{6}$$

implying that either the electric dipole moment is zero or it violates both the P and T symmetries, and under the CPT conservation assumption, it also violates CP-symmetry.

[a] e-mail: yannis@kaist.ac.kr, for the Storage ring EDM collaboration

© The Authors, published by EDP Sciences. This is an open access article distributed under the terms of the Creative Commons Attribution License 4.0 (http://creativecommons.org/licenses/by/4.0/).

1.2 EDM sensitivity level

In 1967, after the discovery of CP-violation in weak interactions in 1964[1] at Brookhaven National Laboratory, Sakharov published his paper[2] on the conditions required to create a matter dominated universe from an originally symmetric one after the big bang:

1. Baryon number violation
2. C and CP-violation
3. Interactions out of thermal equilibrium

The current measurement of the ratio of baryons to photons is $n_B/n_\gamma \approx (6.08 \pm 0.14) \times 10^{-10}$, whereas the amount permitted by the SM through baryogenesis is only $n_B/n_\gamma \approx 10^{-18}$ indicating the existence of a new, much stronger source of CP-violation. An alternative process based on leptogenesis involving neutrinos at the earlier stages of the big bang could also provide a viable solution, but here we will consider only baryogenesis.

It turns out that due to the particular structure in the SM the weak interactions, with only one CP-violating phase, only enters at high order quantum loops as a contribution to EDMs and their values can only be $< 10^{-31}\, e\cdot$cm for hadrons, $< 10^{-38}\, e\cdot$cm for the electrons and $< 10^{-36}\, e\cdot$cm for the muons. Physics beyond the SM, e.g., SUSY with more than 40 CP-violating phases, contributes much more abundantly to EDMs and their contribution could be in the neighborhood of the current experimental limits. For the neutron[3] the experimental limit is $3 \times 10^{-26}\, e \cdot$ cm. Since EDMs require a chirality flip, the contribution to EDM depends to first order on the mass of the particles, i.e., $d_e : d_q : d_\mu = 1 : 20 : 200$, with different, higher mass dependence also possible.

It is not true that the only contribution to the SM comes from the weak interactions CP-violating phase. Another potential source with a much stronger EDM contribution is the so-called θ-term in the QCD Lagrangian:

$$L_{\rm QCD} = \bar{\theta}\frac{\alpha_s}{8\pi}G\tilde{G} \quad (7)$$

with G the gluon field and $\bar{\theta}$ an angle with a value of order one. From dimensional analysis the expected neutron EDM value is

$$d_n(\bar{\theta}) \approx \bar{\theta}\frac{q}{m_n}\frac{m_*}{\Lambda_{\rm QCD}} \approx \bar{\theta}\cdot\left(6\times 10^{-17}\right) e\cdot {\rm cm} \quad (8)$$

with $m_* = \frac{m_u m_d}{m_u+m_d}$ the reduced up and down quark mass. When done more carefully[4] it becomes

$$d_n(\bar{\theta}) \approx -d_p(\bar{\theta}) \approx \bar{\theta}\cdot\left(3.6\times 10^{-16}\right) e\cdot {\rm cm} \quad (9)$$

With $\bar{\theta}$ an angle of order one, it is expected that the neutron and proton EDMs should be of order of $3.6 \times 10^{-16}\, e \cdot$ cm, several orders of magnitude larger than the current experimental limit on the neutron EDM. It implies that the value of $\bar{\theta} \leq 2 \times 10^{-10}$, thus prompting Peccei and Quinn to invoke a new dynamic mechanism[5] that reduces the value of $\bar{\theta}$ below the current experimental limit. Due to breakdown of that symmetry, Weinberg[6] and Wilczek[7] suggested that there should be an axion particle present in vacuum due to the oscillating $\bar{\theta}$ around its average value of very close to zero. The oscillation frequency determines the axion mass, but this mechanism has yet to be detected.

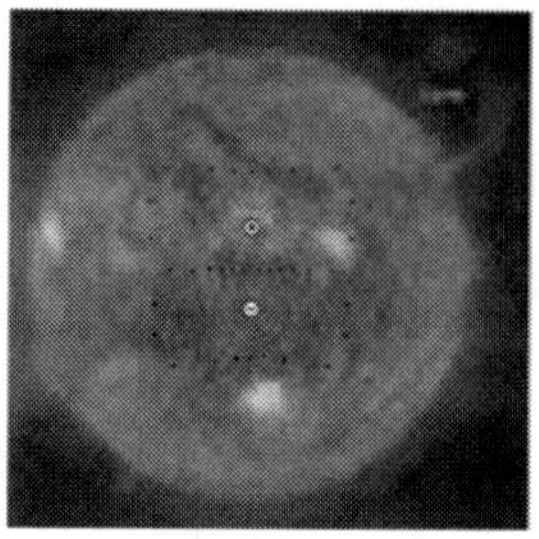

Figure 1. Expanding the proton (approximately 1fm in diameter) to the size of our sun (approximately 1 million km in diameter) would still require to resolve a separation of the quantum of charges by a mere 0.1 μm, demonstrating the sensitivity of the method.

However, the value of $\bar{\theta}$ in principle can be anything and indeed small but finite, so it can contribute to the neutron and proton EDMs at the present levels of current experimental limits. Should the neutron EDM be found to be non-zero we would want to find out the proton EDM. If the value has the same sign then it is unlikely to be due to $\bar{\theta}$, but most likely due to new physics like SUSY. Information from the deuteron is also invaluable. The deuteron, as it has a negative anomalous magnetic moment it requires a combination of magnetic and electric fields for a sensitive EDM experiment and is under development at Jülich, Germany by the JEDI collaboration[8].

It has been estimated[9] that a proton EDM with $d_p < 10^{-29}\, e \cdot$ cm has a physics reach of

$$d_p \approx 0.01(m_p/\Lambda)^2 \tan\phi^{\rm NP}\, q/2m_p, \quad (10)$$

which after substituting for the parameter values

$$d_p \approx 10^{-22}(1{\rm TeV}/\Lambda_{\rm NP})^2\, \tan\phi^{\rm NP}\, e\cdot {\rm cm}. \quad (11)$$

If $\phi^{\rm NP}$ is of order 1, then $\Lambda_{\rm NP} \approx 3000$ TeV is probed, while if $\Lambda_{\rm NP} \approx 1$ TeV, then $\phi^{\rm NP} \approx 10^{-7}$ rad is probed, demonstrating the exquisite potential sensitivity of the method.

For equal but opposite charges $\pm q$, separated by a distance $\vec{r}$ the EDM is classically defined as $\vec{d} = q\vec{r}$. If we take the proton, which has an approximate diameter of 1fm, and blow it up to the size of our sun, with approximately diameter of 1 million km, an EDM $d = 10^{-29}\, e\cdot$cm means that the charge separation would be of order of 0.1 μm, see Figure 1.

2 Experimental method

It is known that electric dipole moments couple to electric fields and magnetic dipole moments to magnetic fields,

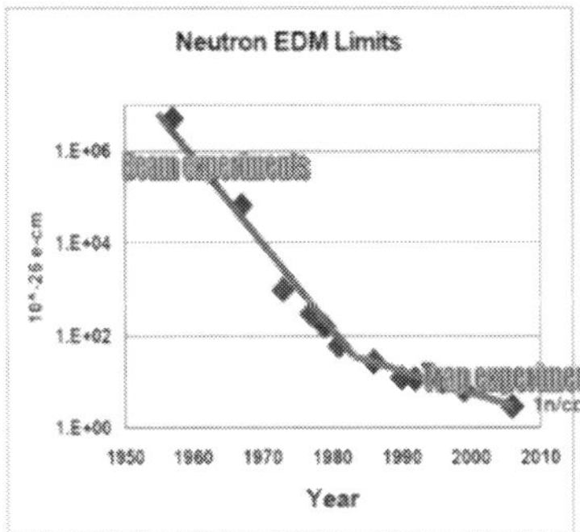

Figure 2. The neutron EDM as a function of publication year, from[10]. There is a clear kink in the EDM sensitivity owing to the potential systematic error due to motional magnetic field when neutrons move in an electric field region.

thus in order to probe the electric dipole moment of a fundamental particle the particle has to feel the presence of an electric field, which when stationary can only happen by applying an electric field to it. The spin precession rate due to this electric field is given by

$$\frac{d\vec{s}}{dt} = \vec{d} \times \vec{E} \Rightarrow \frac{d\theta}{dt} = \frac{2dE}{\hbar} \Rightarrow \omega = 3\text{nrad/s} \tag{12}$$

assuming $d = 10^{-29}\, e \cdot \text{cm}$ and $E = 10$ MV/m. The spin precession rate is very small and a large electric field is required. Ramsey and Parcel decided to use neutrons instead of charged particles in order to avoid the complications due to the application of the electric field. This method required the application of a small magnetic field so that the neutron spin precession could be observed with high sensitivity. Then a large as possible collinear electric field is applied and the spin precession rate is compared to that when the electric field direction is flipped. Since the signal is compatible with zero it is a small phase shift in the precession frequency that is mostly hoped for, i.e., the observed quantity is a scalar. To gain statistical sensitivity a large number of energetic neutrons are guided to pass between two electrically charged plates where there is a strong electric field present. The electric field direction is flipped periodically and the sought signal is a spin precession phase shift correlated with the electric field direction. This method, which had adequate statistics, ran into systematic errors and had to be abandoned for UCN traps, which unfortunately have very limited statistical sensitivity, see Figure 2. The main reason for abandoning the neutron beam experiments was due to the fact that moving neutrons in strong electric fields also feel the presence of significant magnetic fields in their rest frame, depending on their velocity, precessing their spin. When the electric field direction is flipped the neutron velocity distribution could not be controlled with the required accuracy and the method was limited by systematic errors. The storage ring proton EDM experimental method, however, makes use of the motional magnetic field and instead of a systematic error this becomes a feature of the experiment.

2.1 Precession in a storage ring

The spin precession rate of a particle at rest in the presence of magnetic and electric fields is given by

$$\frac{d\vec{s}}{dt} = \vec{\mu} \times \vec{B} + \vec{d} \times \vec{E} \tag{13}$$

For particles in motion the corresponding equation becomes

$$\left(\frac{d\vec{s}}{dt}\right)^{*} = \vec{\mu} \times \vec{B}^{*} + \vec{d} \times \vec{E}^{*} \tag{14}$$

where the rest frame spin precession needs to be transferred to the rest frame by taking into account the accelerated frame due to a possible particle rotation and consequently radial acceleration (Thomas precession). For a particle stored in a circular storage ring the relative spin precession compared to the momentum precession is given by (assuming $\vec{\beta} \cdot \vec{B} = \vec{\beta} \cdot \vec{E} = 0$):

$$\vec{\omega}_a = -\frac{q}{m}\left(a\vec{B} - \left[a - \left(\frac{mc}{p}\right)^2\right]\frac{\vec{\beta} \times \vec{E}}{c}\right) \tag{15}$$

which simplifies to

$$\vec{\omega}_a = -a\frac{q}{m}\vec{B} \tag{16}$$

for $p = mc/\sqrt{a}$, the so-called "magic" momentum, making it independent of the electric field value. This is indeed the principle of the muon $g-2$ experiment used at CERN[11], BNL[12], and next at FNAL[13], where a uniform magnetic field is used to probe the anomalous magnetic moment of the muon with high accuracy. An electric quadrupole field is used to provide vertical focusing without affecting (to first order) $\vec{\omega}_a$, see Figure 3.

In the presence of an EDM the total spin precession rate is given by

$$\vec{\omega} = -\frac{q}{m}\left[a\vec{B} + \frac{\eta}{2}\left(\vec{\beta} \times \vec{B}\right)\right] = \vec{\omega}_a + \vec{\omega}_{\text{edm}} \tag{17}$$

causing a tilt in the total angular precession vector $\vec{\omega}$ with respect to the vertical, with $\tan\theta = \omega_{\text{edm}}/\omega_a$, see Figure 4. In general the sensitivity of the experiment to the muon EDM is limited[15] especially compared to the electron EDM limit[16] due to the finite muon lifetime, the weak muon beam intensities and the diffused muon phase space available. However, the muons are the only second generation particle that we have a chance of reaching better than roughly $10^{-21}\, e \cdot \text{cm}$ in the FNAL[13] and J-PARC[14] experiments in an indirect way. It is possible to improve that sensitivity, especially the systematic errors, by a couple more orders of magnitude in a dedicated storage ring experiment by using a combination of electric and magnetic fields to freeze the muon spin in the horizontal plane[17, 18].

2.2 Storage ring proton EDM

The dedicated storage ring proton EDM experiment based on a dedicated electric field storage ring is the simplest

 DOI: 10.1051/epjconf/201611801032

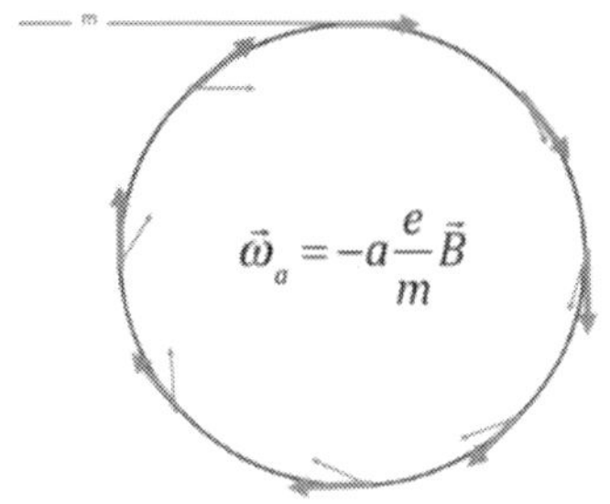

Figure 3. The muon spin precession slightly faster than the muon momentum in a uniform magnetic field ring with electric focusing and the muons at their magic momentum.

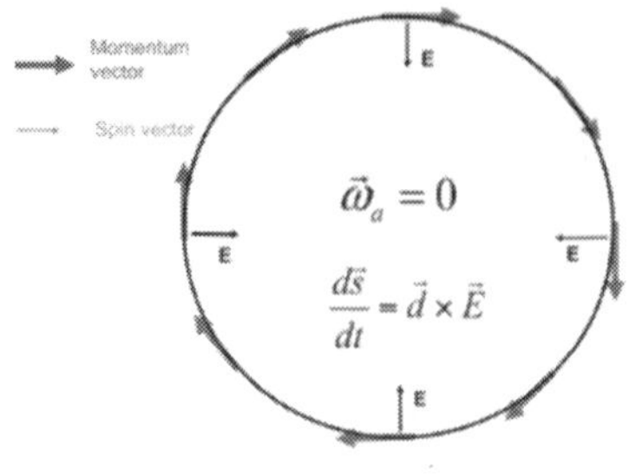

Figure 5. The proton spin follows the momentum exactly, on average, in the horizontal plane ($\vec{\omega}_a = 0$), while the EDM signal ($d\vec{s}/dt = \vec{d} \times \vec{E}$) accumulates for the duration of the storage time causing the spin to precess out of plane.

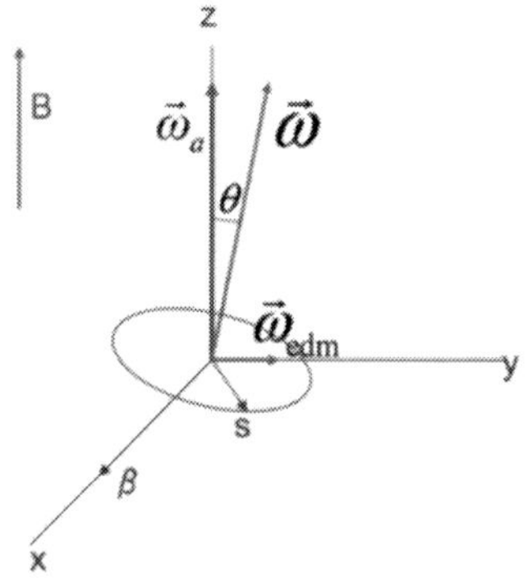

Figure 4. The total muon spin precession plane is slightly tilted from the vertical due to the small contribution by the muon EDM.

way to accomplish high sensitivity in hadronic EDMs. High intensity (more than 10^{11} per cycle) with high beam polarization (more than 80%) are readily available at several labs around the globe, e.g., BNL in the USA and COSY in Jülich/Germany. The anomalous magnetic moment of the proton is $G = 1.79$ and the magic momentum becomes $p = 700$ MeV/c. The spin and momentum precession rate coincide at the magic momentum value with the radial E-field being always on the same side of the EDM vector and the EDM signal is the spin precession into the vertical plane, acting for the duration of the storage time, see Figure 5. The EDM signal causes a spin precession in the vertical plane, perpendicular to the magnetic dipole moment precession plane in the more familiar $g - 2$ case and can therefore be distinguished.

An all electric ring with a radius of $R = 40$M requires a radial E-field of $E \approx 10$MV/m and can be accomplished at a reasonable cost. In a realistic lattice there is a need for beam injection as well as beam diagnostics. In an all-electric ring, counter-rotating beams can be stored simultaneously. This feature is critically important since the main systematic error source that is a non-zero radial B-field, will split the counter-rotating beams vertically. The vertical separation depends on the strength of the vertical focusing, encouraging a rather weak vertical focusing. This vertical separation creates a radial B-field, which is modulated at 1-10 kHz by modulating the strength of the vertical focusing by 1-10%. A realistic (to first order) ring lattice is given in Figure 6. The distortion of the closed orbit in the vertical direction for a single beam is given by

$$y(\theta) = \sum_{N=0}^{\infty} \frac{\beta c R_0 B_{r,N}}{E_0\left(Q_y^2 - N^2\right)} \cos\left(N\theta + \phi_N\right) \quad (18)$$

where E_0 is the radial E-field strength, $Q_y = Q_{y0}\left[1 + A_m \times \cos\left(\omega_m t + \phi_m\right)\right]$ is the modulated vertical tune, and N is the radial B-field harmonic around the ring azimuth.

Only the $N = 0$ radial B-field harmonic is a systematic error source. However, the counter-rotating beams may sense a non-zero effect originating from a higher harmonic and could be a potential systematic error. The ratio of the strength of a non-zero N^{th} harmonic to the $N = 0$ case for the time dependent part is proportional to $(Q_y/N)^4$, which for small Q_y, e.g. $Q_y = 0.3$ and somewhat large value for N, e.g., $N = 10$, the effect diminishes by six orders of magnitude. Therefore, if the SQUID-based BPMs are sensitive up to the 10th harmonic of the radial B-field, reducing the field everywhere below 0.1nT level is adequate. The oscillating radial magnetic field is sensed using SQUID-based magnetometers with single sensor sensitivity of about $3fT/\sqrt{Hz}$, much better than needed for $10^{-29}\,e \cdot$ cm overall sensitivity. The requirement is to have approximately one dozen SQUID-based beam position monitors (BPMs) around the ring, with the BPM segments of the ring to be at cryogenic temperatures of about 4 K, shown in Figure 7.

Another potential systematic error is the geometrical phase, again related to magnetic fields and the ring plate alignment. It turns out, by using high precision beam/spin dynamics simulations[19–21], that in an all-electric storage ring those systematic errors are easier to handle than in tradition trap experiments. The counter-rotating beams can be sensed separately and the traditional BPMs provide enough information to control this error below the statistical sensitivity as long as the absolute level of the B-fields is less than 0.5 nT and the magnetic field gradient is below 0.1 nT/m. These requirements ordinarily would require an

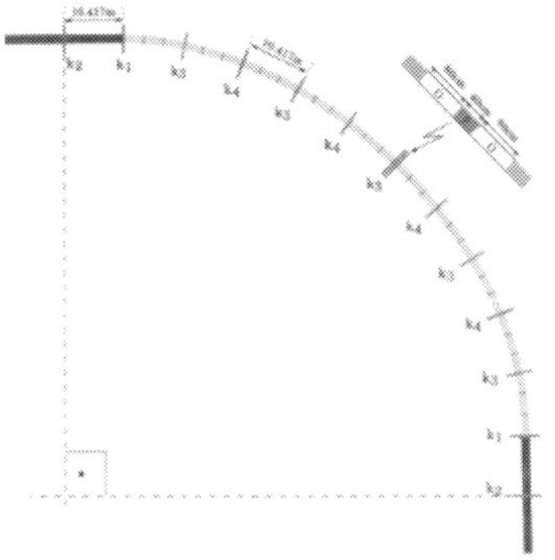

Figure 6. The proton ring lattice has several straight sections accommodating injection sections, polarimeter sections and SQUID-based beam position sections (see text). Alternate focusing is used with the horizontal being as strong as possible to increase the ring admittance and the vertical being rather weak to increase sensitivity to the radial B-field. The radial B-field splits the counter-rotating beams depending on the vertical focusing strength.

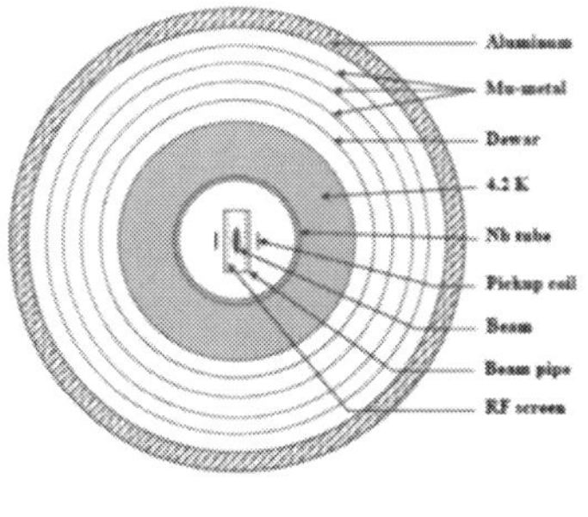

Figure 7. A net radial B-field in the ring would be a systematic error. However, it also splits the counter-rotating beams which depends on the strength of the vertical focusing. The beam splitting creates a radial B-field, which we modulate by modulating the vertical focusing at 1-10 kHz making it possible to be sensed using high sensitivity SQUID-based magnetometers.

elaborate mu-metal shielding around the ring, which could be very expensive. However, recent developments[22] allow for an affordable mu-metal shielding scheme by applying a new degaussing (or equilibration as it is now called) technique. Figure 8 shows a cut-out of the two layer shielding that is deemed to be sufficient to achieve the required shielding.

The statistical sensitivity of the experiment is given by

$$\sigma_d = \frac{2\hbar}{E_R P A \sqrt{N_c f \tau_p T_{\text{tot}}}} \tag{19}$$

with $P = 0.8$ the beam initial polarization, $\tau_p = 10^3$s the beam polarization lifetime, $A = 0.6$ the average left/right asymmetry observed by the polarimeter, $N_c = 10^{11}$ the total number of particles stored per cycle, $T_{\text{tot}} = 10^7$s the total running time per year, $f = 1\%$ the total useful event rate fraction, and $E_R = 7$MV/m the average radial electric

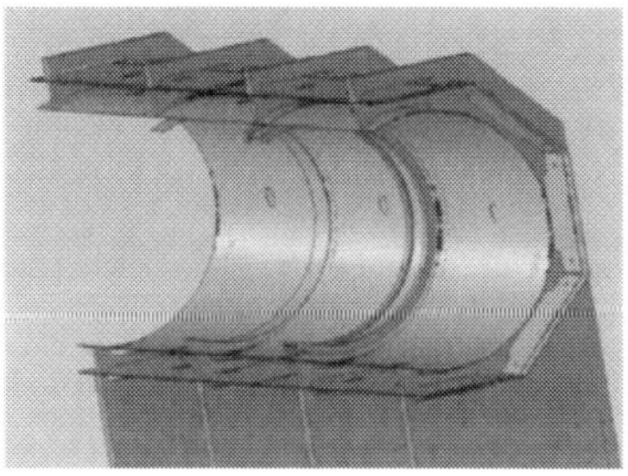

Figure 8. A two layer shielding with special equilibration coils is expected to provide adequate shielding (see text). A prototype of the shielding has been purchased by the Technical University of Munich and is under testing at CAPP/IBS in the KAIST campus.

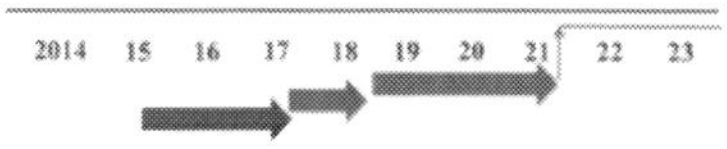

Figure 9. A technically driven schedule is shown here. At CAPP/IBS we are well on our way with the final R&D items required to obtain the necessary information for the final ring optimization. After that time we should be able to finish the ring design in one year and by 2021 we could complete the ring construction.

field around the ring. As a result, the estimated $\sigma_d = 1 \times 10^{-29}\, e \cdot$ cm/year.

2.3 Plan for the storage ring EDM method

A good EDM experiment should have good statistical sensitivity:

1. High intensity beams

2. Long beam lifetime

3. Long spin coherence time

and in addition it needs to have a good way of canceling major systematic errors:

1. Control or cancel the magnetic field effects

2. Small sensitivity to geometrical phase effects and ways to cancel them

3. Detector related systematic errors

4. Negligible dark current from the electric field plates

The proton EDM method has all those stated features and indeed the collaboration is confident that it can put together a successful team to build it and run it within a reasonable cost and schedule. The rough schedule as outlined at the present time is shown in Figure 9

2.4 A storage ring electron EDM: Does it make sense?

The magic momentum of the electron is only 15 MeV/c, making it an ideal candidate for a demonstration storage ring EDM experiment and indeed a sensitive probe for a direct measurement of the electron EDM. Yuri Orlov first pointed this out around 2004 at which time the storage ring EDM collaboration tried to find a sensitive and efficient electron polarimeter for it. It turns out that this is a very difficult electron energy and at the time there was nothing available that could be used for an experiment. William Morse tried to resurrect the idea in 2013 after we came up with the all-electric storage ring idea for the proton and the simultaneous counter-rotating beams. Recently, in 2015 Richard Talman published a new idea[23] where by modifying the observational principle of the experiment and by developing a super-conducting resonance technique one may indeed be possible to put together a sensitive electron EDM experiment.

It is worthwhile to investigate these new options and come to a conclusion whether or not it makes sense to proceed. Even in the worst case, such a ring will become a testing ground for all the relevant, high risk items of the proton EDM experiment and it's probably a good idea to proceed.

2.5 Conclusions and acknowledgements

The current status in physics is very intriguing and indeed puzzling. The available phase-space for SUSY is closing in and it would be nice to have the option of building a machine with much larger energy reach. This may materialize some day, but it's going to take at least a decade to develop even if we had the funding. Storage ring EDM experiments with their very high New Physics mass scale reach can fill the void and provide guidance for the next HEP machine.

I am grateful to my colleagues in the storage ring EDM collaboration for all their ingenuity and hard work over the years and to Octavio Escalante for reading the document very carefully and for his excellent comments. This work was supported in part by IBS of South Korea (IBS-R017-D1-2016-a00).

References

[1] J.H. Christenson, J.W. Cronin, V.L. Fitch, and R. Turley, Phys. Rev. Lett. **13**, 138 (1964)
[2] A.D. Sakharov, JETP Lett. **5**, 24 (1967)
[3] C.A. Baker et al., Phys. Rev. Lett. **97**, 131801 (2006)
[4] M. Pospelov, A. Ritz, Ann. Phys. **318**, 119 (2005)
[5] R.D. Peccei and H.R. Quinn, Phys. Rev. Lett. **38**, 1440 (1977)
[6] S. Weinberg, Phys. Rev. Lett. **40**, 223 (1978)
[7] F. Wilczek, Phys. Rev. Lett. **40**, 279 (1978)
[8] JEDI collaboration, http://collaborations.fz-juelich.de/ikp/jedi/collaboration/collaboration.shtml
[9] W. Marciano, private communication
[10] W. Morse, private communication
[11] J. Bailey et al., Nucl. Phys. **B150**, 1-75 (1979)
[12] G. Bennett et al., (Muon G-2 Collaboration), Phys. Rev. **D73**, 072003 (2006), hep-ex/0602035
[13] J. Grange et al., Muon $(g-2)$ Technical Design Report: FERMILAB-FN-0992-E
[14] N. Saito and J-PARC $(g-2)$/EDM Collaboration, AIP Conf. Proc. 1467 (2012) 45. doi: 10.1063/1.4742078
[15] G.W. Bennett et al., Phys. Rev. **D80**, 052008 (2009)
[16] J. Baron et al., Science **343** 269 (2014)
[17] F.J.M. Farley et al., Phys. Rev. Lett. **93** 052001 (2004)
[18] A. Silenko et al., J-PARC Letter of Intent: Search for a Permanent Muon Electric Dipole Moment at the $10^{-24}\,e\cdot$cm Level, Jan. 2003
[19] V. Anastassopoulos et al., "A Storage Ring Experiment to Detect a Proton Electric Dipole Moment", arXiv:1502.04317, 2015
[20] E.M. Metodiev et al., Phys. Rev. ST Accel. Beams **17** (2014) 7, 074002
[21] E.M. Metodiev et al., Nucl. Instrum. Meth. **A797** 311 (2015)
[22] I. Altarev et al., J. Appl. Phys. **117** 233903 (2015). doi: 10.1063/1.4922671
[23] R. Talman, "Frequency Domain Storage Ring Method for Electric Dipole Moment Measurement", arXiv:1508.04366, 2015

Higher order hadronic and leptonic contributions to the muon $g-2$

Alexander Kurz[1,3a], Tao Liu[2b], Peter Marquard[3c], Alexander V. Smirnov[4d], Vladimir A. Smirnov[5e], and Matthias Steinhauser[1f]

[1] Institut für Theoretische Teilchenphysik, Karlsruhe Institute of Technology (KIT), 76128 Karlsruhe, Germany
[2] Department of Physics, University of Alberta, Edmonton AB T6G 2J1, Canada
[3] Deutsches Elektronen Synchrotron DESY, Platanenallee 6, 15738 Zeuthen, Germany
[4] Scientific Research Computing Center, Moscow State University, 119991, Moscow, Russia
[5] Skobeltsyn Institute of Nuclear Physics of Moscow State University, 119991, Moscow, Russia

Abstract. In this contribution we discuss next-to-next-to-leading order hadronic and four-loop QED contributions to the anomalous magnetic moment of the muon.

1 Introduction

The anomalous magnetic moment of the muon, a_μ, is among the most precise measured quantities in particle physics. It is measured to a precision of 0.54 parts per million which matches the precision of the Standard Model theory prediction [1, 2]. However, since many years one observes a discrepancy of about three to four standard deviations which survives persistent all improvements. This concerns both the experimental data and theoretical calculations entering the prediction.

Currently a new experiment is built at FERMILAB with the aim to increase the accuracy of the measured value by about a factor four [3, 4]. In the upcoming years also improvements on the theory side can be expected. On the one hand this is connected to improved measurements of $R(s)$ at low energies (see, e.g., Refs. [5–7]). On the other hand it can be expected that within the next few years results from lattice simulations become available both for the hadronic vacuum polarization and hadronic light-by-light contributions (see, e.g., Refs. [8–12]).

The by far dominant numerical contribution to a_μ originates from QED corrections which are known to five-loop order [13]. Note, however, that the four- and the five-loop corrections have only been computed by a single group.[1] For this reason we have recently started to systematically check the four-loop results of [13]. In Ref. [17] analytic results for the gauge-invariant subsets with two or three closed electron loops have been obtained neglecting power corrections of the form m_e/m_μ. All contributions involving a τ lepton have been computed in Ref. [20]. After including three (analytic) expansion terms in m_μ^2/m_τ^2 a better precision has been obtained than in the numerical approach of Ref. [13]. The numerically most important QED contributions at four-loop level arise from light-by-light-type diagrams (i.e. the external photon does not couple to the external muon line) containing a closed electron loop. This well-defined subset has been considered in Ref. [21] where an asymptotic expansion for $m_e \ll m_\mu$ has been performed to compute four expansion terms.

We adopt the notation from Ref. [13] and parametrize the anomalous magnetic moment in the form

$$a_\mu = \sum_{n=1}^{\infty} a_\mu^{(2n)} \left(\frac{\alpha}{\pi}\right)^n , \qquad (1)$$

where the four-loop contribution can be written as

$$a_\mu^{(8)} = A_1^{(8)} + A_2^{(8)}(m_\mu/m_e) + A_2^{(8)}(m_\mu/m_\tau) + A_3^{(8)}(m_\mu/m_e, m_\mu/m_\tau) . \qquad (2)$$

$A_1^{(8)}$ contains only contributions from photons and muons, $A_2^{(8)}(m_\mu/m_e)$ and $A_2^{(8)}(m_\mu/m_\tau)$ involve closed electron or tau loops, and each Feynman diagram which contributes to $A_3^{(8)}(m_\mu/m_e, m_\mu/m_\tau)$ contains all three lepton flavours simultaneously. In Sections 2 and 3 we describe the calculation of the light-by-light type QED contribution to $A_2^{(8)}(m_\mu/m_e)$ (see also [21]) and the computation of $A_2^{(8)}(m_\mu/m_\tau)$ (see also [20]), respectively. Afterwards we summarize in Section 4 the computation of the next-to-next-to-leading order (NNLO) hadronic vacuum polarization contribution published in Ref. [22]. A brief summary and an outlook is given in Section 5.

2 Four-loop electron contribution

The numerically most important contribution to $a_\mu^{(8)}$ originates from diagrams involving a closed electron loop (de-

[a] e-mail: alexander.kurz2@kit.edu
[b] e-mail: ltao@ualberta.ca
[c] e-mail: peter.marquard@desy.de
[d] e-mail: asmirnov80@gmail.com
[e] e-mail: smirnov@theory.sinp.msu.ru
[f] e-mail: matthias.steinhauser@kit.edu

[1] Partial four-loop corrections have been obtained in Refs. [14–19].

© The Authors, published by EDP Sciences. This is an open access article distributed under the terms of the Creative Commons Attribution License 4.0 (http://creativecommons.org/licenses/by/4.0/).

$A_2^{(8)}\left(\frac{m_\mu}{m_e}\right)$	[21]	[13, 28]
IV(a0)	116.76 ± 0.02	116.759183 ± 0.000292
IV(a1)	2.69 ± 0.14	2.697443 ± 0.000142
IV(a2)	4.33 ± 0.17	4.328885 ± 0.000293
IV(a)	123.78 ± 0.22	123.78551 ± 0.00044
IV(b)	-0.38 ± 0.08	-0.4170 ± 0.0037
IV(c)	2.94 ± 0.30	2.9072 ± 0.0044

Table 1. Summary of the final results for the individual four-loop light-by-light-type contributions and their comparison with results presented in Refs. [13, 28].

noted by $A_2^{(8)}(m_\mu/m_e)$ in Eq. (2). This contribution contains a gauge invariant subset where the external photon does not couple to the external muon line but to a closed fermion loop, the so-called leptonic light-by-light-type diagrams. Due to Furry's theorem such diagrams do not contribute at two but only start at three loops where four photons can be attached to the closed fermion loop. Here we discuss the four-loop result which can be sub-divided into three gauge invariant and finite contributions which we denote by IV(a), IV(d) and IV(c). Sample Feynman diagrams are shown in Fig. 1. Case IV(a) can be further subdivided according to the flavour of the leptons in the closed fermion loops. The contribution with two electron loops is denoted by IV(a0), the one with one muon and one electron loop and the coupling of the external photon to the electron by IV(a1), and the remaining one with one muon and one electron loop by IV(a2). We do not consider the case with two muon loops since this contribution is part of $A_1^{(8)}$.

The light-by-light-type diagrams are numerically dominant and provide about 95% of the four-loop electron loop contribution. The main reason for this are $\log(m_e/m_\mu)$ terms which are even present in the limit $m_e \to 0$. In fact, IV(a0) even has quadratic logarithms which makes this part the most important one.

Our calculation is based on an asymptotic expansion [23, 24] for $m_e \ll m_\mu$ which is implemented with the help of `asy` [25, 26] and in-house `Mathematica` programs. Similar to the hard mass procedure applied in Section 3 we obtain a factorization of the original two-scale integrals into products of one-scale integrals. The latter are either vacuum or on-shell integrals or integrals containing eikonal propagators of the form $1/(p \cdot q)$ (see Ref. [21] for more details). For each integral class we perform a reduction to master integrals and obtain analytic results expressed as a linear combination of about 150 so-called master integrals. About 50% of them we know analytically or to high numerical precision. The remaining ones are computed with the help of the package `FIESTA` [27] which is the source of the numerical uncertainty in our final result. We would like to stress that in our approach a systematic improvement is possible if it is required to improve the accuracy.

For all five cases we compute terms up to order $(m_e/m_\mu)^3$ (i.e. four expansion terms) and check that the cubic corrections only provide a negligible contribution. Our final results can be found in Tab. 1 where we compare to the findings of Refs. [13, 28]. Note that results for IV(a0) have also been obtained in Refs. [29, 30], though with significantly larger uncertainty. In all cases good agreement is found with [13, 28]. Although our numerical uncertainty, which amounts to approximately $0.4\times(\alpha/\pi)^4 \approx 1.2\times10^{-11}$, is larger, the final result is nevertheless sufficiently accurate as can be seen by the comparison to the difference between the experimental result and theory prediction which is given by

$$a_\mu(\text{exp}) - a_\mu(\text{SM}) \approx 249(87)\times10^{-11}\,. \tag{3}$$

This result is taken from Ref. [13]. Note that the uncertainty in Eq. (3) receives approximately the same amount from experiment and theory. Even after a projected reduction of the uncertainty by a factor four both in a_μ(exp) and a_μ(SM) our numerical precision is a factor ten below the uncertainty of the difference.

3 Four-loop tau lepton contribution

In this section we discuss the gauge invariant and finite subset of Feynman diagrams involving a closed heavy tau lepton loop. In the limit of infinitely heavy m_τ this contribution has to vanish. Thus $A_2^{(8)}(m_\mu/m_\tau)$ has a parametric dependence m_μ^2/m_τ^2 which is of order 10^{-3}. Note, that $\alpha/\pi \approx 2\cdot10^{-3}$ and thus one can expect that the four-loop tau lepton contribution is of the same order as the universal five-loop result [13].

We compute this contribution by applying an asymptotic expansion in the limit $m_\tau^2 \gg m_\mu^2$. This is realized with the help of the program `exp` [31, 32] which is written in C++. As a result the two-scale four-loop integrals factorize into one-scale vacuum (m_τ) and on-shell (m_μ) integrals. Both integral classes are well studied in the literature (for references see [20]). This concerns both the reduction to master integrals and the analytic evaluation of the latter.

In the first line of Fig. 2 a sample Feynman diagram is shown where the thick solid lines represent the tau leptons. Rows two and three of Fig. 2 show the result of the asymptotic expansion where the graphs left of the symbol $\otimes$ have to be expanded in all small quantities, i.e., the external momenta and the muon mass. Thus, the only mass scale of the remaining vacuum integral is the tau lepton mass. The result of the Taylor expansion is inserted into the effective vertex (thick blob) present in the diagram to the right of $\otimes$. Afterwards the remaining loop integrations, which are of on-shell type, are performed.

As a final result we obtain an expansion in m_μ^2/m_τ^2 with analytic coefficients containing $\log(m_\mu^2/m_\tau^2)$ terms. Note that with the help of this method a better accuracy has been obtained than with the numerical approach of Ref. [13]. Inserting numerical values for the lepton masses leads to

$$A_{2,\mu}^{(8)}(m_\mu/m_\tau) = 0.0421670 + 0.0003257 + 0.0000015 + \ldots\,, \tag{4}$$

where the ellipsis indicates terms of order $(m_\mu^2/m_\tau^2)^4$ which are expected to contribute at order 10^{-8} to $A_{2,\mu}^{(8)}(m_\mu/m_\tau)$.

Figure 1. Sample light-by-light-type Feynman diagrams contributing to a_μ. The external solid line represents the muon and at least one of the internal solid loops denotes electrons. In the case of IV(a) the second fermion loop can either be an electron or muon loop. Wavy lines represent photons.

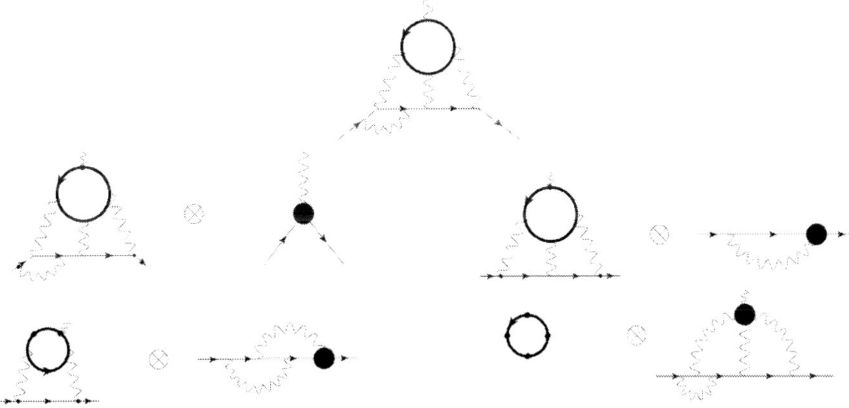

Figure 2. Graphical example for the application of the asymptotic expansion at four loops. Thick solid, thin solid and wavy lines represent taus, muons and photons, respectively. Only four representative sub-diagrams are shown; altogether there are eight contributions for the diagram in the first row.

a_μ receives contribution from τ lepton loops starting at two-loop order. Their numerical impact is given by

$$10^{11} \times a_\mu\Big|_{\tau\text{loops}} = 42.13 + 0.45 + 0.12\,, \qquad (5)$$

where the numbers on the right-hand side correspond to the two, three and four loops. It is interesting to note that the three-loop term is only less than a factor four larger than the four-loop counterpart. Furthermore, it is worth comparing the numbers in Eq. (5) to the universal contributions contained in A_1 which read [13]

$$10^{11} \times a_\mu\Big|_{\text{univ.}} = 116\,140\,973.21 - 177\,230.51 + 1\,480.42 - 5.56 + 0.06\,, \qquad (6)$$

where the individual terms on the right-hand side represent the results from one to five loops. Note that the four-loop tau lepton term is twice bigger than the five-loop photonic contribution.

4 NNLO hadronic contribution

The LO hadronic contribution to the anomalous magnetic moment of the muon is obtained from diagram (a) in Fig. 3. One parametrizes the hadronic contribution (represented by the blob) by the polarization function $\Pi(q^2)$ which appears as a factor in the integrand of the one-loop diagram. In a next step one exploits analyticity of $\Pi(q^2)$ and uses a dispersion integral to introduce its imaginary part,

$$R(s) = \frac{\sigma(e^+e^- \to \text{hadrons})}{\sigma_{pt}}\,, \qquad (7)$$

with $\sigma_{pt} = 4\pi\alpha^2/(3s)$. Note that $\sigma(e^+e^- \to \text{hadrons})$ does note include initial state radiative or vacuum polarization corrections. At that point the loop integration and the dispersion integral are interchanged and one obtains

$$a_\mu^{(1)} = \frac{1}{3}\left(\frac{\alpha}{\pi}\right)^2 \int_{m_\pi^2}^\infty \mathrm{d}s \frac{R(s)}{s} K^{(1)}(s)\,, \qquad (8)$$

A convenient integral representation for the kernel function $K^{(1)}(s)$, which is the result of the loop integration, is given by

$$K^{(1)}(s) = \int_0^1 \mathrm{d}x \frac{x^2(1-x)}{x^2 + (1-x)\frac{s}{m_\mu^2}}\,. \qquad (9)$$

At one-loop order it is possible to obtain analytic results (see Refs. [33, 34]). Nevertheless, it is promising to consider $K^{(1)}(s)$ in the limit $m_\mu^2 \ll s$ which is justified since the lower integration limit in Eq. (8) is m_π^2 which is bigger than m_μ^2. The expansion of $K^{(1)}(s)$ is easily obtained by remembering that it originates from the vertex diagram similar to Fig. 3(a) where the hadronic blob (including the external photon lines) is replaced by a massive photon with mass $\sqrt{s}$. The expansion $m_\mu^2 \ll s$ is easily implemented with the help of the program `exp` [31, 32] which implements the rules of asymptotic expansions involving a large internal mass (see, e.g., Ref. [24]). As a result the original two-scale integral is represented as a sum of one-scale integrals which are simple to compute. Using this approach several expansion terms in m_μ^2/s can be computed. One observes that an excellent approximation for $a_\mu^{(1)}$ is obtained by including terms up to order $(m_\mu^2/s)^5$.

The approach described in detail for the one-loop diagram can also be applied at two and three loops where exact calculations of the kernel functions are either very difficult or even impossible. In Ref. [22] four expansion terms have been computed which provides an approximation at the per mil level.

A slight complication arises for the contributions involving more than one hadronic insertion, see Figs. 3(d,h,i,j,l). In case they are present in the same photon line formulas similar to Eq. (9) can be derived with two- and three-dimensional integrations. Diagrams of type (3c) in Fig. 3 are more involved. Here, we apply a multiple asymptotic expansion in the limits $s \gg s' \gg m_\mu^2$, $s \approx s' \gg m_\mu^2$ and $s' \gg s \gg m_\mu^2$ (s and s' are the integration variables) and construct an interpolating function by combining the results from the individual limits.

The LO result for the hadronic vacuum polarization contribution to a_μ can be found in Refs. [35–38] and NLO analyses have been performed in Refs. [36, 39–41]. Our NLO results for the three contributions read

$$\begin{aligned} a_\mu^{(2a)} &= -20.90 \times 10^{-10}\,, \\ a_\mu^{(2b)} &= 10.68 \times 10^{-10}\,, \\ a_\mu^{(2c)} &= 0.35 \times 10^{-10}\,, \end{aligned} \tag{10}$$

which leads to

$$a_\mu^{\rm had,NLO} = -9.87 \pm 0.09 \times 10^{-10}\,, \tag{11}$$

in a good agreement with Refs. [36, 41]. Note that in our analyses no correlated uncertainties are taken into account. Such a rough treatment should not be done at LO but is certainly acceptable at NNLO.

For the individual NNLO contributions we obtain the results

$$\begin{aligned} a_\mu^{(3a)} &= 0.80 \times 10^{-10}\,, \\ a_\mu^{(3b)} &= -0.41 \times 10^{-10}\,, \\ a_\mu^{(3b,\rm lbl)} &= 0.91 \times 10^{-10}\,, \\ a_\mu^{(3c)} &= -0.06 \times 10^{-10}\,, \\ a_\mu^{(3d)} &= 0.0005 \times 10^{-10}\,, \end{aligned} \tag{12}$$

which leads to

$$a_\mu^{\rm had,NNLO} = 1.24 \pm 0.01 \times 10^{-10}\,. \tag{13}$$

It is interesting to note that similar patterns are observed at two and three loops: multiple hadronic insertions are small and the contributions of type (b) involving closed electron two-point functions reduce the contributions of type (a) by about 50%. However, at three-loop order there is a new type of diagram where the external photon couples to a closed electron loop ($a_\mu^{(3b,\rm lbl)}$) which provides the largest individual contribution. This is in analogy to the three-loop QED corrections where the light-by-light type diagrams dominate the remaining contributions. In fact, due to $a_\mu^{(3b,\rm lbl)}$ the NNLO hadronic vacuum polarization contribution has a non-negligible impact. It has the same order of magnitude as the current uncertainty of the leading order hadronic contribution and should thus be included in future analyses.

An important contribution to a_μ is provided by the so-called hadronic light-by-light diagrams where the external photon is connected to the hadronic blob. The NLO part of this contribution is of the same perturbative order as the corrections in Eq. (13). A first-principle calculation of this part is currently not available, however, in [42] it has been estimated to $a_\mu^{\rm lbl-had,NLO} = 0.3 \pm 0.2 \times 10^{-10}$.

We want to mention that there is a further hadronic contribution where four internal photons couple to the hadronic blob and the external photon couples to the muon line ("internal hadronic light-by-light"). This contribution, which is formally of the same perturbative order as $a_\mu^{\rm had,NNLO}$, is currently unknown.

5 Summary and conclusions

For more than a decade the measured and predicted results for the anomalous magnetic moment of the muon show a discrepancy of three to four standard deviations. This circumstance has triggered many publications which try to interpret the deviation with the help of beyond-SM theories. However, before drawing definite conclusions it is necessary to cross check the experimental result by performing an independent high-precision determination of a_μ. Furthermore, all ingredients of the theory prediction should be computed by at least two groups independently.

In this contribution we describe the calculation of two classes of four-loop QED contributions to a_μ, which up to date only have been computed by one group: the contribution involving tau leptons and the one involving light-by-light-type closed electron loops. Good agreement with the results in the literature is found. To complete the cross check of the four-loop result the non-light-by-light electron contribution, the diagrams involving simultaneously electrons and taus, and the pure-muon contribution have to be computed. From the technical point of view the missing diagram classes have the same complexity as those described in Sections 2 and 3.

As a further topic we have discussed in Section 4 the calculation of the NNLO hadronic vacuum polarization contribution.

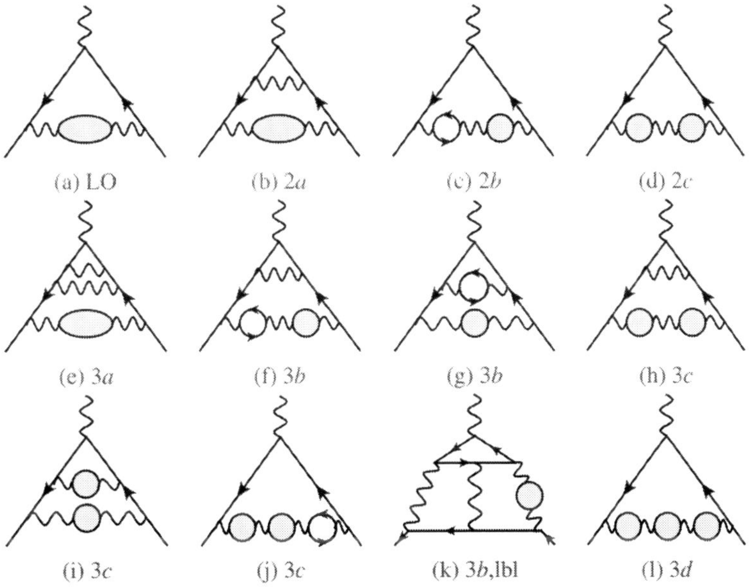

Figure 3. Sample LO, NLO and NNLO Feynman diagrams contributing to $a_\mu^{\rm had}$. The external fermions are muons and the fermions in the closed loops represent electrons.

Acknowledgments

M.S. would like to thank the organizers of "Flavour changing and conserving processes 2015" for the pleasant atmosphere during the conference. We thank the High Performance Computing Center Stuttgart (HLRS) and the Supercomputing Center of Lomonosov Moscow State University [43] for providing computing time used for the numerical computations with FIESTA. P.M. was supported in part by the EU Network HIGGSTOOLS PITN-GA-2012-316704. This work was supported by the DFG through the SFB/TR 9 "Computational Particle Physics". The work of V.S. was supported by the Alexander von Humboldt Foundation (Humboldt Forschungspreis).

References

[1] G. W. Bennett *et al.* [Muon g-2 Collaboration], Phys. Rev. D **73** (2006) 072003 [hep-ex/0602035].

[2] B. L. Roberts, Chin. Phys. C **34** (2010) 741 [arXiv:1001.2898 [hep-ex]].

[3] R. M. Carey *et al.*, FERMILAB-PROPOSAL-0989.

[4] D. Hertzog, these proceedings.

[5] B. Aubert *et al.* [BaBar Collaboration], Phys. Rev. Lett. **103** (2009) 231801 [arXiv:0908.3589 [hep-ex]].

[6] F. Ambrosino *et al.* [KLOE Collaboration], Phys. Lett. B **700** (2011) 102 [arXiv:1006.5313 [hep-ex]].

[7] S. Eidelman, these proceedings.

[8] M. Della Morte, B. Jager, A. Juttner and H. Wittig, JHEP **1203** (2012) 055 [arXiv:1112.2894 [hep-lat]].

[9] F. Burger *et al.* [ETM Collaboration], JHEP **1402** (2014) 099 [arXiv:1308.4327 [hep-lat]].

[10] T. Blum, S. Chowdhury, M. Hayakawa and T. Izubuchi, Phys. Rev. Lett. **114** (2015) 1, 012001 [arXiv:1407.2923 [hep-lat]].

[11] T. Blum, N. Christ, M. Hayakawa, T. Izubuchi, L. Jin and C. Lehner, arXiv:1510.07100 [hep-lat].

[12] C. Lehner, these proceedings.

[13] T. Aoyama, M. Hayakawa, T. Kinoshita and M. Nio, Phys. Rev. Lett. **109** (2012) 111808 [arXiv:1205.5370 [hep-ph]].

[14] S. Laporta, Phys. Lett. B **312** (1993) 495 [hep-ph/9306324].

[15] P. A. Baikov and D. J. Broadhurst, [hep-ph/9504398].

[16] J. -P. Aguilar, D. Greynat and E. De Rafael, Phys. Rev. D **77** (2008) 093010 [arXiv:0802.2618 [hep-ph]].

[17] R. Lee, P. Marquard, A. V. Smirnov, V. A. Smirnov and M. Steinhauser, JHEP **1303** (2013) 162 [arXiv:1301.6481 [hep-ph]].

[18] P. A. Baikov, K. G. Chetyrkin, J. H. Kuhn, C. Sturm, Nucl. Phys. B **867** (2013) 182 [arXiv:1207.2199 [hep-ph]].

[19] P. A. Baikov, A. Maier and P. Marquard, Nucl. Phys. B **877** (2013) 647 [arXiv:1307.6105 [hep-ph]].

[20] A. Kurz, T. Liu, P. Marquard and M. Steinhauser, Nucl. Phys. B **879** (2014) 1 [arXiv:1311.2471 [hep-ph]].

[21] A. Kurz, T. Liu, P. Marquard, A. V. Smirnov, V. A. Smirnov and M. Steinhauser, arXiv:1508.00901 [hep-ph].

[22] A. Kurz, T. Liu, P. Marquard and M. Steinhauser, Phys. Lett. B **734** (2014) 144 [arXiv:1403.6400 [hep-ph]].

[23] M. Beneke and V. A. Smirnov, Nucl. Phys. B **522** (1998) 321 [hep-ph/9711391].

[24] V. A. Smirnov, *Analytic tools for Feynman integrals*, Springer Tracts Mod. Phys. **250** (2012) 1.

[25] A. Pak and A. Smirnov, Eur. Phys. J. C **71** (2011) 1626 [arXiv:1011.4863 [hep-ph]].

[26] B. Jantzen, A. V. Smirnov and V. A. Smirnov, Eur. Phys. J. C **72** (2012) 2139 [arXiv:1206.0546 [hep-ph]].

[27] A. V. Smirnov, Comput. Phys. Commun. **185** (2014) 2090 [arXiv:1312.3186 [hep-ph]].

[28] T. Kinoshita and M. Nio, Phys. Rev. D **70** (2004) 113001 [hep-ph/0402206].

[29] J. Calmet and A. Peterman, Phys. Lett. B **56** (1975) 383.

[30] C. Chlouber and M. A. Samuel, Phys. Rev. D **16** (1977) 3596.

[31] R. Harlander, T. Seidensticker and M. Steinhauser, Phys. Lett. B **426** (1998) 125, arXiv:hep-ph/9712228.

[32] T. Seidensticker, arXiv:hep-ph/9905298.

[33] S. J. Brodsky and E. De Rafael, Phys. Rev. **168** (1968) 1620.

[34] S. Eidelman and F. Jegerlehner, Z. Phys. C **67** (1995) 585 [hep-ph/9502298].

[35] M. Davier, A. Hoecker, B. Malaescu and Z. Zhang, Eur. Phys. J. C **71** (2011) 1515 [Erratum-ibid. C **72** (2012) 1874] [arXiv:1010.4180 [hep-ph]].

[36] K. Hagiwara, R. Liao, A. D. Martin, D. Nomura and T. Teubner, J. Phys. G **38** (2011) 085003 [arXiv:1105.3149 [hep-ph]].

[37] F. Jegerlehner and R. Szafron, Eur. Phys. J. C **71** (2011) 1632 [arXiv:1101.2872 [hep-ph]].

[38] M. Benayoun, P. David, L. DelBuono and F. Jegerlehner, Eur. Phys. J. C **73** (2013) 2453 [arXiv:1210.7184 [hep-ph]].

[39] B. Krause, Phys. Lett. B **390** (1997) 392 [hep-ph/9607259].

[40] D. Greynat and E. de Rafael, JHEP **1207** (2012) 020 [arXiv:1204.3029 [hep-ph]].

[41] K. Hagiwara, A. D. Martin, D. Nomura and T. Teubner, Phys. Rev. D **69** (2004) 093003 [hep-ph/0312250].

[42] G. Colangelo, M. Hoferichter, A. Nyffeler, M. Passera and P. Stoffer, Phys. Lett. B **735** (2014) 90 [arXiv:1403.7512 [hep-ph]].

[43] V. Sadovnichy, A. Tikhonravov, V. Voevodin, and V. Opanasenko, "'Lomonosov': Supercomputing at Moscow State University." In Contemporary High Performance Computing: From Petascale toward Exascale (Chapman & Hall/CRC Computational Science), pp.283-307, Boca Raton, USA, CRC Press, 2013.

Muon $(g-2)$ in the Standard Model and supersymmetric extensions

Markus Bach[1], Dominik Stöckinger[1], Hyejung Stöckinger-Kim[1], and Jae-hyeon Park[2]

[1] *Institut für Kern- und Teilchenphysik, Technische Universität Dresden, Dresden, Germany*
[2] *Departament de Física Teòrica and IFIC, Universitat de València-CSIC, 46100, Burjassot, Spain*

Abstract. Recent new results on the SM and the SUSY prediction for the muon $(g-2)$ are briefly reviewed, and a SUSY scenario with particularly large contributions is discussed.

1 Introduction

In spite of the discovery of a Higgs boson, the first run of the LHC has not revealed any significant sign for physics beyond the Standard Model (SM). On the other hand, in spite of tremendous scrutiny in the past years, the gap between the SM-prediction and the experimental value of the muon anomalous magnetic moment has not been reduced. As discussed in many other talks at this meeting, recent progress on the hadronic contributions to $(g-2)_\mu$ confirm a deviation of more than 3 standard deviations or approximately 3×10^{-9}. Hence the question remains whether this deviation is due to physics beyond the SM.

In the following we first explain the key ingredients of the more precise evaluation of the electroweak SM contributions in Ref. [1]. Then we argue that precise predictions are also necessary in models different from the SM, and we discuss the status of the prediction in the minimal supersymmetric standard model (MSSM). Finally we show that within the framework of the MSSM, very large contributions to $(g-2)_\mu$ are possible even if the masses of all new particles are in the TeV region.

2 The electroweak SM contribution

The electroweak contributions to the SM prediction for $(g-2)_\mu$ have been computed for the first time at the two-loop level in Refs. [2, 3]. Two simplifications were used in these computations: (1) the fermion-loop contributions were evaluated using elementary light quarks. (2) the dependence on the Higgs boson mass was approximated by taking the Higgs boson mass either as equal to the top-quark mass, or taking it to infinity or to zero.

Point (1) was improved in a series of papers which evaluated the hadronic contributions using hadronic models and non-perturbative information; the currently best result and a discussion of other references has been given in Ref. [4]. Ref. [4] also evaluated the leading three-loop logarithmic contributions.

Point (2) was improved in Refs. [5, 6], where in particular the bosonic electroweak two-loop contributions have been evaluated taking into account the exact Higgs boson mass dependence.

However, the results of these references were never combined consistently, because the employed renormalization schemes were different. Hence, in Ref. [1] the two-loop computation of the electroweak contributions has been repeated in the appropriate renormalization scheme, so that a combination with the hadronic and the leading three-loop results of Ref. [4] became possible. Fig. 1 shows sample Feynman diagrams and numerical results of the bosonic and the Higgs-dependent fermionic two-loop contributions.

Furthermore, the remaining sources of parametric and theory uncertainties have been analyzed. The final result obtained in Ref. [1] for the SM electroweak contributions is

$$a_\mu^{\rm EW} = (153.6 \pm 1.0) \times 10^{-11}. \qquad (1)$$

Here the parametric uncertainty due to the input values of the Higgs boson, W- and Z-boson and top-quark is negligible; the uncertainty is dominated by the one of the hadronic contributions and has been taken over from Ref. [4].

3 Contributions in supersymmetric models

3.1 Motivation

Figure 2 provides motivation to consider a_μ as a constraint on the SUSY parameter space. In red it shows the prediction for $a_\mu^{\rm SUSY}$ for the SPS benchmark points [7]. As can be seen these well-motivated points lead to very different contributions, and independently of the resulting value, the future a_μ experiments will exclude most points. In green the figure shows "degenerate" points found in Ref. [8], which correspond to distinct SUSY parameter scenarios which however lead to the same LHC signatures. Combining LHC-measurements with a_μ will allow to select between these points.

© The Authors, published by EDP Sciences. This is an open access article distributed under the terms of the Creative Commons Attribution License 4.0 (http://creativecommons.org/licenses/by/4.0/).

DOI: 10.1051/epjconf/201611801034

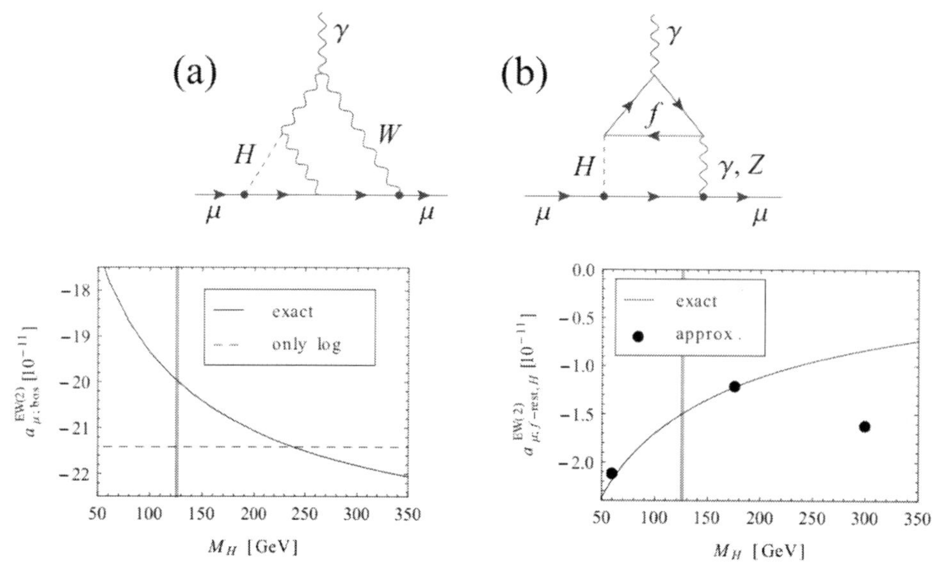

Figure 1. Sample Feynman diagrams and numerical results for the full bosonic electroweak two-loop contributions $a_{\mu;\text{bos}}^{\text{EW}(2)}$ (left) and for the Higgs-dependent fermion-loop electroweak two-loop contributions $a_{\mu;\text{f-rest,H}}^{\text{EW}(2)}$ (right), both as a function of the Higgs boson mass. The vertical band indicates the measured value of M_H. The dashed line in the left plot corresponds to the leading logarithmic approximation as defined in Ref. [5]. The fat dots in the right plot correspond to the approximations for $M_H = 60$ GeV, m_t, 300 GeV given in Ref. [2]. Figures from Ref. [1].

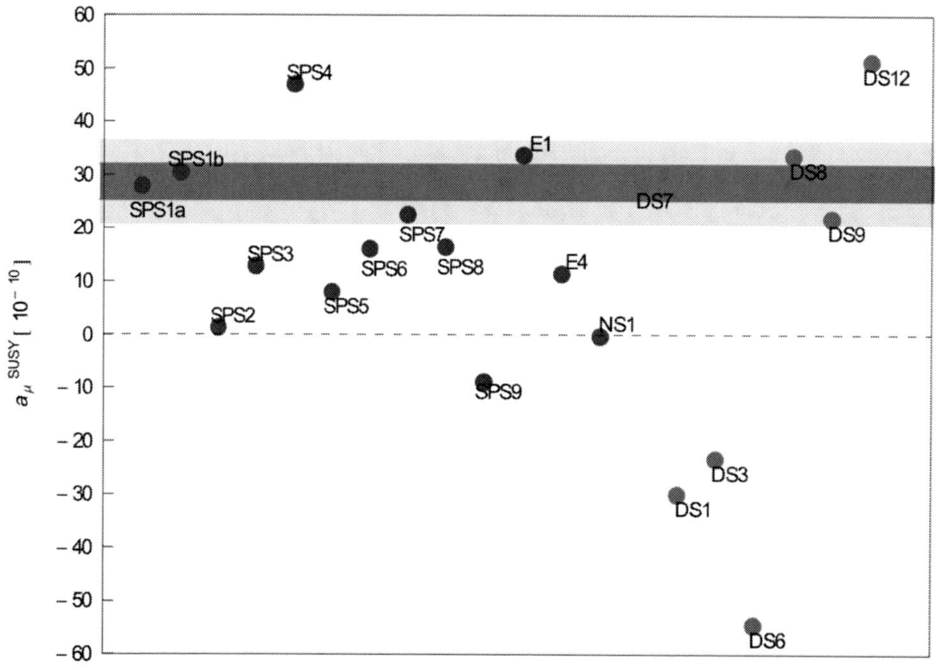

Figure 2. The prediction for a_μ^{SUSY} for the case of the SPS benchmark points [7] and for the "degenerate" points identified in Ref. [8]. The bands indicate the current deviation between experiment and the SM prediction, and the expected precision of future measurements, assuming the same central value. Figure from Ref. [9].

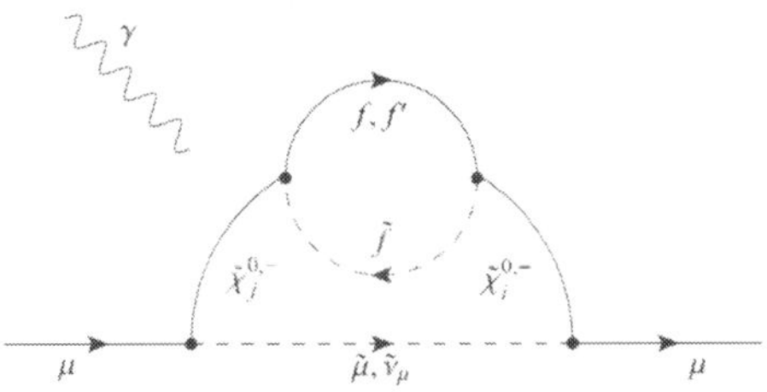

Figure 3. Sample two-loop diagram contributing to the fermion/sfermion-loop contributions of Refs. [12, 13].

3.2 Precision prediction for $a_\mu^{\rm SUSY}$

In view of these prospects and of the expected precision of the future experiments, it is also motivated to aim for a SUSY prediction for a_μ with a theory uncertainty which is as small as possible, ideally smaller than the future experimental uncertainty. This requires the computation of the full two-loop contributions to $a_\mu^{\rm SUSY}$, see Ref. [10].

This goal has not been reached, but significant progress has been achieved in the past years. A computer program has been published in Ref. [11], which implements the currently best prediction of $a_\mu^{\rm SUSY}$, including all corrections discussed below.

Here we highlight in particular the computation of the fermion/sfermion-loop contributions in Refs. [12, 13]. A sample diagram is shown in Fig. 3. This computation has two interesting qualitative consequences. The first is that it reduces the theory uncertainty resulting from the possibility to choose different renormalization schemes for the finestructure constant α and the weak mixing angle $\sin\theta_W$. The one-loop contributions directly depend on these input parameters and thus on the scheme choice. The leading differences between the usually considered schemes originate from fermion or sfermion loops in gauge boson self energies. These terms are precisely included in the counterterm computation of Refs. [12, 13]. The sum of the one-loop contributions and the two-loop contributions of these references is essentially scheme independent (up to 3-loop effects and effects from non-fermion/sfermion-loop contributions).

A second qualitative outcome of that computation is that the contributions from fermion/sfermion loops involve potentially large logarithms between different SUSY particle masses. Particularly very heavy stops in the multi-TeV region can lead to non-decoupling corrections of the SUSY one-loop contribution to a_μ, which can amount to more than 10%. These non-decoupling logarithms also allow to obtain a compact approximation of the fermion/sfermion-loop contributions.

3.3 $a_\mu^{\rm SUSY}$ for $\tan\beta \to \infty$

It is well known that the leading SUSY contributions to a_μ are proportional to $\tan\beta$. However, this proportionality is not exact — in fact for large $\tan\beta$ higher-order effects modify the behaviour, and the limit $\tan\beta \to \infty$ exists and is viable [14, 15], and it is intriguing to study how large a_μ can become in this limit. Here we briefly review the study of Ref. [16]. In order to understand the behaviour it is useful to write the muon mass and the muon magnetic moment as

$$m_\mu = y_\mu v_d + y_\mu v_u \Delta_\mu^{\rm red} + \dots, \tag{2}$$

$$a_\mu^{\rm SUSY} = \frac{y_\mu v_u a_\mu^{\rm red}}{m_\mu} + \dots, \tag{3}$$

where m_μ and y_μ are the muon pole mass and Yukawa coupling, $v_{u,d}$ is the up/down-type Higgs vacuum expectation value, and the dots denote terms proportional to v_d instead of v_u. The limit $\tan\beta \to \infty$ is equivalent to $v_d \to 0$, and in this limit we obtain

$$a_\mu^{\rm SUSY} = \frac{a_\mu^{\rm red}}{\Delta_\mu^{\rm red}}. \tag{4}$$

Here $a_\mu^{\rm red}$ and $\Delta_\mu^{\rm red}$ are given by calculable, finite loop diagrams, which depend (at the one-loop level) on the gaugino, Higgsino, and smuon mass parameters $M_{1,2}, \mu, m_{L,R}$.

Figure 4 shows the resulting $a_\mu^{\rm SUSY}$ as a function of two mass ratios. The gray area shows the region with negative $a_\mu^{\rm SUSY}$, which cannot explain the currently observed deviation between experiment and SM-prediction. The case with universal SUSY masses (all mass ratios equal to one) belongs to the region. However, if there are mass splittings (either Higgsino mass or left-handed smuon mass much larger than the other masses), the resulting $a_\mu^{\rm SUSY}$ becomes positive.

Two sample parameter choices (with mass values given in TeV) and the results for $a_\mu^{\rm SUSY}$ are as follows:

μ	M_1	M_2	m_L	m_R	$a_\mu/10^{-9}$
15	1	−1	1	1	3.01
1.3	1.3	−1.3	26	1.3	2.90

The table and the figure show that the limit $\tan\beta \to \infty$ provides a promising example of a scenario which could explain the deviation in a_μ even though all new particle masses are at or above the TeV-scale.

References

[1] C. Gnendiger, D. Stöckinger and H. Stöckinger-Kim, Phys. Rev. D **88** (2013) 053005.
[2] A. Czarnecki, B. Krause and W. J. Marciano, Phys. Rev. D **52** (1995) 2619.
[3] A. Czarnecki, B. Krause and W. J. Marciano, Phys. Rev. Lett. **76** (1996) 3267.
[4] A. Czarnecki, W. J. Marciano, A. Vainshtein Phys.Rev.D **67** (2003) 073006, Erratum-ibid.D**73** (2006) 119901.
[5] S. Heinemeyer, D. Stöckinger and G. Weiglein, Nucl. Phys. B **699** (2004) 103.
[6] T. Gribouk and A. Czarnecki, Phys. Rev. D **72** (2005) 053016.

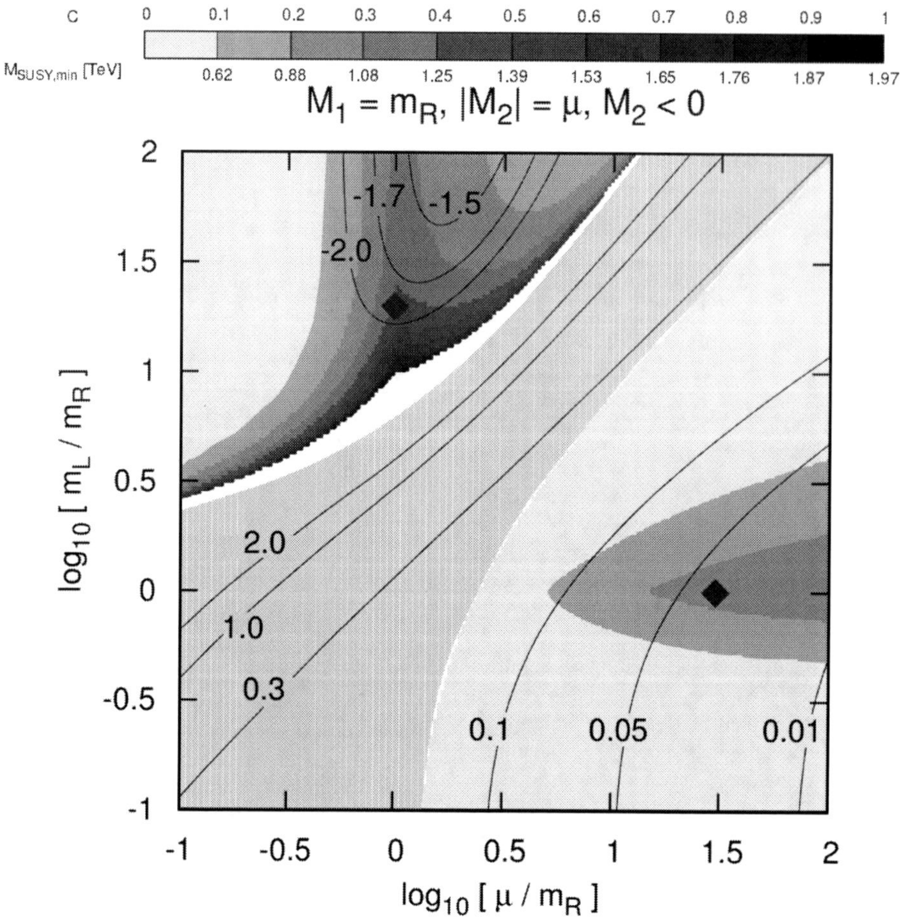

Figure 4. The minimum SUSY mass $M_{\text{SUSY,min}} = \min(M_1, |M_2|, m_L, m_R, \mu)$ for which a_μ^{SUSY} fully explains the current deviation between experiment and SM in the limit $\tan\beta \to \infty$. Equivalently the value C, defined by $a_\mu^{\text{SUSY}} = Cm_\mu^2/M_{\text{SUSY,min}}^2$, is given. The black lines denote values of the muon Yukawa coupling, and the blue squares denote benchmark points. The figure has been taken from Ref. [16].

[7] B. C. Allanach *et al.*, "The Snowmass points and slopes: Benchmarks for SUSY searches," in *Proc. of the APS/DPF/DPB Summer Study on the Future of Particle Physics (Snowmass 2001)* ed. N. Graf, Eur. Phys. J. C **25** (2002) 113 [eConf **C010630** (2001) P125] [arXiv:hep-ph/0202233].

[8] C. Adam, J. -L. Kneur, R. Lafaye, T. Plehn, M. Rauch and D. Zerwas, Eur. Phys. J. C **71** (2011) 1520 [arXiv:1007.2190 [hep-ph]].

[9] J. P. Miller, E. de Rafael, B. L. Roberts and D. Stöckinger, Ann. Rev. Nucl. Part. Sci. **62** (2012) 237.

[10] D. Stöckinger, J. Phys. G G **34** (2007) R45.

[11] P. Athron *et al.*, arXiv:1510.08071 [hep-ph].

[12] H. G. Fargnoli, C. Gnendiger, S. Paßehr, D. Stöckinger and H. Stöckinger-Kim, Phys. Lett. B **726** (2013) 717 [arXiv:1309.0980 [hep-ph]].

[13] H. G. Fargnoli, C. Gnendiger, S. Paßehr, D. Stöckinger and H. Stöckinger-Kim, JHEP **1402** (2014) 070 [arXiv:1311.1775 [hep-ph]].

[14] B. A. Dobrescu and P. J. Fox, Eur. Phys. J. C **70** (2010) 263 [arXiv:1001.3147 [hep-ph]].

[15] W. Altmannshofer and D. M. Straub, JHEP **1009** (2010) 078 [arXiv:1004.1993 [hep-ph]].

[16] M. Bach, J. h. Park, D. Stöckinger and H. Stöckinger-Kim, arXiv:1504.05500 [hep-ph].

EPJ Web of Conferences **118**, 01035 (2016) DOI: 10.1051/epjconf/201611801035
FCCP2015

Evaluating the leading hadronic corrections to the muon g-2: a new approach *

Luca Trentadue[1,2,a]

[1] *Dipartimento di Fisica e Scienze della Terra "Macedonio Melloni" Universitá di Parma, Parma*
[2] *INFN Sezione di Milano Bicocca, Milano, Italy*

Abstract. A novel approach to determine the leading hadronic corrections to the muon g-2 is proposed. It consists in a measurement of the effective electromagnetic coupling in the space-like region. This method may become feasible at flavor factories, resulting in a determination potentially competitive with the dispersive approach via time-like data.

1 Introduction

The motivation of this work [1] is due to a long-standing discrepancy between experiment and the Standard Model (SM) prediction for a_μ, the muon anomalous magnetic moment. For this reason the hadronic corrections have been kept under close scrutiny [2–6]. The hadronic contribution represents the largest uncertainty of the SM value and is comparable with the experimental one. When the new results from the g-2 experiments at Fermilab and J-PARC will reach the unprecedented precision of 0.14 parts per million (or better) [7–9], the uncertainty of the hadronic corrections will become the main limitation of this formidable test of the SM.
Vacuum Polarization makes α_{em} running assuming a well defined effective value at any scale. Vacuum polarization and the effective charge are defined by:

$$e^2 \to e^2(q^2) = \frac{e^2}{1 + (\Pi(q^2) - \Pi(0))} \tag{1}$$

and

$$\alpha(q^2) = \frac{\alpha(0)}{1 - \Delta\alpha} \quad \Delta\alpha = -Re(\Pi(q^2) - \Pi(0)). \tag{2}$$

$\Delta\alpha$ takes contributions from leptonic and hadronic elementary states among these the non-perturbative $\Delta\alpha_{had}$

$$\Delta\alpha = \Delta\alpha_{lept} + \Delta\alpha_{had} + \Delta\alpha_{top} \tag{3}$$

$$\Delta\alpha^{(5)}_{had}(M_z^2) = -\frac{\alpha\, M_z^2}{3\pi} Re \int_{4m_\pi^2}^{\infty} ds \frac{R(s)}{s(s - M_z^2 - i\epsilon)} \tag{4}$$

*This talk is based on a work done in collaboration with C. Carloni Calame, M. Passera and G. Venanzoni and is dedicated to the memory of our friend and colleague Eduard A. Kuraev.

[a] e-mail: luca.trentadue@unipr.it

A direct measurement of the running of $\alpha_{em}(s/t)$ in space/time-like regions can show the running of α_{em}. It can provide a test of duality far away form resonances and ita has been done in the past by few experiments at e^+e^- colliders by comparing well known QED processes with some reference (either data or MonteCarlo one) normalization:

$$(\frac{\alpha(q^2)}{\alpha(q_0^2)})^2 \simeq \frac{N_{signal}(q^2)}{N_{norm}(q_0^2)}. \tag{5}$$

N_{signal} can be any QED process as muon pairs production or others similar ones and N_{norm} can be the Bhabha process, a pure QED channel as $\gamma\gamma$ pair production as well as theory, or any other reference process. q_0 represents a scale chosen as starting point for the running.

2 a_{mu}^{HLO} calculation

The leading-order hadronic contribution to the muon g-2 is given by the well-known formula [6, 13]

$$a_\mu^{\rm HLO} = \frac{\alpha}{\pi^2} \int_0^\infty \frac{ds}{s} K(s)\, {\rm Im}\Pi_{\rm had}(s + i\epsilon), \tag{6}$$

where $\Pi_{\rm had}(s)$ is the hadronic part of the photon vacuum polarization, $\epsilon > 0$,

$$K(s) = \int_0^1 dx \frac{x^2(1-x)}{x^2 + (1-x)(s/m_\mu^2)} \tag{7}$$

is a positive kernel function, and m_μ is the muon mass. As the total cross section for hadron production in low-energy e^+e^- annihilations is related to the imaginary part of $\Pi_{\rm had}(s)$ via the optical theorem, the dispersion integral in eq. (6) is computed integrating experimental time-like ($s > 0$) data up to a certain value of s [3, 17, 18]. The high-energy tail of the integral is calculated using perturbative QCD [19].

© The Authors, published by EDP Sciences. This is an open access article distributed under the terms of the Creative Commons Attribution License 4.0 (http://creativecommons.org/licenses/by/4.0/).

Alternatively, if we exchange the x and s integrations in eq. (6) we obtain [20]

$$a_\mu^{\rm HLO} = \frac{\alpha}{\pi}\int_0^1 dx\,(x-1)\,\overline{\Pi}_{\rm had}[t(x)]\,, \qquad (8)$$

where $\overline{\Pi}_{\rm had}(t) = \Pi_{\rm had}(t) - \Pi_{\rm had}(0)$ and

$$t(x) = \frac{x^2 m_\mu^2}{x-1} < 0 \qquad (9)$$

is a space-like squared four-momentum. If we invert eq. (9), we get $x = (1-\beta)\,(t/2m_\mu^2)$, with $\beta = (1-4m_\mu^2/t)^{1/2}$, and from eq. (8) we obtain

$$a_\mu^{\rm HLO} = \frac{\alpha}{\pi}\int_{-\infty}^0 \overline{\Pi}_{\rm had}(t)\left(\frac{\beta-1}{\beta+1}\right)^2 \frac{dt}{t\beta}. \qquad (10)$$

Equation (10) has been used for lattice QCD calculations of $a_\mu^{\rm HLO}$ [21]; while the results are not yet competitive with those obtained with the dispersive approach via time-like data, their errors are expected to decrease significantly in the next few years [22].

The effective fine-structure constant at squared momentum transfer q^2 is defined above and is $\Delta\alpha(q^2) = -{\rm Re}\overline{\Pi}(q^2)$. The purely leptonic part, $\Delta\alpha_{\rm lep}(q^2)$, can be calculated order-by-order in perturbation theory – it is known up to three loops in QED [23] (and up to four loops in specific q^2 limits [24]). As ${\rm Im}\overline{\Pi}(q^2) = 0$ for negative q^2, eq. (8) can be rewritten in the form [25]

$$a_\mu^{\rm HLO} = \frac{\alpha}{\pi}\int_0^1 dx\,(1-x)\,\Delta\alpha_{\rm had}[t(x)]\,. \qquad (11)$$

Equation (11), involving the hadronic contribution to the running of the effective fine-structure constant at space-like momenta, can be further formulated in terms of the Adler function [26], defined as the logarithmic derivative of the vacuum polarization, which, in turn, can be calculated via a dispersion relation with time-like hadroproduction data and perturbative QCD [25, 27]. We will proceed differently, proposing to calculate eq. (11) by measurements of the effective electromagnetic coupling in the space-like region. The hadronic contribution to the running of α in the space-like region, $\Delta\alpha_{\rm had}(t)$ (see eq. (3)), can be extracted comparing Bhabha scattering data to Monte Carlo (MC) predictions. The LO Bhabha cross section receives contributions from t- and s-channel photon exchange amplitudes. At NLO in QED, it is customary to distinguish corrections with an additional virtual photon or the emission of a real photon (photonic NLO) from those originated by the insertion of the vacuum polarization corrections into the LO photon propagator (VP). let us consider a few simple points. In fig. 1 (left) we plot the integrand $(1-x)\Delta\alpha_{\rm had}[t(x)]$ of eq. (11) using the output of the routine `hadr5n12` [30] (which uses time-like hadroproduction data and perturbative QCD). The range $x \in (0,1)$ corresponds to $t \in (-\infty, 0)$, with $x = 0$ for $t = 0$. The peak of the integrand occurs at $x_{\rm peak} \simeq 0.914$ where $t_{\rm peak} \simeq -0.108$ GeV^2 and $\Delta\alpha_{\rm had}(t_{\rm peak}) \simeq 7.86 \times 10^{-4}$ (see fig. 1 (right)). Such relatively low t values can be explored at e^+e^- colliders with center-of-mass energy $\sqrt{s}$ around or below 10 GeV where

$$t = -\frac{s}{2}(1-cos\theta)\left(1-\frac{4m_e^2}{s}\right), \qquad (12)$$

θ is the electron scattering angle and m_e is the electron mass. Depending on s and θ, the integrand of eq. (11) can be measured in the range $x \in [x_{\rm min}, x_{\rm max}]$, as shown in fig. 2 (left). Note that to span low x intervals, larger θ ranges are needed as the collider energy decreases. In this respect, $\sqrt{s} \sim 3$ GeV appears to be very convenient, as an x interval [0.30, 0.98] can be measured varying θ between $\sim 2°$ and 28°. It is also worth remarking that data collected at flavor factories, such as DAΦNE (Frascati), VEPP-2000 (Novosibirsk), BEPC-II (Beijing), PEP-II (SLAC) and SuperKEKB (Tsukuba), and possibly at a future high-energy e^+e^- collider, like FCC-*ee* (TLEP) [31] or ILC [32], can help to cover different and complementary x regions.

Furthermore, given the smoothness of the integrand, values outside the measured x interval may be interpolated with some theoretical input. In particular, the region below $x_{\rm min}$ will provide a relatively small contribution to $a_\mu^{\rm HLO}$, while the region above $x_{\rm max}$ may be obtained by extrapolating the curve from $x_{\rm max}$ to $x = 1$, where the integrand is null, or using perturbative QCD.

The analytic dependence of the MC Bhabha predictions on $\alpha(t)$ (and, in turn, on $\Delta\alpha_{\rm had}(t)$) is not trivial, and a numerical procedure has to be devised to extract it from the data.This was not the case for example in [14, 15]: there $\alpha(t)$ was extracted from Bhabha data in the very forward region at LEP, where the t channel diagrams are by far dominant and $\alpha(t)$ factorizes (see for example [16]). In formulae, we have to find a function $\alpha(t)$ such that

$$\left.\frac{d\sigma}{dt}\right|_{\rm data} = \left.\frac{d\sigma}{dt}\big(\alpha(t),\alpha(s)\big)\right|_{\rm MC}, \qquad (13)$$

where we explicitly kept apart the dependence on the time-like VP $\alpha(s)$ because we are only interested in $\alpha(t)$. We emphasise that, in our analysis, $\alpha(s)$ is an input parameter. Being the Bhabha cross section in the forward region dominated by the t-channel exchange diagrams , we checked that the present $\alpha(s)$ uncertainty induces in this region a relative error on the θ distribution of less than $\sim 10^{-4}$ (which is part of the systematic error).

We propose to perform the numerical extraction of $\Delta\alpha_{\rm had}(t)$ from the Bhabha distribution of the t Mandelstam variable. The idea is to let $\alpha(t)$ vary in the MC sample around a reference value and choose, bin by bin in the t distribution, the value that minimizes the difference with data. The procedure can be sketched as follows:

1. choose a reference function returning the value of $\Delta\alpha_{\rm had}(t)$ (and hence $\alpha(t)$) to be used in the MC sample, we call it $\bar{\alpha}(t)$;

2. for each generated event, calculate N MC weights by rescaling $\bar{\alpha}(t) \to \bar{\alpha}(t) + \frac{i}{N}\delta(t)$, where $i \in [-N, N]$ and $\delta(t)$ is for example the error induced on $\bar{\alpha}(t)$ by the error on $\Delta\alpha_{\rm had}(t)$. Being done on an event by event basis, the full dependence on $\alpha(t)$ of the MC differential cross section can be kept;

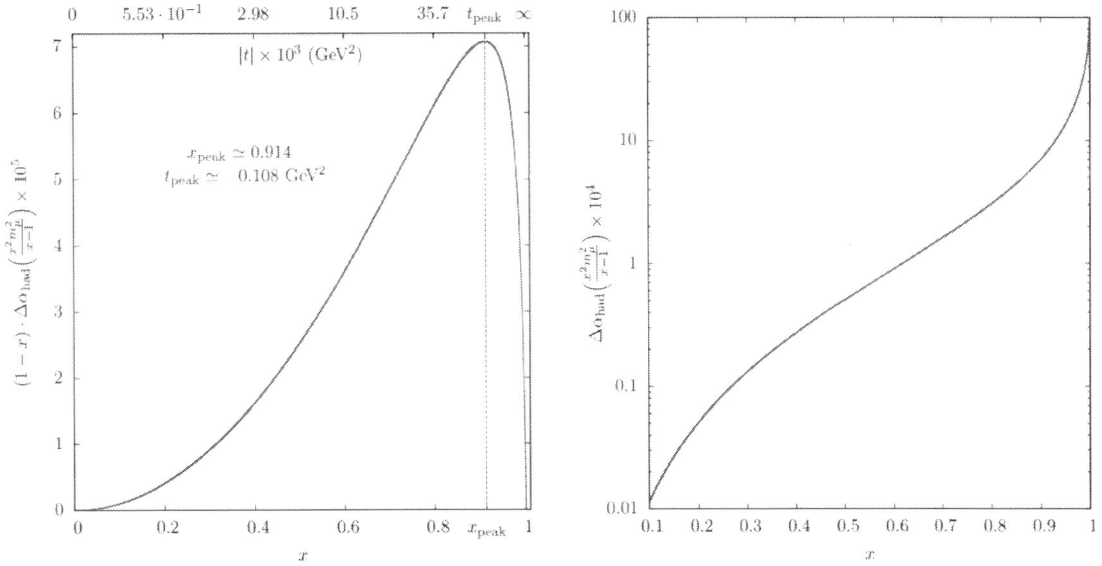

Figure 1. Left: The integrand $(1-x)\Delta\alpha_{\rm had}[t(x)] \times 10^5$ as a function of x and t. Right: $\Delta\alpha_{\rm had}[t(x)] \times 10^4$.

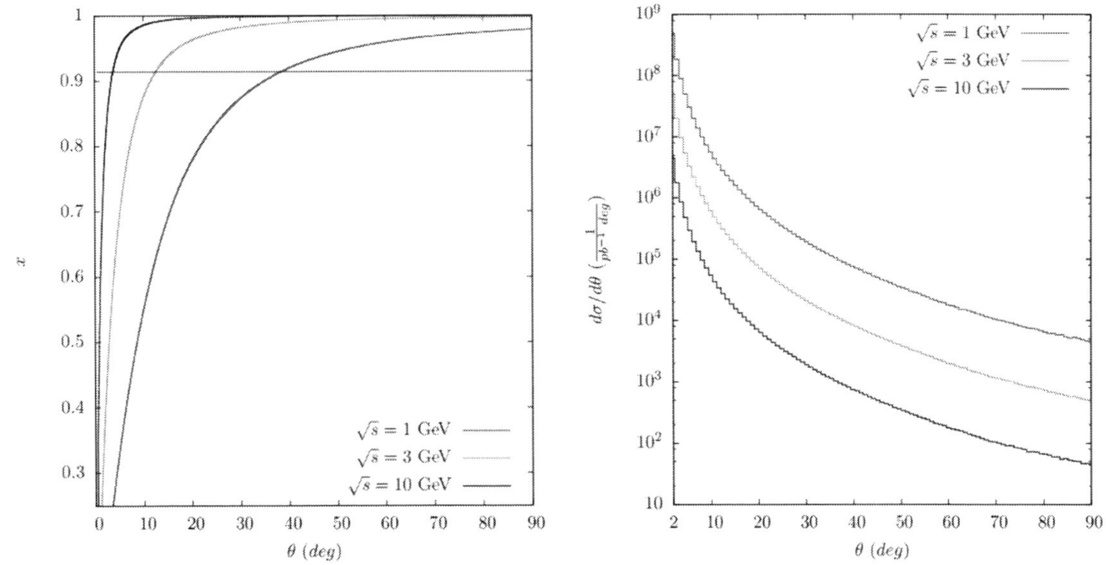

Figure 2. Left: Ranges of x values as a function of the electron scattering angle θ for three different center-of-mass energies. The horizontal line corresponds to $x = x_{\rm peak} \simeq 0.914$. Right: Bhabha differential cross section obtained with BabaYaga [28] as a function of θ for the same three values of $\sqrt{s}$ in the angular range $2° < \theta < 90°$.

3. for each bin j of the t distribution, compare the experimental differential cross section with the MC predictions and choose the i_j which minimizes the difference;

4. $\bar{\alpha}(t_j) + \frac{i_j}{N}\delta(t_j)$ will be the extracted value of $\alpha(t_j)$ from data in the jth bin. $\Delta\alpha_{\rm had}(t_j)$ can then be obtained through the relation between $\alpha(t)$ and $\Delta\alpha_{\rm had}(t)$.

We finally find, for each bin j of the t distribution,

$$\left.\frac{d\sigma}{dt}\right|_{j,\rm data} = \left.\frac{d\sigma}{dt}\Big(\bar{\alpha}(t) + \frac{i_j}{N}\delta(t), \alpha(s)\Big)\right|_{j,\rm MC}. \qquad (14)$$

We remark that the algorithm does not assume any simple dependence of the cross section on $\alpha(t)$, which can in fact be general, mixing s, t channels and higher order radiative corrections, relevant (or not) in different t domains.

In order to test our procedure, we perform a pseudo-experiment: we generate pseudo-data using the parameterization $\Delta\alpha^{I}_{\rm had}(t)$ of refs. [18, 33] and check if we can recover it by inserting in the MC the (independent) parameterization $\Delta\alpha^{II}_{\rm had}(t)$ (corresponding to $\bar{\alpha}(t)$ of eq. 14) of ref. [30] by means of the method described above. For this exercise, we use the generator BabaYaga in its most complete setup, generating events at $\sqrt{s} = 1.02$ GeV, requiring $10° < \theta_\pm < 170°$, $E_\pm > 0.4$ GeV and an acollinearity cut of 15°. We choose $\delta(t)$ to be the error induced on $\alpha(t)$ by the 1-σ error on $\Delta\alpha_{\rm had}(t)$, which is returned by the routine of ref. [30], we set $N = 150$, and we produce distributions with 200 bins. We note that in the present exercise $\alpha(s)$ and all the radiative corrections both in the pseudo-data and in the MC samples are exactly the same, because we are interested in testing the algorithm rather than assessing the achievable accuracy, at least at this stage.

In fig. 3, $\Delta\alpha^{\rm extr}_{\rm had}$ is the result extracted with our algorithm, corresponding to the minimizing set of i_j: the figure shows that our method is capable of recovering the underlying function $\Delta\alpha_{\rm had}(t)$ inserted into the "data". As the difference between $\Delta\alpha^{I}_{\rm had}$ and $\Delta\alpha^{\rm extr}_{\rm had}$ is hardly visible on an absolute scale, in fig. 3 all the functions have been divided by $\Delta\alpha^{II}_{\rm had}$ to display better the comparison between $\Delta\alpha^{I}_{\rm had}$ and $\Delta\alpha^{\rm extr}_{\rm had}$.

In order to assess the achievable accuracy on $\Delta\alpha_{\rm had}(t)$ with the proposed method, we remark that the LO contribution to the cross section is quadratic in $\alpha(t)$, thus we have

$$\frac{1}{2}\frac{\delta\sigma}{\sigma} \simeq \frac{\delta\alpha}{\alpha} \simeq \delta\Delta\alpha_{\rm had} \qquad (15)$$

Equation (15) relates the *absolute* error on $\Delta\alpha_{\rm had}$ with the *relative* error on the Bhabha cross section. Let us stress here that:

1. From the theoretical point of view, the present accuracy of the MC predictions [29] is at the level of about 0.05%, which implies that the precision that our method can, at best, set on $\Delta\alpha_{\rm had}(t)$ is $\delta\Delta\alpha_{\rm had}(t) \simeq 2\cdot 10^{-4}$. Any further improvement requires the inclusion of the NNLO QED corrections into the MC codes (see Ref. [29]).

2. Experimentally a measurement of $a_\mu^{\rm HLO}$ from space-like data competitive with the current time-like evaluations would require an $O(1\%)$ accuracy. Statistical considerations show that a 3% fractional accuracy on the $a_\mu^{\rm HLO}$ integral can be obtained by sampling the integrand $(1-x)\Delta\alpha_{\rm had}[t(x)]$ in ~ 10 points around the x peak with a fractional accuracy of 10%. Given the value of $O(10^{-3})$ for $\Delta\alpha_{\rm had}$ at $x = x_{\rm peak}$, this implies that the cross section must be known with relative accuracy of $\sim 2\times 10^{-4}$. Such a statistical accuracy, although challenging, can be obtained at flavor factories, as shown in fig. 2 (right). With an integrated luminosity of $O(1)$, $O(10)$, $O(100)$ fb^{-1} at $\sqrt{s} = 1$, 3 and 10 GeV, respectively, the angular region of interest can be covered with a 0.01% accuracy per degree. The experimental systematic error must match the same level of accuracy.

A source of experimental systematic errors comes from the machine luminosity, which can be normalized by calculating a theoretical cross section in principle not depending on $\Delta\alpha_{\rm had}$. We devise two possible options for the normalization process:

1. To use the $e^+e^- \to \gamma\gamma$ process, which has no dependence on $\Delta\alpha_{\rm had}$, at least up to NNLO order;

2. To use the Bhabha process at $t \sim 10^{-3}$ GeV^2 ($x \sim 0.3$), where the dependence on $\Delta\alpha_{\rm had}$ is of $O(10^{-5})$ and can be safely neglected.

It is worth quoting that a detailed analysis of the systematic errors involved in the measurement of the luminosity has been carried out at LEP by the OPAL collaboration reaching the final accuracy of $O(10^{-4})$ [15, 34].

3 Conclusions

We presented a novel approach to determine the leading hadronic correction to the muon g-2 based on measurements of the running of $\alpha(t)$ in the space-like region from Bhabha scattering data.

This approach, even if challenging, may become feasible by using data collected at present flavor factories as well as those at future high-energy e^+e^- colliders.

The proposed determination can become competitive with the accuracy of the present results obtained with the dispersive approach via time-like data.

An alternative formula for a_μ^{HLO} in the space-like region has been studied in detail. It emphasizes low values of $t \leq 1 GeV^2$ which can be explored at low energies machines.

We have also argued that this requires a measurement of the Bhabha cross section, at relatively small angles, with an accuracy of the order 10^{-4} or better.

Reaching such an accuracy demands a dedicated experimental and theoretical work for the next few years.

References

[1] C. Carloni Calame, M. Passera, G. Venanzoni, L. Trentadue, Phys. Lett. B **746** (2015) 325-329.

[2] G.W. Bennett *et al.* [Muon g-2 Collaboration], Phys. Rev. D **73** (2006) 072003.

[3] F. Jegerlehner, A. Nyffeler, Phys. Rept. **477** (2009) 1.

[4] F. Jegerlehner, these proceedings.

[5] T. Blum *et al.*, arXiv:1311.2198 [hep-ph]; K. Melnikov, A. Vainshtein, Springer Tracts Mod. Phys. **216** (2006) 1; M. Davier, W.J. Marciano, Ann. Rev. Nucl. Part. Sci. **54** (2004) 115; M. Passera, J. Phys. G **31** (2005) R75; M. Knecht, Lect. Notes Phys. **629** (2004) 37.

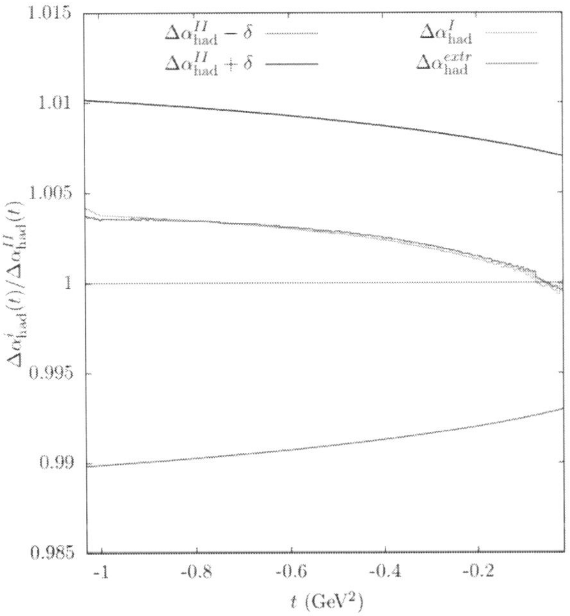

Figure 3. The extracted function $\Delta\alpha^{\rm extr}_{\rm had}(t)$ compared to the function $\Delta\alpha^{I}_{\rm had}(t)$ used in the pseudo-data (see text). The functions $\Delta\alpha^{II}_{\rm had}(t) \pm \delta(t)$ are shown to display the range spanned by the MC samples. All functions have been divided by $\Delta\alpha^{II}_{\rm had}(t)$. The tiny difference between $\Delta\alpha^{I}_{\rm had}$ and $\Delta\alpha^{\rm extr}_{\rm had}$ is due to the binning discretization.

[6] F. Jegerlehner, "The anomalous magnetic moment of the muon," Springer Tracts Mod. Phys. **226**, 2008.

[7] J. Grange *et al.* [Muon g-2 Collaboration], arXiv:1501.06858 [physics.ins-det].

[8] G. Venanzoni [Muon g-2 Collaboration], arXiv:1411.2555 [physics.ins-det].

[9] N. Saito [J-PARC g-2/EDM Collaboration], AIP Conf. Proc. **1467** (2012) 45.

[10] G. Venanzoni, Nuovo Cim. C **037** (2014) 02, 165; G. Venanzoni, Frascati Phys. Ser. **54** (2012) 52.

[11] G.V. Fedotovich [CMD-2 Collaboration], Nucl. Phys. Proc. Suppl. **181-182** (2008) 146.

[12] G. Colangelo, M. Hoferichter, A. Nyffeler, M. Passera, P. Stoffer, Phys. Lett. B **735** (2014) 90.

[13] C. Bouchiat, L. Michel, J. Phys. Radium 22 (1961) 121; L. Durand, Phys. Rev. **128** (1962) 441 [Erratum-ibid. **129** (1963) 2835]; M. Gourdin, E. De Rafael, Nucl. Phys. B **10** (1969) 667.

[14] A.B. Arbuzov, D. Haidt, C. Matteuzzi, M. Paganoni and L. Trentadue, Eur. Phys. J. C **34** (2004) 267.

[15] G. Abbiendi *et al.* [OPAL Collaboration], Eur. Phys. J. C **45** (2006) 1.

[16] A.B. Arbuzov, V.S. Fadin, E.A. Kuraev, L.N. Lipatov, N.P. Merenkov, L. Trentadue, Nucl. Phys. B**485** (1997) 457-502.

[17] M. Davier, A. Hoecker, B. Malaescu, Z. Zhang, Eur. Phys. J. C **71** (2011) 1515 [Erratum-ibid. C **72** (2012) 1874].

[18] K. Hagiwara, R. Liao, A.D. Martin, D. Nomura, T. Teubner, J. Phys. G **38** (2011) 085003.

[19] R.V. Harlander, M. Steinhauser, Comput. Phys. Commun. **153** (2003) 244.

[20] B.E. Lautrup, A. Peterman, E. de Rafael, Phys. Rept. **3** (1972) 193.

[21] C. Aubin, T. Blum, Phys. Rev. D **75** (2007) 114502; P. Boyle, L. Del Debbio, E. Kerrane, J. Zanotti, Phys. Rev. D **85** (2012) 074504; X. Feng, K. Jansen, M. Petschlies, D.B. Renner, Phys. Rev. Lett. **107** (2011) 081802; M. Della Morte, B. Jager, A. Juttner, H. Wittig, JHEP **1203** (2012) 055.

[22] T. Blum, M. Hayakawa, T. Izubuchi, PoS LATTICE **2012** (2012) 022.

[23] M. Steinhauser, Phys. Lett. B **429** (1998) 158.

[24] P.A. Baikov, K.G. Chetyrkin, J.H. Kuhn and C. Sturm, Nucl. Phys. B **867** (2013) 182; C. Sturm, Nucl. Phys. B **874** (2013) 698; P.A. Baikov, A. Maier, P. Marquard, Nucl. Phys. B **877** (2013) 647.

[25] F. Jegerlehner, in Proceedings of "Fifty years of electroweak physics: a symposium in honour of Professor Alberto Sirlin's 70th birthday", New York University, 27-28 October 2000, J. Phys. G **29** (2003) 101.

[26] S.L. Adler, Phys. Rev. D **10** (1974) 3714.

[27] S. Eidelman, F. Jegerlehner, A. L. Kataev, O. Veretin, Phys. Lett. B **454** (1999) 369.

[28] G. Balossini, C.M. Carloni Calame, G. Montagna, O. Nicrosini, F. Piccinini, Nucl. Phys. B **758** (2006) 227.

[29] S. Actis *et al.*, Eur. Phys. J. C **66** (2010) 585.

[30] S. Eidelman, F. Jegerlehner, Z. Phys. C **67** (1995) 585; F. Jegerlehner, Nucl. Phys. Proc. Suppl. **181-182** (2008) 135.

[31] M. Bicer *et al.* [TLEP Design Study Working Group Collaboration], JHEP **1401** (2014) 164 [arXiv:1308.6176 [hep-ex]].

[32] G. Aarons *et al.* [ILC Collaboration], Physics at the ILC, arXiv:0709.1893 [hep-ph].

[33] K. Hagiwara, A.D. Martin, D. Nomura, T. Teubner, Phys. Lett. B **649** (2007) 173; Phys. Rev. D **69** (2004) 093003.

[34] G. Abbiendi et al. [OPAL collaboration], Eur. Phys. J. C**14** (2000) 373-425.

EPJ Web of Conferences **118**, 01036 (2016) DOI: 10.1051/epjconf/201611801036
FCCP2015

Review of Recent Calculations of the Hadronic Vacuum Polarisation Contribution

Zhiqing Zhang[1,a]

[1] *Laboratoire de l'Accélérateur Linéaire, Univ. Paris-Sud, CNRS/IN2P3, Université Paris-Saclay, Orsay, France*

Abstract. Recent calculations of the hadronic vacuum polarisation contribution are reviewed. The focus is put on the leading-order contribution to the muon magnetic anomaly involving e^+e^- annihilation cross section data as input to a dispersion relation approach. Alternative calculation including tau data is also discussed. The τ data are corrected for various isospin-breaking sources which are explicitly shown source by source.

1 Introduction

The hadronic vacuum polarisation (HVP) contribution to the muon magnetic anomaly $a_\mu \equiv (g-2)/2$ may be decomposed into three parts as $a_\mu^{\rm Had} = a_\mu^{\rm Had,LO} + a_\mu^{\rm Had,HO} + a_\mu^{\rm had,LBL}$ corresponding to the leading-order (LO), higher-order (HO) and light-by-light (LBL) scattering contribution, respectively. The corresponding representative numerical values are 692.3 ± 4.2 [1], -9.79 ± 0.09 [2] and 10.5 ± 2.6 [3], in units of 10^{-10} (the same units will implicitly be used for all following quoted a_μ numbers). Therefore the LO term is the dominant hadronic contribution and has the largest uncertainty not only among the three hadronic terms but also among all the electromagnetic, weak and hadronic sectors. This is why the improvement of the uncertainty of the LO HVP has been one of the main research activities by a number of groups and individuals since several decades (Fig. 1).

The HVP involving strongly interacting particles can be computed at large energy scales but not at low scales due to the non-perturbative nature of QCD at large distance. It is possible to overcome this problem by means of a dispersion relation technique involving experimental data on the cross section for e^+e^- annihilation into hadrons. This will be the focus of this writeup. For alternative evaluations such as model dependent or Lattice QCD based calculations, we refer to [5] and [6]. However the precision of the current Lattice calculation is still far from competitive with that of the dispersion relation approach.

This writeup is organised as follows. In Sec. 2, the state-of-the-art techniques used for the e^+e^- based $a_\mu^{\rm Had,LO}$ calculation are reviewed. In Sec. 3, we discuss an alternative evaluation by including τ data and taking into account the known isospin-breaking corrections, followed by a summary in Sec. 4.

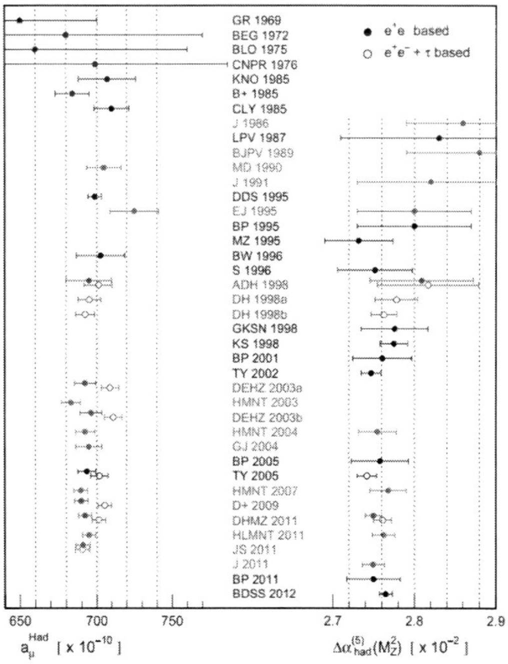

Figure 1. Summary of the evolution as function of time for $a_\mu^{\rm Had}$ and $\Delta\alpha^{(5)}_{\rm had}(M_Z^2)$ (figure taken from [4]).

2 Review of the e^+e^- based $a_\mu^{\rm Had}$ calculations

The LO HVP contribution to $a_\mu^{\rm had}$ is calculated using the dispersion relation [7] as

$$a_\mu^{\rm Had,LO} = \frac{1}{3}\left(\frac{\alpha}{\pi}\right)^2 \int_{m_\pi^2}^{\infty} ds \frac{K(s)}{s} R^{(0)}(s), \qquad (1)$$

[a] e-mail: zhang@lal.in2p3.fr

© The Authors, published by EDP Sciences. This is an open access article distributed under the terms of the Creative Commons Attribution License 4.0 (http://creativecommons.org/licenses/by/4.0/).

DOI: 10.1051/epjconf/201611801036

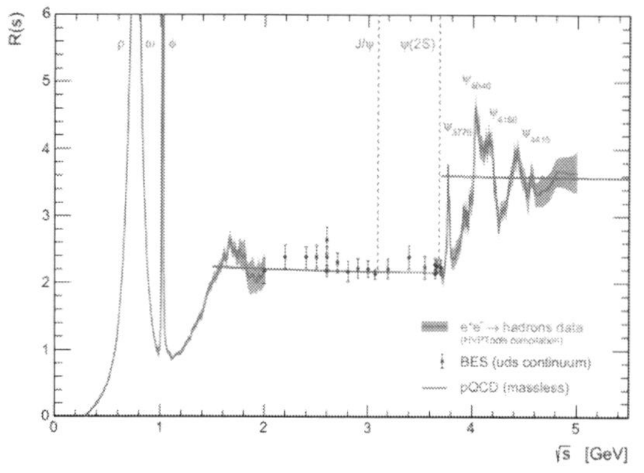

Figure 2. Cross section for the process $e^+e^- \to$ hadrons versus centre-of-mass energy $\sqrt{s}$. The blue band represents the combined experimental measurements with their uncertainty. The red line shows the perturbative QCD prediction, the data points show the inclusive measurements from the BES experiment [9] (figure taken from [4]).

where $K(s)$ is a QED kernel function [8] and $R^{(0)}(s)$ represents the ratio of the bare cross section for e^+e^- annihilation into hadrons to the point-like muon-pair cross section for a given centre-of-mass energy $\sqrt{s}$. The function $K(s) \sim 1/s$ gives a strong weight to the low energy part of the integral. Therefore, $a_\mu^{\rm Had,LO}$ is dominated by the $\rho(770) \to 2\pi$ resonance.

The computation of the dispersion integral requires the knowledge of $R(s)$ at any scale s. Figure 2 shows an overview of the input data used in the evaluation of [1]. In the low energy range between the $\pi^0\gamma$ threshold and 1.8 GeV, the evaluation uses about 22 exclusive channels that are measured, and a few other channels that are (partially) estimated using isospin relations. In the continuum region between 1.8 GeV and 3.7 GeV, the prediction based on four-loop perturbative QCD calculation is used, which is in good agreement with the results from direct inclusive measurements by BES [9]. The same QCD prediction is also used at higher energy scale above 5 GeV. In the region just above the heavy quark thresholds and below 5 GeV, the evaluation uses also e^+e^- annihilation data to hadrons.

The evaluation of [10] uses a similar strategy. But there are a number of differences. For instance, the perturbative QCD predictions are only used between 2.6 and 3.73 GeV and above 11.09 GeV. The data treatment is also different. The evaluation of [1] is based on HVPTools [11] whereas that of [10] using a clustering technique. A simplified comparison is shown in Table 1.

It is instructive to look at Table 2 (Table 4 from [10]). The difference between the two evaluations are often larger than or comparable with the quoted uncertainties. There are also important difference on the quoted uncertainties. These differences could be a reflection of the different data and uncertainty treatment mentioned earlier. The use of different inputs in particular the old data sets could be another source though these old data sets are often imprecise and therefore their relative weight in the combination should be in general small. It is important that these differences could be better understood and reduced.

Table 2 confirms that the $\pi^+\pi^-$ channel is by far the dominant one both in the central value and in the uncertainty. The corresponding inputs measured by different experiments are shown in Fig. 3. The inconsistency between Babar and KLOE in particular is clearly visible. Such inconsistency results in sizeable $\chi^2/n_{\rm dof}$ values (Fig. 4 (left)) preventing currently the expected precision improvement in the combination [11]. The relative weight of each experiment is shown in Fig. 4 (right).

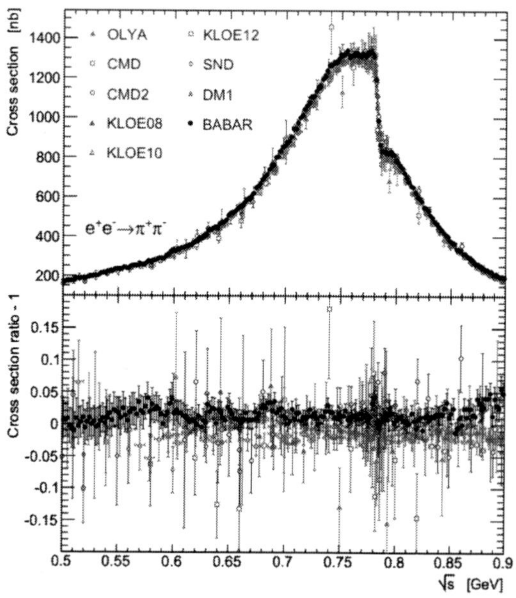

Figure 3. Cross section for the process $e^+e^- \to \pi^+\pi^-$ measured by different experiments (top) and their relative difference with respect to the combined cross section (bottom). The uncertainty bars contain both the statistical and systematic components added in quadrature (figure taken from [4]).

Table 2 also shows that $\pi^+\pi^-2\pi^0$, $\pi^+\pi^-\pi^0$, K^+K^- and $2\pi^+2\pi^-$ are important subleading channels contributing to the uncertainty budget. The corresponding measurements are shown in Fig. 5. For the first channel, the Babar measurements dominate, which are, however, still preliminary and the final results are expected soon and hopefully will improve the precision of the channel. For the $\pi^+\pi^-\pi^0$, the measurements in the ω and ϕ resonances are from the energy-scan experiments whereas the measurements above the resonances come almost exclusively from Babar. There is clearly room for improvement for this channel. The most precise measurements for the ϕ resonance in the K^+K^- channel used to be from CMD-2 experiment. A new measurement from Babar has recently been published [12], which contributes to a_μ with 22.93 ± 0.28 in the energy range from threshold to 1.8 GeV, to be compared with the values in Table 2. There are many measure-

Table 1. Comparison of data interpolation and combination between [11] and [10] where ISR, VP, FSR and n_{dof} stand for initial state radiation, vacuum polarisation, final state radiation correction and the number of degrees of freedom, respectively.

	HVPTools [11]	Clustering method [10]
Common points	Use bare cross sections (remove ISR, VP and add FSR to some early data) Combine different experiments in a same channel before integration	
Data interpolation	Using 2nd order polynomials	Trapezoidal rule (linear interpolation)
Normalisation	Pseudo-MC generation fluctuates a data point along the original measurement taking into account correlation	Use directly measured data allowing the normalisation to float within quoted uncertainty
Bin size	Using small bins (1 MeV) covered by the data point	Using varying bin/cluster size depending on data density
Average	Weighted average and covariance matrix calculated in the small bins	Average obtained with non-linear χ^2 minimisation
Error inflation	Error scaled locally in a bin with $\sqrt{\chi^2/n_{\rm dof}}$ if it is greater than 1	

Table 2. Contributions to a_μ in the energy region from 0.305 to 1.8 GeV from exclusive channels (Table 4 from [10]).

Channel	HLMNT (11) [10]	DHMZ (10) [1]	Difference
$\eta\pi^+\pi^-$	0.88 ± 0.10	1.15 ± 0.19	−0.27
K^+K^-	22.09 ± 0.46	21.63 ± 0.73	0.46
$K_S^0K_L^0$	13.32 ± 0.16	12.96 ± 0.39	0.36
$\omega\pi^0$	0.76 ± 0.03	0.89 ± 0.07	−0.13
$\pi^+\pi^-$	505.65 ± 3.09	507.80 ± 2.84	−2.15
$2\pi^+2\pi^-$	13.50 ± 0.44	13.35 ± 0.53	0.15
$3\pi^+3\pi^-$	0.11 ± 0.01	0.12 ± 0.01	−0.01
$\pi^+\pi^-\pi^0$	47.38 ± 0.99	46.00 ± 1.48	1.38
$\pi^+\pi^-2\pi^0$	18.62 ± 1.15	18.01 ± 1.24	0.61
$\pi^0\gamma$	4.54 ± 0.14	4.42 ± 0.19	0.12
$\eta\gamma$	0.69 ± 0.02	0.64 ± 0.02	0.05
$\eta 2\pi^+2\pi^-$	0.02 ± 0.00	0.02 ± 0.01	0.00
$\eta\omega$	0.38 ± 0.06	0.47 ± 0.06	−0.09
$\eta\phi$	0.33 ± 0.03	0.36 ± 0.03	−0.03
$\phi(\to$ unaccounted)	0.04 ± 0.04	0.05 ± 0.00	−0.01
Sum of isospin channels	5.98 ± 0.42	6.06 ± 0.46	−0.08
Total	634.28 ± 3.53	633.93 ± 3.61	0.35

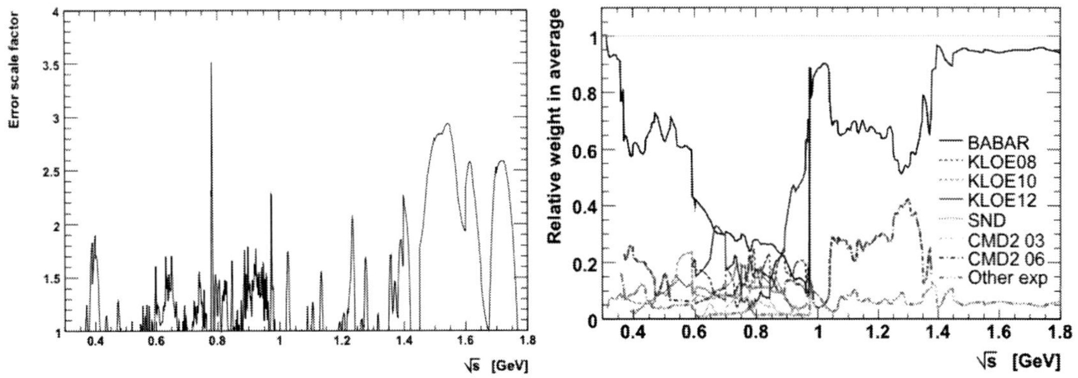

Figure 4. Rescaling factor accounting for inconsistencies among experiments (left, Fig. 1 from [11]) and relative local averaging weight per experiment (right, Fig. 3 from [1]) versus $\sqrt{s}$ in $e^+e^- \to \pi^+\pi^-$.

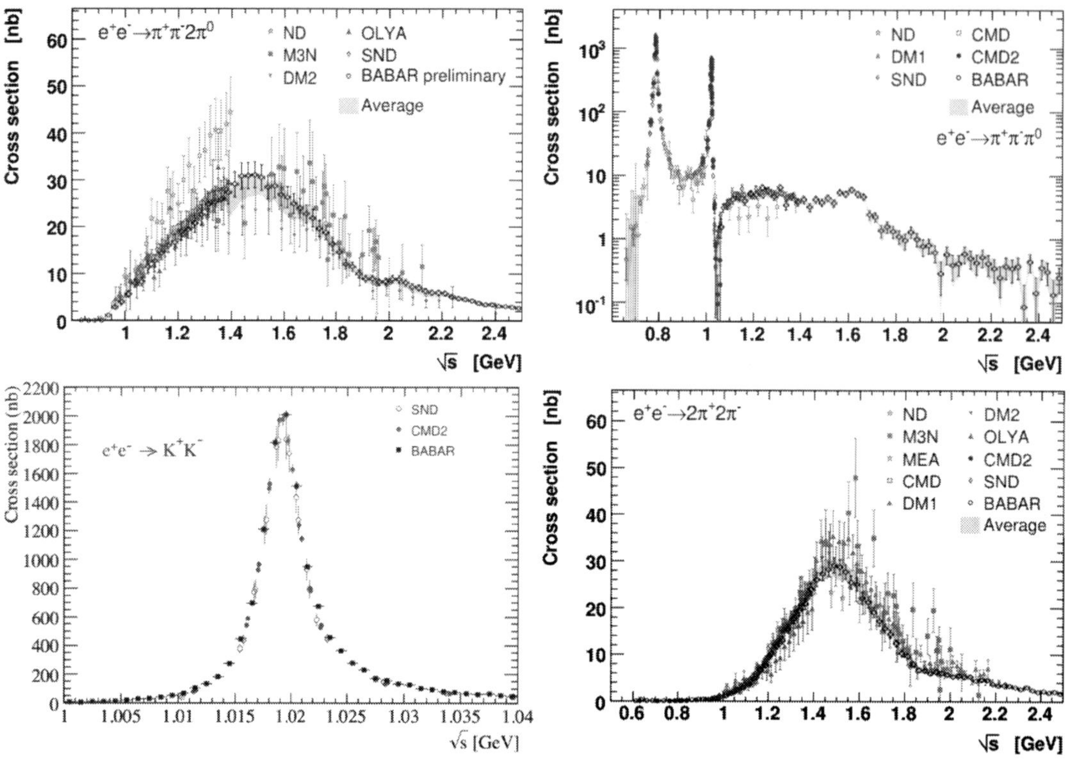

Figure 5. Cross section inputs for the subleading channels $\pi^+\pi^-2\pi^0$ (top left), $\pi^+\pi^-\pi^0$ (top right), K^+K^- (bottom left) and $2\pi^+2\pi^-$ (bottom right). The figures with the green error bands are taken from [1] and the figure for the $K^+K^-(\gamma)$ channel from [12]. The error bands correspond to the 1σ combination uncertainty obtained with the HVPTools.

ments for the $2\pi^+2\pi^-$ channel with the new Babar measurements [13] dominating in the precision of the combination. The updated result is 13.64 ± 0.36, which improves the numbers quoted in Table 2.

3 Alternative calculation of a_μ^{Had}

In Fig. 1, some of the predictions are $e^+e^- + \tau$ based. The use of tau data of semi-leptonic τ decays in the evaluation of a_μ^{Had} and $\Delta\alpha_{\mathrm{had}}^{(5)}$ was originally proposed in Ref. [14]. It is based on the fact that in the limits of isospin invariance, the spectral function of the vector current decay $\tau^- \to X^-\nu_\tau$ is related to the $e^+e^- \to X^0$ cross section of the corresponding isovector final state X^0 (so-called the conserved vector current (CVC) relation),

$$\sigma_{X^0}^{I=1}(s) = \frac{4\pi\alpha^2}{s} v_{1,X^-}(s), \qquad (2)$$

where s is the centre-of-mass energy-squared or equivalently the invariant mass-squared of the τ final state X, α is the electromagnetic fine structure constant, and v_{1,X^-} is the non-strange, isospin-one vector spectral function given by

$$v_{1,X^-}(s) = \frac{m_\tau^2}{6|V_{ud}|^2} \frac{\mathcal{B}_{X^-}}{\mathcal{B}_e} \frac{1}{N_X} \frac{dN_X}{ds} \times \left(1 - \frac{s}{m_\tau^2}\right)^2 \left(1 + \frac{2s}{m_\tau^2}\right)^{-1} \frac{R_{\mathrm{IB}}(s)}{S_{\mathrm{EW}}}, \qquad (3)$$

with

$$R_{\mathrm{IB}}(s) = \frac{\mathrm{FSR}(s)}{G_{\mathrm{EM}}(s)} \frac{\beta_0^3(s)}{\beta_-^3(s)} \left| \frac{F_0(s)}{F_-(s)} \right|^2 . \qquad (4)$$

In Eq. (3), m_τ is the τ mass, $|V_{ud}|$ the CKM matrix element, $\mathcal{B}_{X^-}$ and $\mathcal{B}_e$ are the the branching fractions of $\tau^- \to X^-(\gamma)\nu_\tau$ (final state photon radiation is implied for τ branching fractions) and of $\tau^- \to e^-\bar{\nu}_e\nu_\tau$, $(1/N_X)dN_X/ds$ is the normalised invariant mass spectrum of the hadronic final state, and R_{IB} represents all the s-dependent isospin-breaking (IB) corrections and S_{EW} corresponds to short-distance electroweak radiative effects [15].

The first term in Eq. (4) is the ratio $\mathrm{FSR}(s)/G_{\mathrm{EM}}(s)$, where $\mathrm{FSR}(s)$ refers to the final state radiative corrections [16] in the $\pi^+\pi^-$ channel, and $G_{\mathrm{EM}}(s)$ denotes the long-distance radiative corrections of order α to the photon-inclusive $\tau^- \to \pi^-\pi^0\nu_\tau$ spectrum [15]. The energy dependent corrections used in [15] are compared in Fig. 6 with those by [17]. The smaller correction of G_{EM} in [15] is due to the exclusion of the contributions involving the $\rho\omega\pi$ vertex.

The second correction term in Eq. (4), $\beta_0^3(s)/\beta_-^3(s)$, arises from the $\pi^\pm - \pi^0$ mass splitting and is important only close the threshold (see Fig. 6).

The third IB correction term involves the ratio of the electromagnetic to weak form factors $|F_0(s)/F_-(s)|^2$ and is the most delicate one. Below 1 GeV, the pion form fac-

 DOI: 10.1051/epjconf/201611801036

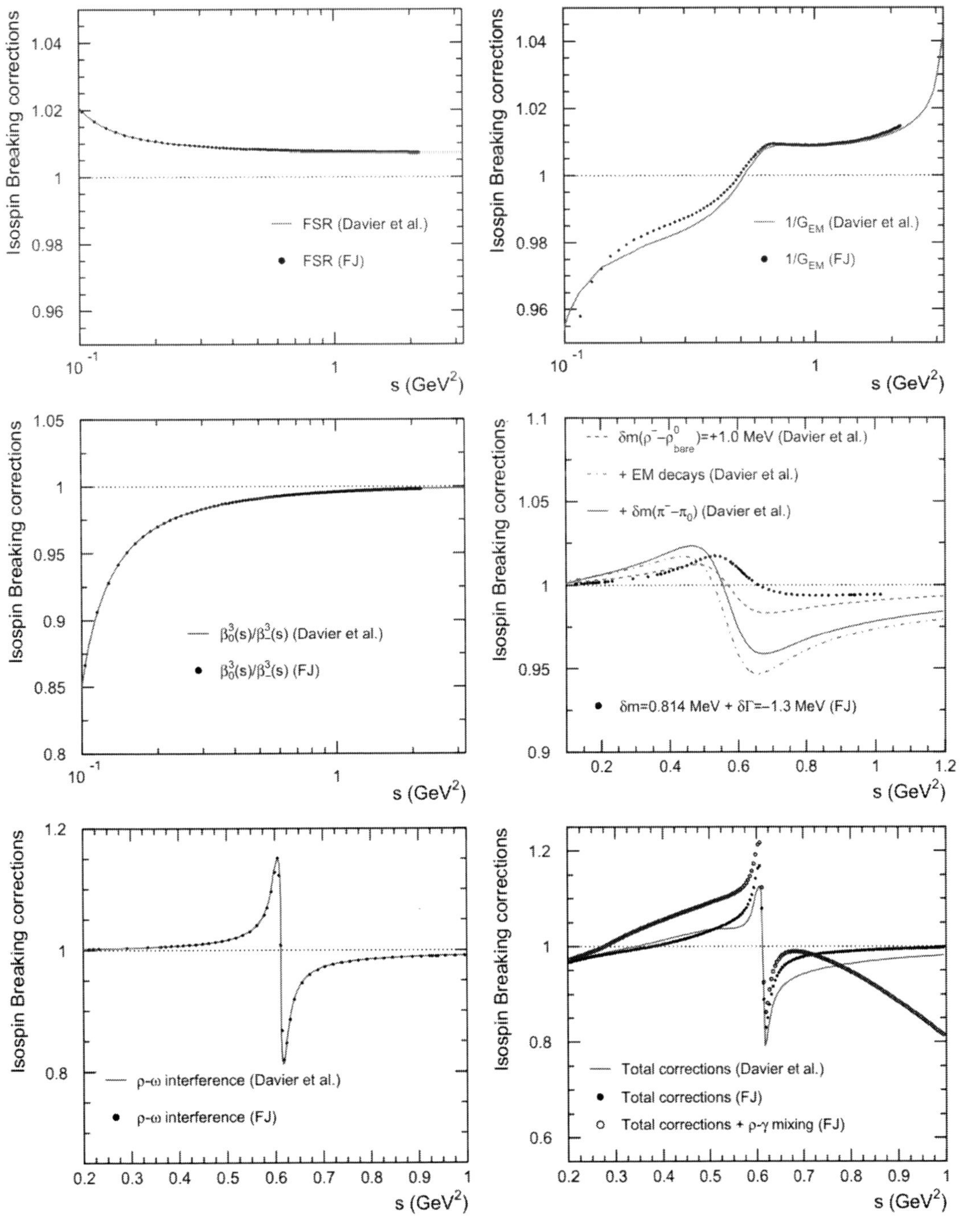

Figure 6. Comparison of isospin-breaking corrections versus s used by Davier et al. [15] and by JS [17]. The different plots correspond to FSR (top left), $1/G_{\rm EM}$ (top right), β_0^3/β_-^3 ratio term (middle left), the effect of the ρ mass and width difference in the $|F_0/F_-|^2$ term (middle right), the effect of the $\rho-\omega$ interference in the $|F_0/F_-|^2$ term (bottom left) and the total corrections (bottom right). The difference between the open blue points and the solid black one in the last plot stems from the $\rho-\gamma$ mixing corrections proposed in [18].

DOI: 10.1051/epjconf/201611801036

tors are dominated by the ρ meson resonance, such that IB effects mainly stem from the mass and width differences between the $\rho^{\pm}$ and ρ^0 mesons, and from $\rho^0 - \omega$ mixing. The difference between the corrections used in [15] and those of [17, 18] is mainly due to different width differences considered. The width difference $\delta\Gamma_\rho = \Gamma_{\rho^0} - \Gamma_{\rho^-}$ used in [15] was based on [19]

$$\delta\Gamma_\rho(s) = \frac{g^2_{\rho\pi\pi}\sqrt{s}}{48\pi}\left[\beta_0^3(s)(1+\delta_0) - \beta_-^3(s)(1+\delta_-)\right], \quad (5)$$

where $g_{\rho\pi\pi}$ is the strong coupling of the isospin-invariant $\rho\pi\pi$ vertex and $\delta_{0,-}$ denote radiative corrections for photon-inclusive $\rho \to \pi\pi$ decays, which include $\rho \to \pi\pi\gamma$. Contrary the expression

$$\delta\Gamma_\rho = \frac{g^2_{\rho\pi\pi}}{48\pi}\left(\beta_0^3 M_{\rho^0} - \beta_-^3 M_{\rho^-}\right), \quad (6)$$

has been used in [18]. The numerical values of Eqs. (5) at $M_\rho = 775$ MeV and (6) are +0.76 MeV and −1.3 MeV, respectively. Another small difference which contributes to the IB difference is from $\delta M_\rho = M_{\rho^-} - M_{\rho^0}$ of 1.0 ± 0.9 MeV [15] and 0.814 MeV [18].

The effects of the IB corrections applied to $a_\mu^{\rm Had,LO}$ using τ data in the dominant $\pi\pi$ channel is shown in Table 3 [15] for the energy range between the 2π mass threshold and 1.8 GeV. The first source corresponds to the effect of $S_{\rm EW} = 1.0235 \pm 0.003$ [15]. The uncertainty of $G_{\rm EM}$ corresponds to the difference of two $G_{\rm EM}$ corrections shown in Fig. 6. The quoted 10% uncertainty on the FSR and $\pi\pi\gamma$ electromagnetic corrections is an estimate of the structure-dependent effects (pion form factor) in virtual corrections and of intermediate resonance contributions to real photon emission [15]. The systematic uncertainty assigned to the $\rho - \omega$ interference contribution accounts for the difference in $a_\mu^{\rm Had,LO}$ between two phenomenological fits, where the mass and width of the ω resonance are either left free to vary or fixed to their world average values. Some of the IB corrections depend on the form factor parametrisation used and the values quoted in Table 3 corresponds to those of Gounaris-Sakurai (GS) parametrisation [20] but the total uncertainty includes the full difference between the GS parametrisation and that of the Kühn-Santamaria (KS) parametrisation [15]. The total correction of −16.07 ± 1.85 is to be compared with the previous correction of −13.8 ± 2.4 [21] thus resulting a net change of −6.9.

In Table 3, the effects of the IB corrections to the CVC prediction of $\mathcal{B}_{\pi\pi^0}$ are also shown. The prediction for the branching fraction of a heavy lepton decaying into a *G*-parity even hadronic final state, X^-,

$$\mathcal{B}_X^{\rm CVC} = \frac{3}{2}\frac{\mathcal{B}_e|V_{ud}|^2}{\pi\alpha^2 m_\tau^2}\int_{s_{\rm min}}^{m_\tau^2} ds\, s\,\sigma^I_{X^0}(s) \times \left(1-\frac{s}{m_\tau^2}\right)^2\left(1+\frac{2s}{m_\tau^2}\right)\frac{S_{\rm EW}}{R_{\rm IB}(s)}, \quad (7)$$

is derived from the vector current using the CVC relation with $s_{\rm min}$ being the threshold of the invariant mass-squared

Table 3. Contributions to $a_\mu^{\rm Had,LO}[\pi\pi,\tau]$ ($\times 10^{-10}$) and $\mathcal{B}^{\rm CVC}_{\pi^-\pi^0}$ ($\times 10^{-2}$) from the isospin-breaking corrections. Corrections shown correspond to the Gounaris-Sakurai (GS) parametrisation [15]. The total uncertainty includes the difference with the Kühn-Santamaria (KS) parametrisation quoted as δ(GS − KS).

Source	$\Delta a_\mu^{\rm Had,LO}[\pi\pi,\tau]$	$\Delta\mathcal{B}^{\rm CVC}_{\pi^-\pi^0}$
$S_{\rm EW}$	−12.21 ± 0.15	+0.57 ± 0.01
$G_{\rm EM}$	−1.92 ± 0.90	−0.07 ± 0.17
FSR	+4.67 ± 0.47	−0.19 ± 0.02
ρ–ω interference	+2.80 ± 0.19	−0.01 ± 0.01
$m_{\pi^\pm} - m_{\pi^0}$ effect on σ	−7.88	+0.19
$m_{\pi^\pm} - m_{\pi^0}$ effect on Γ_ρ	+4.09	−0.22
$m_{\rho^\pm} - m_{\rho^0_{\rm bare}}$	$0.20^{+0.27}_{-0.19}$	+0.08 ± 0.08
$\pi\pi\gamma$, electrom. decays	−5.91 ± 0.59	+0.34 ± 0.03
δ(GS − KS)	−0.67	−0.03
Total	−16.07 ± 1.85	+0.69 ± 0.22

of the final state X^0 in e^+e^- annihilation. CVC comparisons of τ branching fractions are of special interest because they are essentially insensitive to the shape of the τ spectral function, hence avoiding experimental difficulties, such as the mass dependence of the π^0 detection efficiency and feed-through, and biases from the unfolding of the raw mass distributions from acceptance and resolution effects.

Despite the improved IB corrections, there is still a sizeable difference between the e^+e^- based prediction of 692.3 ± 4.2 and the τ based one of 703.0 ± 4.4 [22, 23]. The difference amounts to 10.7 ± 4.9 corresponding to a deviation of 2.2σ. The shape of the combined τ spectral function after the IB corrections is also found different from the one from e^+e^- data (Fig. 7). The discrepancy is further reflected in the τ branching fractions (Fig. 8). This used to be one of the open issues on the subject [22].

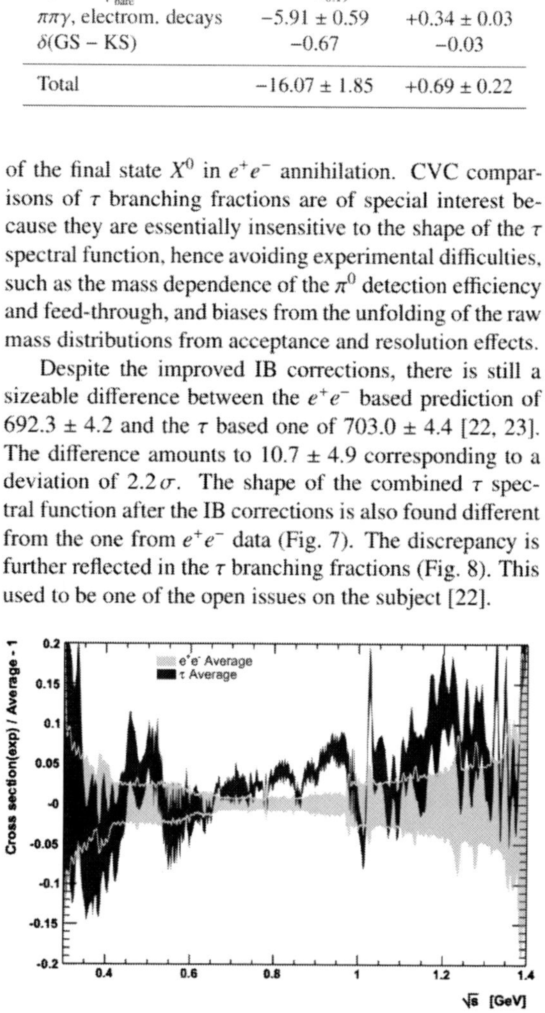

Figure 7. Relative comparison between the combined τ after the IB corrections and e^+e^- spectral functions (figure taken from [11]).

Recently, a model-dependent $\rho - \gamma$ mixing effect, which is absent in the τ data, was proposed in [18] to explain the $e^+e^- - \tau$ discrepancy. The proposed correction corresponds to the difference between the open blue points

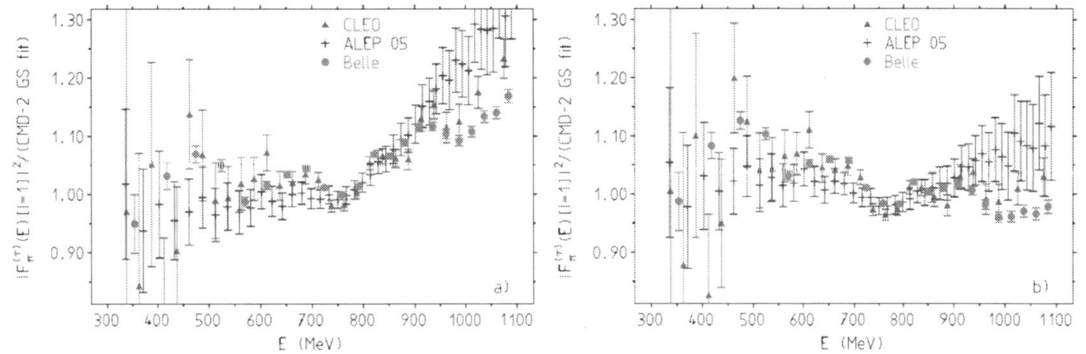

Figure 9. $|F_\pi(E)|^2$ ratio τ versus $e^+e^- I = 1$ (CMD-2 GS fit): left: uncorrected for $\rho - \gamma$ mixing; right: the same after correcting for it (figures taken from [18]).

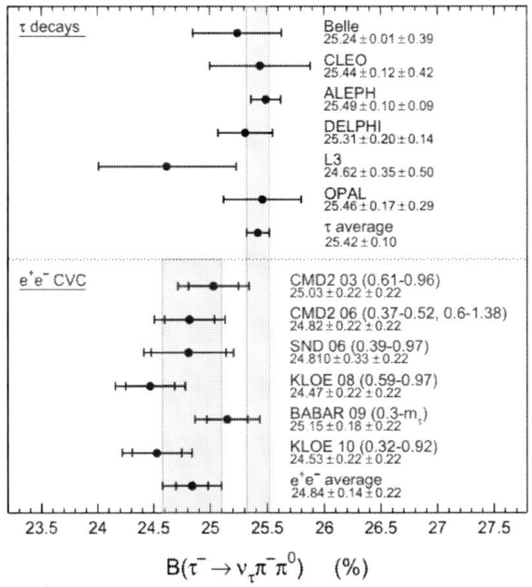

Figure 8. The measured branching fractions for $\tau^- \to \pi^-\pi^0\nu_\tau$ compared to the predictions from the $e^+e^- \to \pi^+\pi^-$ spectral functions, applying the isospin-breaking corrections. The long and short vertical error bands correspond to the τ and e^+e^- averages.

and the solid black points in Fig. 6 (bottom right). The effect of the correction is shown in Fig. 9. After the correction, the $e^+e^- - \tau$ difference above the rho peak looks indeed reduced, on the other hand, the agreement at the peak and below seems to become worse.

4 Summary

The evaluation of the hadronic vacuum polarisation contributions to the muon magnetic anomaly a_μ (and the running of the QED coupling $\alpha(s)$) has a long history. The precision has been steadily improving thanks to more precise and complete cross section measurements of e^+e^- annihilation into hadrons and of tau spectral functions on the one hand and the application of more advanced data interpolation and combination techniques in the dispersion relation approach on the other hand.

Some inconsistencies exist nevertheless among different e^+e^- data sets in particular between Babar and KLOE which limit the accuracy of the combined results. There is also a discrepancy between the e^+e^- data and the corresponding tau data after correcting for all known isospin-breaking effects. The $\rho - \gamma$ mixing effect is suggested to reduce the discrepancy. However, unlike for the analogous $\gamma - Z$ mixing, the correction here is model-dependent because of the ρ hadronic structure.

The prospect is however good as improved or final measurements from Babar are expected for both the dominant $\pi\pi$ channel and the few other significant processes. In addition new data will be collected by CMD-3 and SND-2 at VEPP-2000, BES III at BEPC2, KLOE-2 at DAΦNE, and Belle II at superKEK-B in the next years. The new data and measurements should allow to improve the current precision of the leading-order hadronic vacuum polarisation contribution by a factor of at least two. Together with the expected improvement on the experimental side from Fermilab and J-PARC, the future of the muon magnetic anomaly will be very exciting.

Acknowledgment

I am grateful for the fruitful collaboration with my colleagues and friends in particular M. Davier, A. Hoecker, G. Lopez Castro and B. Malaescu and I thank also F. Jegerlehner for useful discussions on the isospin-breaking corrections.

References

[1] M. Davier, A. Hoecker, B. Malaescu and Z. Zhang, Eur. Phys. J. C **71**, 1515 (2011) [Eur. Phys. J. C **72**, 1874 (2012)] [arXiv:1010.4180 [hep-ph]].

[2] K. Hagiwara, R. Liao, A. D. Martin, D. Nomura and T. Teubner, J. Phys. G **38**, 085003 (2011) [arXiv:1105.3149 [hep-ph]].

[3] J. Prades, E. de Rafael and A. Vainshtein, Adv. Ser. Direct. High Energy Phys. **20**, 303 (2009) [arXiv:0901.0306 [hep-ph]].

[4] M. Davier, A. Hoecker, B. Malaescu and Z. Zhang, *Standard Theory Essays in the 60th Anniversary of CERN* (World Scientific, 2015)

[5] M. Benayoun, proceedings of this workshop.

[6] M. Petschlies, proceedings of this workshop.

[7] C. Bouchiat and L. Michel, J. Phys. Radium **22**, 121 (1961).

[8] S.J. Brodsky and E. de Rafael, Phys. Rev. **168**, 1620 (1968).

[9] J. Bai *et al.*, Phys. Rev. Lett. **84** (2000) 594; Phys. Rev. Lett. **88**, 101802 (2002); M. Ablikim *et al.* Phys. Lett. **B641** (2006) 145; Phys. Lett. **B677**, 239 (2009).

[10] K. Hagiwara, R. Liao, A.D. Martin, D. Nomura and T. Teubner, J. Phys. **G38**, 085003 (2011).

[11] M. Davier, A. Hoecker, B. Malaescu, C. Yuan and Z. Zhang, Eur. Phys. J. C **66**, 1 (2010).

[12] J. P. Lees *et al.* [BaBar Collaboration], Phys. Rev. D **88**, 032013 (2013) [arXiv:1306.3600 [hep-ex]].

[13] J. P. Lees *et al.* [BaBar Collaboration], Phys. Rev. D **85**, 112009 (2012) [arXiv:1201.5677 [hep-ex]].

[14] R. Alemany, M. Davier, and A. Hoecker, Eur. Phys. J. C **2**, 123 (1998).

[15] M. Davier *et al.*, Eur. Phys. J. C **66**, 127 (2010).

[16] J.S. Schwinger, *Particles, Sources and Fields*, vol. **3** (Addison- Wesley, Reading, 1989). See also, M. Drees, K. Hikasa, Phys. Lett. B **252**, 127 (1990).

[17] F. Jegerlehner, private communication.

[18] F. Jegerlehner and R. Szafron, Eur. Phys. J. C **71**, 1632 (2011) [arXiv:1101.2872 [hep-ph]].

[19] F. Flores-Baez, G. Lopez Castro, G. Toledo Sanchez, Phys. Rev. D **76**, 096010 (2007).

[20] G.J. Gounaris, J.J. Sakurai, Phys. Rev. Lett. **21**, 244 (1968).

[21] M. Davier, S. Eidelman, A. Hoecker, Z. Zhang, Eur. Phys. J. C **27**, 497 (2003)

[22] Z. Zhang, *Review of recent leading-order hadronic vacuum polarisation calculation* (Proceedings of PHOTON 2013 conference, Paris), arXiv:1312.7549 [hep-ph].

[23] M. Davier, A. Hoecker, B. Malaescu, C. Z. Yuan and Z. Zhang, Eur. Phys. J. C **74**, 2803 (2014) [arXiv:1312.1501 [hep-ex]].

AUTHOR INDEX